Lecture Notes in Bioinformatics

4175

Edited by S. Istrail, P. Pevzner, an

Editorial Board: A. Apostolico
T. Lengauer   S. Miyano   G. Myer
R. Shamir   T. Speed   M. Vingron

Subseries of Lecture Notes in Computer Science

Philipp Bücher   Bernard M.E. Moret (Eds.)

Algorithms in Bioinformatics

6th International Workshop, WABI 2006
Zurich, Switzerland, September 11-13, 2006
Proceedings

 Springer

Series Editors

Sorin Istrail, Brown University, Providence, RI, USA
Pavel Pevzner, University of California, San Diego, CA, USA
Michael Waterman, University of Southern California, Los Angeles, CA, USA

Volume Editors

Philipp Bücher
Ecole Polytechnique Fédérale de Lausanne, Switzerland
E-mail: Philipp.Bucher@isrec.unil.ch

Bernard M.E. Moret
Ecole Polytechnique Fédérale de Lausanne, Switzerland
E-mail: bernard.moret@epfl.ch

Library of Congress Control Number: 2006932026

CR Subject Classification (1998): F.1, F.2.2, E.1, G.1-3, J.3

LNCS Sublibrary: SL 8 – Bioinformatics

ISSN   0302-9743
ISBN-10   3-540-39583-0 Springer Berlin Heidelberg New York
ISBN-13   978-3-540-39583-6 Springer Berlin Heidelberg New York

Springer is a part of Springer Science+Business Media

springer.com

© Springer-Verlag Berlin Heidelberg 2006

Typesetting: Camera-ready by author, data conversion by Scientific Publishing Services, Chennai, India
Printed on acid-free paper   SPIN: 11851561   06/3142   5 4 3 2 1 0

Preface

We are very pleased to present the proceedings of the *Sixth Workshop on Algorithms in Bioinformatics (WABI 2006)*, which took place in Zürich on September 11-13, 2006, under the auspices of the *International Society for Computational Biology (ISCB)*, the *European Association for Theoretical Computer Science (EATCS)*, and the *Eidgenössische Technische Hochschule Zürich (ETHZ)*.

The *Workshop on Algorithms in Bioinformatics* covers research on all aspects of algorithmic work in bioinformatics. The emphasis is on discrete algorithms that address important problems in molecular biology, that are founded on sound models, that are computationally efficient, and that have been implemented and tested in simulations and on real datasets. The goal is to present recent research results, including significant work-in-progress, and to identify and explore directions of future research. Specific topics of interest include, but are not limited to:

- Exact, approximate, and machine-learning algorithms for genomics, sequence analysis, gene and signal recognition, alignment, molecular evolution, population genetics and nucleotide polymorphism, structure determination or prediction, gene expression and gene networks, proteomics, functional genomics, and drug design.
- Methods, software and dataset repositories for the development and testing of such algorithms and their underlying models.
- High-performance approaches to computationally hard problems in bioinformatics, particularly optimization problems.

A major goal of the workshop is to bring together researchers spanning the range from abstract algorithm design to biological dataset analysis, so as to enable a dialogue between application specialists and algorithm designers, mediated by algorithm engineers and high-performance computing specialists. We believe that such a dialogue is necessary for the progress of computational biology, inasmuch as application specialists cannot analyze their datasets without fast and robust algorithms and, conversely, algorithm designers cannot produce useful algorithms without being conversant with the problems faced by biologists.

Part of this mix has been achieved for all six *WABI* events to date by collocating *WABI* with the *European Symposium on Algorithms (ESA)*, along with other occasional conferences or workshops, so as to form the interdisciplinary scientific meeting known as *ALGO*. This year, *ALGO 2006* comprised the *14th European Symposium on Algorithms (ESA 2006)*, the *6th Workshop on Algorithms in Bioinformatics (WABI 2006)*, the *4th Workshop on Approximation and Online Algorithms (WAOA 2006)*, the *2nd International Workshop on Parameterized and Exact Computation (IWPEC 2006)*, and the *6th Workshop on Algorithmic Methods and Models for Optimization of Railways (ATMOS 2006)*.

We received 100 submissions in response to our call for *WABI 2006* and were able to accept 36 of them, ranging from mathematical tools to experimental studies of approximation algorithms and reports on significant computational analyses. Numerous biological problems are dealt with, including genetic mapping, sequence alignment and sequence analysis, phylogeny, comparative genomics, and protein structure. This year was the first in which *WABI* also called for machine-learning approaches along with combinatorial optimization, and we are delighted to feature five contributions from this area.

We would like to thank all authors for submitting their work to the workshop and all the presenters and attendees for their participation. We were particularly fortunate in enlisting the help of a very distinguished panel of researchers for our program committee, which undoubtedly accounts for the large number of submissions and the high quality of the presentations. Our heartfelt thanks go to all:

Vincent Berry (U. Montpellier)
Rita Casadio (U. di Bologna)
Phoebe Chen (Deakin U.)
Nadia El-Mabrouk (U. Montréal)
Raffaele Giancarlo (U. di Palermo)
David Gilbert (U. Glasgow)
Roderic Guigo (U. Pompeu Fabra)
Vasant Honavar (Iowa State U.)
Daniel Huson (U. Tübingen)
Jens Lagergren (KTH Stockholm)
C. Randal Linder (U. Texas Austin)
Joao Meidanis (U. Campinas)
Satoru Miyano (Tokyo U.)
Gene W. Myers (HHMI Janelia Farm)
Luay Nakhleh (Rice U.)
Cedric Notredame (CNRS Marseilles)
Sven Rahmann (U. Bielefeld)
Knut Reinert (Freie U. Berlin)
Mikhail Roytberg (Russian Academy of Sciences)
Marie-France Sagot (U. Claude Bernard)
David Sankoff (U. Ottawa)
Joao Setubal (U. Campinas)
Adam Siepel (Cornell U.)
Jijun Tang (U. South Carolina)
Olga Troyanskaya (Princeton U.)
Alfonso Valencia (CNB-CSIC)
Jaak Vilo (Egeen Inc.)
Tandy Warnow (U. Texas Austin)
Lusheng Wang (City U. Hong Kong)
Tiffani Williams (Texas A&M U.)
Louxin Zhang (National U. Singapore)

We were fortunate to attract Ron Shamir, from Tel Aviv University, to address the joint conferences on topics in computational biomedicine, along with other distinguished speakers lecturing in more classical algorithmic areas: Erik Demaine (Massachusetts Institute of Technology), Lisa Fleischer (IBM T.J. Watson Research Labs), László Lovász (Eőtvős Loránd University and Microsoft Research), and Kurt Mehlhorn (Max-Planck-Institute Saarbrücken).

Last but not least, we thank Michael Hoffman and his colleagues Angelika Steger, Emo Welzl, and Peter Widmayer, all at ETHZ, for doing a superb job of organizing the joint conferences.

We hope that you will consider contributing to future *WABI* events, through a submission or by participating in the workshop.

September 2006   Phillip Bücher and Bernard M.E. Moret
                                              WABI'06 Program Co-Chairs

Table of Contents

Measures of Codon Bias in Yeast, the tRNA Pairing Index and Possible
DNA Repair Mechanisms 1
   Markus T. Friberg, Pedro Gonnet, Yves Barral,
   Nicol N. Schraudolph, Gaston H. Gonnet

Decomposing Metabolomic Isotope Patterns 12
   Sebastian Böcker, Matthias C. Letzel, Zsuzsanna Lipták,
   Anton Pervukhin

A Method to Design Standard HMMs with Desired Length Distribution
for Biological Sequence Analysis 24
   Hongmei Zhu, Jiaxin Wang, Zehong Yang, Yixu Song

Efficient Model-Based Clustering for LC-MS Data 32
   Marta Łuksza, Bogusław Kluge, Jerzy Ostrowski,
   Jakub Karczmarski, Anna Gambin

A Bayesian Algorithm for Reconstructing Two-Component
Signaling Networks 44
   Lukas Burger, Erik van Nimwegen

Linear-Time Haplotype Inference on Pedigrees
Without Recombinations 56
   M.Y. Chan, Wun-Tat Chan, Francis Y.L. Chin, Stanley P.Y. Fung,
   Ming-Yang Kao

Phylogenetic Network Inferences Through Efficient Haplotyping 68
   Yinglei Song, Chunmei Liu, Russell L. Malmberg, Liming Cai

Beaches of Islands of Tractability: Algorithms for Parsimony
and Minimum Perfect Phylogeny Haplotyping Problems 80
   Leo van Iersel, Judith Keijsper, Steven Kelk, Leen Stougie

On the Complexity of SNP Block Partitioning Under the Perfect
Phylogeny Model 92
   Jens Gramm, Tzvika Hartman, Till Nierhoff, Roded Sharan,
   Till Tantau

How Many Transcripts Does It Take to Reconstruct the Splice Graph? ... 103
   Paul Jenkins, Rune Lyngsø, Jotun Hein

Multiple Structure Alignment and Consensus Identification
for Proteins 115
    Jieping Ye, Ivaylo Ilinkin, Ravi Janardan, Adam Isom

Procrastination Leads to Efficient Filtration for Local
Multiple Alignment 126
    Aaron E. Darling, Todd J. Treangen, Louxin Zhang, Carla Kuiken,
    Xavier Messeguer, Nicole T. Perna

Controlling Size When Aligning Multiple Genomic Sequences
with Duplications 138
    Minmei Hou, Piotr Berman, Louxin Zhang, Webb Miller

Reducing Distortion in Phylogenetic Networks 150
    Daniel H. Huson, Mike A. Steel, Jim Whitfield

Imputing Supertrees and Supernetworks from Quartets 162
    Barbara Hollan, Glenn Conner, Katharina T. Huber,
    Vincent Moulton

A Unifying View of Genome Rearrangements 163
    Anne Bergeron, Julia Mixtacki, Jens Stoye

Efficient Sampling of Transpositions and Inverted Transpositions
for Bayesian MCMC 174
    István Miklós, Timothy Brooks Paige, Péter Ligeti

Alignment with Non-overlapping Inversions in $O(n^3)$-Time 186
    Augusto F. Vellozo, Carlos E.R. Alves, Alair Pereira do Lago

Accelerating Motif Discovery: Motif Matching on Parallel Hardware 197
    Geir Kjetil Sandve, Magnar Nedland, Øyvind Bø Syrstad,
    Lars Andreas Eidsheim, Osman Abul, Finn Drabløs

Segmenting Motifs in Protein-Protein Interface Surfaces 207
    Jeff M. Phillips, Johannes Rudolph, Pankaj K. Agarwal

Protein Side-Chain Placement Through MAP Estimation
and Problem-Size Reduction 219
    Eun-Jong Hong, Tomás Lozano-Pérez

On the Complexity of the Crossing Contact Map Pattern
Matching Problem 231
    Shuai Cheng Li, Ming Li

A Fuzzy Dynamic Programming Approach to Predict RNA
Secondary Structure... 242
    Dandan Song, Zhidong Deng

Landscape Analysis for Protein-Folding Simulation in the H-P Model.... 252
    Kathleen Steinhöfel, Alexandros Skaliotis, Andreas A. Albrecht

Rapid ab initio RNA Folding Including Pseudoknots Via Graph
Tree Decomposition 262
    Jizhen Zhao, Russell L. Malmberg, Liming Cai

Flux-Based vs. Topology-Based Similarity of Metabolic Genes 274
    Oleg Rokhlenko, Tomer Shlomi, Roded Sharan, Eytan Ruppin,
    Ron Y. Pinter

Combinatorial Methods for Disease Association Search
and Susceptibility Prediction 286
    Dumitru Brinza, Alexander Zelikovsky

Integer Linear Programs for Discovering Approximate Gene Clusters 298
    Sven Rahmann, Gunnar W. Klau

Approximation Algorithms for Bi-clustering Problems 310
    Lusheng Wang, Yu Lin, Xiaowen Liu

Improving the Layout of Oligonucleotide Microarrays:
Pivot Partitioning... 321
    Sérgio A. de Carvalho Jr., Sven Rahmann

Accelerating the Computation of Elementary Modes Using
Pattern Trees... 333
    Marco Terzer, Jörg Stelling

A Linear-Time Algorithm for Studying Genetic Variation 344
    Nikola Stojanovic, Piotr Berman

New Constructive Heuristics for DNA Sequencing by Hybridization 355
    Christian Blum, Mateu Yábar Vallès

Optimal Probing Patterns for Sequencing by Hybridization............. 366
    Dekel Tsur

Gapped Permutation Patterns for Comparative Genomics 376
    Laxmi Parida

Segmentation with an Isochore Distribution 388
   Miklós Csűrös, Ming-Te Cheng, Andreas Grimm,
   Amine Halawani, Perrine Landreau

Author Index 401

Measures of Codon Bias in Yeast, the tRNA Pairing Index and Possible DNA Repair Mechanisms

Markus T. Friberg[1], Pedro Gonnet[1], Yves Barral[2],
Nicol N. Schraudolph[3,4], and Gaston H. Gonnet[1]

[1] Institute of Computational Science, ETH Zurich, 8092 Zurich, Switzerland
[2] Institute of Biochemistry, Department of Biology, ETH Zurich, Switzerland
[3] Statistical Machine Learning, National ICT Australia, Canberra ACT 2601, Australia
[4] RSISE, Australian National University, Canberra ACT 0200, Australia

Abstract. Protein translation is a rapid and accurate process, which has been optimized by evolution. Recently, it has been shown that tRNA reusage influences translation speed. We present the tRNA Pairing Index (TPI), a novel index to measure the degree of tRNA reusage in any gene. We describe two variants of the index, how to combine various such indices to a single one and an efficient algorithm for their computation. A statistical analysis of gene expression groups indicate that cell cycle genes have high TPI. This result is independent of other biases like GC content and codon bias. Furthermore, we find an additional unexpected codon bias that seems related to a context sensitive DNA repair.

1 Introduction

Protein translation is a rapid and accurate process, despite the need to discriminate between many possible incoming and competing tRNAs. One can assume that the process has been optimized by evolution. It has been shown that tRNA availability is both a limiting step and a regulatory parameter during translation [1,2]. Recently, through an experiment with synthesized GFP genes, it was shown that tRNA reusage (codon order) influences translation speed in yeast [3]. Here we describe the tRNA Pairing Index (TPI), an index that measures the degree of tRNA reusage in any gene.

By a statistical analysis of the TPI and gene expression, we show that genes that change their expression level rapidly (and thus require the most rapid translation) have a (statistically significant) higher TPI. Specifically, genes involved in cell cycle and DNA damage have a high TPI. These genes are regulated in the most dynamic manner, i.e. they are most rapidly turned on and off in response to intra- or extra-cellular activities.

The TPI distribution over all yeast coding sequences is biased towards positive values, indicating that there is a general tendency of tRNA reusage in the yeast genome.

Codon bias has been extensively studied previously [4,5,6,7,8,9,10]. However, to the best of our knowledge, the problem of measuring tRNA reusage in a gene has not been addressed before. The general analysis of codon autocorrelation suffers from the bias that may be induced by different base frequencies in different parts of the genome. It is known that some parts of the genome are GC-rich while other parts are GC-poor. Such

P. Bücher and B.M.E. Moret (Eds.): WABI 2006, LNBI 4175, pp. 1–11, 2006.

long-stretched biases induce an autocorrelation in the codons, which could be significant. Our first version of the TPI can measure autocorrelation without being affected by this kind of bias.

2   Methods

The TPI is an index which is computed for each protein and measures the autocorrelation (positive or negative) of its codons. Depending on how the background distribution is chosen, it is possible to make TPI completely independent of the frequencies of the amino acids, tRNAs, codons or bases, so that it will not suffer from any of the common sources of bias.

We measure the autocorrelation independently of everything else by analyzing the usage of tRNA in each amino acid of a protein as a combinatorial problem on symbols. For example, suppose that we are considering an amino acid which occurs 7 times in the protein in question and can be translated by two different tRNAs, A and B (e.g. 3 A's and 4 B's). We will extract the tRNAs from our sequence and represent them as a sequence of 7 symbols, e.g. AABABBB.

Highly autocorrelated cases are AAABBBB and BBBBAAA. A highly negatively autocorrelated case is BABABAB. This autocorrelation can be quantified by the number of identical pairs in the sequence or, conversely, by the number of changes C as we read from left to right. Notice that for a sequence of length n, the number of identical pairs plus the number of changes is $n - 1$. The mathematics is completely analogous for the number of pairs or number of changes. We call these breaks in the sequences changes, with the thought that if a tRNA molecule is doing the translation for one particular amino acid, when these breaks happen, this tRNA will have to be changed for another molecule. The first two examples have 1 change each, the last example has 6 changes. The TPI measures how high the actual number of pairs are, or how low C is, compared to all possible permutations of the sequence of tRNAs.

We present two different background distributions: one (TPI$_1$) based on codon frequencies given by the actual gene/genome under study, i.e. all possible orders considered equally likely (2.1) and another one (TPI$_2$) based on variable codon frequencies extracted from the entire genome (2.3).

2.1   TPI$_1$: Constant Codon Frequencies

Computation of the Probability of the Number of Changes. We will now describe the function to compute the probability and cumulative distribution of a given number of changes x.

It is easy to observe that the probability of the number of changes $C(x, n_1, n_2, ..., n_k)$ does not depend on what the symbols are, but rather on how many symbols there are of each kind $(n_1, n_2, ..., n_k)$. C is a (symmetric) function of the number of each kind of different symbols. It is difficult to write a recursion based on C, so instead we will base its computation on another function, called C_r, which does the recursive part of the computation. $C_r(x, n_1, n_2, ..., n_k)$ assumes that we are not at the beginning of the sequence, but rather that the last symbol observed is known (Fig. 1). To identify this known symbol (all symbols are otherwise equivalent), we will make it the first of the

```
C(x,4,3)
```

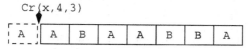

(probability of x changes with 4 and 3 symbols)

```
Cr(x,4,3)
```

(probability of x changes with 4 and 3 when
the previous symbol is of the first kind)

Fig. 1. C and C_r

arguments. Our function C_r assumes that it is called with a symbol of the first class preceding the rest of the symbols (Fig. 1). We explain C_r for $k = 2$ symbols in detail.

$$C_r(x, n_1, n_2) = \begin{cases} 0 & \text{if } n_1 < 0 \text{ or } n_2 < 0 \text{ or } x < 0 \text{ or } x > n_1 + n_2 \\ 1 & \text{if } n_1 = n_2 = 0 \ (x \text{ must be } 0) \\ \frac{1}{n_1+n_2}(n_1 C_r(x, n_1 - 1, n_2) + n_2 C_r(x - 1, n_2 - 1, n_1)) \end{cases} \quad (1)$$

The first symbol is either from the class of n_1 (no change) or from the class of n_2, in which case the preceding symbol now is of the second class and we invert the arguments: $Cr(x - 1, n_2 - 1, n1)$.

The extension of this function to higher k is simple. Supplementary material (http://www.biorecipes.com/TPI/appendix/Cr.M) shows a production quality version of this procedure which takes into account more refined border conditions. $C(x, n_1, n_2)$ can be expressed in two forms in terms of C_r. First, if we allow an arbitrary number of symbols we use

$$C(x, n_1, n_2) = C_r(x + 1, [0, n_1, n_2]) \quad (2)$$

i.e., we create an artificial first symbol (of which we have 0 left) and allow for one more change. Else we can expand based on the first symbol:

$$C(x, n_1, n_2) = \frac{n_1}{n_1 + n_2} C_r(x, n_1 - 1, n_2) + \frac{n_2}{n_1 + n_2} C_r(x, n_2 - 1, n_1) \quad (3)$$

The code for C_r as written above, is exponential. We can use dynamic programming, or we could use something equivalent to *option remember* in Maple [11] to make it polynomial in the product of the n_i.

To estimate how rare a given number of changes is, we need to compute its cumulative distribution. Since the distribution is over the integers, we will take the cumulative distribution which adds one half of the probability at the point.

$$C_{cum}(x, n_1, n_2, ..., n_k) = \sum_{i=0}^{x-1} C(i, n_1, n_2, ..., n_k) + \frac{1}{2} C(x, n_1, n_2, ..., n_k) \quad (4)$$

Our TPI is $1 - 2C_{cum}$, which is more intuitive to use than C_{cum}.

Expected Values and Moments of the Number of Changes. The expected value of the number of changes can be expressed in terms of the symmetric functions S_1 and S_2 on the arguments:

$$\mu'_1 = \sum_{i=1}^{\infty} i \times C(i, n_1, n_2, ..., n_k) = \frac{S_1^2 - S_2}{S_1} \quad \text{where} \quad S_i = \sum_{j=1}^{k} n_j^i \quad (5)$$

The derivation of this formula is not trivial in its general form (for an arbitrary k). However if we observe that all the probabilities are sums of binomial coefficients, then we can conclude that the result (expected value or higher moments) must be a polynomial expression divided an appropriate descending factorial. Since all the moments are symmetric in all the arguments, the moments must be functions of the symmetric polynomials derived from the n_i. Hence by symbolic interpolation we can determine all the moments in a much easier (and safer) way. Of interest are the expected value and the variance. This is because we will attempt a normal approximation to the distribution.

$$\mu_2 = \sum_{i=1}^{\infty} (i - \mu'_1)^2 C(i, n_1, n_2, ..., n_k) = \frac{S_1 S_2 - S_1^3 - 2S_1 S_3 + S_2 S_1^2 + S_2^2}{S_1^2 (S_1 - 1)} \quad (6)$$

Unfortunately, despite the simplicity of the formulas resulting from this approach, they do not resolve our problem completely. The normal approximation gives a good approximation of the cumulative distribution around the average (for large values of S_1) and very good approximations when $\min(n_i)$ is high. However, it gives poor approximations at the tails when some of the n_i are small, which is an important case.

Computing the Distribution of C in Practice. The recursion in C_r, although simple, swaps its arguments, which makes it almost impossible to handle with the standard techniques. Even dynamic programming becomes very difficult to express. In this section we find a mechanism to rewrite the recursion in a way that the argument order is maintained.

Since the function is totally symmetric in its arguments (and C_r is totally symmetric in its arguments but the first) we can sort the arguments in increasing order guaranteeing a time of $O(n_1 n_2 .... n_k)$. This makes the recursion marginally acceptable for real problems (for yeast $k \leq 4$ and for most other genomes $k \leq 5$). This ordering is partly ruined by the swapping of arguments in the recursion (1). Each recursive call to C_r uses a different argument as second argument.

To resolve this problem we find recursions which (while maybe more complicated) do not jumble the arguments. We can illustrate this by doing the transformation on the simplest recursion, $k = 2$. For further simplicity, we will use the auxiliary function $H(x, n_1, n_2) = C_r(x, n_1, n_2) \binom{n_1 + n_2}{n_1}$. As expected, the recursion on $H(x, n_1, n_2)$ is significantly simpler.

$$H(x, n_1, n_2) = H(x, n_1 - 1, n_2) + H(x - 1, n_2 - 1, n_1) \quad (7)$$

We now apply this formula to the shifted arguments

$$-H(x, n_1, n_2 - 1) = -H(x, n_1 - 1, n_2 - 1) - H(x - 1, n_2 - 2, n_1) \qquad (8)$$

$$H(x - 1, n_2 - 1, n_1) = H(x - 1, n_2 - 2, n_1) + H(x - 2, n_1 - 1, n_2 - 1) \qquad (9)$$

Adding these three equations results in

$$H(x, n_1, n_2) = H(x, n_1 - 1, n_2) + H(x, n_1, n_2 - 1) - H(x, n_1 - 1, n_2 - 1) + H(x - 2, n_1 - 1, n_2 - 1)$$

$$(10)$$

Notice that we have managed to obtain a recursion for which all the arguments (n_1, n_2) are in the same order. The new recursion with four terms instead of two is a bit more complicated, but this is an insignificant cost when we observe that in this form it is easy to write a recursive program to compute it. The computation can be done over the space of $n_1 x n_2$ for increasing x, having to keep two copies of the older H.

Transformations for up to $k = 5$ were obtained by doing a Knuth-Bendix style elimination procedure among all shifts of the basic recurrence. This was done in Maple and required some careful and extensive manipulations. Table 1 shows the summary of the results.In the supplementary material, http://www.biorecipes.com/TPI/ appendix/recursions, we show the recursions for $k = 2$ to $k = 5$. With these recursions it was possible to write a C program that can compute all the TPI values for a genome like yeast in about 6 hours. Previous attempts failed after weeks of computing in very large machines.

Table 1. Recursions

k	terms	eq. used	max shift x	max shift $n_1, n_2, ...$
2	4	3	-2	-1
3	12	37	-3	-1
4	32	657	-4	-1
5	80	19125	-5	-1

Analytic Solution for two Symbols. The case with two symbols can be resolved explicitly (unfortunately, we were not able to find closed forms for higher k, and conjecture that no simple forms exist). **Theorem:**

$$H(x, n_1, n_2) = \binom{n_1}{\lfloor \frac{x}{2} \rfloor} \binom{n_2 - 1}{\lfloor \frac{x-1}{2} \rfloor} \qquad (11)$$

This is easily proved by plugging the recursion that defines $H(x, n_1, n_2)$ and separating the case when x is even and when x is odd. For example if x is even then $x = 2w$ and the recursion becomes:

$$\binom{n_1}{w}\binom{n_2 - 1}{w - 1} = \binom{n_1 - 1}{w}\binom{n_2 - 1}{w - 1} + \binom{n_2 - 1}{w - 1}\binom{n_1 - 1}{w - 1} \qquad (12)$$

By removing the common factor $\binom{n_2 - 1}{w - 1}$ we get

$$\binom{n_1}{w} = \binom{n_1 - 1}{w}\binom{n_1 - 1}{w - 1} \tag{13}$$

which is a well-known identity of binomial coefficients [12].

2.2 Scale of the TPI

The previous subsection showed how to compute the cumulative distribution which is needed to compute the TPI. These formulas were applied to each individual amino acid. Now comes the question of how to express, with a single index, the joint distribution for all amino acids.

We use convolution of the cumulative distributions, since the measure is the same for each individual amino acid. To facilitate our understanding, we use $\text{TPI}_1 = 1 - 2C_{cum}$ to scale this convolved cumulative distribution to $-1..1$, so that we can talk about negative TPI (more tRNA changes than expected) and positive TPI (fewer changes than expected).

2.3 TPI$_2$: Variable Codon Frequencies

Some highly expressed genes (e.g., YGR192C) use basically only one tRNA for each amino acid. Although the number of tRNA changes will be practically zero, so will the distribution of possible changes. Hence, the TPI_1 as computed above will be neutral (close to 0) for these genes. This is not desirable, since these highly expressed genes are highly optimized, and have almost the maximum tRNA reusage possible. In some sense, TPI_1 is too independent of the codon usage distribution.

The problem occurs because we have assumed that the choice of codon in a gene is constant, and only consider different orderings of these codons. To resolve this problem, we suggest an alternative TPI_2, where only the choice of amino acids (not codons) is fixed. The test statistic is the same as for TPI_1 (number of tRNA changes). However, the background distribution is estimated from the set of all possible genes resulting in the same protein, where the contribution from each gene is proportional to the probabilities of its codons according to the global codon frequencies in the genome.

For the TPI_2 we will assume that the codon frequency for each amino acid is given. It can be the global distribution or some localized (extracted from a group of genes) distribution. Contrary to the case for the TPI_1, we have an efficient method of computing the exact distribution (although not a closed form).

We use a similar notation, $C(x, n, \mathbf{P})$ will give us the probability that we find x changes in a sequence of n symbols which appear with probabilities $\mathbf{P} = (p_1, p_2, ..., p_k)$. Similarly $C_i(x, n, \mathbf{P})$ will denote the probability of x changes among the next n symbols when the last symbol is the ith.

$$C(x, n, \mathbf{P}) = \sum_{i=1}^{k} p_i C_i(x, n - 1, \mathbf{P}) \tag{14}$$

$$C_i(x, 0, \mathbf{P}) = \delta_{x0} \tag{15}$$

$$C_i(x, n, \mathbf{P}) = p_i C_i(x, n-1, \mathbf{P}) + \sum_{k \neq i} p_k C_k(x-1, n-1, \mathbf{P}) \quad (16)$$

This recursion is fundamentally simpler than the ones for TPI_1, as it is in two variables ($\mathbf{P}$ is a constant). Furthermore, we can compute the functions for increasing n requiring space $O(nk)$ and time $O(n^2 k)$. The computation of TPI_2 for all yeast genes is done in approximately 10 minutes on a modern desktop computer.

The moments of $C(x, n, \mathbf{P})$ have simple expressions, but in this case we can compute the exact distribution for all practical cases, so there is no point in using them.

3   Results

In a recent experiment with synthesized GFP genes, it was shown that genes with maximum tRNA reusage were translated faster than genes with minimum tRNA reusage [3]. To complement that biochemistry experiment of one gene, a bioinformatics analysis is provided here. We examine whether different gene groups have higher or lower average TPI than what can be expected by chance. Gene groups from expression connection, http://db.yeastgenome.org/cgi-bin/expression/expressionConnection.pl were used to evaluate TPI_1 and TPI_2 (Table 2). For comparison, we also included the CAI index [5].

Table 2. Average TPI's (scale: -1..1) and codon adaptation index (CAI) for groups of genes that are up-regulated in different experiments. Consider as an example the first row: average TPI's and CAI were computed for the group of genes with at least 5-fold up-regulation when subjected to glucose limitation. The same was done for 100000 groups of (equally many) randomly picked genes, and p-values were computed by comparing the real average TPI's and CAI to their respective distribution of random values.

Experiment	avg(TPI_1)	p(TPI_1)	avg(TPI_2)	p(TPI_2)	avg(CAI)	p(CAI)
glucose limitation 5x	0.142	0.463	0.512	0.0108	0.240	0.0285
sporulation 20x	0.028	0.942	-0.045	0.9831	0.140	0.9987
diauxic shift 10x	0.215	0.222	0.428	0.0062	0.209	0.0583
DNA damage 10x	0.048	0.861	0.466	0.00001	0.223	0.0015
alpha-factor over time 80x	0.144	0.337	-0.037	0.9992	0.138	1.0000
alpha-factor (var. conc.) 20x	0.192	0.148	0.173	0.207	0.157	0.8808
stress response 50x	0.163	0.251	0.236	0.0267	0.174	0.3849
histone depletion 50x	0.140	0.440	0.207	0.1681	0.170	0.4894
zinc levels 10x	0.198	0.236	0.461	0.0015	0.246	0.0017
phosphate pathway 10x	0.328	0.048	0.215	0.2410	0.180	0.3233
cell cycle 5x	0.175	0.153	0.310	0.00017	0.194	0.0152
cell cycle 10x	0.365	0.013	0.465	0.0026	0.21	0.0900

Overall, TPI_1 does not show a clear correlation with gene expression. The exception is for cell cycle genes, a group that requires fast translation. Cell cycle genes that increase expression at least 10-fold have an average TPI_1 of 0.365, which is significant with p-value 0.013. This is an indication that there is a selection for high tRNA reusage in cell cycle genes.

TPI$_2$, which is also correlated with CAI, shows a clear correlation with several gene expression groups. Genes from the cell cycle, zinc levels, DNA damage, diauxic shift and glucose limitation experiments have a significantly high average TPI$_2$. This is an additional indication for high tRNA reusage in cell cycle genes.

4 Pairwise Codon Bias Suggests Context-Dependent DNA Repair in *Saccharomyces Cerevisiae*

We observe a very significant bias in the selection of synonymous codons in the coding DNA of *S. cerevisiae*. The bias appears in DNA, its reverse complement and in each amino acid individually which strongly suggests that it is neither a translational nor RNA effect, but a feature of the DNA. Isolating this bias from other possible sources, we conjecture that this is caused by a context dependent DNA mismatch correction mechanism.

Given the complete set of coding sequences of the *S. cerevisiae* genome (release 36 of the *S. cerevisiae* genome from EMBL [13]), we can calculate the observed frequencies of amino acids ($f_{obs}^a(A)$) and codons ($f_{obs}^c(C)$) as well as the observed pairwise amino acid ($f_{obs}^{aa}(A_1, A_2)$) and codon frequencies ($f_{obs}^{cc}(C_1, C_2)$). Using f_{obs}^{aa} and f_{obs}^c, we can calculate the *expected* pairwise codon frequencies f_{exp}^{cc} under the hypothesis of independent codon pairing.

We compute the difference between the observed value and the expected value, in units of a standard deviation and get differences of up to 25 standard deviations. If we group the codon pairs according to the nucleotides at the frame border, that is, for a codon pair $x_1x_2x_3$ and $y_1y_2y_3$, the nucleotides x_3 and y_1, we get even more significant results (up to 83.10 standard deviations, Table 3).

Table 3. Difference in standard deviations between expected and observed counts of pairwise codons grouped by the nucleotides at the frame border

$x_3 \backslash y_1$	A	C	G	T
A	19.74	-4.68	-23.46	5.79
C	**77.90**	-11.21	**-61.09**	-19.50
G	-0.67	30.52	2.84	-29.46
T	**-83.10**	-11.20	**65.06**	31.50

A look at the most biased codon pair patterns (Table 4) quickly reveals that the reverse complement of each pattern is similarly biased. This indicates that whatever is causing the observed bias is probably not dependent on the reading-sense of the sequences, since it also appears in the reverse-complement strand in identical form.

Furthermore, if we re-compute the values in Table 3 for each amino acid with more than one codon, using the third nucleotide in each codon for the position x_3 and the

Table 4. The 20 most significantly biased patterns and their reverse complement. The indented patterns are subsets of an already listed pattern. The bias is given in standard deviations.

Pattern	bias	Reverse	bias	Pattern	bias	Reverse	bias
T A	-83.10	self		**C G**	-61.09	self	
C A	77.90	**T G**	65.06	*TC A*	59.07	**T GA*	25.51
**T AA*	-70.59	*TT A**	-69.96	*CC A*	54.69	**T GG*	32.07
*TT A**	-69.96	**T AA*	-70.59	**T A*A	-52.78	T*T A**	-45.06
T G	65.06	**C A**	77.90	**T A*G	-52.61	C*T A**	-51.43
**C AA*	62.39	*TT G**	36.10	**T AG*	-51.57	*CT A**	-48.28
TT AA	-62.02	self		C*T A**	-51.43	**T A*G	-52.61

observed and expected pairwise codon frequencies for each amino acid, the resulting biases (although not all pairs are present for each amino acid) match those in Table 3.

In summary, the pairwise codon bias observed seems to be caused by some mechanism that operates directionally on DNA since it is independent of the coding strand and of the individual amino acids. This excludes as possible mechanisms post-translational events.

We postulate that the bias is caused by a context sensitive DNA mismatch correction mechanism (CSCM). Although DNA replication is in general a very faithful process, errors in the form of mismatches can arise. The most common form of mismatches, and the only type we will consider here, are the purine·pyrimidine mismatches C·A and T·G. Such a mismatch can occur at any given position in the sequence. To avoid DNA packing and transcription problems, such mismatches need to be corrected to either T·A or C·G. In organisms which do not mark the original strand during DNA replication (*e.g. S. cerevisiae*), corrections could be made according to:

- a static, context insensitive correction rule, always correcting to either T·A or C·G. Such a mechanism would create an obvious T·A or C·G bias.
- a random correction scheme, which would not produce a C·G-bias, however, it would not produce the observed codon pair frequency bias either.

In the absence of a strong C·G bias we assume that a more refined context sensitive mechanism exists. We also assume that such a mechanism would produce *better* corrections than the static or random models. For our study we will define a *good* correction as a correction that does not alter the primary structure of the sequence, since it is the only type of error we can quantify.

We define a CSCM as a scheme in which the mismatch correction depends on the mismatch itself and its two neighbouring nucleotides. Considering mismatches in a strand-independent (symmetric) correction scheme, we define the local context as the nucleotides upstream and downstream of the pyrimidine in the mismatch. We shall call these nucleotides x and y. We represent the CSCM as a Table in which for each x and y an A·C or T·G mismatch is corrected to either A·T or C·G (Table 5).

We define the quality of a CSCM as the percentage of correct corrections (PCC). This is the relative number of mismatch corrections which do not induce a change in the primary structure of the sequence or the insertion or deletion of a stop-codon (note that mutations to synonymous codons are accepted).

Table 5. A·C portion of the CSCM. The T·G portion is shown to be identical for optimal correction schemes.

x/y	A	C	G	T
A	A·T	A·T	A·T	A·T
C	C·G	A·T	A·T	A·T
G	C·G	A·T	A·T	A·T
T	C·G	A·T	A·T	A·T

The optimal CSCM (*i.e.* the CSCM with the highest PCC of 74.18% as opposed to 67.12% for random corrections) for the *S. cerevisiae* genome is shown in Table 5. Due to its rather simple structure, we can define, for each nucleotide, *protected* and *vulnerable* contexts:

- **A** preceded by T is *vulnerable*, since after correction it would change to G.
- **C** followed by A is *protected*, since after correction it would remain a C.
- **G** preceded by T is *protected*, since after correction it would remain a G.
- **T** followed by A is *vulnerable*, since after correction it would change to C.

These four contexts are exactly the four most biased nucleotide combinations observed over the frame border (Table 3). Note that the observed bias has very little influence on the choice of the optimal CSCM: the optimal CSCM computed over the *expected* pairwise codon frequencies is identical. **The optimal CSCM could therefore create the observed codon bias.**

5   Conclusions

We have defined and presented two versions of the TPI. The first one assumes constant codon frequencies (codon shuffling). This makes it insensitive to codon bias and different GC content in different parts of the genome. Still, gene expression analysis indicates that tRNA reusage is important to cell cycle genes. The advantage of TPI_1 is that it is constructed to be independent of codon usage, GC content etc, which indicates that the observed signal indeed is due to tRNA reusage.

The second TPI version only assumes constant amino acid sequence, and uses the global codon frequencies to estimate the background distribution. The result is an index that is more similar to the codon adaptation index (CAI). Also, overall TPI_2 p-values correlate quite well with CAI. However, for cell cycle regulated genes, the p-value is much more significant than for CAI. This is an additional indication that tRNA reusage is important for cell cycle genes. Also, the DNA damage experiments show much higher correlation with TPI than with CAI.

The two versions of TPI complement each other. TPI_1 shows that tRNA reusage exists, since it measures tRNA reusage independently of other biases. However, assuming a constant codon population in a gene reduces the measurable tRNA reusage signal. Naturally, if tRNA reusage is a way to optimize translation speed, one can assume that evolution can optimize codon bias for this purpose. By not assuming a constant codon population in a gene, TPI_2 measures tRNA reusage more accurately as it happens in the cell, which is confirmed by the TPI_2 p-values for the cell cycle genes.

In addition to the tRNA reusage bias, we observe an additional unexpected codon bias in *S. cerevisiae*. Isolating this bias from other possible sources, we conjecture that this is caused by a context dependent DNA mismatch correction mechanism.

References

1. Elf, J., Nilsson, D., Tenson, T., Ehrenberg, M.: Selective charging of tRNA isoacceptors explains patterns of codon usage. Science **300** (2003) 1718–1722
2. Dittmar, K.A., Sorensen, M.A., Elf, J., Ehrenberg, M., Pan, T.: Selective charging of tRNA isoacceptors induced by amino-acid starvation. EMBO Rep. **6** (2005) 151–157
3. Barral, Y., Faty, M., von Rohr, P., Schraudolph, N.N., Friberg, M.T., Roth, A.C., Gonnet, P., Cannarozzi, G.M., Gonnet, G.H.: tRNA recycling in translation and its influence on the dynamics of gene expression in yeast. in preparation (2006)
4. Bennetzen, J.L., Hall, B.D.: Codon selection in yeast. J. Biol. Chem. **257** (1982) 3026–3031
5. Sharp, P.M., Li, W.H.: The codon adaptation index–a measure of directional synonymous codon usage bias, and its potential applications. Nucleic Acids Res. **15** (1987) 1281–1295
6. Wright, F.: The 'effective number of codons' used in a gene. Gene **87** (1990) 23–29
7. Boycheva, S., Chkodrov, G., Ivanov, I.: Codon pairs in the genome of *Escherichia coli*. Bioinformatics **19** (2003) 987–998
8. Carbone, A., Zinovyev, A., Kepes, F.: Codon adaptation index as a measure of dominating codon bias. Bioinformatics **19** (2003) 2005–2015
9. Greenbaum, D., Colangelo, C., Williams, K., Gerstein, M.: Comparing protein abundance and mRNA expression levels on a genomic scale. Genome Biol. **4** (2003) 117
10. Friberg, M., von Rohr, P., Gonnet, G.: Limitations of codon adaptation index and other coding DNA-based features for prediction of protein expression in saccharomyces cerevisiae. Yeast **21** (2004) 1083–1093
11. Char, B.W., Geddes, K.O., Gonnet, G.H., Leong, B.L., Monagan, M.B., Watt, S.M.: Maple V Language Reference Manual. Springer-Verlag (1991)
12. Abramowitz, M., Stegun, I.: Handbook of Mathematical Functions, chapter 24.1.1. Dover, New York (1972)
13. Kulikova, T., Aldebert, P., Althorpe, N., Baker, W., Bates, K., Browne, P., van den Broek, A., Cochrane, G., Duggan, K., Eberhardt, R., Faruque, N., Garcia-Pastor, M., Harte, N., Kanz, C., Leinonen, R., Lin, Q., Lombard, V., Lopez, R., Mancuso, R., McHale, M., Nardone, F., Silventoinen, V., Stoehr, P., Stoesser, G., Tuli, M.A., Tzouvara, K., Vaughan, R., Wu, D., Zhu, W., Apweiler, R.: The EMBL nucleotide sequence database. Nucleic Acids Res. **32** (2004) D27–D30

Decomposing Metabolomic Isotope Patterns

Sebastian Böcker[1], Matthias C. Letzel[2],
Zsuzsanna Lipták[3], and Anton Pervukhin[3]

[1] Lehrstuhl für Bioinformatik, Friedrich-Schiller-Universität Jena
Ernst-Abbe-Platz 2, 07743 Jena, Germany
`boecker@minet.uni-jena.de`
[2] Organische Chemie I, Fakultät für Chemie, Universität Bielefeld, PF 100 131, 33501
Bielefeld, Germany
`matthias.letzel@uni-bielefeld.de`
[3] AG Genominformatik, Technische Fakultät, Universität Bielefeld, PF 100 131,
33501 Bielefeld, Germany
`zsuzsa, apervukh@CeBiTec.uni-bielefeld.de`

Abstract. We present a method for determining the sum formula of
metabolites solely from their mass and isotope pattern. Metabolites, such
as sugars or lipids, participate in almost all cellular processes, but the
majority still remains uncharacterized. Our input is a measured isotope
pattern from a high resolution mass spectrometer, and we want to find
those molecules that best match this pattern.

Determination of the sum formula is a crucial step in the identification of an unknown metabolite, as it reduces its possible structures to a
hopefully manageable set. Our method is computationally efficient, and
first results on experimental data indicate good identification rates for
chemical compounds up to 700 Dalton.

Above 1000 Dalton, the number of molecules with a certain mass
increases rapidly. To efficiently analyze mass spectra of such molecules,
we define several additive invariants extracted from the input and then
propose to solve a joint decomposition problem.

1   Introduction

High resolution mass spectrometry (HR-MS) allows determining the mass of
sample molecules with very high accuracy (up to 10^{-3} Dalton), and has become
one preferred method of analyzing metabolites. As with most analysis techniques
in the life sciences, not one but millions of copies of the same molecule are needed.
The output of a mass spectrometer, after preprocessing, consists of peaks that
ideally correspond to the masses of the sample molecules and their abundance,
i.e., the number of sample molecules with this mass. This brings into play the
natural isotopic distributions of the elements: Several peaks in the output correspond to the same type of sample molecule, reflecting its isotope pattern. In
this paper, we make use of this isotope pattern to identify the sample molecule
by determining its sum formula, i.e., the number of atoms of each element.

P. Büchler and B.M.E. Moret (Eds.): WABI 2006, LNBI 4175, pp. 12–23, 2006.
© Springer-Verlag Berlin Heidelberg 2006

The term "metabolite" is usually restricted to small molecules that are intermediates and products of the metabolism. These small molecules participate in almost all cellular processes such as signal transduction, stress response, catabolism, or anabolism. It is widely accepted that every species hosts several thousand metabolites; however, the overwhelming majority of these metabolites is yet uncharacterized. The majority of metabolites have mass below 1000 Dalton: 96.5 % of sum formulas in the KEGG LIGAND database fall into this mass range [6]. Today, metabolites are usually identified through fragmenting the metabolite using electron impact ionization, and subsequent database lookup in a chemical compound library [9]. This method is limited to identifying metabolites and chemical compounds that have been included in some library.

Our input is a list of masses $M_0, \ldots, M_K$ with intensities $f_0, \ldots, f_K$, normalized such that $\sum_i f_i = 1$. We assume that these have been extracted from a high-resolution mass spectrum in a preprocessing step, and that they correspond to the isotope pattern of a sample molecule. Note that, for molecular mixtures, separating isotopic peaks that belong to different molecules is mostly trivial in this case. Our goal then is to find the molecule, or rather its sum formula, whose isotope pattern best matches the input. In the following, we use "molecule" and "sum formula" interchangeably.

To tackle this problem, we proceed as follows: First, we compute all sum formulas that share one or more features (such as monoisotopic mass) with the input mass spectrum. Next, for every such candidate molecule we simulate its isotope pattern, and match and rank it against the input isotope pattern. As the number of candidate molecules is usually large, it is essential to find methods for fast simulation and ranking of isotope patterns. The two studies reported in literature for identifying molecules from their isotope patterns [4,10], both proceed in the same manner but focus on the experimental side of the problem.[1]

Contributions. First, we show how to use integer decomposition techniques introduced in [2] for decomposing real valued molecule masses in Sec. 3, with large improvements over naïve methods that are currently best known for this problem [3]. Second, we present a method for rapid computation of isotope distributions and mean masses of isotope peaks in Sec. 4, improving on best-known results in [12]. Fast simulation of isotope patterns is vital due to the large search space. Third, we show how to rapidly match and rank such simulated spectra against the measured spectrum in Sec. 5. In Sec. 7, we report on the application of our method to high resolution mass spectra.

Fourth, we present methods to further reduce the search space: The number of molecules with a certain monoisotopic mass increases rapidly for large masses [2]. To this end, we introduce the problem of jointly decomposing a set of queries, comparable to the multiple integer knapsack problem [7]. These queries are not the input masses $M_0, \ldots, M_K$, but other values derived from the input such as intensities or average mass. We introduce a method to efficiently generate

---

[1] We stress that this method cannot be used as-is to identify peptides or amino acid compositions, because even short peptides usually have non-unique sum formulas.

all solutions of a multiple integer knapsack problem in Sec. 6, and our simulations show how this technique efficiently cuts down decomposition runtimes, see Sec. 7.

2   Isotopes, Isotope Species, and Isotope Patterns

Atoms are composed of electrons, protons, and neutrons. The number of protons (the atomic number) defines what element the atom is. The elements most abundant in living beings are hydrogen (symbol H) with atomic number 1, carbon (C, 6), nitrogen (N, 7), oxygen (O, 8), phosphor (P, 15), and sulfur (S, 16). The number of neutrons, on the other hand, can vary: Atoms with the same number of protons but different numbers of neutrons are called *isotopes* of the element. Each of these isotopes occurs in nature with a certain abundance. The superscript preceding the symbol denotes the *mass number* of the atom: the number of protons plus the number of neutrons.

The *mass* of an atom is measured in Dalton (Da), which is defined as one twelfth of the mass of a ^{12}C isotope. An atom's mass is roughly but not exactly equal to its mass number, the difference being due to the binding energy in the atom's nucleus. The masses of the different isotopes and their abundance are known up to very high precision [1]; for example, ^{1}H has mass 1.007825 Da with abundance 99.985%, and ^{2}H mass 2.014102 Da with abundance 0.015%.

The *nominal mass* (also called *nucleon number*) of a molecule is the sum of protons and neutrons of the constituting atoms. The *mass* of the molecule is the sum of masses of these atoms. Clearly, nominal mass and mass depend on the isotopes the molecule consists of, thus on the *isotope species* of the molecule. The isotope species where each atom is the isotope with the lowest nominal mass is called *monoisotopic*. Likewise, the mass of the monoisotopic species is called the *monoisotopic mass* of the molecule.

The number of distinct isotope species for a molecule with i_H hydrogen, i_C carbon, i_N nitrogen, i_O oxygen, i_P phosphor, and i_S sulfur atoms is $(i_C+1)(i_H+1)(i_N+1)\binom{i_O+2}{2}\binom{i_S+3}{3}$. This follows because for an element E with r isotope types, a molecule E_l consisting of l atoms of the element has $\binom{l+r-1}{r-1}$ different isotope species. The probability that a certain isotope species occurs can be computed by multiplying the probabilities of the underlying isotopes.

For each element $E \in \Sigma$ we define two discrete random variables, denoted X_E and Y_E, representing the mass and the mass number, respectively. For example, X_C with state space $\{12, 13.003355\}$ and Y_C with state space $\{12, 13\}$ and $\mathbb{P}(X_C = 12) = \mathbb{P}(Y_C = 12) = 0.98890$, $\mathbb{P}(X_C = 13.003355) = \mathbb{P}(Y_C = 13) = 0.01110$ are the random variables of carbon. Given a molecule consisting of l atoms, we assign to the ith atom, $i = 1, \dots, l$, two random variables X_i and Y_i, where $X_i \sim X_E$ and $Y_i \sim Y_E$, with E being the corresponding element. Now we can represent the molecule's *mass distribution* by the random variable $X := X_1 + \dots + X_l$, and its nominal mass distribution, or *isotopic distribution*, by $Y := Y_1 + \dots + Y_l$. Note that X and Y are correlated, since X_E can be viewed as a function of Y_E and E.

In an ideal mass spectrum, normalized peak intensities correspond to the isotopic distribution of the molecule. For ease of exposition, the peak at monoisotopic mass is also called monoisotopic, the following peaks are referred to as $+1$, $+2$, ... peaks. What is the mass of such a superposition peak? It is reasonable to assume that its mass is the mean mass of all isotope species that add to its intensity [12]: For a molecule with monoisotopic nominal mass N, let $X = X_1 + \ldots + X_l$ be the mass distribution and $Y = Y_1 + \cdots + Y_l$ be the isotopic distribution. The mean peak mass of the $+k$ peak is then $m_k = \mathbb{E}(X \mid Y = N + k)$. We refer to the isotopic distribution together with the mean peak masses as the molecule's *isotope pattern*.

3   Decompositions of Real Value Numbers

We want to find all molecules with (monoisotopic) mass in the interval $[l, u] \subseteq \mathbb{R}$ where $l := M_0 - \varepsilon$ and $u := M_0 + \varepsilon$ for some measurement inaccuracy ε. Formally, we search for all solutions of the integer knapsack equation [7]

$$a_1 c_1 + a_2 c_2 + \cdots + a_n c_n \in [l, u] \tag{1}$$

where a_j are real-valued monoisotopic masses of elements satisfying $a_j \geq 0$. We search for all solution vectors $c = (c_1, \ldots, c_n)$ such that all c_j are non-negative integers. We may assume $a_1 < a_2 < \cdots < a_n$.

A straightforward solution is to generate all vectors c with $c_1 = 0$ and $\sum_j a_j c_j \leq u$, and next to test if there is some $c_1 \geq 0$ such that $\sum_j a_j c_j \in [l, u]$. This results in $O(m^{n-1})$ runtime where $m := M_0 / a_2$. Alternatively, we can compute all potential decompositions up to some upper bound U during preprocessing, sort them with respect to mass and use binary search; this results in $O(U^n)$ space requirement. These approaches are unfavorable in theoretical complexity as well as in practice: For the alphabet CHNOPS there exist more than $7 \cdot 10^8$ sum formulas with mass below $1000\,\mathrm{Da}$.

In case of integer coefficients, one can use dynamic programming to compute all solutions efficiently, following the line of thought of [7, Sec. 8.3]. The main disadvantage of this approach is the rather large memory requirement. An alternative method for finding all solutions is given in [2], using a table of size $O(na_1)$, see Sec. 6 for more details. Every solution is constructed in time $O(na_1)$ independent of the input l, u. Regarding the application of decomposing molecule masses, the latter approach uses only $1/15$ of memory and shows better runtimes.

Reconsider the original integer knapsack problem with real-valued coefficients. Choosing a *blowup factor* $b \in \mathbb{R}$, corresponding to precision $1/b$, we can round coefficients by $\phi(a) := \lceil ba \rceil$, so $a'_j := \phi(a_j)$ and $l' := \phi(l)$, $u' := \phi(u)$ form a Diophantine equation. Now, certain solutions c of the integer coefficient knapsack are no solutions of the real-valued coefficient knapsack, and vice versa. We can easily sort out false positive solutions checking (1), resulting in additional runtime. But first, we concentrate on the more intriguing problem of false negative solutions that are missed by the integer coefficient knapsack.

Clearly $\sum_j a_j c_j \geq l$ implies $\sum_j a'_j c_j \geq l'$ since all a'_j are integer. We have to increase the upper bound u' to guarantee that all solutions of (1) are generated.

We define relative rounding errors $\Delta_j = \Delta_j(b) := \frac{\lceil ba_j \rceil - ba_j}{a_j}$ for $j = 1, \ldots, n$, where $0 \le \Delta_j \le \frac{1}{a_j}$, and set $\Delta = \Delta(b) := \max\{\Delta_j\}$. If c satisfies $\sum_j a_j c_j \le u$ then $\sum_j a'_j c_j \le bu + \Delta u$: Clearly, $\sum_j a'_j c_j \le bu + \sum_j (a'_j - ba_j)c_j$ and our claim follows from

$$0 \le \sum_j (a'_j - ba_j)c_j = \sum_j \frac{\lceil ba_j \rceil - ba_j}{a_j} a_j c_j \le \sum_j \Delta_j a_j c_j \le \Delta \sum_j a_j c_j \le \Delta u.$$

One can easily check that this bound is tight. So, we re-define the integer interval by $u' := \lfloor bu + \Delta u \rfloor$. Then, we have to decompose Δu integers in addition to the $(u - l)b$ integers we expect without rounding errors. We stress that the runtime of this approach is dominated by the number of *decompositions* of these integers, and not by the number of integers itself.

As an example, consider the alphabet CHNOPS and blowup factor $b = 10^5$, then $\Delta = \Delta_H = 0.492936$, so for $M_0 = 1000$ we have to decompose an additional 492 integers.

4   Simulating Isotope Patterns

We first observe that for CHNOPS, all resulting molecules have isotopic distributions that decrease rapidly with increasing mass. In particular, we can restrict ourselves to computing the first K non-zero values of the distribution, for rather small K such as $K = 10$. For example, for the molecule C_{166} with nominal mass 1992, the intensities of $+10, +11, \ldots$ peaks sum up to less than 0.00003.

The atoms hydrogen, carbon, and nitrogen have only two (natural) isotopes. Thus, the isotopic distribution of a molecule E_l consisting of l identical atoms of type E with $E \in \{H, C, N\}$ follows a binomial distribution: Let q_k denote the probability that E_l has nominal mass $N + k$, where N is the monoisotopic nominal mass of E_l. Then, $q_k = \binom{l}{k} p^{l-k}(1-p)^k$ where p is the probability of the monoisotopic isotope. The values of the q_k can be computed iteratively, since $q_{k+1} = \frac{l-k}{k+1} \cdot \frac{1-p}{p} q_k$ for $k \ge 0$, thus computation time is $O(K)$.

Where an element E has $r > 2$ isotopes (such as oxygen and sulfur), the isotopic distribution of E_l can be computed as follows: Let p_i for $i = 0, \ldots, r$ denote the probability of occurrence of the ith isotope. Then, the probability that E_l has nominal mass $N + k$ is $\sum \binom{l}{l_0, l_1, \ldots, l_r} \cdot \prod_{i=0}^r p_i^{l_i}$, where the sum runs over all $l_0, \ldots, l_r \ge 0$ satisfying $\sum_{i=0}^r l_i = l$ and $\sum_{i=1}^r i \cdot l_i = k$ [5]. The tuples $(l_0, \ldots, l_r)$ satisfying $\sum_i i \cdot l_i$ are the integer partitions of k into at most r parts, which can be computed recursively with a greedy approach. However, the number of partitions grows rapidly, at least with a polynomial in k of degree $r - 1$.

4.1   Folding Isotope Patterns

Given two discrete random variables Y and Y' with state spaces $\Omega, \Omega' \subseteq \mathbb{N}$, we can compute the distribution of the random variable $Z := Y + Y'$ by folding the distributions, $\mathbb{P}(Z = N) = \sum_k \mathbb{P}(Y = k) \cdot \mathbb{P}(Y' = N - k)$. If we restrict ourselves to the first K values of this sum, we can compute this distribution in time $O(K^2)$. Kubinyi [8] suggests to compute the isotopic distributions of oxygen O_l and sulfur S_l by successive folding of the respective distribution: Using a Russian multiplication scheme for the folding, this results in an algorithm with runtime $O(K^2 \log l)$. In applications, we do not compute these distributions on the fly but during preprocessing, for all $l \leq L$ fixed. This results in $O(KL)$ memory for every such element, where L is small: 64 oxygen atoms already have mass of about $1024\,\mathrm{Da}$, exceeding the relevant mass range. For molecules consisting of different elements, we first compute the isotopic distributions of the individual elements, and then combine these distributions by folding in $O(|\Sigma| \cdot K^2)$ time.

We now come to the more challenging problem of efficiently computing the mean peak masses of a distribution. Doing so using the definition $m_k = \mathbb{E}(X \mid Y = N + k)$ is highly inefficient, because we have to sum up over all isotope species, so pruning strategies have been developed that lead to a loss of accuracy [12, 14]. But there exists a simple recurrence for computing these masses analogous to the folding of distributions, generalizing and improving on results in [12]:

Let $Y = Y_1 + \cdots + Y_l$ and $Y' = Y'_1 + \cdots + Y'_L$ be isotopic distributions of two molecules with monoisotopic nominal masses N and N', respectively. Let $p_k := \mathbb{P}(Y = N + k)$ and $q_k := \mathbb{P}(Y' = N' + k)$ denote the corresponding probabilities, m_k and m'_k the mean peak masses of the $+k$ peaks. Consider the random variable $Z = Y + Y'$ with monoisotopic nominal mass $\tilde{N} = N + N'$.

Theorem 1. *The mean peak mass $\tilde{m}_k$ of the $+k$ peak of the random variable $Z = Y + Y'$ can be computed as:*

$$\tilde{m}_k = \frac{1}{\sum_{j=0}^{k} p_j q_{k-j}} \cdot \sum_{j=0}^{k} p_j q_{k-j} \left(m_j + m'_{k-j} \right) \tag{2}$$

Note that $\sum_{j=0}^{k} p_j q_{k-j} = \mathbb{P}(Z = \tilde{N} + k)$. Since by independence, $\mathbb{P}(Y_1 = N_1, \ldots, Y_l = N_l) = \prod_i \mathbb{P}(Y_i = N_i)$, the theorem follows by rearranging summands, we omit the formal proof.

The theorem allows us to "fold" mean peak masses of two distributions to compute the mean peak masses of their sum. This implies that we can compute mean peak masses as efficiently as the distribution itself. This improves on the previously best known method [12], replacing the linear runtime dependence on the number of atoms by its logarithm.

5   Scoring Candidate Molecules

We want to discriminate between (tens of thousands of) candidate molecules generated by decomposing the monoisotopic mass. To this end, we compare the simulated isotopic distribution with the measured peaks. Matching peak pairs between the spectra is trivial for this application.

Zhang et al. [15] and Zhang and Chait [16] suggest to use Bayesian Statistics to evaluate mass spectra matches:

$$\mathbb{P}(\mathcal{M}_j|\mathcal{D}, \mathcal{B}) = \frac{\mathbb{P}(\mathcal{M}_j|\mathcal{B})\,\mathbb{P}(\mathcal{D}|\mathcal{M}_j, \mathcal{B})}{\sum_i \mathbb{P}(\mathcal{M}_i|\mathcal{B})\,\mathbb{P}(\mathcal{D}|\mathcal{M}_i, \mathcal{B})} \qquad (3)$$

where $\mathcal{D}$ is the data (the measured spectrum), $\mathcal{M}_i$ are the models (the candidate molecules), and $\mathcal{B}$ stands for any prior background information. In particular, we set the prior probability $\mathbb{P}(\mathcal{M}_j|\mathcal{B})$ to zero for all molecules but the decompositions of the monoisotopic mass. We assign prior probability zero to sum formulas that cannot correspond to a molecule, because of chemical considerations: For any molecule, the *degree of unsaturation* $DU = -\frac{v_1}{2} + \frac{v_3}{2} + v_4 + 1$ [11] is a non-negative integer, where v_1, v_3, v_4 denote the number of monovalent atoms (hydrogen), trivalent atoms (nitrogen, phosphor), and tetravalent atoms (carbon) if we assume that all elements are in their lowest valency state.

Next, we assign probabilities to the observed masses and intensities. Assuming independence (in particular from background information) we calculate $\mathbb{P}(\mathcal{D}|\mathcal{M}, \mathcal{B}) = \prod_j \mathbb{P}(M_j|m_j) \prod_j \mathbb{P}(f_j|p_j)$. Here, $\mathbb{P}(M_j|m_j)$ is the probability to observe peak j at mass M_j when its true mass is m_j, and $\mathbb{P}(f_j|p_j)$ is the probability to observe peak j with intensity f_j when its true intensity is p_j. Clearly, the independence of peak intensities is violated because these intensities sum to one, but this product can be seen as a rough estimate of the true probability.

We empirically estimate distributions of mass and intensity differences, as follows: We analyze the 69 mass spectra as described in Sec. 7 and, in addition, spectra of 33 molecules with mass above 1000 Da. We know the correct sum formulas for all of these mass spectra, so we can simulate the ideal isotope patterns and compare with masses and intensities of measured spectra.

Regarding peak masses, our data shows a systematic mass shift due to calibration inaccuracies. This can be eliminated for all masses but the monoisotopic mass: We do not compare masses of the $+1, \ldots$ peaks directly but instead, the difference to the monoisotopic peak, $M_j - M_0$ vs. $m_j - m_0$ for $j \geq 1$. In accordance with expert knowledge, mass differences increase with increasing mass of the molecule, so we use relative mass differences, $(M_0 - m_0)/m_0$ and $(M_j - M_0 - m_j + m_0)/m_j$. We determine mean and variance of these quantities. For intensities, our data indicates that ratios between measured and predicted peak intensity f_j/p_j follow a log normal distribution, so we determine mean and variance of $\log_{10} f_j - \log_{10} p_j$.

We want to estimate the probability that, given a peak with true mass m_j, a peak at mass M_j is observed in the measured spectrum: More precisely, the probability of observing a mass difference of $|M_j - m_j|$ or larger. For simplicity, we assume that relative mass differences follow a Gaussian distribution with parameters $(\bar{\mu}, \bar{\sigma})$. We can compute this probability using the complementary error function "erfc":

$$\mathbb{P}(M_j|m_j) = \mathrm{erfc}\left(\frac{|x_j - \bar{\mu}_j|}{\sqrt{2}\,\bar{\sigma}_j}\right) = \frac{2}{\sqrt{2\pi}} \int_z^\infty e^{-t^2/2} dt \quad \text{with } z := \frac{|x - \bar{\mu}|}{\bar{\sigma}} \qquad (4)$$

where $x_0 = (M_0 - m_0)/m_0$ and $x_j = (M_j - M_0 - m_j + m_0)/m_j$ for $j \geq 1$. Analogous computations can be executed for intensity differences. Regarding peaks missing in the measured spectrum, we can use the smallest intensity of any peak detected in the spectrum as an upper bound, and use a one-sided test for the peak intensity to be below this threshold.

6   Additive Invariants and Joint Decompositions

The mass of the monoisotopic peak is an *additive invariant* of the decompositions we are searching for: Given any solution, the sum of monoisotopic masses of all elements is the input mass M_0. We now present other additive invariants resulting from the observed isotope pattern. We consider a theoretical molecule, where i_E denotes the multiplicity of element E in the molecule, $E \in \Sigma$.

Given the observed normalized intensities $f_0, \ldots, f_K$ and peak masses $M_0, \ldots, M_K$, we easily estimate the *average mass of the molecule* as $M_{\text{av}} := \sum_i f_i M_i$. The average mass of an element E can be estimated by $\mu_1(E) := \mathbb{E}(X_E)$; we decompose the number M_{av} over the set $\{\mu_1(E) \mid E \in \Sigma\}$.

For every element E, let p_E denote the probability that an isotope of this element is monoisotopic. The *intensity of the monoisotopic peak* f_0 should equal the probability p^* that the molecule has monoisotopic mass, which implies that all atoms must have monoisotopic mass, thus $p^* = \prod_{E \in \Sigma} p_E^{i_E}$. Taking the logarithm we find $\sum_{E \in \Sigma} i_E \cdot \log p_E = \log f_0$. Defining a third set $\mu_2(E) := -\log p_E$, we can decompose $-\log f_0$ over the set $\{\mu_2(E) \mid E \in \Sigma\}$.

We have identified two more additive invariants resulting from the molecules' isotope pattern: the relative intensity of the +1 peak f_1/f_0, and the weighted mass difference of the +1 peak $\frac{f_1}{f_0}(M_1 - M_0)$.

For the current problem, we need to find *joint* decompositions for two or more masses $m_1, \ldots, m_k$ where each mass is decomposed over a different set of numbers of the same size. Formally, we state the

JOINT DECOMPOSITION PROBLEM. Let $\{a_{1,1}, \ldots, a_{1,n}\}, \ldots, \{a_{k,1}, \ldots, a_{k,n}\}$ be k sets of non-negative integers. Let $m_1, \ldots, m_k \in \mathbb{N}$. Find all joint decompositions c of $m_1, \ldots, m_k$, i.e., all $c = (c_1, \ldots, c_n) \in \mathbb{N}_0^n$ such that $Ac = m$, where $A = (a_{ij})_{i=1,\ldots,k,j=1,\ldots,n}$ and $m = (m_1, \ldots, m_k)$.

The problem is clearly related to the multidimensional integer knapsack problem [7]. In general, it is NP complete to decide if there exists at least one solution when the matrix has integer entries. At the other extreme, if we have n many equations, then A is a square matrix, and if its rows are linearly independent, we can compute its inverse A^{-1}. We then only need to check whether $c = A^{-1}m$ has only non-negative integer entries; if this is the case, then c is a joint decomposition of $m_1, \ldots, m_n$.

A naïve approach to solve the joint decomposition problem is to generate all decompositions c of m_1 and then test whether $\sum_i c_i a_{j,i} = m_j$ for all $j = 2, \ldots, k$. However, this involves generating many decompositions unnecessarily. Another approach is to construct Extended Residue (ER) tables for each set of numbers [2]: For one set $\{b_1, \ldots, b_m\}$, entry $\text{ER}(r, i)$ of this table, for $r = 0, \ldots, b_1 - 1$ and $i = 1, \ldots, m$, is the smallest number congruent r modulo b_1 which is decomposable over the set $\{b_1, \ldots, b_i\}$. Then, while running the algorithm on the ER table for $\{a_{1,1}, \ldots, a_{1,n}\}$, in each step of the recursion, we check whether there is still a feasible solution for all m_j, $j = 2, \ldots, k$, as well. If the answer is negative for one j, we terminate the current recursion step and continue with the next candidate. Note that this is a runtime heuristic only, since there may exist decompositions over each alphabet, but they may contradict each other.

Consider matrix A of dimension $k \times n$. By Gaussian elimination, we can find a lower triangular matrix $L \in \mathbb{R}^{k \times k}$ of full rank, and an upper triangular matrix $R \in \mathbb{N}^{k \times n}$ such that $A = LR$. Then, $Ac = m$ if and only if $Rc = m'$, where we can compute $m' = L^{-1}m$. In particular, c must satisfy the *bottom equation* of $Rc = m'$ which has at most $n - k + 1$ non-zero coefficients. If all coefficients of R are non-negative integers, we have a new instance of the joint decomposition problem, which we can solve iteratively, beginning with the bottom equation: We build ER tables for each (new) set of numbers, run the decomposition algorithm on the bottom one, checking in each step of the recursion whether the solution is feasible over all sets. When having computed a decomposition of m'_n over $r_{k,k}, \ldots, r_{k,n}$, we continue with the next equation, which has one variable more. Schur's Theorem [13] states that the number of decompositions of an integer M over $\{b_1, \ldots, b_m\}$ grows with a polynomial in M of degree $m - 1$. Since the ER-algorithm of [2] runs in time proportional to the number of decompositions, and the number of solutions of the bottom equation is considerably lower than of any of the original equations, we improve on runtime.

However, even though we can guarantee that all entries of R are integers, some could be negative, yielding infinitely many solutions. In order to avoid negative entries, one needs to exchange columns; details will be described elsewhere. We refer to this algorithm as Dimension Reduction (DR) algorithm. In Sec. 7, we will see that the DR algorithm yields a significant improvement over the approach of simultaneously decomposing over the individual sets.

7   Results

Data set. Our data set consists of 69 mass spectra with single charge from several organic (macro)molecules, composed of the elements CHNOPS. For every such spectrum, the sum formula of the sample molecule is known. The spectra were acquired over the last two years; the molecules range in mass from 284 to 960 Da. Electrospray ionization (ESI) experiments were performed using a Fourier Transform Ion Cyclotron Resonance (FT-ICR) mass spectrometer APEX III (Bruker Daltonik GmbH, Bremen, Germany) equipped with a 7.0 T, 160 mm bore superconducting magnet (Bruker Analytik GmbH – Magnetics, Karlsruhe, Germany), infinity cell, and interfaced to an external (nano)ESI ion source. Peak detection and estimation of peak masses and intensities are conducted using vendor software.

Identification accuracy and runtimes. Every input "mass spectrum" consists of masses $M_0, \ldots, M_k$ and intensities $f_0, \ldots, f_k$. For every such spectrum, we compute all molecules such that the monoisotopic mass m_0 has relative mass

Table 1. Number of correct sum formulas at certain positions of the output list, for several mass ranges. Runtimes in seconds per spectrum. See text for details.

mass range	no. of spectra	rank in output list					average no. sum formulas			runtime
		1	2	3–5	6–10	11+	int.	real	chem.	
200–300	3	3	0	0	0	0	60.7	26.3	5	0.0006
300–400	20	18	2	0	0	0	165.3	70.1	6.4	0.0012
400–500	25	13	5	5	2	0	560.3	236.4	17.8	0.0043
500–600	1	0	1	0	0	0	1956	833	51	0.0164
600–700	2	1	0	1	0	0	2204	934.5	30.5	0.0190
700–800	5	0	2	1	0	2	7548.6	3205.2	167.6	0.0706
800–900	8	0	1	0	1	6	12521	5325.9	340.6	0.1217
900–1000	5	0	0	0	0	5	23443	9972.8	770	0.2338

difference of at most 2 ppm, $|M_0 - m_0|/m_0 \leq 2 \cdot 10^{-6}$. To do so, we decompose integer masses with some blowup $b \in \mathbb{R}$, see Sec. 3, and discard molecules with real mass outside the mass interval. Next, we discard molecules that have negative or non-integer degree of unsaturation DU. For every such molecule, we compute its theoretical isotopic distribution (with $K = 10$) and compare it to the measured isotopic distribution, see Sec. 4 and 5. We rank the molecules according to resulting probabilities. We do not use any other background information to identify the molecule, in order to evaluate the discriminative power of isotope patterns.

Out of the 69 mass spectra, 35 result in a correct identification; in 81 % of the mass spectra, the correct interpretation is found in the top 10 interpretations. There is a clear correlation between mass and identification accuracy, as seen in Table 1. For mass spectra below 700 Da, the correct interpretation is always found in the top 10 interpretations.

We analyzed all 69 mass spectra on a Pentium M 1.5 GHz processor with blowup $b = 5 \cdot 10^4$, using only a few Megabyte of memory. This results in runtimes of less than 1/4 second per spectrum for the complete analysis of one mass spectrum. Clearly, runtimes depend on molecule masses, see Table 1. Increasing the blowup beyond $5 \cdot 10^4$ increased runtimes, presumably because the smaller table can be kept in the processor cache.

For every mass range, we also report in Table 1 the number of integer decompositions (int.), the number of real decompositions (real), cf. Sec. 3, and the number of sum formulas with non-negative integer degree of unsaturation (chem.). These numbers are averages over all molecules in the mass range.

Joint decomposition algorithm. As a first evaluation of our algorithms for decomposing metabolite isotope patterns, we decomposed molecular masses over the CHNOPS alphabet, using data from the KEGG LIGAND database [6]: We extracted 10 300 sum formulas over the alphabet CHNOPS, which reduced to 5 627 non-redundant sum formulas. We computed the monoisotopic and average masses and used these as input for our algorithms using precision $\delta = 10^{-4}$ Da.

Runtimes on a Sun Fire 880 with 900-MHz UltraSPARC-III-CPU, 32 GB RAM, are shown in Fig. 1: (i) computing all decompositions of the monoisotopic mass, (ii) doing the same respecting decompositions of the average mass of the

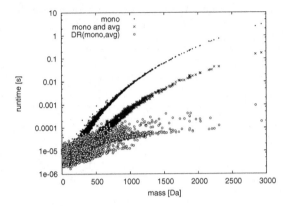

Fig. 1. Runtimes of decomposition algorithms in comparison (logarithmic scale): decomposing the monoisotopic mass with no additional information (dots), simultaneously decomposing the monoisotopic and average masses (crosses), and the DR-algorithm on monoisotopic and average masses (circles)

molecule, and (iii) using the DR algorithm on the monoisotopic and average masses. Runtimes for $\delta = 10^{-3}$ Da are similar (data not shown). Our experiments show that using dimension reduction as is done by the DR algorithm is greatly superior to using additive invariants directly. It is realistic to expect that results will carry over straightforwardly to decompositions of real masses, since the runtimes of all algorithms are effected in the same way.

8 Conclusion

We presented an approach to determine the sum formula of an unknown metabolite solely from its high resolution isotope pattern. Our approach allows us to reduce the number of potential sum formulas to only a few candidates; in many cases we were able to determine the correct sum formula. The approach is time and memory efficient and can be executed on a regular desktop PC. Results on experimental data show its potential, in particular for metabolites below 700 Da. We further presented methods for the efficient simulation of isotope patterns, as well as an approach to significantly reduce the search space of molecule candidates (additive invariants, joint decomposition). Both are vital in ranges above 1000 Da, where the search space increases rapidly.

Nevertheless, our results are only a first step towards automated determination of sum formula from high resolution mass spectrometry data. We want to conduct further studies regarding mass and intensity variations and are currently gathering an independent test set of about 100 sample spectra. Note that we have deliberately ignored some information available in the data, in order to evaluate the discriminative power of a single isotope pattern, such as different charge states, prior probability of the elements, or neutral losses. We are also evaluating the impact of increased mass accuracy on our algorithms.

Acknowledgments

AP supported by Deutsche Forschungsgemeinschaft (BO 1910/1), ZsL by Alexander von Humboldt Foundation and the Bundesministerium für Bildung und Forschung, within the group "Combinatorial Search Algorithms in Bioinformatics." Additional programming by Marcel Martin, whom we thank for his unfailing support, and by Marco Kortkamp.

References

1. G. Audi, A. Wapstra, and C. Thibault. The AME2003 atomic mass evaluation (ii): Tables, graphs, and references. *Nucl. Phys. A*, 729:129–336, 2003.
2. S. Böcker and Zs. Lipták. Efficient mass decomposition. In *Proc. of ACM Symposium on Applied Computing (ACM SAC 2005)*, pages 151–157, 2005.
3. A. Fürst, J.-T. Clerc, and E. Pretsch. A computer program for the computation of the molecular formula. *Chemom. Intell. Lab. Syst.*, 5:329–334, 1989.
4. A.H. Grange, M.C. Zumwalt, and G.W. Sovocool. Determination of ion and neutral loss compositions and deconvolution of product ion mass spectra using an orthogonal acceleration time-of-flight mass spectrometer and an ion correlation program. *Rapid Commun. Mass Spectrom.*, 20(2):89–102, 2006.
5. C.S. Hsu. Diophantine approach to isotopic abundance calculations. *Anal. Chem.*, 56(8):1356–1361, 1984.
6. M. Kanehisa, S. Goto, M. Hattori, K.F. Aoki-Kinoshita, M. Itoh, S. Kawashima, T. Katayama, M. Araki, and M. Hirakawa. From genomics to chemical genomics: new developments in KEGG. *Nuc. Acid Res.*, 34:D354–D357, 2006.
7. H. Kellerer, U. Pferschy, and D. Pisinger. *Knapsack Problems*. Springer, 2004.
8. H. Kubinyi. Calculation of isotope distributions in mass spectrometry: A trivial solution for a non-trivial problem. *Anal. Chim. Acta*, 247:107–119, 1991.
9. F.W. McLafferty. *Wiley Registry of Mass Spectral Data*. John Wiley & Sons, 7th edition with NIST 2005 spectral data edition, 2005.
10. S. Ojanperä, A. Pelander, M. Pelzing, I. Krebs, E. Vuori, and I. Ojanperä. Isotopic pattern and accurate mass determination in urine drug screening by liquid chromatography/time-of-flight mass spectrometry. *Rapid Commun. Mass Spectrom.*, 20(7):1161–1167, 2006.
11. V. Pellegrin. Molecular formulas of organic compounds: the nitrogen rule and degree of unsaturation. *J. Chem. Educ.*, 60(8):626–633, 1983.
12. A.L. Rockwood, J.R. Van Orman, and D.V. Dearden. Isotopic compositions and accurate masses of single isotopic peaks. *J. Am. Soc. Mass Spectr.*, 15:12–21, 2004.
13. H. Wilf. *Generating Functionology*. Academic Press, 1990.
14. J.A. Yergey. A general approach to calculating isotopic distributions for mass spectrometry. *Int. J. Mass Spectrom. Ion Phys.*, 52(2–3):337–349, 1983.
15. N. Zhang, R. Aebersold, and B. Schwikowski. ProbID: a probabilistic algorithm to identify peptides through sequence database searching using tandem mass spectral data. *Proteomics*, 2(10):1406–1412, 2002.
16. W. Zhang and B.T. Chait. ProFound: an expert system for protein identification using mass spectrometric peptide mapping information. *Anal. Chem.*, 72(11):2482–2489, 2000.

A Method to Design Standard HMMs with Desired Length Distribution for Biological Sequence Analysis

Hongmei Zhu, Jiaxin Wang, Zehong Yang, and Yixu Song

Tsinghua National Laboratory for Information Science and Technology,
Department of Computer Science and Technology, Tsinghua University,
Beijing, 100084, China
zhuhongmei00@mails.tsinghua.edu.cn

Abstract. Motivation: Hidden Markov Models (HMMs) have been widely used for biological sequence analysis. When modeling a phenomenon where for instance the nucleotide distribution does not change for various length of DNA, there are two popular approaches to achieve a desired length distribution: explicit or implicit modeling. The implicit modeling requires an elaborately designed model structure. So far there is no general procedure available for designing such a model structure from the training data automatically.

Results: We present an iterative algorithm to design standard HMMs structure with length distribution from the training data. The basic idea behind this algorithm is to use multiple shifted negative binomial distributions to model empirical length distribution. The negative binomial distribution is obtained by an array of n states, each with the same transition probability to itself. We shift this negative binomial distribution by using a serial of states linearly connected before the binomial model.

1 Introduction

Biological sequences, such as DNA, usually contain some homogeneous regions with uncertain length where the composition distribution does not change. When hidden Markov models are used for biological sequence analysis, there are typically one or more states modeling this kind of regions. The simplest model design is to make a state with transition to itself with probability p [1]. After entering the state there is a probability $1 - p$ of leaving it, so the probability of staying in the state for l residues is

$$P(l) = (1 - p)p^{l-1}. \tag{1}$$

Here the emission probabilities are disregarded. In other words, this model generates a sequence of length l with probability $P(l)$. This exponentially decaying distribution on lengths (called a geometric distribution) can be inappropriate in some applications.

P. Bücher and B.M.E. Moret (Eds.): WABI 2006, LNBI 4175, pp. 24–31, 2006.

A more subtle way of obtaining a non-geometric length distribution is to use an array of n states, each with a transition to itself of probability p and a transition to the next of probability $1 - p$ (see Figure 1(a)). For any given state sequence (called path) of length l through the model, the probability of all its transition is $p^{l-n}(1-p)^n$. The number of possible paths of length l through the model is $\binom{l-1}{n-1}$, so the total probability summed over all possible paths is

$$P(l) = \binom{l-1}{n-1} p^{l-n}(1-p)^n. \tag{2}$$

This distribution is called a negative binomial. This model thus generates sequences with lengths varying according to this distribution, as shown in Figure 2 for $p = 0.99$ and $n \leq 5$.

The length distribution can be modeled in an explicit way, if the states of HMMs are allowed to emit more than one symbol [2]. This extension to HMMs is called hidden semi-Markov models (HSMMs). The "super" states once emit a sequence of length with probability from the desired length distribution. This model allows an accurate modeling of the length at the cost of computation time. If no further heuristic is used the computation time of the typical algorithms (Viterbi, forward algorithm) is at least proportional to the maximal possible length of the state [3]. In some situations, these homogeneous regions can be very long and it is practically infeasible to explicitly model the whole length distribution by HSMMs.

In this work we focus on modeling length distribution implicitly and accurately by using standard HMMs. A general method is presented to design model structure to meet the desired length distribution. The negative binomial distribution model is used as basic structure element. By placing a serial of states linearly connected before the binomial model, the negative binomial distribution is shifted (see Figure 1(b)). This method searches for several shifted negative binomial distributions and combines them to fit the desired length distribution.

In the following we explain how to design a standard HMM with desired length distribution from the very beginning, the training data. Then the experiments on Ciona intestinalis intron data and rice exon data are showed.

a. Negative binomial model

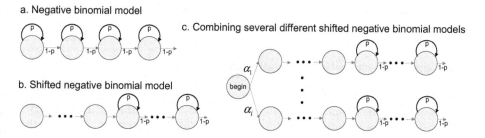

c. Combining several different shifted negative binomial models

b. Shifted negative binomial model

Fig. 1. Model structures used to model the length distribution in this work

2   Methods

In this section, first we describe the shifted negative binomial model more concretely. Then we give the procedures of designing a standard HMM structure with desired length distribution estimated from the training data.

2.1   Shifted Negative Binomial Distribution and the Combination

As described above, by using an array of n states, each with a transition to itself of probability p and a transition to the next of probability $1 - p$, we obtain a negative binomial length distribution like (2). A serial of preceding states linearly connected will shift this binomial length distribution. If the number of this serial states is m, the resulting length distribution would be

$$P(l) = \binom{l - m - 1}{n - 1} p^{l-m-n}(1 - p)^n. \tag{3}$$

Figure 3 shows the binomial distribution for $n = 3$ and $m = 100$.

These shifted negative binomial distribution models are used as basic structure elements in the following structure design. They are combined in the way shown in Figure 1(c), each with a coefficient α_i.

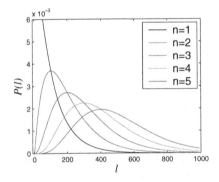

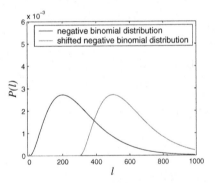

Fig. 2. The probability distribution over lengths for models in Figure 1(a) with $p = 0.99$ and n ranging from 1 to 5

Fig. 3. The probability distribution over lengths for negative binomial distribution and shifted negative binomial distribution with $p = 0.99$ and $n = 3$

2.2   Length Distribution Modeling Algorithm

Suppose the training dataset contains some homogeneous sequences of length within a wild range. First we calculate the length distribution (called empirical distribution) of training sequences based on length frequencies. This length distribution usually is very rough because of the relatively small sample in the training set. Then it is smoothed by some smoothingprocedure, which gives the smoothed distribution. The shifting range is determined based on the empirical distribution. After that the shifted negative binomial distribution candidates are decided. We choose the most fit candidate for the smoothed distribution based on their correlation coefficient. The least squares fitting of this candidate to the smoothed distribution is solved as a quadratic programming problem. Then the fitting part is subtracted from the smoothed distribution, which results in the "residue distribution". In the next iteration, the most fit candidate for the residue distribution is found, and the least squares fitting of these two selected distribution elements to the smoothed distribution is solved. After each iteration the residue distribution is updated and the iteration stops until the exit condition is reached.

More concretely, the modeling algorithm works like this:

1. Calculate the length frequencies: $F(l)$. The empirical distribution is given by

$$P^e(l) = F(l)/ \sum_{Lmin}^{Lmax} F(i). \qquad (4)$$

2. Find the length L_{perc1} that satisfies $P(l > L^{perc1}) = 1\%$ and length L_{perc10} that satisfies $P(l > L^{perc10}) = 10\%$.

3. Replace the probability mass $P^e(l)$ at position $l(l = L_{min} \cdots L^{perc1})$ by a "discretized" normal density with mean $\mu = l$ and variance $\sigma^2 = 2l/F(l)$, scaled so that the total mass is $P^e(l)$ [4]. The resulting distribution is the smoothed distribution denoted as $P^s(l)(l = L_{min} \cdots L^{perc1})$.

4. The minimum length the model is allowed to capture is denoted by L_{min}. Then the minimum shifting value $m_{min} = L_{min} - 1$. The maximum shifting value $m_{max} = L^{perc10}$. The shifted negative binomial candidates are all these distribution elements $E_{m,n,p}$ with $m = m_{min}, \cdots, m_{max}$, $n = 1, \cdots, 5$ and $p = 0.90, 0.91, \cdots, 0.99$. All the distributions $E_{m,n,p}(l)(L = L_{min}, \cdots, L^{perc1})$ with $m = 0$ are calculated before the iteration algorithm.

5. Set $j = 0$ and the residue distribution $R^0 = P^s$.

6. Set $j = j+1$. Calculate the correlation coefficient r of each candidate distribution and the residue distribution R^{j-1}. Choose the distribution candidate that gives the maximum r as the current selected element $E_j(l)$. The least squares fitting of $E_1(l), \cdots, E_j(l)$ to the smoothed distribution is solved as a quadratic programming problem. The solution minimizes

$$S^j = \frac{1}{2}\|\Phi^j \alpha^j - P^s\|^2 + C^j\|\alpha^j\| \qquad (5)$$

With respect to α^j under the constraints of $\alpha_i^j \geq 0(i = 1, \cdots, j)$. Here $\Phi^j = [E_1 \cdots E_j]$ and C^j is used to control the over-fitting phenomenon. After α^j is determined, the residue distribution is given by

$$R^j = P^s - \Phi^j \bullet \alpha^j. \qquad (6)$$

7. If the increment in the correlation coefficient of $\Phi^j \bullet \alpha^j$ and P^s is larger than a predetermined value and j does not exceed a predetermined value, go to step 6.

8. Normalize α^j so that $\sum_1^j \alpha_i^j = 1$. Use the shifted negative binomial models $E_1(l), \cdots, E_j(l)$ and their corresponding coefficients to construct a HMM for these sequences in the training set.

2.3   State *Tying* and the Most Probable Labeling

This implicit length modeling, in other words, has a few basic states appearing again and again to achieve a certain length distribution. A basic states and all its copies are usually supposed to have the same emission probabilities. This is called *tying* of states [5]. During parameter estimation, emission probabilities are needed not to be computed for every state, but just for every *tying* group of states.

There's another feature about the implicit length modeling. The probability of a particular length is distributed across many different state paths. In practice, the states can be labeled and different state paths may give the same 'labeling' path. The most probable labeling path is often more significant than the most probable state path identified by the Viterbi algorithm. Krogh present a so-called 1-best algorithm to find this most probable labeling path [5].

2.4   Allow a State to Emit a Certain Number of Symbols

There may be a lot of states used to shift the negative binomial models. As the number of states is increased, the time and space complexity of typical HMM algorithms is increased.

The time complexity of standard HMM is given as $O((B + K)NT)$ [3]. Here B denotes the number of operations to compute an observation likelihood for one symbol. K denotes average number of predecessor states. N denotes the total number of unique states in HMM set. T is the number of symbols in the sequence being analyzed. The space complexity of standard HMM is given as $O(NT)$. After the shifting states are used, N is increased by several times. Since all the linearly connected states have only one predecessor state, the time complexity is not increased significantly. The space complexity is proportional to the number of states.

If we use one single state to replace these linearly connected states, it will largely reduce the space complexity. This modification is doable and will not lose any useful information, since all the internal transmission probabilities are 1. Let N_1 denote the number of states emitting a certain number (more than one) of symbols, N_2 denote all the other state, and D denote the maximum length one state emits. Then we obtain the time complexity of this HMM as $O((B + K)N_2T + (DB + K)N_1T)$ and the space complexity as $O((N_1 + N_2)T)$. N_1 is typically less than 10 and as we can see this modification largely reduces the space complexity. The time complexity of typical HSMMs is $O((B + D^sK)NT)$ and D^s is the maximum length one state emits in HSMMs and usually much larger than D.

3   Experiments

3.1   Modeling the Length Distribution of Ciona Intestinalis Introns

Ciona intestinalis genes were downloaded from the ftp.ensembl.org site provided by the Ensembl project. After the 'sanity' checks there are 21,164 genes left, from which 44,053 introns were extracted as our intron training dataset.

Figure 4 shows the smoothed length distribution and Figure 5 shows the length distribution given by the HMM structure designed by the methods described above. All the selected shifted negative binomial elements are listed in Table 1. After six iterations we obtained a correlation coefficient $r = 0.99$ of the length distribution given by the combined model and the smoothed distribution.

Table 1. Selected negative binomial elements for ciona intron length distribution (α—coefficients corresponding to model elements; m—the number of shifting states; n—the number of states in the negative binomial models; p—the transition probability from a state to itself in the binomial model)

index	α	m	n	p
1	0.52	10	3	0.99
2	0.08	47	2	0.90
3	0.28	248	2	0.99
4	0.04	206	5	0.90
5	0.06	479	5	0.99
6	0.02	192	2	0.90

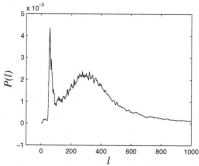

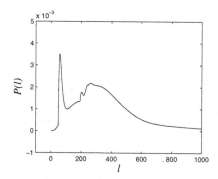

Fig. 4. The smoothed length distribution for Ciona intestinalis introns

Fig. 5. The modeled length distribution for Ciona intestinalis introns

3.2   Modeling the Length Distribution of Rice Exons

Rice genes are from the rice annotation dataset assembled by TIGR Rice Genome Project [6], at http://www.tigr.org/tdb/e2k1/osa1s. After the 'sanity' checks, there are there 25,074 genes left, from which 78,170 internal exons were extracted as our exon training dataset.

Nucleotide distribution in Exons depends on the three codon positions. To make this experiment simple, we disregard non-homogeneousness here. Figure 6 shows the smoothed length distribution and Figure 7 shows the length distribution given by the HMM structure designed by the methods described above. All the selected shifted negative binomial elements are listed in Table 2. After four iterations we obtained a correlation coefficient $r = 0.95$ of the length distribution given by the combined model and the smoothed distribution.

Table 2. Selected negative binomial elements for rice internal exon length distribution (the denotations are the same as in Table 1)

index	α	m	n	p
1	0.81	29	2	0.98
2	0.03	68	1	0.90
3	0.14	155	3	0.99
4	0.02	4	2	0.90

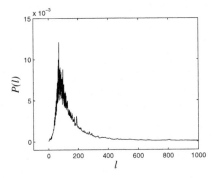

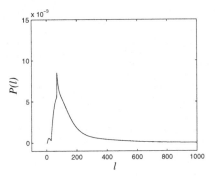

Fig. 6. The smoothed length distribution for rice internal exons

Fig. 7. The modeled length distribution for rice internal exons

For rice exons, the modeled length distribution is not as accurate as for the ciona introns according to the final correlation coefficients. This may be mostly caused by the rougher smoothed distribution of rice exon length. However the true distribution of lengths is likely to be much smoother than this. We do not expect an equally accurate fitting to this rougher distribution.

4   Conclusion

We present a simple iteration algorithm to design standard HMMs with desired length distribution. To reduce the complexity of typical HMM-related algorithm, a state is allowed to emit a certain number of symbols. The resulting HMMs have the capability to accurately model the length distribution and are not as complex as hidden semi-Markov models in computation time. The design method can work without any manual help, which make it applicable to on-line model design.

Acknowledgements

We thank Zheng Danian and Liu Hongbo, PhD students in the Department of Computer Science and Technology, Tsinghua University, for helpful discussions.

References

1. Durbin, R., Eddy, S., Krogh, A., Mitchison, G.: Biological sequence analysis: probabilistic models of proteins and nucleic acids. Tsinghua University Press, Beijing (2002)
2. Rabiner, L.R.: A tutorial on hidden markov models and selected applications in speech recognition. In: Proceedings of the IEEE. Volume 77. (1989) 257–286
3. Michael, T.J.: Capacity and complexity of hmm duration modeling techniques. IEEE signal processing letters **12**(5) (2005) 407–410
4. Burge, C.: Identification of genes in human genomic DNA. PhD thesis, CA: Stanford University (1997)
5. Krogh, A.: Two methods for improving performance of an hmm and their application for gene-finding. In: proceedings of the 5th international Conference on Intelligent Systems for Molecular Biology, Menlo Park, CA, AAAI Press (1997) 179–186
6. Yuan, Q., Ouyang, S., Liu, J., Suh, B., Cheung, F., Sultana, R., Lee, D., Quackenbush, J., Buell, C.R.: The TIGR rice genome annotation resource: Annotating the rice genome and creating resources for plant biologists. Nucleic Acids Research **31**(1) (2003) 229–233

Efficient Model-Based Clustering
for LC-MS Data*

Marta Łuksza[1], Bogusław Kluge[1], Jerzy Ostrowski[2], Jakub Karczmarski[2],
and Anna Gambin[1]

[1] Institute of Informatics, Warsaw University, Banacha 2, 02-097 Warsaw, Poland
[2] Department of Gastroenterology, Medical Center for Postgraduate Education and
Maria Sklodowska-Curie Memorial Cancer Center and Institute of Oncology,
Roentgena 5, 02-781, Warsaw, Poland

Abstract. Proteomic mass spectrometry is gaining an increasing role
in diagnostics and in studies on protein complexes and biological systems.
The issue of high-throughput data processing is therefore becoming more
and more significant. The problems of data imperfectness, presence of
noise and of various errors introduced during experiments arise.

In this paper we focus on the peak alignment problem. As an alternative to heuristic based approaches to aligning peaks from different mass
spectra we propose a mathematically sound method which exploits the
model-based approach. In this framework experiment errors are modeled
as deviations from real values and mass spectra are regarded as finite
Gaussian mixtures. The advantage of such an approach is that it provides
convenient techniques for adjusting parameters and selecting solutions of
best quality. The method can be parameterized by assuming various constraints. In this paper we investigate different classes of models and select
the most suitable one. We analyze the results in terms of statistically significant biomarkers that can be identified after alignment of spectra.

1   Introduction

In the 1990s mass spectrometry arose as a powerful tool for protein analysis [1]
becoming one of the main technologies used in proteomics. The previously used
techniques, such as the two-dimensional gel electrophoresis, suffered from the
lack of sensitive methods for protein identification in a reasonable time. To the
contrary, mass spectrometric experiments, which measure amounts of molecules
present in a sample, are fast and allow for high-throughput data processing.

Multidimensional mass spectrometry. Mass spectrometry combined with
chromatographic technologies produces more detailed, multidimensional spectra.
Liquid chromatography mass spectrometry (LC-MS) separates compounds with
accordance to their hydrophobicity before they are introduced into the ion source
and mass spectrometer. With the usage of the HPLC column it produces a

---

* The research described in this paper was partially supported by Polish Ministry of
Education and Science grants KBN-8 T11F 021 28 and PBZ-KBN-088/P04/2003.

P. Bücher and B.M.E. Moret (Eds.): WABI 2006, LNBI 4175, pp. 32–43, 2006.

mass spectrum for each eluted fraction of molecules. Each detected peptide is characterized by two coordinates: its molecular mass over charge (m/z) ratio and elution time value (also called *retention* time).

Using computational methods of mass spectra analysis, composition of a sample can be almost uniquely identified. This ability can be exploited in diagnostics, while searching for known factors that cause or are related to a disease. On the other hand, even if such proteins are unknown, a significant insight can be acquired by a comparative analysis of samples. Mass spectra obtained from the diseased samples may show some characteristic patterns in peak intensities that distinguish them from the rest of the samples. Detection of such patterns is possible again with mathematical and statistical analysis and by application of various data mining techniques and algorithms (see [2,3]).

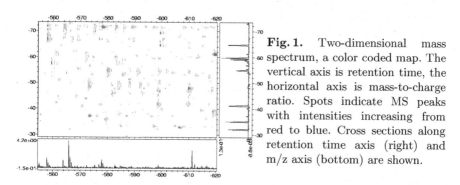

Fig. 1. Two-dimensional mass spectrum, a color coded map. The vertical axis is retention time, the horizontal axis is mass-to-charge ratio. Spots indicate MS peaks with intensities increasing from red to blue. Cross sections along retention time axis (right) and m/z axis (bottom) are shown.

Peak alignment problem. One of the crucial points in the comparative analysis is the assumption that the mass spectra obtained from different samples are properly aligned, i. e. the corresponding peptide signals are matched with each other and the related peptide intensities are correctly identified in every sample. In the ideal case the same molecules detected in spectra for different samples should have the same m/z and retention time values. Unfortunately, chemical experiments are very sensitive to external conditions and instruments responsible for measurements are imperfect [4]. Any contamination of the HPLC column causes changes in the elution time, thereby only identical experiment conditions can guarantee that a peptide is eluted at exactly the same time. One can observe quite significant drifts, especially in the retention time dimension.

Related research. The detailed description of the crucial preprocessing phase is often neglected in studies on mass spectrometry data that focus mainly on further stages of analysis [3]. Most of the solutions that have been proposed concern aligning peaks of one dimensional spectra. To this end, Tibshirani et al. [2] apply complete-linkage hierarchical clustering to align MALDI-MS spectra, where distances are computed along log m/z axis and each cluster represents one peptide signal. Wong et al. [5] construct a reference, consensus spectrum, which is aligned with the remaining spectra with insertion and deletion of data points. Prakash et al. [6] approach the problem of LC-MS spectra alignment with a dynamic programming algorithm for aligning a pair of mass spectrum runs and a heuristic score function for assessing spectra similarity. In the algorithm proposed by Smith et al. [7] (included in the xcms package) two-dimensional spectra are sliced into narrow overlapping m/z segments, within which a kernel density estimator is used to detect the so called "meta-peaks," gathering peaks of similar values of retention time.

Mass spectra peak alignment is a clustering problem: finding groups of related peaks in a data set. Clustering based on probability models emerges as an alternative to the heuristic based approaches to grouping. In the model-based approach [8] it is assumed that the data is generated by a finite mixture of underlying multivariate Gaussian distributions. This method has been successfully applied in biological context by Yeung et al. [9] for clustering gene expression data. Its great advantage is that it provides convenient mathematical techniques for determining the number of clusters and other model parameters.

Our results. We propose a novel approach to mass spectrum peak alignment with takes an advantage of the clustering based on probability models. The implemented process, presented in Figure 2, involves a two-stage clustering of peak

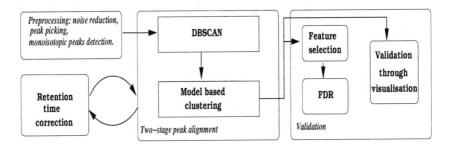

Fig. 2. The control flow through different phases of the proposed method

data: in the first step the data is partitioned into preliminary groups with the DBSCAN algorithm [10], in the second step the model-based procedure is applied. The problem of peak alignment is very challenging due to large size of the data and the large number of expected clusters. Thanks to the proposed strategy, the overall method is both efficient and it retains the advantages of the model-based approach. The quality of an alignment can be validated with a visual analysis. Besides, we also propose a feature selection based method for this purpose. We demonstrate the results on a colorectal cancer mass spectrum data set.

Overview of the paper. Section 2 discusses the model-based clustering in detail, covering the problem of model selection and the description of our method. Experiments and results are presented and discussed in Section 3. Conclusions and directions for further research are given in Section 4.

2   Methods

LC-MS data sets. From the mathematical point of view, a mass spectrum peak $p \in \mathbb{R}^3$ is a tuple (p_m, p_{rt}, p_{int}), where p_m is the m/z value, p_{rt} is the retention time and p_{int} is the intensity.

Each peptide has many isotopic compositions, moreover it may get different charges when processed by the mass spectrometer. This results in the whole family of peaks in the spectrum corresponding to the same peptide. To reduce the redundancy, we use an automatic MS interpretation tool described in [11] to generate the list of monoisotopic peaks from each spectrum.

The problem of alignment of a set of LC-MS peaks can be modeled as a Gaussian mixture, where each component corresponds to peaks from the same peptide with the same charge in different samples. As mentioned in Section 1, variances are determined by a measurement error of the spectrometer and various chemical factors. The dimensions taken into account during the clustering are the mass-to-charge ratio and the retention time (not the intensity).

In the ideal case each peptide signal has one corresponding peak in every sample. However, in some samples the signal may not be detected. Hence the clusters are expected to have sizes about the number of samples or smaller. The total number of clusters is often very large, with ideally one cluster corresponding to each peptide signal.

Model-based clustering. The drawback of the heuristic based approaches is that they do not provide means for evaluating clustering quality and for setting the right algorithm parameters, for instance the number of clusters. Furthermore, the statistical properties of these methods are unspecified and hence no formal inference can be applied. These problem can be overcome if a probability model underlying the data is assumed. In the following section we discuss the use of finite Gaussian mixtures to model clusters of MS peaks.

The idea underlying the model-based clustering [8] is that the data is sampled independently from many component distributions, but the components' labels are missing. Assume that $Y = \{y_i | i = 1, \ldots, n\}$ is a set of independent, p-dimensional observations. Let G be the number of components and let f_k, $k \in \{1, \ldots, G\}$ be the density function of the k-th component and θ_k be the set of parameters describing f_k. The probability that observation y is from the k-th component is τ_k. The probability density function of the mixture of the distributions is:

$$f(y) = \sum_{k=1}^{G} \tau_k f_k(y | \theta_k)$$

In our applications f_k is the 2-dimensional Gaussian density ϕ with mean μ_k and covariance matrix Σ_k:

$$\phi(y | \mu_k, \Sigma_k) = \frac{\exp\left(-\frac{1}{2}(y - \mu_k)^T \Sigma_k^{-1}(y - \mu_k)\right)}{\sqrt{(2\pi)^p \det \Sigma_k}}$$

The *complete data likelihood* of the observed data Y is given by the formula:

$$\mathcal{L}(\tau_1, \ldots, \tau_G, \mu_1, \ldots, \mu_G, \Sigma_1, \ldots, \Sigma_G) = \prod_{i=1}^{n} \sum_{k=1}^{G} \tau_k \phi(y_i | \mu_k, \Sigma_k) \tag{1}$$

Computing the parameters $\tau_1, \ldots, \tau_G, \mu_1, \ldots, \mu_G, \Sigma_1, \ldots, \Sigma_G$ that maximize the likelihood (1) is rather problematic in case of Gaussian mixture models. The *Expectation-Maximization* (EM) algorithm proposed by Dempster et al. [12] was designed to solve this kind of problems. The basic idea of the algorithm is to introduce latent (hidden) variables, knowledge of which would simplify the maximization. In case of the clustering problem, these variables indicate which component each observation belongs to.

Convergence of the EM algorithm may be very slow [13]. Furthermore, it does not guarantee to return the global maximum of the likelihood. The algorithm is very sensitive to initialization conditions, therefore is often started with a cluster assignment from a model-based hierarchical clustering algorithm. Both the model-based algorithms, agglomerative hierarchical and EM, can be parameterized by adding constraints on the covariance matrices of the mixture components. Note that while the mean μ_k is the center around which the elements of the k-th component are distributed, the shape of the cluster is determined by the covariance matrix Σ_k.

Banfield and Raftery [14] proposed to decompose the covariance matrix as $\Sigma_k = \lambda_k D_k B_k D_k^T$, where λ_k is a scalar responsible for the cluster volume, D_k is the matrix of orthogonal eigenvectors that determines the cluster orientation and B_k, where $\det B_k = 1$ is a diagonal matrix which determines the cluster shape. This parameterization allows for imposing constraints on cluster shapes when one possesses prior knowledge or beliefs about what the clusters should look like.

In case of a mass spectrum peak data one does not expect to observe any correlation between errors on the m/z and the retention time dimensions. Hence, we have decided to focus our attention only on a class of diagonal models, i.e. the ones that can be noted as $\Sigma_k = \lambda_k B_k$, where B_k is a diagonal matrix and $\det B_k = 1$. Since we knew what range of standard deviations we should expect, we also investigated the more restrictive models with covariance matrices set manually. The standard deviations were imposed either only on the m/z dimension or on both dimensions. In the first case we let the retention time deviation vary among clusters. The reasoning for this is that variations in retention time are far less predictable than the ones of mass-to-charge ratios. We tried to fit the following models:

- $[\lambda_k B_k]$ – the most general diagonal model, where the volumes and shapes are allowed to vary between clusters,
- $[\lambda B]$ – a model of identical clusters,
- $[\lambda_k B]$ – a model in which clusters have identical shape but their volumes may vary,
- $[\lambda B_k]$ – a model, where all the clusters have the same volume but their shapes may vary,

- [**manually set variances, equal volume**] – a model with $\Sigma_k = \begin{pmatrix} c_1 & 0 \\ 0 & c_2 \end{pmatrix}$, where $c_1, c_2 > 0$ are constants. As mentioned before, we also investigated the diagonal models where variance of only one of the dimensions was set and the second one was to be estimated,
- [**manually set variances, varying volumes**] – a model similar to the previous one apart from the fact that the cluster volumes along the dimensions of the unset variances are allowed to vary; for example, $\Sigma_k = \begin{pmatrix} c & 0 \\ 0 & \sigma_k \end{pmatrix}$, where $c > 0$ is a constant.

We used the mclust R package [15] for clustering in the first four models and wrote our own procedures for the other two models.

Following [8] we used *Bayesian Information Criterion* (BIC) [16,17] to select the number of clusters. BIC is defined as:

$$BIC = 2\ln P(D|\hat{\theta}, M) - r\ln n,$$

where D is the data, $\hat{\theta}$ are the estimated parameters of the model M and r is the number of independent parameters estimated in the model M (it depends on the number of clusters). We pick the number of clusters with the greatest BIC.

DBSCAN algorithm.(*A Density-Based Algorithm for Discovering Clusters in Large Spatial Databases with Noise*) [10] was designed for finding clusters in spatial data, such as satellite images and protein structure data. This algorithm applies a local cluster criterion. Clusters are regarded as dense regions in the data space which are separated by regions of low object density, called *noise*. Each cluster can have an arbitrary shape and the objects inside a cluster region may be arbitrarily distributed.

Let $N(p, \varepsilon)$ denote the ε-neighborhood of point p and $|N(p, \varepsilon)|$ the number of points from the input dataset in this neighborhood. Points p such that $|N(p, \varepsilon)| < minPts$ are considered noise. Two points p and q are assigned to the same cluster if there exists a sequence $p_1, \ldots, p_k$ of non-noise points, such that $p_1 = p$, $p_k = q$ and $p_{i+1} \in N(p_i, \varepsilon)$ for $i = 1, \ldots, k - 1$. In our application $\varepsilon = (\varepsilon_m, \varepsilon_{rt})$ and $q \in N(p, \varepsilon)$ if and only if $|p_m - q_m| < \varepsilon_m$ and $|p_{rt} - q_{rt}| < \varepsilon_{rt}$.

Parameter ε can be chosen so as to account for our knowledge about the range of expected mass spectrometer accuracy errors and retention time drifts. Moreover, depending on the $minPts$ parameter, some of the peaks in very sparse regions can be sieved out.

The algorithm was implemented so as to assure that sizes of the resulting subsets are not greater than the threshold parameter, $maxSize$. If for initial value of parameter ε some resulting subsets are of greater size, the algorithm is rerun for each of those subsets separately with ε decreased. This step is repeated until all of the subsets have sizes lower than $maxSize$.

Two-stage clustering. The model-based clustering cannot be directly applied to the entire set of peaks due to efficiency reasons. We propose the following two-stage procedure:

1. preliminary partitioning the data set into non-overlapping subsets of moderate sizes,
2. application of the model-based clustering to each subset separately.

In the first step the DBSCAN procedure is used, in the second step different models are fitted.

Retention time correction. Retention time deviations are often too significant for the correct peak alignment to be possible. Due to this fact Smith et al. [7] propose an iterative procedure that comprises multiple steps of peak alignment alternated with the retention time correction. This procedure is implemented in the xcms package [7] and we adopt it to our setting.

Feature selection. A *biomarker* is an indicator of a specific biological process, usually it is a biochemical feature or facet that can be used to measure the progress of disease or the effects of treatment. In case of mass spectrometry experiments we are interested in peptide signals, presence, lack or certain level of which somehow corresponds to patient's state.

Mathematical methods can be used to detect biomarkers. In our setting, in terms of data mining, potential biomarkers are the features that allow for the best discrimination between the classes. We exploited the feature selection mechanism in order to compare the clusterings. Two approaches are used in the sequel: the well known T-test procedure and, recently gaining much interest, the Random Forest based feature selection algorithm [18].

False Discovery Rate. Feature selection methods usually score attributes accordingly to their computed importance. By choosing a particular level of the score, one can simply select the attributes exceeding the threshold. However, the issue of selecting an appropriate threshold is problematic itself. It is hard to assess what level yields the really significant selections, i.e. the ones that are lowly probable to occur by chance. We apply here the False Discovery Rate (FDR) [19] which is a meta-test, independent on the feature selection method, designed for choosing appropriate threshold values. The intuition is that the lower the FDR values are, the more significant the features are.

3   Results and Discussion

Data set. Data was provided by the Mass Spectrometry Laboratory from the Institute of Biochemistry and Biophysics of Polish Academy of Sciences. The mass spectrometer used in the experiments was an Electro Spray Ionization Fourier Transform Ion Cyclotron Resonance (ESI-FTICR) coupled with the HPLC retention column.

The data set comprised mass spectra acquired from plasma samples for colorectal cancer patients. Apart from the patients data, control samples were also collected from healthy donors and analyzed with the mass spectrometer. The colorectal cancer data set consisted of 40 spectra, 23 samples corresponding to patients and 17 to healthy donors.

The raw data in mzXML file format was preprocessed (noise reduction and peak picking) using the NMRPipe tool [20]. We used a monoisotopic peak detection program [11] to obtain a list of peak coordinates, i.e. m/z values and retention times of the most abundant molecules. In total, there were 155294 monoisotopic peaks detected in 40 samples. The time range of detected peaks was $922.6 - 4871.3$ seconds ($15.4 - 81.2$ minutes), mass-to-charge ratio range was $300.127 - 1499.33$.

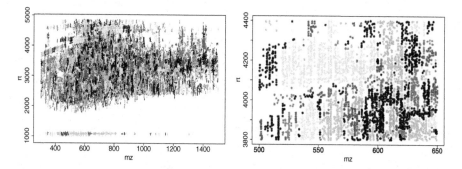

Fig. 3. Colorectal cancer data clustered with the DBSCAN algorithm, $\varepsilon_m = 5$, $\varepsilon_{rt} = 30$, $minPts = 10$. Picture on the right presents fragment of the data in greater detail. The cluster colors are recycled.

Experiments. We tested the model-based approach on the data set described above. The DBSCAN algorithm was run with the following parameters: $\varepsilon = (\varepsilon_m, \varepsilon_{rt}) = (5, 30)$, $minPts = 10$, the upper limit for size of a preliminary cluster was 1000 elements. We made the assumption that the peaks that did not have at least 10 neighbors had noise origins. The algorithm resulted in 8216 preliminary groups and 3076 points were assigned to the *noise* cluster. Figure 3 presents the peak map with preliminary clusters.

All the models presented in Section 2 were fitted within each of the preliminary clusters in the second stage of the algorithm. At one time the same model was assumed in all the preliminary clusters. For a preliminary cluster of size n ($1 \leq n \leq 1000$) clusterings of the number of clusters from interval $[n/40, n/10]$ were compared (40 is the number of samples).

In case of the models with manually set covariance matrices, the standard deviation on mass-to-charge ratios was set to 0.04, which was selected after consultations with experienced mass spectrometer operators. We tested manually set retention time deviations of 50, 100 and 200 seconds. Apart from the mentioned models, we also fitted both models where the retention time deviation was estimated by the algorithm.

The initial cluster assignments that are being improved with the EM algorithm were obtained with the model-based hierarchical algorithm. Implementation from the mclust [15] R package was used. The model assumed in the hierarchical algorithm was λI which stands for identical, spherical clusters. Hence, the dimensions, which are given in different units, had to be properly scaled. It was established with empirical tests that dividing the retention time values by 100 results in reasonable initial clusters.

During the clustering peaks that gained cluster probability assignment lower than 0.3 were sieved out. We also disregarded the clusters that had less than 4 elements from both the colorectal cancel sample group and the healthy donor sample group.

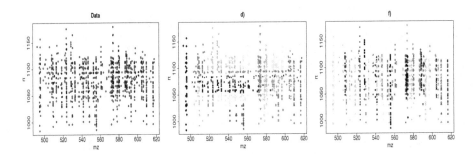

Fig. 4. Example of clustering acquired for one of the preliminary clusters from Figure 3 of size 999. Outcomes for models: d) λB_k; f) manually set deviation of the m/z dimension to 0.04, retention time deviations set to 100. Cluster labels and colors are recycled.

Visual analysis of a subset (model selection). Figure 4 shows exemplary clusters obtained from the DBSCAN algorithm and different clusterings resulting from different parameterizations. Peaks corresponding to the same peptide are expected to form elongated groups along the retention time axis with rather small variance along the m/z axis. One can see that the less constrained models detected clusters that should not occur in nature, i.e. elongated along the m/z axis. In particular, it concerns models $\lambda_k B_k$ and λB_k (see picture d in Figure 4). This artifact did not occur in case of models λB and $\lambda_k B$, but some of the clusters obtained with these models comprised peaks from too wide m/z interval. The remaining models had manually set m/z deviation and hence the clusters look more as expected (e.g for picture f in Figure 4 manually set retention time deviation was 100 seconds). Due to space limitations, the detailed results for the other models are omitted. Please refer to our supplementary web site: http://bioputer.mimuw.edu.pl/papers/clust. Since it is not doable to visually examine all the clusters, we tried to detect some overall tendencies in cluster characteristics and to select most suitable model.

Comparison with XCMS package (FDR test). Besides the visualization, the quality of alignments was also evaluated using the FDR test with the two feature selection methods. The idea underlying this comparison was that properly aligned peptides enable further reasoning, whereas peptides that are aligned

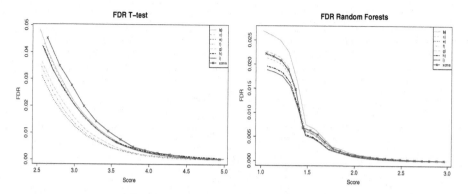

Fig. 5. FDR statistic computed for the T-test and the Random Forest based feature selection. The horizontal axis shows different values of the score assigned to attributes, the vertical axis shows FDR test values. There were 500 permutations performed and compared with the true T-test values. and 100 permutations with the true Random Forest scores. See Table 1 caption for the explanation of fitted model names: b) – i); the outcome of xcms clustering is also plotted.

randomly should not reveal any valuable information. The goal of feature selection is to extract the features, in our case aligned peaks, that best discern classes of samples.

We compared the performance of alignments acquired with our method to the grouping proposed in the xcms package [7]. In the xcms algorithm we set the width of m/z segments to 0.04 (which corresponds to the standard deviation in our method) and we applied the same filtering criterion as in our method: clusters that did not contain at least 4 peaks neither from the colorectal cancer samples nor from the healthy donors samples were sieved out. In all the experiments we performed 4 iterations of the retention time correction.

In Figure 5 we present FDR plots for each of the clusterings for the T-test and Random Forest based feature selection respectively. The closer to zero a plot is, the greater is the number of significant features detected.

Several interesting observations can be made: in general, the models with manually imposed m/z deviation got plots closer to zero. The differences get smaller with the increasing score value. The best among the remaining models is $\lambda_k B$. Unexpectedly, plots of models $\lambda_k B_k$ and λB are very close while the clusters on the visually analyzed subset were very different.

When comparing FDR plots it is visible that our method produces higher quality clusters than xcms. The number of clusters is roughly on the same level, but in case of xcms processing significantly more peaks are filtered out (see Table 1). The running time of our method depends on the selected model and varies from 40 to 90 minutes on a desktop computer (heuristic algorithm implemented in the xcms package is much quicker – 3 minutes).

Table 1. Clustering statistics, reported are: the number of clusters and the number of peaks that were either rejected by the DBSCAN algorithm or after the clusters' filtration. Explanation of model names: a) $\lambda_k B_k$; b) λB; c)$\lambda_k B$; d) λB_k; e)–i) All the models have manually imposed deviation of the m/z dimension to 0.04, retention time deviations are: e) 50 f) 100 g) 200 h) estimated, the same for all clusters i) estimated, varying between clusters.

Model	number of:		Model	number of:	
	clusters	rejected peaks		clusters	rejected peaks
a	5718	20437	f	7768	21794
b	5979	23158	g	7753	21622
c	6374	22662	h	7935	27250
d	5775	19878	i	8168	29050
e	7753	22418	**xcms**	**7996**	**47524**

4 Conclusions

The alignment of corresponding peptide signals across samples is a very crucial stage of mass spectrum data interpretation, errors introduced here are propagated in the further analysis and often prevent its success. We have proposed a novel approach to this problem based on a sound mathematical framework. The method is designed for aligning two dimensional LC-MS spectra, but the extension to analysis of spectra of more dimensions is straightforward.

The comparison with the only accessible tool (xcms) shows the superior performance of the model-based approach. With our method we could select the models that detected a greater number of significant features. The visual analysis of clusters proved that the models with imposed m/z deviations resulted in alignments corresponding to our expectations (see Figure 4).

A great advantage of our approach is not only its performance but also the possibility of the correct model selection offered by the BIC analysis. By applying the "divide and conquer" strategy (the DBSCAN pre-clustering) we also demonstrated the practicability of the model-based approach on inherently large LC-MS data sets.

In the proposed method we took into account only the m/z and retention time values for peak description, disregarding peak intensities. One peptide usually has multiple corresponding peptide signals in a spectrum, for each possible charge value. The distribution of such related peptide signals corresponding to a peptide should be similar across mass spectra. This model requires further study.

References

1. Aebersold, R., Mann, M.: Mass-spectrometry based proteomics. Nature **422** (2003) 198–207
2. Tibshirani, R., Hastie, T., Narasimhan, B., Soltys, S., Shi, G., Koong, A., Le, Q.T.: Sample classification from protein mass spectrometry, by "peak probability contrasts". Bioinformatics **20** (2004) 3034–3044

3. Wu, B., Abbott, T., Fishman, D., McMurray, W., Mor, G., Stone, K., Ward, D., Williams, K., Zhao, H.: Comparison of statistical methods for classification of ovarian cancer using mass spectrometry data. Bioinformatics **19** (2003) 1636–1643
4. Wang, W., Zhou, H., Lin, H., Roy, S., Shaler, T.A., Hill, L.R., Norton, S., Kumar, P., Anderle, M., Becker, C.H.: Quantification of proteins and metabolites by mass spectrometry without isotopic labeling or spiked standards. Analytical Chemistry **75** (2003) 4818–4826
5. Wong, J.W.H., Cagney, G., Cartwright, H.M.: SpecAlign. processing and alignment of mass spectra datasets. Bioinformatics **21** (2005) 2088–2090
6. Prakash, A., Mallick, P., Whiteaker, J., Zhang, H., Paulovich, A., Flory, M., Lee, H., Aebersold, R., Schwikowski, B.: Signal maps for mass spectrometry-based comparative proteomics. Molecular and Cellular Proteomics **5** (2006) 423–432
7. Smith, C.A., Want, E.J., O'Maille, G., Abagyan, R., Siuzdak, G.: XCMS: Processing mass spectrometry data for metabolite profiling using nonlinear peak alignment, matching, and identification. Analytical Chemistry **78** (2006) 779 – 787
8. Fraley, C., Raftery, A.E.: How many clusters? which clustering method? answers via model-based cluster analysis. The Computer Journal **41** (1998) 578–588
9. Yeung, K.Y., Fraley, C., Murua, A., Raftery, A.E., Ruzzo, W.L.: Model-based clustering and data transformations for gene expression data. Bioinformatics **17** (2001) 977–987
10. Ester, M., Kriegel, H.P., Sander, J., Xu, X.: A density-based algorithm for discovering clusters in large spatial databases with noise. In Simoudis, E., Han, J., Fayyad, U., eds.: Second International Conference on Knowledge Discovery and Data Mining, AAAI Press (1996) 226–231
11. Gambin, A., Dutkowski, J., Karczmarski, J., Kluge, B., Kowalczyk, K., J.Ostrowski, Poznański, J., Tiuryn, J., Bakun, M., Dadlez, M.: Automated reduction and interpretation of multidimensional ms data for analysis of complex peptide mixtures. International J. of Mass Spectrometry (2006) *(in press)*.
12. Dempster, A.P., Laird, N., Rubin, D.: Maximum likelihood from incomplete data via the EM algorithm. J. of Royal Statististical Society **Series B** (1977) 1–38
13. Petersen, K.B.: On the slow convergence of EM and VBEM in low-noise linear models. Neural Computation **17** (2005) 1921–1926
14. Banfield, J.D., Raftery, A.E.: Model-based Gaussian and non-Gaussian clustering. Biometrics **49** (1993) 803–821
15. Fraley, C., Raftery, A.E.: MCLUST: Software for model-based clustering, density estimation and discriminant. Technical Report 415R, University of Washington, Department of Statistics (2002)
16. Haughton, D.M.A.: On the choice of a model to fit data from an exponential family. The Annals of Statistics **16** (1988) 342–355
17. Schwarz, G.: Estimating the dimension of a model. The Annals of Statistics **6** (1978) 461–464
18. Breiman, L.: Random forests. Machine learning **45** (2001) 5–32
19. Storey, J., Tibshirani, R.: Statistical significance for genomewide studies. PNAS **100** (2003) 9440–9445
20. Delaglio, F., Grzesiek, S., Vuister, G.W., Zhu, G., Pfeifer, J., Bax, A.: Nmrpipe: a multidimensional spectral processing system based on unix pipes. J. Biomol. NMR **6** (1995) 277–293

A Bayesian Algorithm for Reconstructing Two-Component Signaling Networks

Lukas Burger and Erik van Nimwegen

Biozentrum, University of Basel
Klingelbergstrasse 50/70, CH-4056, Basel, Switzerland
{lukas.burger, erik.vannimwegen}@unibas.ch

Abstract. We present an algorithm, based on a Bayesian network model, for *ab initio* prediction of signaling interactions in bacterial two-component systems. The algorithm uses a large training set of known interacting kinase/receiver pairs to build a probabilistic model of dependency between the amino acid sequences of the two proteins and uses this model to predict which pairs interact. We show that the algorithm can accurately reconstruct cognate kinase/receiver pairs across all sequenced bacteria. We also present predictions of interacting orphan kinase/receiver pairs in the bacterium *Caulobacter crescentus* and show that these significantly overlap with experimentally observed interactions.

1 Introduction

The automated prediction of protein-protein interactions on the basis of their amino acid sequences alone is one of the great challenges in computational biology. Here we present a first attempt at such an algorithm for the large class of bacterial two-component systems. In their simplest form, two-component systems consists of two proteins: a histidine kinase and a response regulator [1]. In many cases the histidine kinase is a membrane-bound protein, with a sensor domain which responds to environmental cues and, on the cytoplasmic side, a kinase domain, which autophosphorylates upon activation of the sensor. The kinase domain very specifically interacts with its cognate response regulator by transferring the phosphate to the regulator's receiver domain. Phosphorylation leads to the activation of the regulator, which often acts as a transcription factor. Since two-component systems are responsible for most signal transduction in bacteria[1, 2], successful computational prediction of two-component system interactions would allow exhaustive reconstruction of signaling networks across all fully sequenced bacterial genomes.

There are several reasons that make two-component systems particularly attractive for computational modeling. Firstly, both the histidine kinase and the receiver domain exhibit a high degree of sequence similarity and they can be easily detected in fully-sequenced genomes using hidden Markov models. Secondly, two-component systems are very abundant in the bacterial and archeal kingdom, with many tens of interacting pairs in some genomes, and thousands of

P. Bücher and B.M.E. Moret (Eds.): WABI 2006, LNBI 4175, pp. 44–55, 2006.

examples across all genomes, providing enough data for relatively subtle statistical modeling. Finally, for a significant fraction of all two-component systems, the interacting partners lie in the same operon on the genome, which allows us to easily extract a large number of examples of "known" interacting pairs.

In this article, we will present an algorithm that uses a statistical model to predict interacting kinase/receiver pairs. We test the performance of the algorithm on reconstructing known cognate pairs from all sequenced bacteria and use it to predict interaction partners for orphan kinases in the Gram-negative bacterium *Caulobacter crescentus*, where orphans play an important role in the cell-cycle progression [3]. We will show that our predictions agree well with the experimental results.

To our knowledge our method is the first computational approach for comprehensive prediction of two-component interactions, and the first to explicitly model dependencies between interacting amino acids in this context. In a previous work it was shown that some two-component interactions can be predicted within the context of a general method for inferring interactions between proteins from the similarity of their phylogenetic trees [4]. However, this method is not applicable beyond small selected subsets of kinase/receiver pairs.

2   Outline of the Algorithm

Our prediction algorithm operates in two steps. Comparison of the kinases from all sequenced bacteria shows that there are 7 major classes of domain architectures. Using a training set of cognate receivers for each class of kinases we build position-specific weight matrices (WMs) for the receivers of each class and use these to classify receivers. This allows us to predict, for each receiver, which type of kinase it will interact with. In the second step of our algorithm we aim to identify which kinase/receiver pairs within a class interact. To this end we again use training sets of cognate kinase/receiver pairs and identify pairs of amino acid positions in kinase and receiver that show significant mutual information. Using a network of such correlated positions we construct statistical models for the joint distribution of amino acids in interacting kinase/receiver pairs. The final "score" for a putative interacting pair is given by the ratio of the likelihood of their sequences given that they are an interacting pair and the likelihood assuming independence of their sequences. In order to reconstruct cognate kinase/receiver pairs genome-wide we use Markov chain Monte-Carlo sampling to sample all ways of assigning kinase/receiver pairs, sampling each assignment in proportion to the likelihood of the sequences of all interacting pairs in the assignment.

3   Classifying Bacterial Two-Component Systems

To gather an exhaustive collection of two-component system proteins, we first collected a set of hidden Markov profiles from the Pfam database [5] that characterize two-component systems. These are the histidine-kinase profiles HisKA, HisKA_2(or H2), HisKA_3(or H3), and HWE_HK, the ATP-binding domain

HATPase_c, the His-containing phosphotransfer domain Hpt, and the response regulator receiver domain Response_reg (or RR). We then collected all bacterial genomes from the NCBI database (ftp.ncbi.nlm.nih.gov/genomes/Bacteria) and searched for matches to all these domains, using the hmmpfam program (http://hmmer.wustl.edu/) with an E-value cutoff of 10^{-4}.

3.1   Cognate Pairs and Orphans

Depending on their position on the DNA, kinases and response regulators were further classified as follows. We defined operons as maximal sets of contiguous genes on the same strand of the DNA with all intergenic regions between consecutive genes less than 100 bps in length. Whenever an operon contains only one kinase and one receiver these two were considered a cognate pair that we assume to interact. Kinases and receivers that occur by themselves in an operon were named orphans. For virtually all of these 'orphan' kinases and receivers it is currently unknown what partners they interact with and one of the major aims of our algorithm is to predict interaction partners for these orphans.

3.2   Kinase Domain Architectures

Whereas the response regulators are characterized by a single receiver profile, the kinases are represented by 6 different Pfam profiles. Although kinases display a large variety of different domain architectures, we find that most domain architectures are very rare, and that almost all kinases fall within the 7 most abundant classes shown in Table 1.

Table 1. Pfam domain combinations of the most abundant kinase architectures and the numbers of their occurence in both cognates and orphans. Both the short and long hybrid architecture can contain one or two receiver domains.

Name	Architecture	no.cognates	no.orphans
HisKA	HisKA, HATPase_c	2165	979
H3	H3, HATPase_c	415	75
Chemotaxis	Hpt, HATPase_c	113	35
Long hybrid	HisKA, HATPase_c, RR, (RR), Hpt	86	151
Short hybrid	HisKA, HATPase_c, RR, (RR)	82	591
HWE	HWE or H2, HATPase_c	22	172
Hpt	Hpt	19	48

3.3   Multiple Alignments

To produce multiple alignments of the receiver domains and of the kinases in each of the 7 classes we first used the program Hmmalign (http://hmmer.wustl.edu/) for each domain. For the HisKA, chemotaxis, HWE and H3 kinase classes we constructed a full alignment by simply concatenating the alignments of the kinase and the HATPase_c domains. For the short hybrids we aligned the HisKA and HATPase_c domains and for the long hybrids only the Hpt domain.

To check the accuracy of the alignments we compared the Hmmalign alignments with alignments made by the ProbCons algorithm [6]. For each class 200 sequences were selected at random (or all if the class has less than 200 sequences) and aligned with ProbCons. We then selected all columns in the hmmalign alignments that contain less than 15% gaps and for which at least 80% of the amino acids in the column also align together in a single column in the ProbCons alignment. We call these columns the 'trusted positions'. Finally, we replaced each alignment with the alignment of only the trusted positions.

3.4   Classification of Response Regulators

We found that response regulators that interact with different types of kinases show distinct amino acid compositions and these differences can be used to predict, for each receiver, what kind of kinase it will interact with.

We divided the multiple alignment of all cognate receivers into 7 sub-alignments corresponding to all receivers that interact with kinases of a particular class. For each of the 7 alignments we then constructed a position specific weight matrix

$$w_{i\alpha}^c = \frac{n_{i\alpha}^c + \lambda}{(n^c + 21\lambda)}. \tag{1}$$

Here $n_{i\alpha}^c$ is the number of times amino acid α occurs in column i of the alignment (gaps are treated as a 21st amino acid) of cognate receivers of class c, n^c is the total number of sequences in the alignment, and λ is the pseudocount resulting from the Dirichlet prior (we used $\lambda = 1/2$). $w_{i\alpha}^c$ is thus the estimated probability of seeing amino acid α in position i of a receiver of class c.

Given a receiver with sequence S we can now determine the posterior probability $P(c|S)$ that it belongs to class c. We have

$$P(c|S) = \frac{P(S|c)P(c)}{\sum_{c'} P(S|c')P(c')} \quad \text{with} \quad P(S|c) = \prod_{i \in \text{TP}} w_{iS_i}^c. \tag{2}$$

Here S_i stands for the amino acid in the ith position of receiver sequence S. Note that the product only runs over all the trusted positions TP. We assumed a uniform prior $P(c) = 1/7$. When classifying a regulator whose sequence was used in the construction of the WM we removed its contribution from the counts $n_{i\alpha}^c$.

The results of the classification are shown in Figure 1. The posterior probabilities for the 7 classes were calculated for each receiver and the receiver was assigned to the class with the highest posterior probability (which is often close to 1). The results show that for the three most abundant types (HisKA, H3, and chemotaxis kinases) the classifier predicts almost perfectly which receivers interact with HisKA kinases, which with H3 kinases, and which with chemotaxis kinases. For the other, rarer classes the classification is still correct in the majority of the cases, except for the very rare Hpt kinases where slightly more than half are misclassified. The lower performance for the rarer classes is presumably due to the fact that the WM models for these classes are based on a relatively small number of examples.

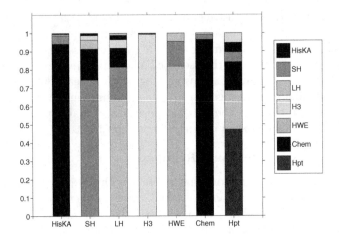

Fig. 1. Predicted classification of receivers. Each bar represents the set of all receivers that are member of a cognate pair with kinases of a particular type. The color distribution in the bar shows what percentages of the receivers are classified with each class. The correspondence between color and kinase type is shown in the legend on the right. SH and LH stand for short and long hybrid, respectively, and Chem stands for chemotaxis.

The types of misclassifications match what is to be expected based on the domain architectures. Both chemotaxis and long hybrid kinases contain an Hpt domain and some of the receivers that interact with a single Hpt domain kinase are mistaken for a receiver that interacts with the Hpt domain of a chemotaxis or long hybrid kinase. Similarly, both long and short hybrids contain an HisKA domain and their receivers are sometimes mistaken for a receiver that interacts with a single HisKA domain kinase. Overall, the WM model predicts the correct type of kinase for 93% of the cognate receivers.

4   Predicting Cognate Interactions

Once we have classified the receivers according to the type of kinase they interact with, the second step of our algorithm consists of predicting, for each class, which pairs of kinases and receivers interact. To do this we make alignments of all cognate kinase/receiver pairs in each class by simply concatenating the respective kinase and receiver alignments. We then build probabilistic "dependent" models for the joint amino acid sequences of cognate kinase/receiver pairs and "independent" models for the kinases and receivers independently. The algorithm then predicts interactions between kinase/receiver pairs whose sequences are more likely under the dependent than under the independent model.

4.1   Quantifying Dependence Between Positions in Kinase and Receiver

Given the joint multiple alignment of kinase/receiver pairs in a particular class we quantify the dependence between all pairs of trusted positions (i, j), where the positions i and j may both be either in the kinase or in the receiver, using a measure closely related to mutual information. For each pair (i, j) we calculate the likelihood of the observed columns of amino acids under a model that assumes the amino acids at the two positions were drawn from two independent distributions and under a model that assumes general dependence between the two amino acids. In particular, for the independent model let p_α denote the probability to observe amino acid α at position i, and let q_β denote the probability to observe amino acid β at position j. For the dependent model, let $w_{\alpha\beta}$ denote the probability to observe the pair of amino acids (α, β) at positions (i, j). Finally, let D_{ij} denote the columns of observed amino acids in the alignments at positions i and j, $n_{\alpha \cdot}$ the number of times α is observed at position i, $n_{\cdot \beta}$ the number of times β is observed at position j, and $n_{\alpha\beta}$ the number of times the pair of amino acids (α, β) is observed at positions (i, j).

Given the joint probability $w_{\alpha\beta}$ the probability of the data D_{ij} is given by

$$P(D_{ij}|w) = \prod_{\alpha\beta} (w_{\alpha\beta})^{n_{\alpha\beta}} \tag{3}$$

and under the independent models p and q the probability of the data is given by

$$P(D_{ij}|p, q) = P(D_i|p)P(D_j|q) = \left[\prod_{\alpha} (p_\alpha)^{n_{\alpha\cdot}}\right]\left[\prod_{\beta} (q_\beta)^{n_{\cdot\beta}}\right]. \tag{4}$$

Since the distributions p, q and the joint distribution w are unknown, they are nuisance parameters that we integrate out of the likelihood for the dependent and independent models. We use Dirichlet priors of the form $P(w) \propto \prod_{\alpha\beta} w_{\alpha\beta}^{\lambda-1}$ and integrate over the simplices $\sum_\alpha p_\alpha = \sum_\beta q_\beta = \sum_{\alpha\beta} w_{\alpha\beta} = 1$. We then obtain for the probability of the data under the dependent model

$$P(D_{ij}|\text{dep}) = \int P(D_{ij}|w)P(w)dw = \frac{\Gamma(21^2\lambda)}{\Gamma(n + 21^2\lambda)} \prod_{\alpha\beta} \frac{\Gamma(n_{\alpha\beta} + \lambda)}{\Gamma(\lambda)}, \tag{5}$$

and similarly for the probability of the data under the independent model

$$P(D_{ij}|\text{indep}) = \frac{\Gamma^2(21\lambda)}{\Gamma^2(n + 21\lambda)} \left[\prod_{\alpha} \frac{\Gamma(n_{\alpha\cdot} + \lambda)}{\Gamma(\lambda)}\right]\left[\prod_{\beta} \frac{\Gamma(n_{\cdot\beta} + \lambda)}{\Gamma(\lambda)}\right], \tag{6}$$

where $\Gamma(n)$ is the Gamma function. Finally, we quantify the amount of dependence between positions i and j by the log-ratio R_{ij} of likelihoods of the dependent and independent models

$$R_{ij} = \log\left[\frac{P(D_{ij}|\text{dep})}{P(D_{ij}|\text{indep})}\right]. \tag{7}$$

For our calculations we used the Jeffreys, or information theory prior with $\lambda = 1/2$. One can think of the quantity R as a finite-size corrected version of the mutual information that takes into account the larger model space of the dependent model [7].

4.2   Probabilities of Kinase/Receiver Pairs Under Interacting and Independent Models

For each class, R_{ij} was calculated for each pair of trusted positions both between kinases and receivers and within the proteins themselves. Since, due to the evolutionary relationship of our sequences, the correlations may be overestimated, we chose a stringent cut-off of $R = 50$ for the HisKA class and, to still obtain a reasonable number of dependent positions, a more lenient cut-off of $R = 0$ for the other classes. For each class we then collected the set of 'significant positions' Ω^c that score over the threshold with at least one other position.

The two-point correlation structure of the significant positions can be represented by a graph in which each node is a significant position and two nodes are connected if R for the two positions scores over the threshold. Interestingly, we find that this graph generally consists of a large connected component containing both kinase and receiver positions, plus a few small connected components containing either only kinase or only receiver positions. Since the positions in these small components do not contain information about the dependence between kinase and receiver we discarded them from the set Ω^c.

We now approximate the joint distribution of the significant positions in interacting kinase/receiver pairs using pairwise conditional probabilities between positions. The procedure is illustrated in Figure 2. The multiple alignments of cognate kinase-receiver pairs are shown at the top with the significant positions as colored columns and the arcs indicating which pairs of columns correlate

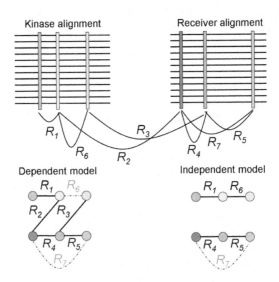

Fig. 2. Multiple alignments of cognate (interacting) kinase/receiver pairs with significant positions shown as colored columns. The arcs show the pairs of positions that are significantly correlated. The correlation structure of the dependent and independent models are shown at the bottom. The edges that are removed by the Chow-Liu algorithm are shown as dotted lines.

significantly. To factorize the joint probability of all significant positions we use a slightly modified version of the Chow-Liu algorithm [8] that reduces the correlation graph to a tree while maximizing the sum over the R values along the remaining edges. For example, in the bottom left of Figure 2 the links 6 and 7 with the lowest R values were removed to yield a tree. Once a root is chosen (arbitrarily) each position i (except for the root) will have exactly one parent $\pi(i)$ and we factor the joint probability by assuming the amino acid at position i is only dependent on the amino acid at position $\pi(i)$. That is, if $S_{K,R}$ denotes the set of significant positions for kinase K and receiver R then the probability $P(S_{K,R}|c)$ of the sequences assuming that they are an interacting pair of class c is given by

$$P_{K,R}(S_{K,R}|c) = \prod_{i \in \Omega^c} P(S^i_{K,R}|S^{\pi(i)}_{K,R}, c) \quad \text{with} \quad P(S^r_{K,R}|S^{\pi(r)}_{K,R}, c) \equiv P(S^r_{K,R}|c),$$

(8)

for the root of the tree r. Here $S^i_{K,R}$ is the amino acid in the ith significant position of the kinase-regulator sequence and $\pi(i)$ is the parent of position i as defined by the tree. The probability $P(\alpha|\beta, c)$ to observe amino acid α at position i given that amino acid β occurs at position $\pi(i)$ is given by

$$P(\alpha|\beta, c) = \frac{n^c_{\alpha\beta} + \lambda w^c_{i\alpha}}{n^c_{\cdot\beta} + \lambda},$$

(9)

where $n^c_{\alpha\beta}$ is the number of times the pair $\alpha\beta$ occurs at positions i and $\pi(i)$ of the cognate kinase-receiver pairs of class c, $n^c_{\cdot\beta}$ is the total number of times that β occurs at position $\pi(i)$, and λ is the pseudo-count of the Dirichlet prior (here we use a much larger $\lambda = 10$ to smooth fluctuations due to the small sample size). Note that we made the prior for the conditional probabilities proportional to the independent probability, i.e. the WM w^c, for class c.

In complete analogy we calculate the independent probabilities $P(S_K|c)$ of the kinase and $P(S_R|c)$ of the receiver, where we now only allow conditional dependence between positions within the kinase and positions within the receiver as in the bottom right of Figure 2. Finally, we assign a "score" $Z(K, R|c)$ to the pair K, R which equals the logarithm of the likelihood ratio

$$Z(K, R|c) = \log \left[\frac{P(S_{K,R}|c)}{P(S_K|c)P(S_R|c)} \right].$$

(10)

4.3   Results on Reconstructing Cognate Pairs

For each genome and each class we collected all kinases in the class together with their cognate receivers. For the two largest classes, the HisKA and H3 class, we randomly divided the genomes into two groups of equal size and used one group as the training set for scoring the other group. For the remaining classes, due to their relatively small size, we followed a leave-one-out strategy, i.e. we scored each pair using all the remaining kinase/receiver pairs as the

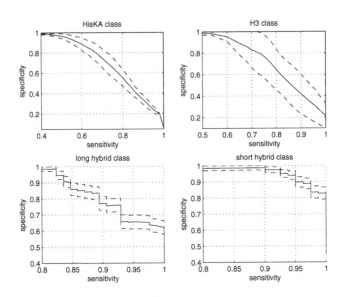

Fig. 3. Sensitivity/Specificity curves for the 4 most abundant kinase classes. The solid lines give the estimated specificity and the dashed lines give one standard error of the estimate. In the case of the HisKA and H3 class, the specificity is averaged over ten different partitions into training and test set.

training set. For each class, we then used Markov chain Monte-Carlo sampling to sample, for every genome, all ways of assigning one kinase to each receiver. Let a denote an assignment and let $R(K, a)$ denote the receiver assigned to kinase K in assignment a. The probability of sampling a is then given by

$$P(a) \propto \exp\left[\sum_K Z(K, R(K, a))\right]. \tag{11}$$

We then measured what fraction of the time $f(R, K)$ during sampling each kinase K was assigned to each receiver R. For different values of f we counted what fraction of true interacting pairs (i.e. cognate pairs) from all genomes have $f(R, K) > f$ (sensitivity) and also what fraction of all pairs that have $f(R, K) > f$ are true interacting pairs (specificity). The resulting sensitivity/ specificity curves for the 4 most abundant kinase classes are shown in Figure 3. For the HisKA and H3 classes, we repeated the random partition into training and test set ten times and calculated, for each given sensitivity, the mean specificity as well as the fluctuations around the mean. As the figure shows, our model very accurately predicts for all classes which kinase interacts with which receiver. For example, more than 50% percent of all cognate pairs for all classes can be predicted at a specificity close to 1. Note that in a genome with n cognate pairs there are only n true interactions among n^2 possible interactions. This explains why the lowest specificities are obtained for the large HisKA class, i.e. the correct interactions have to be discovered in a much larger set of putative interactions for this class. Still, even for HisKA 70% of all true interactions are predicted at a specificity of about 70%.

5   Prediction of Orphan Interactions in *Caulobacter Crescentus*

Although the previous section shows that our algorithm can accurately reconstruct interacting pairs for the cognate kinases and receivers, these predictions are not biologically novel since for cognate pairs the interacting partners could already be determined from their positions on the DNA. Therefore, we next applied our algorithm to predict interaction partners for orphan kinases and receivers. It is difficult to assess the performance of our algorithm in this context since only very few orphan interactions have been experimentally characterized. Moreover, most of the experimental work is done *in vitro* under conditions that are very different from those *in vivo* and it is not clear if observed interactions *in vitro* reliably reflect *in vivo* interactions.

We chose the bacterium *Caulobacter crescentus* as a test case since most experimentally known orphan interactions are from that organism [3, 9]. *C. crescentus* contains 40 orphan kinases of which 6 are in the class of HisKA kinases. Since all but two of the known interactions involve HisKA kinases we decided to focus on the 5 HisKA kinases for which at least one interaction has been ex-

Table 2. Predictions for HisKA orphan kinases of *Caulobacter Crescentus* for which at least one interaction has been experimentally characterized. For each kinase K the receivers are sorted by their score $Z(K, R)$ and the known interactions are indicated. We cut off each list to include the known interactions except for the interaction of DivL with the receiver CtrA, which occurs at position 16 in the list of DivL.

kinase	regulator	interaction score	experimental evidence
DivL	DivK	3.75	yeast two-hybrid screen [10]
PleC	DivK	1.95	*in vitro* phosphorylation [11]
PleC	PleD	-0.47	*in vitro* phosphorylation [11]
CckN	CC1364 (CheYIII protein)	9.28	
CckN	DivK	8.47	yeast two-hybrid screen [10]
CenK	CC1842	7.38	
CenK	CenR	6.39	*in vitro* phosphorylation [11]
DivJ	CC3155 (CheYIII protein)	-0.51	
DivJ	CC0612 (NasT)	-1.75	
DivJ	CC3162	-2.17	
DivJ	CC1842	-2.28	
DivJ	CC3471	-2.38	
DivJ	CC1364 (CheYIII)	-2.65	
DivJ	DivK	-2.65	*in vitro* phosphorylation [11]
DivJ	PleD	-3.52	*in vitro* phosphorylation [11]

perimentally characterized. There are 23 orphan receivers in *C. crescentus* and we determined the score Z for each orphan receiver with each of these 5 HisKA orphan kinases. The results are shown in table 2. As shown in the table, 5 of 7 experimentally observed interactions rank either immediately at the top or at the second position of the ordered list for each kinase. For DivJ the two known interactions with DivK and PleD occur at positions 7 and 8 of the list (of 23 receivers in total). The only known interaction not shown in the table is the interaction of DivL with CtrA which occurs at position 16 of DivL's list. To test the significance of these predictions we calculated p-values under a rank-sum test, i.e. by randomly permuting the ranks of the interaction scores. If we include the "bad" case DivL-CtrA, the probability of getting a set of predictions as good or better in ranks than ours is $p = 5 \cdot 10^{-4}$. Without CtrA, the p-value is $p = 3.5 \cdot 10^{-5}$.

In summary, in spite of the small number of experimentally determined orphan interactions the predictions of our algorithm show a significant overlap with the known interactions.

6   Conclusions

We have presented the first computational method for extensive reconstruction of bacterial signaling networks from knowledge of amino acid sequences only. First, we found that the domain architectures of almost all kinases of bacterial two-component systems fall into 7 distinct classes and that, using position-specific weight matrices, one can accurately predict which of these kinase classes each receiver domain interacts with. Using training sets of known interacting kinase/receiver pairs we determined which positions in the kinase and the receiver show clear evidence of dependence between their amino acids. From this correlation structure we constructed a probabilistic model for the joint distribution of the amino acid sequences of interacting pairs, and 'independent' models for the distributions of amino acids in kinases and receivers separately. Finally, with these probabilistic models we predict kinase/receiver interactions across all sequenced bacterial genomes. We first tested our predictions on the cognate pairs. These tests show that the cognate interactions can be very accurately reconstructed using our model. Second, we predicted interactions between orphan kinase and receivers in *Caulobacter crescentus*, and compared these with the few interactions that have been characterized in the literature. This test showed a significant overlap between the known interactions and the predictions of our algorithm. Given the small number of examples involved we cannot yet assess if the very high performance observed on the cognates generalizes to the orphans but it is highly encouraging that for 4 of the 5 tested kinases observed interactions ranked at the first or second position of our list of predictions. We believe that the large number of orphan interactions predicted by our algorithm across all sequenced genomes already form a valuable data-set for experimental investigation.

References

1. Stock, A., Robinson, V., Goudreau, P.: Two-component signal transduction. Ann. Rev. Biochem. **69** (2000), 183–215
2. Grebe, T., Stock, J.: The histidine protein kinase superfamily. Advances in Microbial Physiology **41** (1999), 139–227
3. Ausmees, N., Jacobs-Wagner, C.: Spatial and temporal control of differentiation and cell cycle progression in *Caulobacter Crescentus*. Ann. Rev. Microbiol. **57** (2003), 225–247
4. Ramani, A., Marcotte, E.: Exploiting the co-evolution of interacting proteins to discover interaction specificity. J. Mol. Biol. **327** (2003), 273–284
5. Bateman, A., Coin, L., Durbin, R., Finn, R., Hollich, V., Griffiths-Jones, S., Khanna, A., Marshall, M., Moxon, S., Sonnhammer, E., Studholme, D., Yeats, C., Eddy, S.: The Pfam protein families database. Nucl. Acids Res. **32** (2004), D138–D141
6. Do, C., Mahabhashyam, M., Brudno, M., Batzoglou, S.: Probcons: Probabilistic consistency-based multiple sequence alignment. Genome Research **15** (2005), 330–340
7. van Nimwegen, E., Zavolan, M., Rajewsky, N., Siggia, E.D.: Probabilistic clustering of sequences: Inferring new bacterial regulons by comparative genomics. Proc. Natl. Acad. Sci. USA **99** (2002), 7323–7328
8. Chow, C., Liu, C.: Approximating discrete probability distributions with dependence trees. IEEE Transactions on Information Theory **IT-14** (1968), 462–467
9. Skerker, J., Laub, M.: Cell-cycle progression and the generation of asymmetry in *Caulobacter crescentus*. Nature Reviews Microbiology **3** (2004), 325–337
10. Ohta, N., Newton, A.: The core dimerization domains of histidine kinases contain specificity for the cognate response regulator. J. Bacteriology **185** (2003), 4424–4431
11. Skerker, J., Prasol, M., Perchuk, B., Biondi, E., Laub, M.: Two-component signal transduction pathways regulating growth and cell cycle progression in a bacterium: a systems-level analysis. PLOS Biol. **3** (,2005) e334

Linear-Time Haplotype Inference on Pedigrees Without Recombinations

M.Y. Chan[1], Wun-Tat Chan[1], Francis Y.L. Chin[1,*],
Stanley P.Y. Fung[2], and Ming-Yang Kao[3]

[1] Department of Computer Science, University of Hong Kong, Hong Kong
{mychan, wtchan, chin}@cs.hku.hk
[2] Department of Computer Science, University of Leicester, Leicester, UK
pyfung@mcs.le.ac.uk
[3] Department of Electrical Engineering and Computer Science, Northwestern University, USA
kao@cs.northwestern.edu

Abstract. In this paper, a linear-time algorithm, which is optimal, is presented to solve the haplotype inference problem for pedigree data when there are no recombinations and the pedigree has no mating loops. The approach is based on the use of graphs to capture SNP, Mendelian and parity constraints of the given pedigree.

1 Introduction

The modeling of human genetic variation is critical to the understanding of the genetic basis for complex diseases. *Single nucleotide polymorphisms* (SNPs) [6] are the most frequent form of this variation, and it is useful to analyze *haplotypes*, which are sequences of linked SNP genetic markers (small segments of DNA) on a single chromosome. In diploid organisms, such as humans, chromosomes come in pairs, and experiments often yield *genotypes*, which blend haplotypes for the chromosome pair. This gives rise to the problem of inferring haplotypes from genotypes.

Before defining our problem, some preliminary definitions are needed. The physical position of a marker on a chromosome is called a *locus* and its state is called an *allele*. Without loss of generality, the alleles of a *biallelic* SNP can be denoted by 0 and 1, and a haplotype with m loci is represented as a length-m string in $\{0,1\}^m$, and a genotype as a length-m string in $\{0,1,2\}^m$. Haplotype pair $\langle h_1, h_2 \rangle$ is *SNP-consistent* with genotype g if where the two alleles of h_1 and h_2 are the same at the same locus, say 0 (respectively 1), the corresponding locus of g is also 0 (1), which denotes a *homozygous* locus; otherwise, where the two alleles of h_1 and h_2 are different, the corresponding locus of g is 2, which denotes a *heterozygous* locus (i.e. SNP). A genotype with s heterozygous loci can have 2^{s-1} SNP-consistent haplotype solutions. For example, genotype

---

* This research was supported by Hong Kong RGC Grant HKU-7119/05E and HKU Strategic Research Team Fund.

P. Bücher and B.M.E. Moret (Eds.): WABI 2006, LNBI 4175, pp. 56–67, 2006.

$g = 012212$ with $s = 3$ has four SNP-consistent haplotype pairs: $\{\langle 01\underline{1}\underline{1}\underline{1}\underline{1},$
$01\underline{0}\underline{0}10\rangle, \langle 01\underline{1}\underline{1}10, 01\underline{0}\underline{0}11\rangle, \langle 01\underline{1}011, 01\underline{0}110\rangle, \langle 01\underline{1}010, 01\underline{0}111\rangle\}$.

A *pedigree* is a fundamental connected structure used in genetics. Figure 1 shows the pictorial representation of a pedigree with 4 nodes, with a square representing a male node and a circle representing a female node and children placed under their parents: in particular, a *father* (node F), a *mother* (node M) and two *children* (son node S and daughter node D). F-M-S (also F-M-D) is a *father-mother-child trio* or simply *trio*. Furthermore, each individual node in the pedigree is associated with a genotype. We assume that there are no *mating loops*, i.e., no marriages between descendants of a common ancestor, in the pedigree.

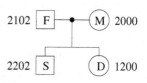

Fig. 1. Example of a pedigree with 4 nodes

A *Consistent Haplotype Configuration* (with no recombinations) for a given pedigree is an assignment of a pair of haplotypes to each individual node such that (i) all the haplotype pairs are SNP-consistent with their corresponding genotypes and (ii) the haplotypes of each child are *Mendelian-consistent*, i.e. one of the child's haplotype is exactly the same as one of its father's and the other is the same as one of its mother's.

Haplotyping Pedigree Data (with No Recombinations) Problem (HPD-NR): Given a pedigree P where each individual node of P is associated with a genotype, find a consistent haplotype configuration (CHC) for P.   □

Wijsman [8] proposed a 20-rule algorithm, and O'Connell [5] described a genotype elimination algorithm, both of which can be used for solving the HPD-NR problem. Li and Jiang [2] formulated the problem as an $mn \times mn$ matrix and solved HPD-NR by Gaussian elimination which could be solved in polynomial time $(O(m^3n^3))$, where n is the number of individuals in the pedigree and m is the number of loci for each individual. Xiao, Liu, Xia and Jiang [9] later improved this to $O(mn^2 + n^3 \log^2 n \log \log n)$. For the case without mating loops, their algorithm runs in $O(mn^2 + n^3)$ time. In this paper, we propose a new 4-stage algorithm that can either find a CHC solution or report "no solution" in optimal $O(mn)$ time when there are no mating loops. Due to space constraints, some proofs are omitted from this version.

2   The Algorithm

2.1   Stage 1

Definition 1. *If there exists a father F, mother M and two children C_1 and C_2 in the pedigree and two locus i and j such that i and j are heterozygous loci for F, M and C_1 but are homozygous and heterozygous, respective, for C_2, then we say that the pedigree has a **family problem**.*   □

Stage 1A - Checking for family problems: Since a pedigree with a family problem has no CHC solution, our algorithm begins by checking for family problems. Only if there are no family problems will the algorithm continue; otherwise, "no solution" is reported.   □

Stage 1B – Generation of vector-pairs: For each trio in the given pedigree, let the respective genotypes of the father F, the mother M and the child C be: $x_1 x_2 \ldots x_m$ and $y_1 y_2 \ldots y_m$ and $z_1 z_2 \ldots z_m$ where x_i, y_i, $z_i \in \{0, 1, 2\}$. We determine a pair of vectors (or vector-pair) each for the father, the mother and the child, namely: $\langle f_1, f_2 \rangle$, $\langle m_1, m_2 \rangle$ and $\langle c_1, c_2 \rangle$, respectively, where $f_1 = x_{1,1} x_{1,2} \ldots x_{1,m}$ and $f_2 = x_{2,1} x_{2,2} \ldots x_{2,m}$; $m_1 = y_{1,1} y_{1,2} \ldots y_{1,m}$ and $m_2 = y_{2,1} y_{2,2} \ldots y_{2,m}$; $c_1 = z_{1,1} z_{1,2} \ldots z_{1,m}$ and $c_2 = z_{2,1} z_{2,2} \ldots z_{2,m}$. The vector-pairs are determined in the following manner.

1. For each locus i, for f_1 and f_2:
   (a) If $x_i = 0$ then $x_{1,i} = x_{2,i} = 0$.
   (b) If $x_i = 1$ then $x_{1,i} = x_{2,i} = 1$.
   (c) If $x_i = 2$ and $z_i = 0$ then $x_{1,i} = 0$ and $x_{2,i} = 1$.
   (d) If $x_i = 2$ and $z_i = 1$ then $x_{1,i} = 1$ and $x_{2,i} = 0$.
   (e) If $x_i = 2$ and $z_i = 2$ and $y_i = 0$ then $x_{1,i} = 1$ and $x_{2,i} = 0$.
   (f) If $x_i = 2$ and $z_i = 2$ and $y_i = 1$ then $x_{1,i} = 0$ and $x_{2,i} = 1$.
   (g) If $x_i = 2$ and $z_i = 2$ and $y_i = 2$ then $x_{1,i} = ?$ and $x_{2,i} = ?$.
2. m_1 and m_2 are similarly determined.
3. We assume C inherits f_1 from F and m_1 from M and thus $\langle c_1, c_2 \rangle = \langle f_1, m_1 \rangle$. Check if $\langle c_1, c_2 \rangle$ is consistent with C's genotype $z_1 z_2 \ldots z_m$. If not, report "no solution".   □

Observe that, if a particular node N in the pedigree belongs to k different trios, then k vector-pairs, or $2k$ vectors, will be created for N in Stage 1. Let $\Phi(N)$ be the multiset comprised of these k vector-pairs. It is sometimes convenient to refer to the vectors rather than the vector-pairs. Thus, we let $\Gamma(N)$ be the multiset of $2k$ vectors, containing the two vectors of each vector-pair in $\Phi(N)$. Note that we can define SNP-consistency and Mendelian-consistency in terms of vector-pairs.

SNP-Consistency Condition: *SNP-consistency is said to be maintained* iff, for all nodes N in the pedigree, each vector-pair in $\Phi(N)$ is SNP-consistent with N's genotype. Vector-pair $\langle h_1, h_2 \rangle$ is said to be *SNP-consistent* with genotype g if h_1 and h_2 are both 0 (respectively 1) at the same locus, the corresponding locus of g is also 0 (1); otherwise, if h_1 is 0 (respectively 1) and h_2 is 1 (0) at the same locus, the corresponding locus of g is 2 (2).   □

Mendelian-Consistency Condition [1, 7]: *Mendelian-consistency is said to be maintained* iff, for all nodes N in the pedigree, if N is a child in a trio comprised of F, M and N, then $\Phi(N)$ contains a vector-pair $\langle c_1, c_2 \rangle = \langle f_1, m_1 \rangle$ where $f_1 \in \Gamma(F)$ and $m_1 \in \Gamma(M)$.   □

Stage 1C - Initial construction of $G = (V, E)$: Let V be the multiset of all the vectors created in Stage 1B and E be the set of red and *brown* edges defined below.

1. A **red** edge will be introduced to join the two vectors of each vector-pair generated in Stage 1A and indicates that a ? appearing at locus i of both vectors must be resolved differently in the later stages of the algorithm (the two vectors can be different or the same at other locus positions depending on whether the genotype has a 2 or not at that locus). *[SNP-consistency]*
2. For each F-M-C trio, let $\langle f_1, f_2 \rangle$, $\langle m_1, m_2 \rangle$ and $\langle c_1, c_2 \rangle$ be vector-pairs in $\Phi(\text{F})$, $\Phi(\text{M})$ and $\Phi(\text{C})$, respectively, associated with this trio. Two **brown** edges will be introduced, one connecting c_1 and f_1, and the other connecting c_2 and m_1. A brown edge between two vectors means that the two vectors must be the same at all locus positions. *[Mendelian-consistency]*   □

Example 1: Consider the pedigree with F (father), M (mother), S (son), D (daughter) shown in Figure 1. Stage 1 produces the following graph G of 12 vertices and 10 edges (6 red and 4 brown), comprised of two connected components.

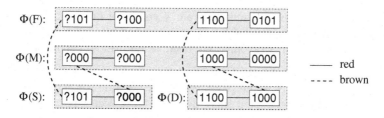

Fig. 2. Graph G for Example 1

Definition 2. *For any loci in a connected component $\mathcal{G}$ of G, we say*
1. *Locus i is* **resolved** *in $\mathcal{G}$ iff all vectors in $\mathcal{G}$ have 0 or 1 at locus i.*
2. *Locus i is* **unresolved** *in $\mathcal{G}$ iff all vectors in $\mathcal{G}$ have ? at locus i.*
3. *Otherwise, locus i is a mix of ? and non-? at i.*   □

In Example 1, the connected component for trio F-M-S has one unresolved locus (locus 1) and three resolved loci (locus 2, 3 and 4). Meanwhile, the component for trio F-M-D has no unresolved loci and four resolved loci (locus 1, 2, 3 and 4).

Lemma 1. *The time complexity of Stage 1 (Stage 1A, 1B and 1C) is $O(mn)$, where n is the number of nodes in the pedigree and m is the number of loci in each genotype. Furthermore, after Stage 1, all loci are either resolved or unresolved in each connected component of G, and G has $O(n)$ nodes and edges.*   □

In Stages 2 and 3, no vector-pairs will be added to or deleted from each $\Phi(\text{N})$ and the 0's and 1's of Stage 1 will remain as they are (unchanged). The unresolved loci of each component of G will become resolved with SNP-consistency and Mendelian-consistency maintained, and components of G will be repeatedly merged with the addition of connecting **green** (added in Stage 2) or **white** (added in Stage 3) edges until G evolves into being a single connected component. Each green or white edge is added between two vectors belonging in the same $\Gamma(N)$. This structured way of adding edges to make G connected can be done given Lemma 2 below.

Lemma 2. *If G has more than one connected component, then there exists a $\Phi(N)$ for some N such that there are two vector-pairs in $\Phi(N)$ which belong to two different components.*

Proof. Suppose to the contrary that, for all N, the vector-pairs in $\Phi(N)$ are all connected. We make use of the fact that the brown edges in G preserve the connectivity of any two nodes in the pedigree, which we have assumed to be connected. Therefore, if vector-pairs in $\Phi(N)$ are all connected for all N, then all vectors are connected together in a single connected component, which contradicts the assumption that G has more than one connected component. ☐

As loci are resolved, each multiset $\Phi(N)$ may contain one or more copies of more than one unique vector-pair. However, by the time all loci are resolved, for all nodes N, each multiset $\Phi(N)$ must contain k copies of one unique vector-pair $\langle h_1, h_2 \rangle$, which represents the haplotype-pair in the CHC for N, where k is the number of trios to which N belongs. We need an additional condition:

Endgame-Consistency Condition: *Endgame-consistency is said to be maintained* iff, for all nodes N is the pedigree, N is Endgame-consistent. Node N is said to be *Endgame-consistent* if there does not exist vector-pairs $\langle u_1, u_2 \rangle$, $\langle v_1, v_2 \rangle \in \Phi(N)$ such that the vector values at some heterozygous locus i and j ($i \neq j$) for u_1, u_2, v_1 and v_2 are a permutation of the four possibilities: 00, 01, 10 and 11; and *Endgame-inconsistent* otherwise. A connected component $\mathcal{G}$ of graph G is said to be *Endgame-consistent* if there does not exist a node N and vector-pairs $\langle u_1, u_2 \rangle$, $\langle v_1, v_2 \rangle$ in both $\Phi(N)$ and $\mathcal{G}$ such that the vector values at some heterozygous locus i and j ($i \neq j$) for u_1, u_2, v_1 and v_2 are a permutation of the four possibilities: 00, 01, 10 and 11; and *Endgame-inconsistent* otherwise. ☐

Our algorithm achieves a solution if, at the end of Stage 4, (a) graph G comprises a single connected component; (b) all loci are resolved in G; and (c) SNP-consistency, Mendelian-consistency and Endgame-consistency are maintained. However, our algorithm might report "no solution" if some N is Endgame-inconsistent before the end of Stage 4.

2.2   Stage 2

We begin by defining an important subroutine called LOCUS_RESOLVE. LOCUS_RESOLVE($\mathcal{G}$, i, u, x) will resolve all ?'s at an unresolved locus i in a connected component $\mathcal{G}$ (of G) starting with resolving the ? at locus i of vector u in $\mathcal{G}$ to $x \in \{0, 1\}$ in a manner consistent with red and non-red edges.

LOCUS_RESOLVE($\mathcal{G}, i, u, x$):

1. Let vector $u = u_1 u_2 \ldots u_m$. Set $u_i \leftarrow x$
2. For each edge $e = (u, v)$:

3.   Let vector $v = v_1 v_2 \dots v_m$.

4.   If $v_i = $? then

5.   If e is a red edge then LOCUS_RESOLVE($\mathcal{G}$, i, v, $1 - x$)

6.   else LOCUS_RESOLVE($\mathcal{G}$, i, v, x)   □

The idea of Stage 2 is to add $O(n)$ green edges to connect components of G together, where green edges are like brown edges requiring that the ?s in the two vectors connected by the edge to be resolved the same. The way in which green edges are added respects Endgame-consistency. In particular, green edges are added to connect two unconnected vectors that have the value 0 at heterozygous locus i.

Stage 2 – Adding Green Edges: For each locus i do the following:
1. For each node N, if locus i is heterozygous in N, (a) let $u = u_1 u_2 \dots u_m$ in $\Gamma(N)$ such that $u_i = 0$ (if any); and (b) for each other vector $v = v_1 v_2 \dots v_m$ in $\Gamma(N)$ such that $v_i = 0$ do the following:
   (a) For each locus j such that $u_j \in \{0,1\}$ and $v_j = $?, run LOCUS_RESOLVE (G_v, j, v, u_j). In so doing, we say that we **use u to resolve all unresolved loci of G_v**.
   (b) Likewise, for each locus j such that $v_j \in \{0,1\}$ and $u_j = $?, run LOCUS_RESOLVE(G_u, j, u, v_j). Thus, we **use v to resolve all unresolved loci of G_u**.
   (c) Add a green edge joining u and v.
2. Make G acyclic, by removing green edges only.   □

Lemma 3. *The time complexity of Stage 2 is $O(mn)$. Furthermore, after Stage 2, all loci are either resolved or unresolved in each connected component of G, and G has $O(n)$ nodes and edges.*

Proof. There are two aspects for the time complexity of Stage 2. Firstly, only unresolved loci in each component are considered, and thus a locus, once resolved, will not be considered again even upon the component's subsequent joining with other components by green edges. In this way, $O(mn)$ time complexity can be achieved. Secondly, when heterozygous locus i is considered, at most $n-1$ green edges will be added to G and thus G will still have $O(n)$ edges. Step 2 is intended to prevent an explosion of green edges by eliminating any cycles among vectors in $\Gamma(N)$ by removing green edges and can be done in $O(n)$ time by a traversal of G and is only done once for each locus. Note that, after Stage 2, there may still exists unconnected vectors u and v in $\Gamma(N)$ with $u_i = v_i = 0$ for some heterozygous locus i in N; such u and v will become properly connected in Stage 3.   □

Stage 2 ensures that each connected component has only resolved and unresolved loci. This property is important. Lemma 4 essentially tells us that we can arbitrarily resolve unresolved loci in any such component of G, and it will not affect Endgame-consistency in the sense that no matter how the unresolved loci are resolved, either Endgame-consistency will be maintained or not maintained within that component. Stage 1A and Stage 2 combined ensure the mother-father property of Lemma 5.

Lemma 4. *If a component $\mathcal{G}$ (of G) has only resolved and unresolved loci, then all possible ways of resolving ?'s in vectors in $\mathcal{G}$ such that SNP-consistency and Mendelian-consistency are maintained will either all make $\mathcal{G}$ Endgame-consistent or all make $\mathcal{G}$ Endgame-inconsistent.*

Proof. Consider a particular resolution of ?'s in the vectors in $\mathcal{G}$ such that SNP-consistency and Mendelian-consistency are maintained. Suppose Endgame-inconsistency occurs at node N, i.e. there exist two vector-pairs $\langle x_1, x_2 \rangle$, $\langle y_1, y_2 \rangle \in \Phi(N)$. We can assume, without loss of generality, that the value at some heterozygous locus i and j ($i \neq j$) for x_1, x_2, y_1 and y_2 are 00, 11, 01 and 10 respectively. Consider the following three cases for the state of locus i and j prior to the resolution:

Case 1: *Suppose locus i and j were both unresolved in $\mathcal{G}$.* Then, for all other possible resolutions, the values at locus i and j for x_1, x_2, y_1 and y_2 would either be 00, 11, 01 and 10 respectively, or 11, 00, 10 and 01 respectively, and Endgame-consistency would also be violated.

Case 2: *Suppose only one of locus i and j was unresolved, say i, in $\mathcal{G}$.* Then, for all other possible resolutions, the values at locus i and j for x_1, x_2, y_1 and y_2 would either be 00, 11, 01 and 10 respectively, or 10, 01, 11 and 00 respectively, and Endgame-consistency would also be violated.

Case 3: *Suppose both locus i and j were not unresolved (i.e., resolved).* Then, the Endgame-inconsistency existed prior to any resolution of ?'s.   □

Lemma 5. *Suppose (a) M and F are the mother and father of two unconnected trios in G after Stage 2 and (b) the given pedigree has no family problems. Then, for all possible way of resolving ?s in vectors in the two trios such that SNP-consistency and Mendelian-consistency are maintained, M and F are either both Endgame-consistent or both Endgame-inconsistent.*

Proof. Suppose F is Endgame-inconsistent. Without loss of generality, let the values at locus i and j for x_1, x_2, y_1 and y_2 be 00, 11, 01 and 10 respectively where $\langle x_1, x_2 \rangle$, $\langle y_1, y_2 \rangle \in \phi(F)$. This means locus i and j are heterozygous loci for F. Since the two trios are unconnected by a green edge, locus i and j are also heterozygous for M also. Let C_1 and C_2 be the two respective children of F connected to $\langle x_1, x_2 \rangle$ and $\langle y_1, y_2 \rangle$ by a brown edge. In the absence of family problems and green edges connecting the two trios, there are only three cases to consider: (i) when locus i and j are both heterozygous for both C_1 and C_2; (ii) when locus i and j are both heterozygous for C_1 and both homozygous for C_2; and (iii) when locus i is heterozygous for C_1 and homozygous for C_2 while locus j is homozygous for C_1 and heterozygous for C_2. It can be readily shown that in all three cases, M would also be Endgame-inconsistent.   □

2.3   Special Case of a Connected Graph

Let us consider the special case where G becomes a connected graph (i.e. a single connected component) after Stage 2. By Lemma 3, we are left only with at most two kinds of locus in G: resolved and unresolved. To resolve all unresolved loci in G (if any), we do the following. Arbitrarily pick a vector u of G. For all unresolved locus i, we simply run LOCUS_RESOLVE(G, i, u, 0). Note that running LOCUS_RESOLVE(G, i, u, 1) would have worked equally well (Lemma 4), the effect being all 1's become 0 and all 0's become 1 at locus i and gives another solution. Finally, we check that that all N are Endgame-consistent and report "no solution" if any N were Endgame-inconsistent. This procedure for dealing with G when G is a single connected component will later be called Stage 4.

Lemma 6. *If G is a connected graph after Stage 2, we can either achieve a solution that represents a CHC for the given pedigree, or report "no solution" when there is no CHC for the pedigree, in $O(mn)$ time.*

Proof. By Lemma 4, we do not have to try all possible resolutions; one will do. The time complexity of resolving the remaining k unresolved loci in the manner described above is $O(kn)$ since LOCUS_RESOLVE runs in $O(n)$ time. Checking all N for Endgame-consistency can be done in $O(mn)$ time.   □

Lemma 7. *Suppose G is a connected graph after Stage 2. If there exists a CHC solution, there are 2^s different CHC solutions, where s is the number of unresolved loci in G, unless every node in the pedigree has exactly s heterozygous loci in which case there are 2^{s-1} different CHC solutions.*

Proof. If there is a CHC solution, it is easy to see that it will remain a solution if all values at a particular unresolved locus were reversed (i.e. 0 changed to 1 and vice versa) because SNP-consistency, Mendelian-consistency and Endgame-consistency will be maintained. Thus, there are 2^s possible CHC solutions altogether, as long as there exists at least one node with more than s heterozygous loci. However, when each node in the pedigree has exactly s heterozygous loci, i.e. all the other loci are homozygous, the number of different CHC solutions is 2^{s-1}.   □

2.4   Stage 3

After Stage 2, suppose G is left with r connected components where $r > 1$, with each component having only resolved and unresolved loci. The idea of Stage 3 is to connect components of G together so that a single connected component results. After G becomes a single connected component, we can continue in the manner described in the previous section for a single connected component. Note that white edges will be treated as "non-red" edges by LOCUS_RESOLVE.

As it turns out, we can connect components in a structured way with the help of a **support graph H**. This we do in Stage 3A.

Stage 3A – Constructing Support Graph H:

1. For each node N in the pedigree, if N is unmarried, $\Gamma(N)$ cannot intersect with more than one connected component of G. Nothing is added to H. Otherwise, suppose N is married to M in the pedigree. Let G_N denote the set of connected components in G that intersect $\Gamma(N)$ but not $\Gamma(M)$. Similarly,

G_M denote those that intersect $\Gamma(M)$ but not $\Gamma(N)$, and G_{MN} denote those that intersect both $\Gamma(M)$ and $\Gamma(N)$. Now,

(a) Pick a vector from $\Gamma(N)$ from each connected component in $G_N \cup G_{MN}$. Connect the k chosen vectors with $k-1$ edges.

(b) Next, pick a vector from $\Gamma(M)$ from each connected component in G_M. Connect them to one of the vectors in $\Gamma(M)$ from a connected component in G_{MN}.

2. Next, we introduce $k'-1$ edges to connect up the k' vectors in H that are in the same component of G, and for each such edge (u, v) introduced, we label the edge with 0 if there is a path with an even number of red edges between u and v in G; otherwise, we label it with 1.

Lemma 8. *If there are no mating loops in the pedigree, H is acyclic.*

Proof. We claim that, if there are no mating loops (cycles) in the pedigree, any two components both intersect the Γ of at most two nodes. Furthermore, if there are two such nodes, they are the parents within two unconnected trios. This being the case, by making sure there are no cycles between a node and its spouse in H, as we have done in Step 1, there are no cycles in H. To prove the claim, we make use of the fact that the brown edges in G preserve and reflect the connectivity of any two nodes in the pedigree. □

Lemma 9. *H has $O(n)$ edges, and can be constructed in $O(n)$ time.* □

The idea is that we will label each edge of support tree H with 0 and 1. Some edges have been labeled in Stage 3A and others have not. We are mainly interested in the label of edge (u, v) in H where u and v are unconnected in G. Such a labeling will be done in Stage 3C. If the label is 0, then we would connect (unconnected) u and v with a white edge in G. Otherwise, we would instead connect u and the vector that is connected to v by a red edge. This is how H is used. Note that, a CHC solution of the pedigree corresponds a labeling of the edges of H. Our challenge is to finding that labeling.

In order to assist the labeling, we construct a **parity constraint graph J**, which is constructed in Stage 3B. One of the essential differences between H and J is that H shows connections between "neighboring" components while J captures all parity constraints between far-apart components.

Stage 3B – Construct parity constraint graph J:

1. Nodes in J are the same as the nodes in H.

2. Add an edge between two vectors u and v in J if (u, v) is labeled in H. Furthermore, the label of this edge in J is the same as its label in H.

3. If there is a path between two vectors u and v in H and a heterozygous locus i such that u and v are resolved (has 0 or 1) at locus i but all other vectors (if any) in the path are unresolved at locus i, add an edge (u, v) labeled L between u and v in J, where L is 1 if u and v are resolved differently at locus i and 0 otherwise, provided there is no such edge already in J. Note that there may still be two edges between any two pairs of vectors u and v in J, one labeled 0 and the other labeled 1, which is an odd cycle.

4. Check that all cycles in J have an even number of edges labeled 1. Report "no solution" and stop if there is a cycle in J with an odd number of edges labeled 1.

5. Let graph K be a copy of graph J. Note that K is not necessarily connected. To make K connected, we add edge (u, v) to K when u and v are in different components in K where (u, v) is an edge in H. This is always possible because H is a connected graph and K and H have the same set of vectors as nodes. We arbitrarily label this edge with 0 and call the corresponding edge in H a **free edge** because we have the freedom to label (u, v) with 1 instead. We continue adding edges until K is connected. □

Lemma 10. *If H has no cycles but J has an odd cycle, then there is no CHC solution.* □

Lemma 11. *K has at most $O(mn)$ edges and can be constructed in $O(mn)$ time.* □

Stage 3C – Complete labeling of H:

1. Traverse K, computing, for each node v in K, whether the number of 1-labeled edges in the path from a fixed node t in K is odd or even, i.e. parity.
2. For each unlabeled edge (u, v) in H: if u and v have same parity in K then label edge (u, v) in H with 0; else with 1. □

Lemma 12. *All edges in H can be labeled with 0 or 1 in $O(mn)$ time in Stage 3C, and the labels in H are consistent with the parity constraints specified in J in the sense that the parity between any two vectors u and v specified in J is consistent with the number of 1-label edges in the path between u and v in H.* □

Lemma 13. *Suppose the pedigree has a CHC solution, which corresponds to a labeling of edges in H where free edge e is labeled $\alpha \in \{0, 1\}$. Then, changing the label on e to $1 - \alpha$ will result in a labeling that also corresponds to a CHC solution.* □

Stage 3D – Adding White Edges to G: For each edge (u, v) in H where u is in say component G_u and v in G_v:

1. If edge is labeled 1 then let $x \leftarrow$ vector adjacent to v by red edge else $x \leftarrow v$.
2. Add white edge between u and x.
3. Use u to resolve unresolved loci in G_v.
4. Use x to resolve unresolved loci in G_u.
5. G now has one less component.

Lemma 14. *Stage 3D can be done in $O(mn)$ time, and after Stage 3D, G will be a single connected component with only unresolved and resolved loci.* □

Lemma 15. *If the pedigree has a CHC solution, Stage 3D maintains Endgame-consistency.*

Proof. Suppose, to the contrary, that some node N becomes Endgame-inconsistent after Stage 3D. Without loss of generality, let the values at locus i and j for x_1, x_2, y_1 and y_2 be 00, 11, 01 and 10, respectively, where $\langle x_1, x_2 \rangle, \langle y_1, y_2 \rangle \in \Phi(N)$. We say that the two vectors are Endgame-inconsistent.

Consider the situation prior to Stage 3D. Since the pedigree has a CHC solution, given Lemma 4, $\langle x_1, x_2 \rangle$ and $\langle y_1, y_2 \rangle$ must belong to the different components. Now suppose $\langle x_1, x_2 \rangle$ and $\langle y_1, y_2 \rangle$ become connected during Stage 3D, in particular, after the addition of a white edge e. Before the addition of white edge e, suppose $\langle x_1, x_2 \rangle$ belonged to component G_1 and $\langle y_1, y_2 \rangle$ belonged to component G_2. There are four cases to consider:

Case 1: *e connects $\langle x_1, x_2 \rangle$ and $\langle y_1, y_2 \rangle$.* White edge e corresponds to an edge in H that is labeled with a unique parity. Suppose e connects x_1 and y_1 and is labeled 0. This white edge will make x_1 and y_1 equal and therefore the value of locus i and j cannot possibly become 00 for x_1 and 01 for y_1.

Case 2: *e connects $\langle x_3, x_4 \rangle$ in G_1 and $\langle y_3, y_4 \rangle$ in G_2 where $\langle x_3, x_4 \rangle$ and $\langle y_3, y_4 \rangle$ $\in \Phi(N)$.* Since pedigree has a CHC solution and G_1 has only resolved and unresolved loci, according to Lemma 4, G_1 must be Endgame-consistent. This implies that $\langle x_1, x_2 \rangle$ and $\langle x_3, x_4 \rangle$, which are in G_1, are Endgame-consistent. Likewise, $\langle y_1, y_2 \rangle$ and $\langle y_3, y_4 \rangle$ must also be Endgame-consistent. Because of the argument in Case 1, $\langle x_3, x_4 \rangle$ and $\langle y_3, y_4 \rangle$ must also be Endgame-consistent. This makes it impossible for $\langle x_1, x_2 \rangle$ and $\langle y_1, y_2 \rangle$ to be Endgame-inconsistent.

Case 3: *e connects $\langle x_3, x_4 \rangle$ in G_1 and $\langle y_3, y_4 \rangle$ in G_2 where $\langle x_3, x_4 \rangle$ and $\langle y_3, y_4 \rangle \in \Phi(M)$ and M is N's spouse.* Suppose $\langle u_1, u_2 \rangle \in \phi(M)$ belongs to the same trio as $\langle x_1, x_2 \rangle$ and suppose $\langle v_1, v_2 \rangle \in \phi(M)$ belongs to the same trio as $\langle y_1, y_2 \rangle$. According to the Lemma 5, $\langle u_1, u_2 \rangle$ and $\langle v_1, v_2 \rangle$ are also Endgame-inconsistent. Thus, we can consider $\langle u_1, u_2 \rangle$ and $\langle v_1, v_2 \rangle$ instead of $\langle x_1, x_2 \rangle$ and $\langle y_1, y_2 \rangle$, and accordingly, apply the arguments of Case 2.

Case 4: *e connects $\langle x_3, x_4 \rangle$ in G_1 and $\langle y_3, y_4 \rangle$ in G_2 where $\langle x_3, x_4 \rangle$ and $\langle y_3, y_4 \rangle \in \Phi(M)$ and M is neither N nor N's spouse.* Assuming no mating loops, this case does not exist.   □

2.5   Stage 4

Now we deal with the single connected component G as described before:

Stage 4 – Dealing with Single Component:
1. Arbitrarily pick a vector u of G. For all unresolved locus i, run LOCUS_RESOLVE($G, i, u, 0$).
2. For all N, check $\Phi(N)$ for Endgame-consistency and report *"no solution"* if it is not maintained.

Theorem 1. *For a given pedigree, we can either achieve a solution that represents a CHC for the given pedigree, or report "no solution" when there is no solution, in $O(mn)$ time where n is the number of nodes in the pedigree and m is the number of loci.*   □

3   Concluding Remarks

In this paper, a linear-time algorithm is presented to solve the haplotype problem for pedigree data when there are no recombinations and the pedigree has no mating loops. We are currently extending the algorithm to handle mating loops.

For the haplotyping problem with recombinations, the problem becomes intractable even when at most one recombination is allowed at each haplotype of a child, or when the problem is to find a feasible haplotype with the minimum number of recombinations (even without mating loops) [4]. However, there is still much scope for further study. For example, in practice, pedigree data often contains a significant amount of missing alleles (up to 14-15% of the alleles belonging to a block could be missing in the pedigree data studied). In some cases, the deduction of the missing information on alleles is possible. The goal is then to devise an efficient algorithm to determine as many missing alleles as possible.

References

1. R. Cox, N. Bouzekri, *et al.* Angiotensin-1-converting enzyme (ACE) plasma concentration is influenced by multiple *ACE*-linked quantitative trait nucleotides. *Hum. Mol. Genet.*, 11:2969–2977, 2002.
2. J. Li and T. Jiang. Efficient rule-based haplotyping algorithms for pedigree data. *RECOMB'03*, pages 197–206, 2003.
3. J. Li and T. Jiang. Efficient inference of haplotypes from genotypes on a pedigree. *J. Bioinfo. Comp. Biol,* 1(1):41–69, 2003.
4. J. Li and T. Jiang. An exact solution for finding minimum recombinant haplotype configurations on pedigrees with missing data by integer linear programming. *RECOMB'04*, pages 20–29, 2004.
5. J.R. O'Connell. Zero-recombinant haplotyping: applications to fine mapping using SNPs. *Genet. Epidemiol.*, 19 Suppl 1:S64–70, 2000.
6. E. Russo *et al.* Single nucleotide polymorphism: Big pharmacy hedges its bets. *The Scientist*, 13, 1999.
7. N. Wang, J.M. Akey, K. Zhang, K. Chakraborty, and L. Jin. Distribution of recombination crossovers and the origin of haplotype blocks: The interplay of population history, recombination, and mutation. *Am. J. Hum. Genet..*, 11:1227–1234, 2002.
8. E.M. Wijsman. A deductive method of haplotype analysis in pedigrees. *Am. J. Hum. Genet.*, 41(3):356–373, 1987.
9. J. Xiao, L. Liu, L. Xia and T. Jiang. Fast Elimination of Redundant Linear Equations and Reconstruction of Recombination-Free Mendelian Inheritance on a Pedigree. *Manuscript.*

Phylogenetic Network Inferences Through Efficient Haplotyping

Yinglei Song[1,*], Chunmei Liu[1], Russell L. Malmberg[2], and Liming Cai[1,*]

[1] Dept. of Computer Science, Univ. of Georgia, Athens GA 30602, USA
{chunmei, song, cai}@cs.uga.edu
[2] Department of Plant Biology, University of Georgia, Athens GA 30602, USA
russell@plantbio.uga.edu

Abstract. The genotype phasing problem is to determine the haplotypes of diploid individuals from their genotypes where linkage relationships are not known. Based on the model of perfect phylogeny, the genotype phasing problem can be solved in linear time. However, recombinations may occur and the perfect phylogeny model thus cannot interpret genotype data with recombinations. This paper develops a graph theoretical approach that can reduce the problem to finding a subgraph pattern contained in a given graph. Based on ordered graph tree decomposition, this problem can be solved efficiently with a parameterized algorithm. Our tests on biological genotype data showed that this algorithm is extremely efficient and its interpretation accuracy is better than or comparable with that of other approaches.

1 Introduction

An important yet challenging problem in human genetics is the study of DNA differences among individuals. The variations of DNA sequences among a population of individuals often provide information on the genetic traits of many complex diseases. Single Nucleotide Polymorphisms (SNPs) are one of the major types of such variations. The chromosome of a diploid individual generally contains two copies of nucleotide sequences that are not completely identical. SNP sites, often called heterozygous sites, are locations where at least two different nucleotides occur in a large percentage of the population. For a region of interest in a chromosome, a description of its nucleotides from a single copy is called a *haplotype*, while that of the conflated data for the two copies is called a *genotype* [8]. In general, haplotypes are more informative on the genetic causes of diseases than genotypes; the goal of the *genotype phasing* problem is thus to determine the haplotypes from their corresponding genotype data.

Both the genotype and haplotypes of an individual can be determined with biological experiments. However, experimental techniques for haplotyping are more expensive. An extensively used procedure for haplotyping is thus to determine the genotypes of a set of individuals experimentally and then infer the haplotypes with computational approaches. Programs based on statistical models [17,7,21,23,28] have been developed to solve the genotype phasing problem. For example, PHASE [28] exhaustively enumerates all possible sets of haplotypes that can resolve the given genotypes; an EM-based

---

* To whom all correspondence should be addressed.

P. Bücher and B.M.E. Moret (Eds.): WABI 2006, LNBI 4175, pp. 68–79, 2006.

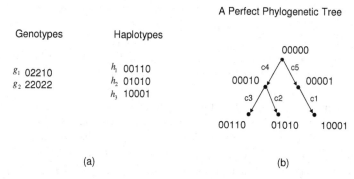

Fig. 1. (a) The genotype matrix and the corresponding haplotypes resolving the genotypes, i.e., h_1, h_2 resolve g_1, and h_2, h_3 resolve g_2. (b) A perfect phylogenetic tree for the genotypes in (a); edges in the tree are labelled with a column in the genotype matrix.

algorithm is then used to select the one with the maximum likelihood. HAPLOTYPER [21] partitions the genotype data into segments and then uses Gibbs sampling to resolve the genotypes in each segment. The haplotypes for all segments are combined to form an overall solution.

Various phylogeny models have been introduced for the inference of phylogenetic relationships among haplotypes as well as haplotypes themselves. The parsimony model assumes that the solution for genotype phasing contains a minimum number of haplotypes. Minimizing the number of haplotypes is NP-hard [25]; a few practically efficient heuristics [4] and optimal algorithms [10,11] have been developed. The perfect phylogeny model [8] considers the evolution of haplotypes; it assumes that the haplotypes resolving the given genotypes are the leaves of a perfect phylogenetic tree (see Figure 1). Based on graph matroid theory, efficient optimal algorithms [6,8,5,15] have been developed for haplotyping under the assumption of perfect phylogeny.

However, the perfect phylogeny model is not entirely satisfactory because recombinations between SNP sites have been observed in a significantly increased amount of genotype data [22]. New approaches are thus needed to include recombinations in models and algorithms for genotype phasing to correctly interpret these genotypes. The first phylogenetic model that includes the recombinations of haplotypes was developed in [29]. Assuming recombinations may not happen very often, efficient algorithms have been developed to construct "galled trees" where the recombination cycles are node disjoint and estimate a lower bound for the number of recombinations needed [12,16,19,20,26]. More recently, a decompsosition theory for phylogenetic networks is developed in [9]. Based on the new concept of "blobbed tree", the underlying maximal tree structure of a phylogenetic network can be efficiently computed from the genotype data.

In this paper, we consider the more general and yet more difficult problem of phylogenetic network inference through haplotyping where recombination cycles are allowed to share nodes and edges. We introduce a new graph theoretical model for genotype phasing. In particular, we use a genotype graph to describe all genotypes and reduce the problem to finding a certain subgraph pattern in the genotype graph. An efficient parameterized algorithm can be developed to solve this problem based on ordered graph tree decomposition. Our algorithm requires the number of heterozygous sites to be small in each taxon. In practice, a genotype data set can be partitioned into contiguous blocks such that the number k of heterozygous sites in a block is a small number (e.g., $k \leq 5$). Our algorithm can be used to resolve the genotypes in these blocks and the the resulted haplotypes can thus be combined to form an overall solution with a dynamic programming approach proposed in [6].

We have implemented this algorithm and compared its performance with that of PHASE and HAPLOTYPER on 192 biological genotype data sets downloaded from the SeattleSNPS database [3]. Our testing results showed that this algorithm is significantly faster and can achieve better interpretation accuracy than both PHASE and HAPLOTYPER on genotype data sets that contain recombinations. In addition to an efficient and accurate solution to the phylogenetic network inference problem, this graph theoretical model can possibly be used to solve a few other problems associated with haplotyping, such as the perfect phylogeny, the galled tree inference [29,9], phylogenetic netoworks with bounded number of recombinations [27], and a few incomplete perfect phylogeny problems [2,14,18].

2   Models and Algorithms

2.1   Problem Description

We use 0 or 1 to represent a *homozygous site* where the two copies of the chromosome contain the same nucleotide and 2 for a heterozygous site. A genotype can thus be described with a string of 0, 1 and 2's. Characters in such a string are also called *alleles*. The input of the genotype phasing problem is assumed to contain m genotypes of length n, which form a *genotype matrix* M. The solution is a *haplotype matrix* N, where each row is a string of length n containing 0 and 1's. In particular, for each genotype g_i in M, there exists two haplotypes h_{i_1} and h_{i_2} in N such that g_i can be resolved by h_{i_1} and h_{i_2}. For example, Figure 1(a) shows two genotypes resolved by three haplotypes.

The perfect phylogeny model assumes that the haplotypes in N are from the leaves of a rooted perfect phylogenetic tree. Internal nodes of the tree are intermediate haplotypes during the evolution and each edge in the tree is labelled by a column of matrix M and represents a mutation event that occurs on the corresponding site. In addition, each column in M can only be used to label at most one edge in the tree. Haplotyping under the framework of perfect phylogeny can be solved in linear time (see Figure 1(b)). However, solutions compatible with the perfect phylogeny model only exist for genotype matrices that do not contain conflicting columns. Two columns i and j in M *conflict* if they contain all of the four possible pairs $(0, 0)$, $(0, 1)$, $(1, 0)$, and $(1, 1)$ [8]; these pairs may exist in a population if a recombination has occurred. Recent study has shown that recombinations are biologically important for explaining genotype matrices that contain conflicting columns. As an example, Figure 2(b) provides a phylogenetic network for the set of genotypes in Figure 2(a).

In general, a recombination event between two haplotypes generates a new haplotype. Each character in this new haplotype is inherited from one of its parent haplotypes. In this paper, we only consider the *single-crossover recombination*. In particular, the resulting haplotype is a sequential combination of a prefix of one parent haplotype and a

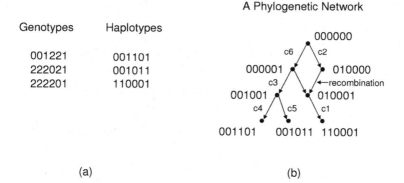

Fig. 2. (a) The genotype matrix and the corresponding haplotypes for two individuals. (b) A phylogenetic network with a single cross-over recombination for the genotypes in (a); edges that represent mutations are labelled with a column in the genotype matrix.

suffix of the other one. Single-crossover recombinations contain only one "cross-over" and are one of the most important types of recombinations [9,12,27].

We thus relax the constraints of the perfect phylogeny to model the recombinations between SNP sites. In particular, a $m \times n$ genotype matrix M can be resolved by a $2m \times n$ haplotype matrix if there exists a phylogenetic network that satisfies the constraints proposed in the following definition:

Definition 1. *Let N be a $2m \times n$ matrix, a phylogenetic network P for N is a directed connected acyclic graph that satisfies the following properties:*

1. *Each vertex in P is labelled by a haplotype of length n;*
2. *Each of the $2m$ rows in N labels one of the vertices in P;*
3. *No vertex in P has more than 2 incoming edges; exactly one vertex in P has zero incoming edges;*
4. *A vertex with exactly one incoming edge is labelled with a haplotype that is the result of mutations (but only from 0 to 1) from the haplotype labeling the precedent vertex. This edge is labelled with some column of N.*
5. *A vertex with exactly two incoming edges, called a* recombined vertex, *is labelled by a haplotype that is the result of a recombination of two haplotypes that respectively label the two precedent vertices. This edge is not labelled by any column of N.*

We propose to solve the genotype phasing problem by finding the phylogenetic network with the minimum number of recombinations. In general, a genotype data set can be partitioned into contiguous blocks, each containing a small number of heterozygous sites [6]. Therefore, we consider a simplified instance for this problem, where the number of heterozygous sites in each taxon is bounded by a small constant k. We thus can efficiently enumerate all 2^k possible haplotypes that can resolve each of the genotypes. Given a set of genotypes, we construct a *genotype graph* as follows. We enumerate the haplotypes for all genotypes and represent each of the enumerated haplotypes with a vertex. A directed edge is generated to connect from vertex u to v if h_u evolves to

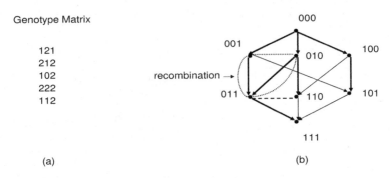

Fig. 3. (a) A genotype matrix with conflicting columns (columns 2 and 3). (b) The genotype graph constructed for the genotype matrix in (a), where thick directed lines represent a phylogenetic network contained in it; dashed and dotted lines represent principal and recombinant edges respectively. For simplicity, these two types of edges are only partially drawn and genotype vertices are not shown in the figure.

h_v by a single 0 to 1 mutation. In addition, we find all vertex triples (u, v, g) in G such that g can be obtained from u and v by a single cross-over recombination. We connect vertices in each triple into a triangle with nondirected *recombinant edges*. We generate an additional *genotype vertex* g_i for the ith genotype, and for each pair of the haplotypes h_{i_1} and h_{i_2} that can explain g_i, we connect the corresponding vertices and g_i into a triangle with nondirected *principal edges*. The resulting graph is a mixed graph.

It is not difficult to see that any phylogenetic network for some haplotype matrix that resolves a given genotype matrix can be obtained from a graphic pattern contained in its genotype graph. To guarantee that every genotype is resolved by a pair of haplotypes in the phylogenetic network, we require that, every genotype vertex g_i is *edge-dominated*, in other words, there exists a triangle that contains g_i as one of its vertices and the other two vertices are both in the phylogenetic network. Figure 3(b) partially shows the genotype graph constructed from the genotype matrix in Figure 3(a).

We would like to point out that the above construction of the genotype graph does not result in easy problems. For example, the haplotyping Maximum Resolution (MR) problem remains difficult after being formulated as a graph theoretic problem by enumerating all haplotypes for each genotype [13]. It remains NP-hard even if the exponential time reduction is assumed cost-free. We suspect that on introduced genotype graphs, the problem of finding a phylogenetic network with the minimum number of recombinations remains intractable.

2.2 Ordered Tree Decomposition

Definition 2. *Let $G = (V, E)$ be a directed acyclic graph, where V is the set of vertices in G, E denotes the set of edges in G. Pair (T, X) is an* ordered tree decomposition *of graph G if it satisfies the following conditions:*

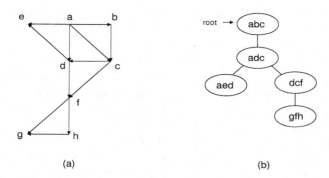

(a)   (b)

Fig. 4. (a) An example of a directed acyclic graph. (b) An ordered tree decomposition for the graph in (a).

1. $T = (I, F)$ defines a rooted tree, the sets of vertices and edges in T are I and F respectively,
2. $X = \{X_i | i \in I, X_i \subseteq V\}$, and $\forall u \in V$, $\exists i \in I$ such that $u \in X_i$,
3. $\forall (u, v) \in E$, $\exists i \in I$ such that $u \in X_i$ and $v \in X_i$,
4. $\forall i, j, k \in I$, if k is on the path that connects i and j in tree T, then $X_i \cap X_j \subseteq X_k$,
5. $\forall X_i \in X$, $X_j \in X$, where i is on the path from the root to j, there do not exist vertices u, v such that $u \in X_j$, $v \in X_i$ and there is a directed path from u to v in G.

The tree width of the ordered tree decomposition (T, X) is defined as $\max_{i \in I} |X_i| - 1$.

Figure 4 shows an example of a directed acyclic graph and an ordered tree decomposition for it. Ordered tree decomposition is a variant of the traditional tree decomposition [24], For directed acyclic graphs, we slightly modify the original definition of tree decomposition to include additional information on the topological order of graph vertices. We show later in the paper that, based on an ordered tree decomposition, a dynamic programming approach can be developed to find graphic patterns that satisfy certain topological constraints in a directed acyclic graph.

There exists an efficient algorithm that can construct an ordered tree decomposion for a directed acyclic graph G. Indeed, we can slightly modify the greedy fill-in heuristic [1] used for the efficient construction of graph tree decompositions. In particular, in the fill-in procedure, we can remove a vertex that contains no outgoing edges and connect its neighbors into a clique with nondirected edges. It is not difficult to verify that the tree decomposition generated this way for G is an ordered one.

2.3   The Algorithm

Without loss of generality, we assume that the given tree decomposition is a binary tree. Each tree node is associated with a dynamic programming table with multiple entries, the table stores the partial optimal solutions for subgraphs induced by vertices in the subtree rooted at that tree node. The algorithm follows a general bottom-up process to fill the dynamic programming tables in all tree nodes, starting with the leaves of the tree [1]. The table for a tree node with vertices $\{v_1, v_2, \cdots, v_t\}$ contains t columns, each column stores the *decision bits* for each vertex in the tree node. A table entry consists of a certain combination of the selection bits for all vertices in the tree node. The decision bit for a haplotype vertex is 1 if it is included in the partial solution and otherwise 0. For a genotype vertex, the decision bit is set to be 1 if it has been edge-dominated and otherwise 0. Two additional columns V and N are also included in the table to store the validity and the number of recombinations in the partial optimal solutions.

For each triangle that represents a recombination event, the algorithm finds the tree nodes that contain all its three vertices and marks the one with the minimum height. To compute the table for a leaf node of the tree, the algorithm enumerates all possible combinations of the selection bits for all its vertices. It also determines the validity and number of recombinations for these valid ones. An entry is invalid if the subgraph induced by the selected vertices does not satisfy the topological constraints for a phylogenetic network. For an internal node X_i with child nodes X_j and X_k, the algorithm needs to query the tables of X_j and X_k to compute the values of V and N for each entry. In particular, for a given entry e_i in the table for X_i, the computation queries only valid entries in the tables for X_j and X_k whose selection bits on vertices in $X_i \cap X_j$ and $X_i \cap X_k$ are *consistent* with e_i. The algorithm checks all possible combinations from queried entries e_j, e_k in the tables for X_j and X_k. e_i is valid if there exists an entry pair (e_j, e_k) such that the vertices selected in both e_j and e_k can form a subgraph that satisfies the topological constraints of a phylogenetic network. The number of recombinations S_{ij} for this entry pair can be computed by adding the N values for e_j and e_k together. S_{ij} is the *potential* for entry pair (e_j, e_k). The algorithm then finds the minimum potential over all queried entry pairs and add it to the number of selected marked triangles in X_i to obtain the N value for e_i.

The vertices with selection bit 1 in e_i and e_j, e_k need to be checked to determine whether the e_i is consistent with them. In particular, for a vertex u in $X_i \cap X_j$, if u represents a haplotype, e_i and e_j must set the same selection bit for u to be consistent. For a genotype vertex $v \in X_i$, the algorithm considers its selection bits in both e_j and e_k together to determine whether e_i and e_j, e_k are consistent on v. The principles for checking this consistency are as follows:

1. e_i, e_j and e_k are consistent on v if its selection bit is 1 in e_i and it is edge-dominated by an edge whose two ends have selection bit 1 in e_i.
2. e_i, e_j and e_k are consistent on v if its selection bit is 0 in e_i and the vertices with selection bit 1 in e_i do not edge-dominate e_i, in addition, selection bits for v in both e_j and e_k are 0.
3. e_i, e_j and e_k are consitent on v if its selection bit is 1 in e_i and the selection bit of v for at least one of e_j and e_k is 1.

Figure 5 provides an example for computing the entries in the table for an internal tree node $X_i = \{a, b, g_1\}$, with two child nodes $X_j = \{b, c, g_1\}$ and $X_k = \{a, d, g_1\}$, where g_1 is a genotype vertex. Since $X_i \cap X_j = \{b, g_1\}$, and $X_i \cap X_k = \{a, g_1\}$, to compute the values of V and N for an entry $e = (1, 1, 1)$ in the table for X_i, the algorithm needs to query all the valid entries with selection bit 1 for b in the table for X_j and those with selection bit 1 for a in the table for X_k. In addition, in cases where

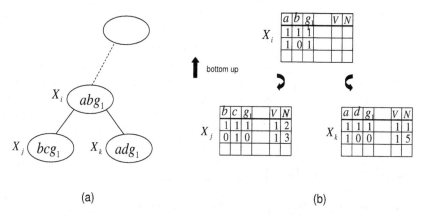

Fig. 5. (a) An ordered tree decomposition, where X_i is an internal node, X_j and X_k are its two child nodes. (b) The dynamic programming tables and the table entries for X_i, X_j and X_k. The algorithm needs to query the tables for X_j and X_k to determine the values of V and N for each entry in the table for X_i.

a and b edge-dominates g_1, the selection bit for g_1 in a queried entry from the table of X_j or X_k can be either 0 or 1. On the other hand, if g_1 is not edge-dominated by a and b, the algorithm needs to consider three possible combinations $(0, 1)$, $(1, 0)$, and $(1, 1)$ for the selection bits of g_1 while querying the entries in tables for X_j and X_k.

After the tables for all tree nodes have been completely determined, the algorithm follows a top-down tracing back procedure to obtain the phylogenetic network with the minimum number of recombinations. Specifically, the algorithm maintains a global array to mark the vertices that are selected. All valid entries in the table for the root are checked; the one with the minimum number of recombinations is selected. The vertices with selection bit 1 in this entry are then marked in the global array. The consistent entries in the tables of its child nodes are then queried. A single entry can be selected from the table for each child node such that the sum of their N values is minimized. The algorithm then proceeds to the child nodes and recursively applies the same procedure on the selected entries for these nodes. In the last stage, the algorithm adds two directed edges pointing to each recombinant vertex in the subgraph obtained from the tracing back procedure. The recombinant edges in the corresponding triangle are then removed.

The time complexity of the algorithm on a genotype graph with ordered tree width t is $O(6^t 2^k m)$, where k is the number of heterozygous sites for each genotype and m is the number of taxa. The correctness of the algorithm is guaranteed by the topological property of an ordered tree decomposition, which does not change the topological order of the vertices in a genotype graph. The algorithm thus only needs to query the tables of the child nodes of a tree node to determine the validity and number of recombinations for each entry in its table. The total number of possible combinations of selection bits for vertices in a tree node is $O(2^t)$. For a genotype vertex v in a tree node X_i, in the worst case, where v is contained in both of its two children X_j and X_k, the algorithm may need to consider three possible combinations of consistent entries from tables for X_j and X_k. The number of genotype vertices in a tree node is bounded by t, the computation time needed to compute the table for a tree node is thus bounded by $O(6^t)$. Since the total number of vertices in a genotype graph is bounded by $O(2^k m)$, the time complexity of the algorithm is $O(6^t 2^k m)$. We thus obtain the following theorem.

Theorem 1. *Given a genotype matrix with m rows and k heterozygous sites for each row. Based on an ordered tree decomposition of its genotype graph with tree width bounded by t, the phylogenetic network with the minimum number of recombinations can be computed in time $O(6^t 2^k m)$.*

3 Experiments and Results

3.1 The Ordered Tree Widths for Genotype Data Sets

We have implemented this algorithm and tested its performance on 192 genotype data sets downloaded from the SeattleSNPs database [3]. A graph trimming heuristic [4] is used to safely remove some of the vertices and edges from the genotype graph. In particular, this heuristic arbitrarily selects a vertex with no incoming edges and computes the set of all vertices that are reachable from this vertex. This vertex can be safely removed from the genotype graph if this set cannot resolve all the genotypes. This procedure can be recursively applied to a genotype graph and practically reduce its ordered tree width to an appropriate number.

Most of the 192 data sets contain missing alleles. The algorithm thus needs to consider all possible alleles on these sites while enumerating the haplotypes for a genotype with missing data. However, a pair of haplotypes that resolve the genotype may have the same allele on these missing sites. Genotypes with missing data can slightly increase the ordered tree width. To avoid processing genotypes with a large number of missing alleles and heterozygous sites, the algorithm partitions each data set into short blocks, the solutions on all blocks are then combined with a dynamic programming approach used in [6] to obtain an overall solution. In particular, this algorithm considers the consistency of haplotyping results for contiguous blocks and a dynamic programming algorithm can be used to find an overal haplotyping results that have minimum conflicts with those of the blocks. Table 1 shows the distribution of tree widths on all the short blocks where each genotype contains up to 5 heterozygous sites and missing alleles. It can be seen from the table that the tree widths for most of the genotype graphs constructed on short blocks are in the range from 5 to 10.

Table 1. The distribution of tree widths of the genotype graphs for all the short blocks of genotype data where each genotype contains up to 5 heterozygous sites and missing alleles in the block, these blocks are obtained from the 192 available data sets

Tree Width	5	6	7	8	9	10	11	< 11
Percentage (%)	33.30	36.21	16.24	5.71	3.36	3.78	1.40	98.6

3.2 The Accuracy and Efficiency of Haplotyping

We used the program to compute the phylogentic network for each genotype data set and the haplotypes that resolve it. We compared the accuracy of the program with that of PHASE and HAPLOTYPER. We used error rates to evaluate the accuracy of the phasing results. In particular, error rate is defined as the total percentage of alleles that are incorrectly interpreted in the haplotyping results. We compare the performance of our program with that of HAPLOTYPER and PHASE. These two software are the only tools that are both available and can be smoothly compiled on our system. Table 2 shows the error rates of the three programs on 192 genotype data sets. It is evident from the table that our program achieves a lower error rate than that of PHASE and HAPLO-TYPER on these testing data sets. In addition, our program is significantly faster than both of them. PHASE in general needs a few hours to compute the haplotypes for a data set, while HAPLOTYPER needs more than 10 seconds for a single run on 95% of the testing data sets. Table 3 shows the distribution of the computation time of our program on the testing data sets. The computation time needed by our program on 99.0% data sets is less than 1.0 second.

Table 2. The error rates and their standard deviations on the 192 genotype data sets for PHASE, HAPLOTYPER and our program

	Error Rate (%)	Standard Deviation (%)
PHASE	9.3%	0.8%
HAPLOTYPER	7.4%	0.6%
Our Program	5.7%	1.7%

Table 3. The cumulative distribution of the computation time needed by our program on the testing data sets

	< 0.01(sec)	< 0.05(sec)	< 0.1(sec)	< 0.5(sec)	< 1.0(sec)
Percentage	53.6%	68.8%	71.4%	93.8%	99.0%

3.3 The Number of Recombinations

The program can also obtain the minimum number of recombinations (single cross-over) needed to construct a phylogenetic netork for each data set.

Table 4 shows the distribution of this number for all testing data sets. It is evident from the table that the perfect phylogeny model can only explain 32.3% ($R = 0$) of testing data sets and there are around 19.3% data sets whose phylogenetic networks

Table 4. The distribution of the minimum number of recombinations (single cross-over) for the phylogenetic networks for the available data sets; R is the number of recombinations

	$R = 0$	$R = 1$	$1 < R \leq 5$	$5 < R \leq 10$	$10 < R \leq 15$	$15 < R \leq 20$	$R > 20$
Percentage	32.3%	11.5%	12.0%	11.5%	12.5%	1.0%	19.3%

contain more than 20 recombination events. This distribution also suggests that recombination events are important for correctly interpreting most of the genotype data sets in practice.

4   Conclusions

We have developped a new parameterized algorithm that can solve the genotype phasing problem and, at the same time, computes the corresponding phylogenetic network with the minimum number of single crossover recombinations. Our method reduces the problem of phylogenetic network construction to finding a certain subgraph pattern contained in a graph that represents the genotype matrices. Based on ordered graph tree decomposition, this problem can be solved with a parameterized algorithm. Experiments on biological data sets have demonstrated the advantage of this method over some other methods in accuracy and efficiency. Moreover, we believe it is possible to apply similar methods to solve a few other problems related to haplotyping such as the galled tree inference [29], and a few incomplete perfect phylogeny problems [2,14,18,27].

Acknowledgement

We thank the constructive comments from the anonymous reviewers on an earlier version of the paper.

References

1. S. Arnborg and A. Proskurowski, "Linear time algorithms for NP-hard problems restricted to partial k-trees," *Discrete Applied Mathematics*, 23:11–24, 1989.
2. R. Cilibrasi, L. Iersel, S. Kelk, and J. Tromp, "On the complexity of several haplotyping problems", *Proc. 5th Workshop on Algorithms in Bioinformatics WABI'05*, 128–139, 2005.
3. D.C. Crawford, C.S. Carlson, M.J. Rieder, D.P. Carrington, Q. Yi, J.D. Smith, M. A. Eberle, L. Kruglyak, and D.A. Nickerson, "Haplotype diversity across 100 candidate genes for inflammation, lipid metabolism, and blood pressure regulation in two populations," *American J. of Human Genetics*, 74:610–622, 2004.
4. A.G. Clark, "Inference of haplotypes from PCR-amplified samples of diploid populations," *Molecular Biology and Evolution*, 7(2):111–122, 1990.
5. Z. Ding, V. Filkov, and D. Gusfield, "A Linear-Time Algorithm for the Perfect Phylogeny Haplotyping (PPH) Problem," *Proc. 10th Int'l Conf. on Research in Comput. Molecular Biol. RECOMB'06,*, 231–245, 2006.
6. E. Eskin, E. Halperin, and R.M. Karp, "Large scale reconstruction of haplotypes from genotype data," *Proc. 7th Int'l Conf. on Research in Comput. Molecular Biol. RECOMB'03*, 104–113, 2003.
7. G. Greenspan and D. Greiger, "Model-based inference of haplotype block variation," *J. of Computational Biology*, 11:493–504, 2004.
8. D. Gusfield, "Haplotyping as perfect phylogeny: conceptual framework and efficient solutions," *Proc. 6th Int'l Conf. on Research in Comput. Molecular Biol. RECOMB'02*, 166–175, 2002.

9. D. Gusfield and V. Bansal, "A fundamental decomposition theory for phylogenetic networks and incompatible characters.," *Proc. 9th Int'l Conf. on Research in Comput. Molecular Biol. RECOMB'05*, 217–232, 2005.

10. D. Gusfield, "Haplotyping by pure pasimony," *Proc. 14th Symp. on Combinatorial Pattern Matching CPM'03*, 144–155, 2003.

11. D. Gusfield, "A practical algorithm for optimal inference of haplotypes from diploid populations," *Proc. 8th Conf. on Intelligent Systems for Molecular Biology ISMB'00*, 183–189, 2000.

12. D. Gusfield, S. Eddhu, and C. Langley, "Optimal efficient reconstruction of phylogenetic networks with constrained recombination," *J. of Bioinformatics and Computational Biology*, 2(1):173–213, 2004.

13. D. Gusfield, "An overview of combinatorial methods for haplotype inference," in *Computational Methods for SNPs and Haplotype Inference*, LNCS 2983, 9–25, 2004.

14. D. Gusfield, "Inference of haplotypes from samples of diploid populations: complexity and algorithms," *J. of Computational Biology*, 8(3):305–324, 2001.

15. E. Halperin and R.M. Karp, "Perfect phylogeny and haplotype assignment," *Proc. 8th Int'l Conf. on Research in Comput. Molecular Biol. RECOMB'04*, 10–19, 2004.

16. J. Hein, "Reconstructing evolution of squences subject to recombination using parsimony," *Mathematical Biosciences*, 98:185–200, 1990.

17. G. Kimmel and R. Shamir, "Maximum likelihood resolution of multi-block genotypes," *Proc. 8th Int'l Conf. on Research in Comput. Molecular Biol. RECOMB'04*, 2–9, 2004.

18. G. Kimmel and R. Shamir, "The incomplete perfect phylogeny haplotype problem," *J. of Bioinformatics and Computational Biology*, 3(2):359–384, 2005.

19. B.M.E. Moret, L. Nakhleh, T. Warnow, C.R. Linder, A. Tholse, A. Padolina, J. Sun, and R. Timme, "Phylogenetic networks: Modeling, reconstructibility, and accuracy," *IEEE/ACM Transactions on Computational Biology and Bioinformatics*, 1:13–23, 2004.

20. S.R. Myers and R.C. Griffiths, "Bounds on the minimum number of recombination events in a sample history," *Genetics*, 163:375–394, 2003.

21. T. Niu, Z. S. Qin, X. Xu, and J.S. Liu, "Bayesian haplotype inference for multiple linked single-nucleotide polymorphisms," *American J. of Human Genetics*, 70(1):157–169, 2002.

22. D. Posada and K. Crandall, "Intraspecific gene genealogies: trees grafting into networks," *Trends in Ecology and Evolution*, 16:37–45, 2001.

23. P. Rastas, M. Koivisto, H. Mannila, and E. Ukkonen, "A hidden Markov technique for haplotype reconstruction," *Proc. 5th Workshop on Algorithms in Bioinformatics WABI'05*, 140–151, 2005.

24. N. Robertson and P.D. Seymour, "Graph Minors II. Algorithmic aspects of tree-width," *J. of Algorithms*, 7:309–322, 1986.

25. R. Sharan, B.V. Halldórsson, and S. Istrail, "Islands of tractability for parsimony haplotyping," *Proc. IEEE Computational Systems Bioinformatics Conf. CSB'05*, 65–72, 2005.

26. Y.S. Song and J. Hein, "On the minimum number of recombination events in the evolutionary history of DNA sequences," *J. of Mathematical Biology*, 48:160–186, 2004.

27. Y.S. Song, Y. Wu, and D. Gusfield, "Algorithms for imperfect phylogeny haplotyping (IPPH) with a single homoplasy or recombination event," *Proc. 5th Workshop on Algorithms in Bioinformatics WABI'05*, 152–164, 2005.

28. M. Stephens, N.J. Smith, and P. Donnelly, "A new statistical method for haplotype reconstruction from population data," *American J. of Human Genetics*, 68:978–989, 2001.

29. L. Wang, K. Zhang and L. Zhang, "Perfect phylogenetic networks with recombination," J. of Computational Biology 8(1):69–78, 2001.

Beaches of Islands of Tractability: Algorithms for Parsimony and Minimum Perfect Phylogeny Haplotyping Problems[*]

Leo van Iersel[1], Judith Keijsper[1], Steven Kelk[2], and Leen Stougie[1,2]

[1] Technische Universiteit Eindhoven (TU/e), Den Dolech 2,
5612 AX Eindhoven, Netherlands
`l.j.j.v.iersel@tue.nl, j.c.m.keijsper@tue.nl`
`http://www.tue.nl`
[2] Centrum voor Wiskunde en Informatica (CWI), Kruislaan 413, 1098 SJ
Amsterdam, Netherlands
`steven.kelk@cwi.nl, leen.stougie@cwi.nl`
`http://www.cwi.nl`

Abstract. The problem *Parsimony Haplotyping* (*PH*) asks for the smallest set of haplotypes which can explain a given set of genotypes, and the problem *Minimum Perfect Phylogeny Haplotyping* (*MPPH*) asks for the smallest such set which also allows the haplotypes to be embedded in a *perfect phylogeny* evolutionary tree, a well-known biologically-motivated data structure. For *PH* we extend recent work of [17] by further mapping the interface between "easy" and "hard" instances, within the framework of (*k, l*)-*bounded instances*. By exploring, in the same way, the tractability frontier of *MPPH* we provide the first concrete, positive results for this problem, and the algorithms underpinning these results offer new insights about how *MPPH* might be further tackled in the future. In both *PH* and *MPPH* intriguing open problems remain.

1 Introduction

The computational problem of inferring biologically meaningful haplotype data from the genotype data of a population continues to generate considerable interest at the interface of biology and computer science/mathematics. A popular underlying abstraction for this model (in the context of diploid organisms) represents a genotype as a string over a $\{0, 1, 2\}$ alphabet, and a haplotype as a string over $\{0, 1\}$. The precise goal depends on the biological model being applied but a common, minimal algorithmic requirement is that, given a set of genotypes, a set of haplotypes must be produced which resolves the genotypes.

In this paper we focus on two different models. The first model, the *parsimony haplotyping* (*PH*) model [10], asks for a smallest (i.e., most parsimonious) set of haplotypes to resolve the input genotypes. To be precise, we are given a *genotype matrix G* with elements in $\{0, 1, 2\}$, the rows of which correspond to genotypes,

---

[*] Supported by the Dutch BSIK/BRICKS project.

P. Bücher and B.M.E. Moret (Eds.): WABI 2006, LNBI 4175, pp. 80–91, 2006.

while its columns correspond to sites on the genome, called SNP's. A *haplotype matrix* has elements from $\{0, 1\}$, and rows corresponding to haplotypes. Haplotype matrix H *resolves* genotype matrix G if for each row g_i of G, containing at least one 2, there are two rows h_{i_1} and h_{i_2} of H, such that $g_i(j) = h_{i_1}(j)$ for all j with $h_{i_1}(j) = h_{i_2}(j)$ and $g_i(j) = 2$ otherwise, in which case we say that h_{i_1} and h_{i_2} resolve g_i, we write $g_i = h_{i_1} + h_{i_2}$, and we call h_{i_1} the *complement* of h_{i_2} with respect to g_i, and vice versa. A row g_i without 2's is itself a haplotype and is uniquely resolved by this haplotype, which therefore has to be contained in H.

The *Parsimony Haplotyping* problem (PH) is given a genotype matrix G to find a haplotype matrix H with a minimum number of rows that resolves G. There is a rich literature in this area, of which recent papers such as [5] give a good overview. The problem is APX-hard [13,17] and the best known approximation algorithms are rather weak, yielding approximation guarantees of 2^{k-1} where k is the maximum number of 2's appearing in a row of the genotype matrix [13,14]. The lack of success in finding strong approximation guarantees has led many authors to consider methods based on Integer Linear Programming (ILP) [5,10,11,13]. A different response to the hardness is to search for "islands of tractability" amongst special, restricted cases of the problem, exploring the frontier between hardness and polynomial-time solvability. In the literature available in this direction [6,14,17], this investigation has specified classes of (k, l)-*bounded instances*: in a (k, l)-*bounded instance* the input genotype matrix G has at most k 2's per row and at most l 2's per column (cf. [17]). If k or l is a "$*$" we mean instances that are bounded only by the number of 2's per column or per row, respectively. This paper aims to supplement this "tractability" literature with mainly positive results, and doing so almost completes the bounded instance complexity landscape.

Next to the PH model we study a related model: the *Minimum Perfect Phylogeny Haplotyping* ($MPPH$) model [2]. Again a minimum-size set of resolving haplotypes is required but this time under the additional, biologically-motivated restriction that the produced haplotypes permit a *perfect phylogeny* i.e., that they can be placed at the leaves of an evolutionary tree within which each site mutates at most once. Haplotype matrices admitting a perfect phylogeny are completely characterised [8,9] by the absence of the forbidden submatrix

$$F = \begin{bmatrix} 1 & 1 \\ 0 & 0 \\ 1 & 0 \\ 0 & 1 \end{bmatrix}.$$

The *Minimum Perfect Phylogeny Haplotyping* problem ($MPPH$) is given a genotype matrix G find a haplotype matrix H with a minimum number of rows that resolves G and admits a perfect phylogeny.

The feasibility question (PPH)—given a genotype matrix G, find any haplotype matrix H that resolves G and admits a perfect phylogeny, or state that no such H exists—is solvable in linear-time [7,19]. Researchers in this area are now moving on to explore the PPH question on phylogenetic *networks* [18].

The $MPPH$ problem, however, has so far hardly been studied beyond an NP-hardness result [2] and occasional comments within PH and PPH literature [4][19][20]. In this paper we thus provide what is one of the first attempts to analyse the parsimony optimisation criteria within a well-defined and widely applicable biological framework. We seek namely to map the $MPPH$ complexity landscape in the same way as the PH complexity landscape: using the concept of (k, l)-boundedness. We write $PH(k,l)$ and $MPPH(k,l)$ for these problems restricted to (k, l)-bounded instances.

In [13] it was shown that $PH(3, *)$ is APX-hard. In [6][14] it was shown that $PH(2, *)$ is polynomial-time solvable. Recently in [17], it was shown (amongst various other results) that $PH(4, 3)$ is APX-hard. In this paper, we bring the boundaries between hard and easy classes closer by showing that $PH(3, 3)$ is APX-hard and that $PH(*, 1)$ is polynomial-time solvable.

As far as $MPPH$ is concerned there have been, prior to this paper, no concrete results beyond the above mentioned NP-hardness result. We show that $MPPH(3, 3)$ is APX-hard and that, like their PH counterparts, $MPPH(2, *)$ and $MPPH(*, 1)$ are polynomial-time solvable (in both cases using a reduction to the PH counterpart.)

For both problems the $(*, 2)$-bounded versions remain the intriguing open case. Analogous to a result from [17] for a subclass of $PH(*, 2)$, we show here that $MPPH(*, 2)$ is solvable in polynomial-time if the *compatibility graph* of the input genotype matrix is a clique. The compatibility graph $C(G)$ of a genotype matrix G has vertices representing the rows (genotypes) of G, and there is an edge between two vertices if the corresponding two genotypes coincide in each column in which none of the two has a 2. Our prediction is that learning the complexity of $PH(*, 2)$ and $MPPH(*, 2)$ in the case where the compatibility graph is a (graph-theoretical) sum of two or three cliques, will reveal the complexity of the full classes $PH(*, 2)$ and $MPPH(*, 2)$.

As explained by Sharan et al. in their "islands of tractability" paper [17], identifying tractable special classes can be practically useful for constructing high-speed subroutines within ILP solvers, but perhaps the most significant aspect of this paper is the analysis underpinning the results, which - by deepening our understanding of how this problem behaves - assists the search for better, faster approximation algorithms and for determining the exact beaches of the islands of tractability. Indeed, the continuing absence of approximation algorithms with strong accuracy guarantees underlines the importance of such work. Furthermore, the fact that (prior to this paper) concrete and positive results for $MPPH$ had not been obtained (except for rather pessimistic modifications to ILP models [5]), means that the algorithms given here for the $MPPH$ cases, and the data structures used in their analysis (e.g. the *restricted compatibility graph* in Section 3), assume particular importance.

Finally, this paper yields some interesting open problems, of which the outstanding $(*, 2)$ case (for both PH and $MPPH$) is only one; prominent amongst these questions (which are discussed at the end of the paper) is the question of whether $MPPH$ and PH instances are inter-reducible, at least within the bounded-instance framework.

The paper is organised as follows. In Section 2 we give the hardness results, in Section 3 we present the polynomial-time solvable cases, and we finish in Section 4 with conclusions and open problems. A full version of the paper including all proofs is available online [12].

2   Hard Problems

Theorem 1. $MPPH(3,3)$ *is APX-hard.*

Proof. The proof in [2] that $MPPH$ is NP-hard uses a reduction from VERTEX COVER. Using the same construction, but reducing instead from the APX-hard problem 3-VERTEX COVER (i.e., where every vertex has at most degree 3) [1][15], gives a (3,3)-bounded instance. In such a case it is not too difficult to show that (for $\epsilon > 0$) a $(1+\epsilon)$ approximation for the constructed $MPPH$ instance can be used to create a $(1+8\epsilon)$ approximation for the size of the minimum vertex cover on the input graph. We defer the details to a full version of the paper [12].   □

Theorem 2. $PH(3,3)$ *is APX-hard.*

Proof. We observe that in the proof that $PH(4,3)$ is APX-hard, by Sharan et al in [17], the leftmost 2 of an *element genotype* is actually only necessary if the element in question appears in fewer than three triples. This slight modification thus yields a (3,3)-bounded instance, and the reduction used in [17] is otherwise unchanged. We defer the proof of correctness to a full version of the paper [12].□

3   Polynomial-Time Solvability

3.1   Parsimony Haplotyping

The following result shows the polynomial-time solvability of PH on (*,1)-bounded instances.

We say that two genotypes g_1 and g_2 are *compatible*, denoted as $g_1 \sim g_2$, if $g_1(j) = g_2(j)$ or $g_1(j) = 2$ or $g_2(j) = 2$ for all j. A genotype g and a haplotype h are *consistent* if h can be used to resolve g, ie. if $g(j) = h(j)$ or $g(j) = 2$ for all j. The *compatibility graph* is the graph with vertices for the genotypes and an edge between two genotypes if they are compatible. Proof of the following two lemmas is omitted.

Lemma 1. *If g_1 and g_2 are rows of a genotype matrix with at most one 2 per column and g_1 and g_2 are compatible then there exists exactly one haplotype that is consistent with both g_1 and g_2.*   □

We use the notation $g_1 \sim_h g_2$ if g_1 and g_2 are compatible and h is consistent with both. We prove that the compatibility graph has a specific structure. A *1-sum* of two graphs is the result of identifying a vertex of one graph with a vertex of the other graph. A 1-sum of $n+1$ graphs is the result of identifying a vertex of a graph with a vertex of a 1-sum of n graphs. See Figure 1 for an example of a 1-sum of three cliques (K_3, K_4 and K_2).

Lemma 2. *If G is a genotype matrix with at most one 2 per column then every connected component of the compatibility graph of G is a 1-sum of cliques, where edges in the same clique are labelled with the same haplotype.* □

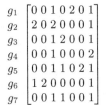
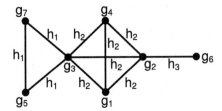

Fig. 1. Example of a genotype matrix and the corresponding compatibility graph, with $h_1 = (0,0,1,1,0,0,1)$, $h_2 = (0,0,1,0,0,0,1)$ and $h_3 = (1,0,0,0,0,0,1)$

From this lemma, it follows directly that in $PH(*,1)$ the compatibility graph is *chordal*, meaning that all its induced cycles are triangles. Every chordal graph has a *simplicial* vertex, a vertex whose (closed) neighbourhood is a clique. Deleting a vertex in a chordal graph gives again a chordal graph (see for example [3] for an introduction to chordal graphs). The following lemma leads almost immediately to polynomial solvability of $PH(*,1)$. We use set-operations for the rows of matrices: thus, e.g., $h \in H$ says h is a row of matrix H, $H \cup h$ says h is added to H as a row, and $H' \subset H$ says H' is a submatrix consisting of rows of H.

Lemma 3. *Given haplotype matrix H' and genotype matrix G with at most one 2 per column it is possible to find, in polynomial time, a haplotype matrix H that resolves G, has H' as a submatrix and has a minimum number of rows.*

Proof. The proof is constructive. Let problem (G, H') denote the above problem on input matrices G and H'. Let C be the compatibility graph of G, which implied by Lemma 2 is chordal. Suppose g corresponds to a simplicial vertex of C. Let h_c be the unique haplotype consistent with any genotype in the closed neighbourhood clique of g. We extend matrix H' to H'' and update graph C as follows.

1. If g has no 2's it can be resolved with only one haplotype $h = g$. We set $H'' = H' \cup h$ and remove g from C.
2. Else, if there exist rows $h_1 \in H'$ and $h_2 \in H'$ that resolve g we set $H'' = H'$ and remove g from C.
3. Else, if there exists $h_1 \in H'$ such that $g = h_1 + h_c$ we set $H'' = H' \cup h_c$ and remove g from C.
4. Else, if there exists $h_1 \in H'$ and $h_2 \notin H'$ such that $g = h_1 + h_2$ we set $H'' = H' \cup h_2$ and remove g from C.

5. Else, if g is not an isolated vertex in C then there exists a haplotype h_1 such that $g = h_1 + h_c$ and we set $H'' = H' \cup \{h_1, h_c\}$ and remove g from C.
6. Otherwise, g is an isolated vertex in C and we set $H'' = H' \cup \{h_1, h_2\}$ for any h_1 and h_2 such that $g = h_1 + h_2$ and remove g from C.

The resulting graph is again chordal and we repeat the above procedure for $H' = H''$ until all vertices are removed from C. Let H be the final haplotype matrix H''. It is clear from the construction that H resolves G.

The proof that H has a minimum number of rows is by induction on the number of genotypes and deferred to a full version of the paper [12]. □

Theorem 3. *The problem $PH(*, 1)$ can be solved in polynomial time.*

Proof. The proof follows from Lemma 3. Construction of the compatibility graph takes $O(n^2 m)$ time, for an n times m input matrix. Finding an ordering in which to delete the simplicial vertices can be done in time $O(n^2)$ (see [16]) and resolving each vertex takes $O(n^2 m)$ time. The overall running time of the algorithm is therefore $O(n^3 m)$. □

3.2   Minimum Pure Parsimony Haplotyping

Polynomial-time solvability of PH on $(2, *)$-bounded instances has been shown in [6] and [14]. We prove it for $MPPH(2, *)$. We start with a definition.

Definition 1. *For two columns of a genotype matrix we say that a reduced resolution of these columns is the result of applying the following rules as often as possible to the submatrix induced by these columns: deleting one of two identical rows and the replacement rules, for $a \in \{0, 1\}$,*

$$[2\ a] \rightarrow \begin{bmatrix} 1 & a \\ 0 & a \end{bmatrix}, \ [a\ 2] \rightarrow \begin{bmatrix} a & 1 \\ a & 0 \end{bmatrix}, \ [2\ 2] \rightarrow \begin{bmatrix} 1 & 1 \\ 0 & 0 \end{bmatrix} \text{ and } [2\ 2] \rightarrow \begin{bmatrix} 1 & 0 \\ 0 & 1 \end{bmatrix}$$

Note that two columns can have more than one reduced resolution if there is a genotype with a 2 in both these columns. The reduced resolutions of a column pair of a genotype matrix G are submatrices of (or equal to) F and represent all possibilities for the submatrix induced by the corresponding two columns of a minimal haplotype matrix H resolving G, after collapsing identical rows.

Theorem 4. *The problem $MPPH(2, *)$ can be solved in polynomial time.*

Proof. We reduce $MPPH(2, *)$ to $PH(2, *)$, which can be solved in polynomial time (see above). Let G be an instance of $MPPH(2, *)$. We may assume that any two rows are different.

Take the submatrix of any two columns of G. If it does not contain a $[2\ 2]$ row, then in terms of Definition 1 there is only one reduced resolution. If G contains two or more $[2\ 2]$ rows then, since by assumption all genotypes are different, G must have $\begin{bmatrix} 2 & 2 & 0 \\ 2 & 2 & 1 \end{bmatrix}$ and therefore $\begin{bmatrix} 2 & 0 \\ 2 & 1 \end{bmatrix}$ as a submatrix, which can only be resolved by a haplotype matrix containing the forbidden submatrix F. It follows that in this case the instance is infeasible. If it contains exactly one [2 2] row, then there are clearly two reduced resolutions. Thus we may assume that for each column pair there are at most two reduced solutions.

Observe that if for some column pair all reduced resolutions are equal to F the instance is again infeasible. On the other hand, if for all column pairs none of the reduced resolutions is equal to F then $MPPH(2, *)$ is equivalent to $PH(2, *)$ because any minimal haplotype matrix H that resolves G admits a perfect phylogeny. Finally, consider a column pair with two reduced resolutions, one of them containing F. Because there are two reduced resolutions there is a genotype g with a 2 in both columns. For any such g, replace g in G by h_1 and h_2, where h_1 and h_2 are the haplotypes that correspond to the resolution of g that does not lead to F. This ensures that a minimal haplotype matrix H resolving G can not have F as a submatrix in these two columns.

Repeating this procedure for every column pair either tells us that the matrix G was an infeasible instance or creates a genotype matrix G' such that any minimal haplotype matrix H resolves G' if and only if H resolves G, and H admits a perfect phylogeny.   □

Theorem 5. *The problem $MPPH(*, 1)$ can be solved in polynomial time.*

Proof. As in the proof of Theorem 4 we reduce $MPPH(*, 1)$ to $PH(*, 1)$. We defer it to a full version of the paper.   □

The open complexity problems in PH and $MPPH$ are now $PH(*, 2)$ and $MPPH(*, 2)$. Unfortunately, we have not found the answer to these complexity questions. However, the borders have been pushed slightly further. In [17] $PH(*, 2)$ is shown to be polynomially solvable if the input genotypes have the complete graph as compatibility graph, we call this problem $PH(*, 2)\text{-}C1$. We will give the counterpart result for $MPPH(*, 2)\text{-}C1$.

Let G be an $n \times m$ $MPPH(*, 2)\text{-}C1$ input matrix. Since the compatibility graph is a clique, every column of G contains only one symbol besides possible 2's. If we replace in every 1-column of G (a column containing only 1's and 2's) the 1's by 0's and mark the SNP corresponding to this column 'flipped', then we obtain an equivalent problem on a $\{0, 2\}$-matrix G'. To see that this problem is indeed equivalent, suppose H' is a haplotype matrix resolving this modified genotype matrix G' and suppose H' does not contain the forbidden submatrix F. Then by interchanging 0's and 1's in every column of H' corresponding to a flipped SNP, one obtains a haplotype matrix H without the forbidden submatrix which resolves the original input matrix G. And vice versa. Hence, from now on we will assume, without loss of generality, that the input matrix G is a $\{0, 2\}$-matrix.

If we assume moreover that $n \geq 3$, which we do from here on, the *trivial haplotype* h_t defined as the all-0 haplotype of length m is the only haplotype consistent with all genotypes in G. We define the *restricted* compatibility graph $C_R(G)$ of G as follows. As in the normal compatibility graph, the vertices of $C_R(G)$ are the genotypes of G. However, there is an edge $\{g, g'\}$ in $C_R(\mathrm{G})$ only if $g \sim_h g'$ for some $h \neq h_t$, or, equivalently, if there is a column where both g and g' have a 2.

Lemma 4. *If G is a feasible instance of $MPPH(*, 2)\text{-}C1$ then every vertex in $C_R(G)$ has degree at most 2.*

Proof. Any vertex of degree higher than 2 in $C_R(G)$ implies the existence in G of submatrix:

$$B = \begin{bmatrix} 2 & 2 & 2 \\ 2 & 0 & 0 \\ 0 & 2 & 0 \\ 0 & 0 & 2 \end{bmatrix}$$

It is easy to verify that no resolution of this submatrix permits a perfect phylogeny. □

Suppose that G has two identical columns. There are either 0, 1 or 2 rows with 2's in both these columns. In each case it is easy to see that any haplotype matrix H resolving G can be modified, without introducing a forbidden submatrix, to make the corresponding columns in H equal as well (simply delete one column and duplicate another). This leads to the first step of the algorithm **A** that we propose for solving $MPPH(*, 2)\text{-}C1$:

Step 1 of A: Collapse all identical columns in G.

From now on, we assume that there are no identical columns. Let us partition the genotypes in G_0, G_1 and G_2, denoting the set of genotypes in G with, respectively, degree 0,1, and 2 in $C_R(G)$. For any genotype g of degree 1 in $C_R(G)$ there is exactly one genotype with a 2 in the same column as g. Because there are no identical columns, it follows that any genotype g of degree 1 in $C_R(G)$ can have at most two 2's. Similarly any genotype of degree 2 in $C_R(G)$ has at most three 2's. Accordingly we define G_1^1 and G_1^2 as the genotypes in G_1 that have one 2 and two 2's, respectively, and similarly G_2^2 and G_2^3 as the genotypes in G_2 with two and three 2's, respectively.

The following lemma states how genotypes in these sets must be resolved if no submatrix F is allowed in the solution. If genotype g has k 2's we denote by $g[a_1, a_2, \ldots, a_k]$ the haplotype with entry a_i in the position where g has its i-th 2 and 0 everywhere else.

Lemma 5. *In a feasible solution to the problem $MPPH(*, 2)\text{-}C1$ all genotypes are resolved in one of the following ways:*

1. *a genotype $g \in G_1^1$ is resolved by $g[1]$ and $g[0] = h_t$;*
2. *a genotype $g \in G_2^2$ is resolved by $g[0, 1]$ and $g[1, 0]$;*
3. *a genotype $g \in G_1^2$ is either resolved by $g[0, 0] = h_t$ and $g[1, 1]$ or by $g[0, 1]$ and $g[1, 0]$; or*
4. *a genotype $g \in G_2^3$ is either resolved by $g[1, 0, 0]$ and $g[0, 1, 1]$ or by $g[0, 1, 0]$ and $g[1, 0, 1]$ (assuming that the two neighbours of g have a 2 in the first two positions where g has a 2).*

Proof. A genotype $g \in G_2^2$ has degree 2 in $C_R(G)$, which implies the existence in G of a submatrix:

$$D = \begin{matrix} g \\ g' \\ g'' \end{matrix} \begin{bmatrix} 2 & 2 \\ 2 & 0 \\ 0 & 2 \end{bmatrix}.$$

Resolving g with $g[0,0]$ and $g[1,1]$ clearly leads to the forbidden submatrix F. Similarly, resolving a genotype $g \in G_2^3$ with $g[0,0,1]$ and $g[1,1,0]$ or with $g[0,0,0]$ and $g[1,1,1]$ leads to a forbidden submatrix in the first two columns where g has a 2. It follows that resolving the genotypes in a way other than described in the lemma yields a haplotype matrix which does not admit a perfect phylogeny.

Now suppose that all genotypes are resolved as described in the lemma and assume that there is a forbidden submatrix F in the solution. Without loss of generality, we assume F can be found in the first two columns of the solution matrix. We may also assume that no haplotype can be deleted from the solution. Then, since F contains $[1\ 1]$, there is a genotype g starting with $[2\ 2]$. Since there are no identical columns there are only two possibilities. The first possibility is that there is exactly one other genotype g' with a 2 in exactly one of the first two columns. Since all genotypes different from g and g' start with $[0\ 0]$, none of the resolutions of g can have created the complete submatrix F. Contradiction. The other possibility is that there is exactly one genotype with a 2 in the first column and exactly one genotype with a 2 in the second column, but these are different genotypes, i.e., we have the submatrix D. Then $g \in G_2^3$ or $g \in G_2^2$ and it can again be checked that none of the resolutions in (2) and (4) leads to the forbidden submatrix. $\qquad\square$

Lemma 6. *Let G be an instance of $MPPH(*, 2)$ and G_1^2, G_2^3 as defined above.*

1. *any nontrivial haplotype is consistent with at most two genotypes in G; and*
2. *A genotype $g \in G_1^2 \cup G_2^3$ must be resolved using at least one haplotype that is not consistent with any other genotype.*

Proof. For the first statement, let h be a nontrivial haplotype. There is a column where h has a 1 and there are at most two genotypes with a 2 in that column. For the second statement, a genotype $g \in G_1^2 \cup G_2^3$ has a 2 in a column that has no other 2's. Hence there is a haplotype with a 1 in this column and this haplotype is not consistent with any other genotypes. $\qquad\square$

A haplotype that is only consistent with g is called a *private haplotype* of g. Based on (1) and (2) of Lemma 5 we propose the next step of **A**:

Step 2 of A: Resolve all $g \in G_1^1 \cup G_2^2$ by the unique haplotypes allowed to resolve them according to Lemma 5. Also resolve each $g \in G_0$ with h_t and the complement of h_t with respect to g. This leads to a partial haplotype matrix H_2^p.

The next step of **A** is based on Lemma 6 (2).

Step 3 of A: For each $g \in G_1^2 \cup G_2^3$ with $g \sim_{h'} g'$ for some $h' \in H_2^p$ that is allowed to resolve g according to Lemma 5, resolve g by adding the complement h'' of h' w.r.t. g to the set of haplotypes, i.e., set $H_2^p := H_2^p \cup \{h''\}$, and repeat this step as long as new haplotypes get added. This leads to partial haplotype matrix H_3^p.

Notice that H_3^p does not contain any haplotype that is allowed to resolve any of the genotypes that have not been resolved in Steps 2 and 3. Let us denote this set of leftover, unresolved haplotypes by GL, the degree 1 vertices among those by $GL_1 \subseteq G_1^2$, and the degree 2 vertices among those by $GL_2 \subseteq G_2^3$. The restricted compatibility graph induced by GL, which we denote by $C_R(GL)$ consists of paths and circuits. We first give the final steps of algorithm **A** and argue optimality afterwards.

Step 4 of A: Resolve each cycle in $C_R(GL)$, necessarily consisting of GL_2-vertices, by starting with an arbitrary vertex and, following the cycle, resolving each next pair g, g' of vertices by haplotype $h \neq h_t$ such that $g \sim_h g'$ and the two complements of h w.r.t. g and g' respectively. In case of an odd cycle the last vertex is resolved by any pair of haplotypes that is allowed to resolve it. Note that h has a 1 in the column where both g and g' have a 2 and otherwise 0. It follows easily that g and g' are both allowed to use h (and its complement) according to (4) of Lemma 5.

Step 5 of A: Resolve each path in $C_R(GL)$ with both endpoints in GL_1 by first resolving the GL_1 endpoints by the trivial haplotype h_t and the complements of h_t w.r.t. the two endpoint genotypes, respectively. The remaining path contains only GL_2-vertices and is resolved according to Step 6.

Step 6 of A: Resolve each remaining path by starting in (one of) its GL_2-endpoint(s), and following the path, resolving each next pair of vertices as in Step 4. In case of a path with an odd number of vertices, resolve the last vertex by any pair of haplotypes that is allowed to resolve it in case it is a GL_2-vertex, and resolve it by the trivial haplotype and its complement w.r.t. the vertex in case it is a GL_1 vertex.

By construction the haplotype matrix H resulting from **A** resolves G. In addition, from Lemma 5 follows that H admits a perfect phylogeny. To argue minimality of the solution, first observe that the haplotypes added in Step 2 and Step 3 are unavoidable by Lemma 5 (1) and (2) and Lemma 6 (2). Lemma 6 tells us moreover that the resolution of a cycle of k genotypes in GL_2 requires at least $k + \lceil \frac{k}{2} \rceil$ haplotypes that can not be used to resolve any other genotypes in GL. This proves optimality of Step 4. To prove optimality of the last two steps we need to take into account that genotypes in GL_1 can potentially share the trivial haplotype. Observe that to resolve a path with k vertices one needs at least $k + \lceil \frac{k}{2} \rceil$ haplotypes. Indeed **A** does not use more than that in Steps 5 and 6. Moreover, since these paths are disjoint, they cannot share haplotypes for resolving their genotypes except for the endpoints if they are in GL_1, which can share the trivial haplotype. Indeed, **A** exploits the possibility of sharing the trivial haplotype in a maximal way, except on a path with an even number of vertices and one endpoint in GL_1. Such a path, with k (even) vertices, is resolved in $\mathbf{A}$ by $3\frac{k}{2}$ haplotypes that can not be used to resolve any other genotypes. The degree 1 endpoint might alternatively be resolved by the trivial haplotype and its complement w.r.t. the corresponding genotype, adding the latter private haplotype, but then for resolving the remaining path with $k - 1$ (odd) vertices only from GL_2 we still need $k - 1 + \lceil\frac{k-1}{2}\rceil$, which together with the private haplotype of the degree 1 vertex gives $3\frac{k}{2}$ haplotypes also (not even counting h_t).

Theorem 6. $MPPH(*, 2)$ *is solvable in polynomial time if the compatibility graph is a clique.* □

4   Postlude

There remain a number of open problems. The complexity of $PH(*, 2)$ and $MPPH(*, 2)$ is still unknown. An approach that might raise the necessary insight is studying $PH(*, 2)$-Ck and $MPPH(*, 2)$-Ck variants of these problems (i.e., where the compatibility graph is the sum of k cliques) for small k.

Another intriguing open question concerns the relative complexity of PH and $MPPH$ instances. Has $PH(k, l)$ always the same complexity as $MPPH(k, l)$, in terms of well-known complexity measurements (polynomial-time solvability, NP-hardness, APX-hardness)? For hard instances, do approximability ratios differ? There do not yet exist any approximation algorithms for $MPPH$ and an immediate question is whether the weak 2^{k-1} approximation ratio for PH can be attained (or improved) for $MPPH$. A related question is whether it is possible to directly encode PH instances as $MPPH$ instances, and/or vice-versa, and if so whether/how this affects the bounds on the number of 2's in columns and rows.

For hard $PH(k, l)$ instances it would also be interesting to determine if the 2^{k-1} approximation ratio can be improved for fixed l. Finally, with respect to $MPPH$, it could be good to explore how parsimonious the solutions are that are produced by the various PPH feasibility algorithms, and whether searching through the entire space of PPH solutions (as proposed in [19]) yields practical algorithms for solving $MPPH$.

References

1. Alimonti, P., Kann, V., Hardness of approximating problems on cubic graphs, Proc. of the 3rd Italian Conf. on Algorithms and Complexity, 288–298 (1997)
2. Bafna, V., Gusfield, D., Hannenhalli, S., Yooseph, S., A Note on Efficient Computation of Haplotypes via Perfect Phylogeny, *J. of Computational Biology*, 11(5), pp. 858–866 (2004)
3. Blair, J.R.S., Peyton, B., An introduction to chordal graphs and clique trees, in *Graph theory and sparse matrix computation*, pp. 1–29, Springer (1993)
4. Bonizzoni, P., Vedova, G.D., Dondi, R., Li, J., The haplotyping problem: an overview of computational models and solutions, *J. of Computer Science and Technology* 18(6), pp. 675–688 (2003)

5. Brown, D., Harrower, I., Integer programming approaches to haplotype inference by pure parsimony, *IEEE/ACM Transactions on Computational Biology and Informatics* 3(2) (2006)
6. Cilibrasi, R., Iersel, L.J.J. van, Kelk, S.M., Tromp, J., On the Complexity of Several Haplotyping Problems, Proc. 5th Workshop on Algorithms in Bioinformatics (WABI 2005), LNBI 3692, Springer Verlag, Berlin, pp. 128–139 (2005)
7. Ding, Z., Filkov, V., Gusfield, D., A linear-time algorithm for the perfect phylogeny haplotyping (PPH) problem, *J. of Computational Biology*, 13(2) pp. 522–533 (2006)
8. Gusfield, D., *Algorithms on Strings, Trees, and Sequences: Computer Science and Computational Biology*, Cambridge University Press (1997)
9. Gusfield, D., Efficient algorithms for inferring evolutionary history, *Networks* 21, pp. 19–28 (1991)
10. Gusfield, D., Haplotype inference by pure parsimony, Proc. 14th Symp. Combinatorial Pattern Matching CPM'03, pp. 144–155 (2003)
11. Halldórsson, B.V., Bafna, V., Edwards, N., Lippert, R., Yooseph, S., Istrail, S., A survey of computational methods for determining haplotypes, Proc. DIMACS/RECOMB Satellite Workshop: Computational Methods for SNPs and Haplotype Inference, pp. 26–47 (2004)
12. Iersel, L.J.J. van, Keijsper, J.C.M., Kelk, S.M., Stougie, L., Beaches of islands of tractability: Algorithms for parsimony and minimum perfect phylogeny haplotyping problems, technical report, http://www.win.tue.nl/bs/spor/2006-09.pdf (2006)
13. Lancia, G., Pinotti, M., Rizzi, R., Haplotyping populations by pure parsimony: complexity of exact and approximation algorithms, *INFORMS J. on Computing* 16(4) pp. 348–359 (2004)
14. Lancia, G., Rizzi, R., A polynomial case of the parsimony haplotyping problem, *Operations Research Letters* 34(3) pp. 289–295 (2006)
15. Papadimitriou, C.H., Yannakakis, M., Optimization, approximation, and complexity classes, *J. Comput. System Sci.* 43, pp. 425–440 (1991)
16. Rose, D.J., Tarjan, R.E., Lueker, G.S., Algorithmic aspects of vertex elimination on graphs, *SIAM J. Comput.* 5, pp. 266–283 (1976)
17. Sharan, R., Halldórsson, B.V., Istrail, S., Islands of tractability for parsimony haplotyping, *IEEE/ACM Transactions on Computational Biology and Bioinformatics*, to appear
18. Song, Y.S., Wu, Y., Gusfield, D., Algorithms for imperfect phylogeny haplotyping (IPPH) with single haploplasy or recombination event, Proc. 5th Workshop on Algorithms in Bioinformatics (WABI 2005), LNBI 3692, Springer Verlag, Berlin, pp. 152–164 (2005)
19. VijayaSatya, R., Mukherjee, A., An optimal algorithm for perfect phylogeny haplotyping, *J. of Computational Biology*, to appear
20. Xian-Sun Zhang, Rui-Sheng Wang, Ling-Yun Wu, Luonan Chen, Models and Algorithms for Haplotyping Problem, *Current Bioinformatics* 1, pp. 105–114 (2006)

On the Complexity of SNP Block Partitioning Under the Perfect Phylogeny Model

Jens Gramm[1], Tzvika Hartman[2], Till Nierhoff[3],
Roded Sharan[4], and Till Tantau[5]

[1] Wilhelm-Schickard-Institut für Informatik, Universität Tübingen, Germany
gramm@informatik.uni-tuebingen.de
[2] Dept. of Computer Science, Bar-Ilan University, Ramat-Gan 52900, Israel
hartmat@cs.biu.ac.il
[3] International Computer Science Institute, Berkeley, USA
[4] School of Computer Science, Tel-Aviv University, Tel-Aviv 69978, Israel
roded@tau.ac.il
[5] Institut für Theoretische Informatik, Universität zu Lübeck, Germany
tantau@tcs.uni-luebeck.de

Abstract. Recent technologies for typing single nucleotide polymorphisms (SNPs) across a population are producing genome-wide genotype data for tens of thousands of SNP sites. The emergence of such large data sets underscores the importance of algorithms for large-scale haplotyping. Common haplotyping approaches first partition the SNPs into blocks of high linkage-disequilibrium, and then infer haplotypes for each block separately. We investigate an integrated haplotyping approach where a partition of the SNPs into a minimum number of non-contiguous subsets is sought, such that each subset can be haplotyped under the perfect phylogeny model. We show that finding an optimum partition is NP-hard even if we are guaranteed that two subsets suffice. On the positive side, we show that a variant of the problem, in which each subset is required to admit a perfect *path* phylogeny haplotyping, is solvable in polynomial time.

1 Introduction

Single nucleotide polymorphisms (SNPs) are differences in a single base, across the population, within an otherwise conserved genomic sequence [21]. SNPs account for the majority of the variation between DNA sequences of different individuals [19]. Especially when occurring in coding or otherwise functional regions, variations in the allelic content of SNPs are linked to medical condition or may affect drug response.

The sequence of alleles in contiguous SNP positions along a chromosomal region is called a *haplotype*. A SNP commonly has two variants, or *alleles*, in the population, corresponding to two of the four genomic letters A, C, G, and T. For diploid organisms, the *genotype* specifies for every SNP position the particular alleles that are present at this site in the two chromosomes. Genotype data contains information only on the combination of alleles at a given site; it does not reveal the association of each allele with one of the two chromosomes. Current

P. Bücher and B.M.E. Moret (Eds.): WABI 2006, LNBI 4175, pp. 92–102, 2006.

technology, suitable for large-scale polymorphism screening, obtains only the genotype information at each SNP site. The actual haplotypes in the typed region can be obtained at a considerably higher cost [19]. Due to the importance of haplotype information in association studies, it is desirable to develop efficient methods for inferring haplotypes from genotype information.

Extant approaches for inferring haplotypes from genotype data include parsimony approaches [3,12], maximum likelihood methods [7], and statistical methods [18,20]. Here we consider a perfect-phylogeny-based technique for haplotype inference, first introduced in a seminal paper by Gusfield [13]. This approach assumes that the underlying haplotypes can be arranged in a phylogenetic tree, so that for each SNP site the set of haplotypes with the same state at this site forms a connected subtree. The theoretical elegance of the perfect phylogeny approach to haplotyping as well as its efficiency and good performance in practice [2,5] have spawned several studies of the problem and its variants [1,5,15]. For more background on perfect phylogeny haplotyping see [14].

A more restricted model is the *perfect path phylogeny* model [9,10], in which the phylogenetic tree is a single long path. The motivation for considering path phylogenies is the discovery that yin-yang (complementary) haplotypes, which imply that in the prefect phylogeny model any phylogeny has to take the form of a path, are very common in human populations [22]. We previously found that over 70% of publicly available human genotype matrices that admit a perfect phylogeny also admit a perfect path phylogeny [9,10]. In the presence of missing data, finding perfect path phylogenies appears to be easier since this problem is fixed-parameter tractable [10], which is not known to be the case for perfect (branching) phylogenies.

The perfect phylogeny assumption is particularly appropriate for short genomic regions that have not undergone recombination events. For longer regions, it is common practice to sidestep the recombination problem by inferring haplotypes only for small blocks of data and then assembling these blocks to obtain the complete haplotypes [6]. Thus, the common approach to large-scale haplotyping consists of two phases: First, partition the data into blocks of SNPs. Then, infer the haplotypes for each block separately using an algorithm based on the perfect phylogeny model. Most existing block-partitioning methods partition the data into contiguous blocks, whereas in real biological data the blocks need not be contiguous [17].

In this paper we study the computational complexity of a combined approach that aims at finding a partition of an input set of SNPs into a minimum number of subsets (not necessarily contiguous), such that the genotype data induced on each subset is amenable to haplotyping under a perfect phylogeny model. We consider several variants of this problem. First, we show that for haplotype data it is possible to check in polynomial time whether there is a perfect phylogeny partition of size at most two (Section 4). However, for size three and more the problem becomes NP-hard. The situation for genotype data is even worse: Coming up with a partition into a constant number of subsets is NP-hard even if we are guaranteed that two sets suffice (Section 5). On the positive side, we show that the partitioning problem under the perfect path phylogeny model can be solved efficiently even for genotype matrices (Section 6). This result implies a novel haplotyping method that integrates the block partitioning phase and the haplotyping phase under this model. Moreover, unlike most block-partitioning techniques, our algorithm does not assume that the blocks are contiguous.

2   Preliminaries and Problem Statement

In this section we provide background on haplotyping via perfect phylogeny and formulate the partitioning problems that are at the focus of this paper.

2.1   Haplotypes, Genotypes, and Perfect Phylogenies

A *haplotype* is a row vector with binary entries. Each position of the vector corresponds to a SNP site, and specifies which of the two possible alleles are present at that position (we consider only bi-allelic SNPs since sites with more alleles are rare). For a haplotype h, let $h[i]$ denote the ith position of h. A *haplotype matrix* is a binary matrix whose rows are haplotypes. A haplotype matrix B *admits a perfect phylogeny* or just *is pp* if there exists a rooted tree T_B such that:

1. Every row of B labels exactly one node of T_B.
2. Each column of B labels exactly one edge of T_B.
3. Every edge of T_B is labeled by at least one column of B.
4. For every two rows h_1 and h_2 of B and every column i, we have $h_1[i] \neq h_2[i]$ iff i lies on the path from h_1 to h_2 in T_B.

A *genotype* is a row vector with entries in $\{0, 1, 2\}$, each corresponding to an SNP site. A 0- or 1-entry in a genotype implies that the two underlying haplotypes have the same entry in this position. A 2-entry in a genotype implies that the two underlying haplotypes differ at that position. A *genotype matrix* is a matrix whose rows are genotypes. Two haplotypes h_1 and h_2 *explain* (or *resolve*) a genotype g if for each position i the following holds: $g[i] \in \{0, 1\}$ implies $h_1[i] = h_2[i] = g[i]$; and $g[i] = 2$ implies $h_1[i] \neq h_2[i]$. Given an $n \times m$ genotype matrix A and a $2n \times m$ haplotype matrix B, we say that B *explains* A if for every $i \in \{1, \ldots, n\}$ the haplotypes in rows $2i - 1$ and $2i$ of B explain the genotype in row i of A. For a genotype g and a value $v \in \{0, 1, 2\}$, the set of columns with value v in g is called the v-*set* of g. Given an $n \times m$ genotype matrix A, we say that it *admits a perfect phylogeny* or just *is pp* if there is a $2n \times m$ haplotype matrix B that explains A and admits a perfect phylogeny. The problem of determining whether a given genotype matrix admits a perfect phylogeny, and if it does, finding the explaining haplotypes, is called *perfect phylogeny haplotyping*.

In general, the haplotype labeling the root of a perfect phylogeny tree can have arbitrary ancestral states (0 or 1) at each site. In the *directed* version of perfect phylogeny haplotyping the ancestral state of every SNP site is assumed to be 0 or, equivalently, the root of the tree corresponds to the all-0 haplotype. As shown in [5], one can reduce the general (undirected) problem to the directed case using a simple transformation of the input matrix: In each column of the genotype matrix search for the first non-2-entry from above; and if this entry is a 1-entry, exchange the roles of 0-entries and 1-entries in this column.

2.2   Perfect Path Phylogenies

A *perfect path phylogeny* is a perfect phylogeny in the form of a path, which means that the perfect phylogeny may have at most two leaves and branching occurs only at the root. If a haplotype/genotype matrix admits a perfect path phylogeny, we say that *it is ppp*.

The motivation for considering path phylogenies in the context of haplotyping is the discovery that yin-yang (complementary) haplotypes are very common in human populations [22]. We previously found, see [10,9], that over 70% of publicly available human genotype matrices that admit a perfect phylogeny also admit a perfect path phylogeny. In the presence of missing data, finding perfect path phylogenies appears to be easier since this problem is fixed-parameter tractable, which is not known to be the case for perfect (branching) phylogenies.

2.3   Partitioning Problems

Given a set C of columns of a haplotype or genotype matrix, define the following functions: $\chi_{pp}(C) = \min\{k \mid \exists C_1, \ldots, C_k : C = C_1 \cup \cdots \cup C_k, \text{each } C_i \text{ is pp}\}$ and $\chi_{ppp}(C) = \min\{k \mid \exists C_1, \ldots, C_k : C = C_1 \cup \cdots \cup C_k, \text{each } C_i \text{ is ppp}\}$. By "$C_i$ is pp" we mean that the matrix formed by the columns in C_i is pp (the pp-property does not depend on the order of the columns). We call a partition $(C_1, \ldots, C_k)$ of C in which each C_i is pp a *pp-partition*. In a slight abuse of notation we write $\chi_{pp}(A)$ for $\chi_{pp}(C)$, when C is the set of columns in the matrix A. The notation for ppp is analogously defined.

Our objective in the present paper is to determine the computational complexity of the functions χ_{pp} and χ_{ppp}, both for haplotype matrices and, more generally, for genotype matrices. The *pp-partition* problem is to compute χ_{pp} and a partition realizing the optimum value, and the *ppp-partition* problem is to compute χ_{ppp} and a corresponding partition.

Similarly to perfect phylogeny haplotyping, there are directed and undirected versions of the pp- and ppp-partition problems, but the above-mentioned transformation of Eskin et al. [5] can again be used to reduce the more general undirected case to the directed case. This shows the both versions are equivalent, allowing us to restrict attention to the directed version in the following.

3   Review of Related Results

In this section we review results from the literature that we use in the sequel. This includes both results on haplotyping as well as results from order theory.

3.1   The Complexity of Perfect Phylogeny Haplotyping

A polynomial-time algorithm for perfect phylogeny haplotyping was first given by Gusfield [13]. A central tool in Gusfield's algorithm and those that followed it, is the concept of *induce*: The *induce* of a genotype matrix A is the set of rows that is common to all haplotype matrices B that explain A. For example, the induce of the genotype matrix $\left(\begin{smallmatrix} 2 & 2 & 1 \\ 1 & 0 & 0 \end{smallmatrix}\right)$ is just $\{100\}$, but the induce of $\left(\begin{smallmatrix} 0 & 2 \\ 1 & 0 \end{smallmatrix}\right)$ is $\{00, 01, 10\}$. A key theorem on perfect phylogenies is the following (cf. [11]):

Theorem 3.1. (Four-Gamete Test) *A haplotype matrix B is pp iff the induce of any pair of its columns has size at most 3.*

For genotype matrices, an induce of size 4 for two columns also means that the matrix admits no perfect phylogeny, but the converse is no longer true and a more elaborate algorithm is needed to check whether a genotype matrix is pp.

3.2   A Partial-Order Perspective on Haplotyping

We now review results from [9] that relate haplotyping to order theory. As shown in [9], though the result is also implicit in [13], one can characterize the genotype matrices that admit a directed perfect phylogeny as follows:

Theorem 3.2. *A genotype matrix A admits a directed perfect phylogeny iff there exists a rooted tree T_A such that:*

1. *Each column of A labels exactly one edge of T_A.*
2. *Every edge of T_A is labeled by at least one column of A.*
3. *For every row r of A: (a) the columns in its 1-set label a path from the root to some node u; and (b) the columns in the 2-set of row r label a path that visits u and is contained in the subtree rooted at u.*

We consider the following partial order $\succeq$ (introduced by Eskin et al. [5]) on the columns of A: Let $1 \succ 2 \succ 0$ and extend this order to $\{0, 1, 2\}$-columns by setting $c \succeq c'$ if $c[i] \succeq c'[i]$ holds for all rows i. The following theorem shows that the existence of a perfect path phylogeny for a matrix A with column set C can be decided based on the properties of $(C, \succeq)$ alone, but we first need a definition.

Definition 3.3. *Two columns are* separable *if each has a 0-entry in the rows where the other has a 1-entry. We say that a set C of $\{0, 1, 2\}$-columns has the* ppp-property *if it can be covered by two (possibly empty) chains $(C_1, \succeq)$ and $(C_2, \succeq)$, so that their maximal elements are separable, if both are non-empty. The pair (C_1, C_2) is called a* ppp-cover *of C.*

Theorem 3.4 ([9]). *A genotype matrix A admits a directed perfect path phylogeny iff its column set has the ppp-property.*

3.3   Colorings of Hypergraphs

A *hypergraph* $H = (V, E)$ consists of a *vertex set* V and a *set E of hyperedges*, which are subsets of V. A hypergraph is *k-uniform* if each edge has exactly k elements. A *legal χ-coloring* of a hypergraph H is a function $f : V \to \{1, \ldots, \chi\}$ such that no edge in E is monochromatic. The *chromatic number of H* is the minimum χ for which there exists a legal χ-coloring of H.

It has been known for a long time that one can check in polynomial time whether a graph (a 2-uniform hypergraph) can be 2-colored and that checking whether it can be χ-colored is NP-hard for every $\chi \geq 3$. This implies that, for every $k \geq 2$ and every $\chi \geq 3$, checking whether a k-uniform hypergraph is χ-colorable is NP-hard. It is even NP-hard to approximate the chromatic number within a factor of n^ϵ, see [16].

4   PP-Partitioning Problems for Haplotype Matrices

In this section we study the complexity of $\chi_{\mathrm{pp}}(B)$ for *haplotype* matrices B. It turns out we can decide in polynomial time whether $\chi_{\mathrm{pp}}(B)$ is 1 or 2, but it is NP-hard to decide whether it is 3 or more. The proofs of these results rely on easy reductions from χ_{pp}, restricted to haplotype matrices, to the chromatic functions for graphs and back.

Theorem 4.1. *There is a polynomial-time algorithm that checks, on input of a haplotype matrix B, whether $\chi_{\mathrm{pp}}(B) \leq 2$.*

Proof. By Theorem 3.1 we can check in polynomial time whether $\chi_{\mathrm{pp}}(B) = 1$ holds. To check whether $\chi_{\mathrm{pp}}(B) \leq 2$, we construct the following graph on the columns of the matrix B: We put an (undirected) edge between every two columns whose induce has size 4. We claim that $\chi_{\mathrm{pp}}(B) \leq 2$ iff the resulting graph can be colored with two colors. To see this, note that if the chromatic number of the graph is larger than 2, then any subset of the columns of B will contain two columns having an induce of size 4. On the other hand, if the graph is 2-colorable, then the two color classes constitute a covering of the matrix B in which no color class contains two columns having an induce of size 4. Hence, by Theorem 3.1, each color class is pp.   □

Theorem 4.2. *For every $k \geq 3$, it is NP-hard to pp-partition a haplotype matrix B into k perfect phylogenies.*

Proof. We prove the claim by presenting a reduction of the NP-hard problem k-COLORING to pp-partitioning a haplotype matrix into k perfect phylogenies.

Reduction. Let a simple undirected graph $G = (V, E)$ be given as input. We map it to the following haplotype matrix B: There is a column for each vertex $v \in V$. The first row in B is an all-0 row. For each vertex v there is one row having a 1 in column v and having 0's in all other column. Finally, for each edge $\{u, v\} \in E$ there a row in B having 1-entries in columns u and v and having 0-entries in all other columns.

Correctness. Consider a coloring of the graph G. This coloring induces a partition of the columns of the matrix B. For any two column of the same class of the partition, the induce will not contain the bit string 11 and, thus, this class is a perfect phylogeny by Theorem 3.1. For the other direction, consider a partition of B into perfect phylogenies. Inside each class the induce of any two different columns must have size at most 3. Since the induce of any two different columns always contains 00, 01, and 10, the induce must be missing 11. Hence, for any two columns in the same class there cannot be an edge in G. Thus, the partition induces a coloring of the graph G.   □

Theorem 4.3. *Unless* P = NP, *the function* χ_{pp} *cannot be approximated within a factor of* n^ϵ *for any* $\epsilon > 0$.

Proof. In the reduction given in the proof of Theorem 4.2 the number of perfect phylogenies directly corresponds to the number of colors in a coloring. The coloring problem for graphs is NP-hard to approximate within a factor of n^ϵ, see [16].   □

5   PP-Partitioning Problems for Genotype Matrices

By the results of the previous section there is little hope of finding (or just coming close to) the minimum number of perfect phylogenies that cover a haplotype matrix. Since haplotype matrices are just restricted genotype matrices (namely, genotype matrices in which no 2-entries occur), the situation for genotype matrices can even be worse. The only hope left is that we might be able to find a partition of the columns of a genotype matrix into exactly two perfect phylogenies whenever this is possible in principle. As we saw before, for haplotype matrices we can find the desired partition in polynomial time.

In the present section we show that for genotype matrices the situation is much worse: even if we *know* that two perfect phylogenies suffice, coming up with a partition into any constant number χ of perfect phylogenies is still "NP-hard." By this we mean that every problem in NP can be reduced to the pp-partitioning problem in such a way that for all genotype matrices A output by the reduction either $\chi_{pp}(A) \le 2$ or $\chi_{pp}(A) > \chi$.

Theorem 5.1. *For every* $\chi \ge 2$, *it is* NP-*hard to come up with a pp-partition of a genotype matrix* A *into* χ *classes, even if we know that* $\chi_{pp}(A) \le 2$ *holds.*

Proof. We reduce from the problem of coloring a 3-uniform, 2-colorable hypergraph with a constant number of colors, which is known to be "NP-hard" in the sense sketched above: In [4] it is shown that every problem in NP can be reduced to this problem in such a way that the hypergraphs output by the reduction are 3-uniform and either 2-colorable or not χ-colorable.

Reduction. Given a 3-uniform hypergraph H, construct A as follows: A has four rows per hyperedge and one column per vertex. For each hyperedge $h = \{u, v, w\}$, the submatrix of A corresponding to the rows for h and to the columns for u,

v, and w is the matrix $S := \begin{pmatrix} 2 & 2 & 2 \\ 1 & 0 & 0 \\ 0 & 1 & 0 \\ 0 & 0 & 1 \end{pmatrix}$. Every entry of A not contained in such a submatrix is 0.

Correctness. We show how to construct a pp-partition of the columns of A into k sets given a k-coloring of H, and how to construct a k-coloring of H given a pp-partition into k sets.

Given a k-coloring of H with color classes $V_1, \ldots, V_k$, let C_i be the columns corresponding to the vertices of V_i. We claim that each C_i is pp. To this end, let A_i denote the submatrix of A that consists of the columns C_i. Each row contains either one 1-entry or up to two 2-entries and otherwise the rows contain only 0-entries: No row can contain three or more 2-entries, because the maximum number of 2-entries per row of A is three and the columns of these entries cannot all be contained in C_i, since V_i does not contain whole hyperedges.

Those rows that do not contain any 2-entries are resolved trivially by having two copies of these rows in the haplotype matrix. Those containing 2-entries are replaced by two haplotype rows as follows: If they contain at most one 2-entry, they are replaced by two copies in which the 2-entry is substituted by a 0- and a 1-entry. If they contain two 2-entries, in the first copy the 2-entries are replaced by a 0- and a 1-entry (in this order), in the second copy they are replaced by 1- and 0-entry (in this order). Other than 2-entries, these rows only contain 0-entries; so the haplotypes they are replaced by have only one 1-entry.

This way of resolving the genotypes in A_i into haplotypes leaves at most one 1-entry per row, which implies that the haplotype matrices are pp by the four-gamete test (Theorem 3.1).

Given a pp-partition $(C_1, \ldots, C_k)$ of the columns of A, let V_i contain the vertices corresponding to the set C_i. We claim that no V_i contains a complete hyperedge in H. Assume for a contradiction that $u, v, w \in C_i$ for some i and that $h = \{u, v, w\}$ is an edge in H. Then, by the reduction, the submatrix A_i, consisting of the columns C_i, contains the submatrix S. Consider a replacement of the first row with a consistent haplotype pair. One of the haplotypes has to contain two 1-entries and, consequently, there is a pair of columns that induces all four gametes, a contradiction. $\square$

6 A Polynomial-Time Algorithm for PPP-Partitioning Genotype Matrices

Our result on the positive side, which we prove in this section, is a polynomial-time algorithm for ppp-partitioning genotype matrices. The algorithm is based on reducing the problem to bipartite matching, which can be solved in polynomial time.

Let A be a genotype matrix and let C be the set of columns of A. Let $C' := \{c' \mid c \in C\}$ and $C'' := \{c'' \mid c \in C\}$. Let $E_1 := \{\{c', d''\} \mid c \succ d\}$ and

---

algorithm PPP-PARTITIONING
    let $G \leftarrow (C' \cup C'', E_1 \cup E_2)$
    let $M \leftarrow$ maximal_matching(G).
    let $G \leftarrow (C' \cup C'', M)$
    foreach $c \in C$ **do**
        let $G \leftarrow G$ with the pair $\{c', c''\}$ contracted to a single vertex
    foreach connected component X of G **do**
        output the perfect path phylogeny corresponding to X

Fig. 1. A polynomial-time algorithm for finding a ppp-partition

let $E_2 := \{\{c', d'\} \mid c$ and d are separable$\}$. Fulkerson's reduction of Dilworth's Theorem to the König–Egerváry Theorem consists mainly of the observation that the matchings M in the bipartite graph (C', C'', E_1) correspond one-to-one to the partitions of $(C, \succeq)$ into $|C| - |M|$ chains (see [8] for more details). Our method for computing $\chi_{\mathrm{ppp}}(A)$ relies on the following modification of that observation:

Theorem 6.1. *The matchings M of the graph $G = (C' \cup C'', E_1 \cup E_2)$ correspond one-to-one to the partitions of the set of columns C into $k = |C| - |M|$ subsets that admit a directed perfect path phylogeny.*

Proof. Let M be a matching of G. Contract all pairs of vertices $\{c', c''\}$ to a single vertex c. The resulting graph (C, M) has maximum degree 2 and contains no cycles. We claim that each vertex set of a component of (C, M) has the ppp-property. Then, as $\{c', c''\}$ is not an edge for any c, there are $|C| - |M|$ components, and their vertex sets are a partition into $|C| - |M|$ subsets of C that have the ppp-property. Indeed, each component of (C, M) can contain at most one edge from E_2. If it does not contain one, the vertices are a chain and thus have the ppp-property. If it contains an edge from E_2, then all other vertices are on two chains below the end vertices of that edge. So the vertices are covered by two chains whose maximal elements form an edge in E_2 and are therefore separable. Thus, also in this case, the vertex set has the ppp-property and, by Theorem 3.4, the corresponding set of columns admits a directed perfect path phylogeny.

Let $C_1, \ldots, C_k$ be a partition of C into subsets that have the ppp-property. Each C_i gives rise to a matching of size $|C_i| - 1$ in the induced subgraph $G[C_i' \cup C_i'']$. The union of these matchings is disjoint and, therefore, a matching of size $|C| - k$. □

The polynomial-time algorithm for ppp-partitioning is summarized in Figure 1. We now arrive at our main result:

Corollary 6.2. *The ppp-partition problem can be solved in polynomial time.*

7   Concluding Remarks

In this paper we studied the complexity of SNP block partitioning under the perfect phylogeny model. We showed that although the partitioning problems are NP-hard for the perfect phylogeny model, they are tractable for the more restricted perfect path phylogeny model. The contribution is two-fold. On the theoretical side, this demonstrates again the power of the perfect path phylogeny model. On the practical side, we present a block partitioning protocol that integrates the block partitioning phase and the haplotyping phase. We note, however, that there may be an exponential number of minimal partitions, and thus, in order to choose the most biologically meaningful solution we might need to consider also some other criteria for block partitioning. Future directions may include testing the algorithm on real data, and comparing this method with other block partitioning methods. Also, it would be interesting to explore the space of optimal solutions in order to find the most relevant one.

Acknowledgments. JG was supported by a grant for the DFG project *Optimal solutions for hard problems in computational biology*. JG, TN and TT were supported through a postdoc fellowship by the DAAD. TT was supported by a grant for the DFG project *Complexity of haplotyping problems*. RS was supported by an Alon Fellowship.

References

1. V. Bafna, D. Gusfield, G. Lancia, and S. Yooseph. Haplotyping as perfect phylogeny: A direct approach. *J. of Computational Biology*, 10(3–4):323–340, 2003.
2. R.H. Chung and D. Gusfield. Empirical exploration of perfect phylogeny haplotyping and haplotypers. In *Proc. 9th Conf. on Computing and Combinatorics CPM'03*, volume 2697 of *LNCS*, pages 5–19. Springer, 2003.
3. A.G. Clark. Inference of haplotypes from PCR-amplified samples of diploid populations. *J. of Molecular Biology and Evolution*, 7(2):111–122, 1990.
4. I. Dinur, O. Regev, and C.D. Smyth. The hardness of 3-uniform hypergraph coloring. In *Proc. 43rd Symposium on Foundations of Computer Science*, pages 33–42, 2002.
5. E. Eskin, E. Halperin, and R.M. Karp. Efficient reconstruction of haplotype structure via perfect phylogeny. *J. of Bioinformatics and Computational Biology*, 1(1):1–20, 2003.
6. E. Eskin, E. Halperin, R. Sharan. Optimally phasing long genomic regions using local haplotype predictions. In: *Proc. 2nd RECOMB Satellite Workshop on Computational Methods for SNPs and Haplotypes*, Pittsburgh, Pennsylvania, 2004, pp. 13–26.
7. L. Excoffier and M. Slatkin. Maximum-likelihood estimation of molecular haplotype frequencies in a diploid population. *Molecular Biology and Evolution*, 12(5):921–927, 1995.
8. S. Felsner, V. Raghavan, and J. Spinrad. Recognition algorithms for orders of small width and graphs of small Dilworth number. *Order*, 20:351–364, 2003.
9. J. Gramm, T. Nierhoff, R. Sharan, and T. Tantau. Haplotyping with missing data via perfect path phylogenies. *Discrete Applied Mathematics*, 2006. In press.
10. J. Gramm, T. Nierhoff, and T. Tantau. Perfect path phylogeny haplotyping with missing data is fixed-parameter tractable. In *Proc. 2nd International Workshop on Parameterized and Exact Computation IWPEC'04*, volume 3162 of *LNCS*, pages 174–186. Springer-Verlag, 2004.

11. D. Gusfield. Efficient algorithms for inferring evolutionary trees. *Networks*, 21:19–28, 1991.

12. D. Gusfield. Inference of haplotypes from samples of diploid populations: complexity and algorithms. *J. of Computational Biology*, 8(3):305–323, 2001.

13. D. Gusfield. Haplotyping as perfect phylogeny: Conceptual framework and efficient solutions. In *Proc. 6th Conf. on Computational Molecular Biology RECOMB'02*, pages 166–175. ACM Press, 2002.

14. D. Gusfield and S.H. Orzack. Haplotype Inference. In *CRC Handbook on Bioinformatics*, 2005.

15. E. Halperin and R.M. Karp. Perfect phylogeny and haplotype assignment. In *Proc. 8th Conf. on Computational Molecular Biology RECOMB'04*, pages 10–19. ACM Press, 2004.

16. C. Lund and M. Yannakakis. On the hardness of approximating minimization problems. *J. of the ACM*, 45(5):960–981, 1994.

17. C.S. Carlson, M.A. Eberle, L. Kruglyak and D.A. Nickerson. Mapping complex disease loci in whole-genome association studies *Nature*, 429:446–452, 2004.

18. T. Niu, S. Qin, X. Xu, and J. Liu. Bayesian haplotype inference for multiple linked single nucleotide polymorphisms. *American J. of Human Genetics*, 70(1):157–69, 2002.

19. N. Patil, A.J. Berno, D.A. Hinds, et al. Blocks of limited haplotype diversity revealed by high-resolution scanning of human chromosome 21. *Science*, 294(5547):1719–1723, 2001.

20. M. Stephens, N. Smith, and P. Donnelly. A new statistical method for haplotype reconstruction from population data. *American J. of Human Genetics*, 68(4):978–989, 2001.

21. D.G. Wang, J.B. Fan, C.J. Siao, A. Berno, P.P. Young, et al. Large-scale identification, mapping, and genotyping of single nucleotide polymorphisms in the human genome. *Science*, 280(5366):1077–1082, 1998.

22. J. Zhang, W.L. Rowe, A.G. Clark, and K.H. Buetow. Genomewide distribution of high-frequency, completely mismatching SNP haplotype pairs observed to be common across human populations. *American J. of Human Genetics*, 73(5):1073–1081, 2003.

How Many Transcripts Does It Take to Reconstruct the Splice Graph?

Paul Jenkins, Rune Lyngsø, and Jotun Hein

Dept. of Statistics, Oxford University, Oxford, OX1 3TG, United Kingdom
{jenkins, lyngsoe, hein}@stats.ox.ac.uk

Abstract. Alternative splicing has emerged as an important biological process which increases the number of transcripts obtainable from a gene. Given a sample of transcripts, the alternative splicing graph (ASG) can be constructed—a mathematical object minimally explaining these transcripts. Most research has so far been devoted to the reconstruction of ASGs from a sample of transcripts, but little has been done on the confidence we can have in these ASGs providing the full picture of alternative splicing. We address this problem by proposing probabilistic models of transcript generation, under which growth of the inferred ASG is investigated. These models are used in novel methods to test the nature of the collection of real transcripts from which the ASG was derived, which we illustrate on example genes. Statistical comparisons of the proposed models were also performed, showing evidence for variation in the pattern of dependencies between donor and acceptor sites.

1 Introduction

Alternative splicing allows the creation of multiple mRNA transcripts from a single gene. Splicing takes place after the initial transcription of DNA into precursor (pre-) mRNA and before its translation. The process modifies pre-mRNA by discarding certain regions—known as *introns*—and retaining the rest. The resulting strand of ligated *exons*—retained sections—composes the mature mRNA, and by ligating different combinations of exons multiple mRNAs can be synthesised. Studies suggest that in many eukaryotes it is highly prevalent: as many as 74% of human genes undergo alternative splicing [1], with some genes able to produce a large number of different transcripts. Around 5% of human genes may each provide more than 100 putative transcripts [2]. Alternative splicing can therefore account for a number of otherwise unresolved problems, such as the discrepancy between the size of the human proteome and the smaller genome from which it is derived. It is also thought that alternative pre-mRNA splicing is a central mode of genetic regulation in higher eukaryotes (e.g., [3])—one well characterized example is the sexual identity of *Drosophila melanogaster* [4, 5]. Alternative splicing is therefore of central importance, and can now be studied in more depth thanks to the development of tools such as expressed sequence tags (ESTs) and, in recent years, microarray analyses [1, 6].

P. Bücher and B.M.E. Moret (Eds.): WABI 2006, LNBI 4175, pp. 103–114, 2006.

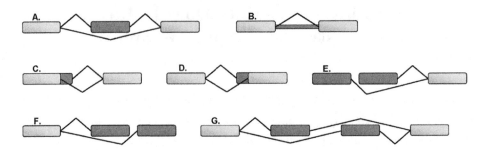

Fig. 1. Basic patterns of alternative splicing. Exons are shown as rectangles. Constitutive exons are in light grey, regions which may be spliced out in dark grey. Black lines represent paths of translation, from left to right. **A.** Cassette exon. **B.** Retained intron. **C.** Alternative 5' site. **D.** Alternative 3' splice site. **E.** Alternative promoter site. **F.** Alternative polyadenylation site. **G.** Mutually exclusive exons.

Exons can be spliced in different ways. Most exons are *constitutive*, that is, always retained in the mRNA. Exons either fully omitted or fully included are called *cassette exons*. Alternative splice sites are also found within individual exons, known as alternative 5' or 3' splice sites. The mRNAs themselves may have alternative 5' or 3' ends, with alternative selection of the 5'-most or 3'-most exons. Finally a *retained intron* denotes an intron flanked by exons that is also included in the final mRNA. These 'building blocks' are illustrated in Figure 1 (see also [4]). Some or all of these patterns may be observed from translations of a gene's mRNA, leading to potentially complex overall splicing patterns (Figure 2). Traditionally the transcriptome of a gene has been represented by an exhaustive list of its splice variants. However, as the prevalence of alternative splicing has become apparent, a need for more concise notation has emerged. Heber *et al.* [7] introduce the idea of the *alternative splice graph* (ASG), which enables the set of possible transcripts to be represented in a single graph, avoiding the error-prone nature of case-by-case transcript reconstruction. Denote the ASG as $G = (V, E)$, defined as follows. Let $\{s_1, ..., s_n\}$ be the set of RNA transcripts of the gene of interest, with each s_k corresponding to a sequence of genomic positions V_k ($V_i \neq V_j$ for $i \neq j$). Define $V := \bigcup_i V_i$, the set of all transcribed positions, and $E := \{(v, w) : v \text{ and } w \text{ form consecutive positions in at least one transcript } s_i\}$. Hence the ASG G is a directed graph and a putative transcript is any path in G. The graph is also acyclic, since the exons present in any spliced transcript are retained in the correct 5' to 3' linear order [8, 9]. Finally strings of consecutive vertices with *indegree = outdegree = 1* are collapsed into a single vertex. So each exon fragment (i.e. portion of an exon bounded by two splice sites) is represented as a single vertex. This enables the ASG to be illustrated in a similar manner to that shown in Figure 2—the numbered blocks are vertices, and the arcs read from left to right are directed edges. The ASG is a convenient, compact representation of all the splicing events associated with a particular gene, and lends itself to much further investigation.

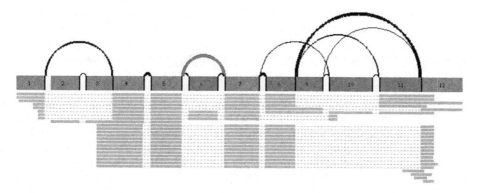

Fig. 2. An example of more complicated splicing patterns: human gene neurexin III-β (Ensembl ID ENSG00000021645, gene not to scale). Splicing events are represented by curved edges. The fragment labelled 6 is a cassette exon. More complicated and nested relationships are also visible. Edge thicknesses are proportional to EST support for that splicing event. ESTs from which the ASG was reconstructed are aligned below. Image derived from the Alternative Splicing Gallery [2].

Note the number of putative transcripts of an ASG equals or exceeds the number of distinct transcripts used to construct the ASG. The assumption that any path through the ASG represents a putative transcript in effect assumes that splicing events are independent. In this paper we propose two ASG based Markovian models of isoform generation to investigate this independence assumption. We further introduce simulation based and graph theoretical algorithms to investigate the question of whether the existing transcripts associated with a given gene are likely to have come from a random sample or from a strongly pruned subset of the transcriptome, either through non-independence of exons or through other effects such as ascertainment bias.

2   Transcript Generation Models

We propose two simple models of transcript generation, each utilising different parameter spaces. The process in our models is Markov in the sense that if we reach a particular exon fragment, the following fragment to be included does not depend on any other earlier decisions upstream. There exists a probability distribution over the transcriptome of a given gene, which can be modified conditional on additional knowledge, such as a cell's tissue type. For now we will not assume such further knowledge, which in many cases this will not unduly affect the distribution of interest. Tissue-specific control appears to be restricted to a relatively small number of specialised genes: only 2.2% of alternative splicing relationships have been observed with high confidence to be tissue-specific [10]. Tissue-specific control would cause a higher degree of exon coupling, since transcripts are effectively generated from two overlapping, yet distinct, sub-ASGs.

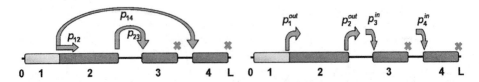

Fig. 3. Transcript generation of a simple example gene under each model. Constitutive exon shown in light grey. Exons which may be spliced out shown in dark grey. In this example label members of V as $1, \ldots, 4$, so that $I = \{1\}$, $p_1^{\text{start}} = 1$, $T = \{3, 4\}$; the read is terminated on reaching the 3' end of exon 3 or 4. (*Left*) Pairwise model. P is the zero matrix other than p_{12}, p_{23} and p_{14} ($p_{23} = 1$). (*Right*) In-out model. Here, $\mathbf{p}^{\text{in}} = (0, 0, p_3^{\text{in}}, p_4^{\text{in}})$ and $\mathbf{p}^{\text{out}} = (p_1^{\text{out}}, p_2^{\text{out}}, 0, 0)$ (with $p_2^{\text{out}} = 1$).

2.1   Model 1: Pairwise Model

We approximate the discrete structure of a gene by an interval of the real line $[0, L]$. Superimposed on the gene is a (fixed) set V of exon fragments, a collection of subsets of the line (as in Figure 2). Based on the underlying ASG there exist pairwise probabilities for each pair of fragments (v_1, v_2) that have been observed to be adjacent in at least one transcript, representing the probability that, as we read through the mRNA's sequence, if it contains v_1 we then jump forward to v_2. These probabilities can be captured as a $|V| \times |V|$ strictly upper triangular probability matrix P, with entries defined by $p_{ij} =$ probability of a transcript jumping from fragment i to fragment j, given that it contains i.

To account for the features of Figure 1, define sets $I, T \subseteq V$ of initiation fragments and terminal fragments, respectively. For each walk, rather than beginning at 0 proceed randomly with probability p_i^{start} from $i \in I$. Similarly, for each walk that reaches a fragment $t \in T$, transcript generation is terminated at the 3' end of that fragment. Thus, the complete model is captured by the collection $(V, P, I, \mathbf{p}^{\text{start}}, T)$—see Figure 3. Intuitively this is the most general model consistent with observed splicing events that assumes independence between splicing events.

2.2   Model 2: In-Out Model

The pairwise model allows the modelling of dependencies between the donor and acceptor sites in a splicing event. As an alternative, we will also consider the most general model consistent with observed splicing events that models 'donation' and 'acceptance' of the splicing event independently. With each exon fragment $x \in V$, associate two probabilities p_x^{in}, p_x^{out}, the probabilities of jumping 'into' and 'out of' the the gene. Conceptually we can imagine travelling along the real line from 0 to L, and as we reach each exon fragment jumping 'in' with probability p^{in} if we are 'out', then jumping 'out' with probability p^{out} if we are 'in'. This in effect models inclusion of isolated exons as independent events, where each exon is included with a probability reflecting the strength of its acceptor site. Note that at most two probabilities are used at each fragment rather than up to $n = |V|$ for each in the pairwise model. The in-out model seeks to explain the splicing events we observe with only $O(n)$ parameters, compared to the $O(n^2)$ parameters of the pairwise model. $I, T \subseteq V$ and $\mathbf{p}^{\text{start}}$ are defined as in the pairwise model. Thus, the complete model is captured by the collection $(V, \mathbf{p}^{\text{in}}, \mathbf{p}^{\text{out}}, I, \mathbf{p}^{\text{start}}, T)$ (Figure 3).

2.3 Hypothesis Testing

The in-out model is nested in the pairwise model; we can represent an in-out model $(S, \mathbf{p}^{\text{in}}, \mathbf{p}^{\text{out}}, I, \mathbf{p}^{\text{start}}, T)$ as a pairwise model $(S, P, I, \mathbf{p}^{\text{start}}, T)$ with $p_{ij} = p_i^{\text{out}} p_j^{\text{in}} \prod_{i<k<j}(1 - p_k^{\text{in}})$. In a similar way, the pairwise model can be emded-ded in what we'll refer to as model 0, that which simply assigns a probability to each putative transcript. Given a gene we can propose the following test for assessing the relative applicability of two models a, b, with $b \subseteq a$. For a given sample of transcripts $\{s_1, \ldots, s_n\}$ define the likelihood ratio statistic $\Lambda = \sup_{\mathbf{q_b}} L_b(\mathbf{q_b}) / \sup_{\mathbf{q_a}} L_a(\mathbf{q_a})$, where $\mathbf{q_a}, \mathbf{q_b}$ are the parameters under each model and L_a, L_b are the likelihoods of the data under each model, assuming independent sampling. The probability of each transcript is the product of the relevant probabilities involved in its generation. For example in Figure 3, $P(1 \cup 2 \cup 3) = p_{12} p_{23}$ under the pairwise model and $P(1 \cup 2 \cup 3) = (1 - p_1^{\text{out}}) p_2^{\text{out}} p_3^{\text{in}}$ under the in-out model. If the in-out model holds then $-2 \ln \Lambda \dot\sim \chi^2(z)$, where z is the number of degrees of freedom.

3   ASG Recovery Tests

3.1   Model Based Tests

Once we have a model describing transcript generation from an ASG, we can address the highly relevant question of the confidence we have in knowing the full true ASG. We propose a bootstrap-like method to assess the likelihood that the full ASG has been reconstructed, or alternatively to detect ascertainment biases in existing transcript databases, using a transcript generation model as follows. Assume that we have reconstructed an ASG from m transcripts. We may then ask what the probability is of drawing m independent samples from the full ASG that covers all edges in the full ASG. This can be computed ex-actly, albeit very inefficiently. Alternatively one can repeatedly sample m tran-scripts from the full ASG and check whether all edges are represented in these transcripts (or, if the in-out model is assumed, whether all choices are repre-sented) to obtain a p-value for the scenario of recovering the full ASG from m transcripts.

Unfortunately, we do not necessarily know the full ASG but only the inferred ASG. So what can we expect if we sample from the inferred ASG? Assume that the inferred ASG is in fact the full ASG, and that the chosen model of transcript generation holds. Then we are indeed sampling from the full ASG and we can expect the rejection rate—i.e. the false negative rate—to equal one minus the p-value computed. If the inferred ASG does not coincide with the full ASG, the acceptance rate—i.e. the false positive rate—cannot be similarly tied to the p-value computed. Indeed if $m = 1$, the inferred ASG will offer only one putative transcript and our sampling test will always accept the inferred ASG. However, as shown in Section 4, the false positive rate does seem to follow the p-value threshold for realistic data. Intuitively, if after m transcripts the ASG is fully recovered, or close to it, then there is a higher probability of some redundancy in the real collection of transcripts—indicating that they do indeed cover the whole ASG. Alternatively, if in general sampling m transcripts does not recover the ASG then there is little redundancy in the collection, and hence a higher probability that there exist other undiscovered edges.

If testing whether a fraction α of the full ASG has been recovered, we are on even less solid ground. Sampling from the inferred ASG and accepting if a fraction α of the inferred ASG has been recovered, not even the false negative rate can be theoretically linked to the p-value computed. Assume that the full ASG offers three possible transcripts and that $m = 2$ and $\alpha = \frac{2}{3}$. With probability $\frac{2}{3}$ the inferred ASG will be based on two different transcripts, i.e. offer two possible transcripts. However, sampling from the inferred ASG we only achieve a p-value of $\frac{1}{2}$ for having recovered a fraction of α of the full ASG. Again we refer to Section 4 for empirical results on the usefulness of our computed p-value on realistic data.

3.2   ASG Based Tests

Without an accepted model for the alternative splicing observed for a gene, we cannot simulate transcript generation. We may however still make a qualitative assessment of the validity of the reconstructed ASG—or alternatively of whether transcripts are fully determined by regulatory factors rather than generated according to the combinatorial model implicit in the ASG representation—in the context of the transcripts used to reconstruct it by considering *informative* transcripts. A transcript is considered informative if it reveals one or more new edges of the ASG. A transcript corresponds to a path through the ASG. So a set of transcripts elucidating the full potential of the ASG uniquely corresponds to a set of paths covering all the edges in the ASG (i.e. a set of paths $\mathbf{P} = \{P_1, \ldots, P_k\}$ such that every edge of the ASG occurs in at least one path P_i in $\mathbf{P}$). For convenience we will assume that all paths have to start at source s and terminate at sink t. This can be realised by amending the ASG with s that has edges to all initiation fragments and t that all terminal fragments have an edge to. If $G = (V, E)$ denotes the ASG, it is a straightforward observation that the maximum number of informative transcripts is

$$2 + |E| - |V| . \tag{1}$$

The minimum number of informative transcripts is equivalent to a minimum path cover, a classic problem related to maximum flow (see e.g. [11]). For reference, algorithm 1 provides a simple augmenting path solution for reducing any path cover to a minimum path cover in time $O((|V| + |E|)\,|\mathbf{P}|)$ where $\mathbf{P}$ is the initial path cover. For each edge $e \in E$, its weight $w(e)$ is initialised to

---

Algorithm 1. Minimum Path Cover

---

  while there is a non-cyclic path π from s to t in G_w **do**
    for all edges $e \in \pi$ **do**
      if $e \in E$ **then**
        $w(e) \leftarrow w(e) - 1$
      else
        $w(e) \leftarrow w(e) + 1$
    Recompute G_w

---

the number of paths covering e in the initial cover. Define $G_w = (V, E_w)$ where $E_w = \{e \in E \mid w(e) > 1\} \cup \{(v, u) \mid (u, v) \in E\}$. I.e. G_w contains all edges covered by more than one path, and the reverse edge of all the edges in G. At termination the minimum path cover size can be determined as the sum of the weights of the edges leaving the source node.

4 Results

We are interested in choosing a model relevant to an ASG constructed from real transcripts. Ideal for obtaining large-scale data on alternative splicing events is microarray technology, but this is still in its infancy, with only a handful of large-scale investigations into exon skipping events [12,13]. Ultimately it is hoped that the ability to attach accurate inclusion rates to individual exons, and even the possibility of sampling full-length mRNA transcripts [14] will be possible. For illustrative purposes we must now content ourselves with using ESTs, whilst being mindful of their limitations [15], e.g. ESTs exhibit a strong bias for the 3' end of the gene. The Alternative Splicing Gallery [2] catalogues EST support for each human gene, from which maximum likelihood estimates (MLEs) for the probabilities associated with each exon fragment can be calculated via a simple transcript counting argument. We apply this to an example gene, Neurexin III-β; alternative splicing in neurexins has been well-characterized [16]. Consider Figure 2. EST support for this gene suggests several distant exon coupling relationships, for example between exons 6 and 10. For convenience extend any partial EST to its full-length counterpart if this can be achieved unambiguously, otherwise omit it. A hypothesis test comparing model 0 against the pairwise model yields a p-value of 0.0026, confirming our suspicion that entirely independent splicing of exons may not be applicable for this gene.

For genes with larger ASGs, the cardinality of the set of all putative transcripts and hence the number of parameters required for use with model 0 can grow exponentially with the number of alternative splice sites, so that a large number of observations are required to accept model 0. At present these are generally lacking (suggesting that in fact the true ASG has not yet been observed—see Section 3.1), so for these genes we must either focus on short alternatively splicing regions, or instead we can test the relative merits of the pairwise and in-out models to provide some measure of the dependence in splicing between different

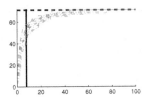 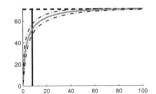 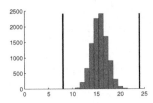

Fig. 4. (*Left*) Ten simulated reconstructions of the ASG for human gene ABCB5, under the pairwise model. Number of sampled transcripts (*x*-axis) is plotted against size of the reconstructed ASG (*y*-axis). Full ASG size shown as a dashed line. Minimal possible number of transcripts annotated as a vertical line. (*Centre*) Mean number of reconstructed edges across 10000 simulations ± 1 standard deviation. (*Right*) Histogram across 10000 simulations of number of informative transcripts. Maximum and minimum number of such transcripts are annotated.

exons. As an example consider the gene ABCB5, one of the 89 human genes known to offer more than 5000 putative transcripts [2]. It is a gene of interest also due to its association with drug resistance in human malignant melanoma, with both functional and non-functional splicing variants [17]. The likelihood ratio test was applied to four regions of the gene observed to exhibit alternative splicing. We make the additional assumption that these regions are bounded by constitutive exons, prohibiting under the models the splicing together of fragments from disparate regions of the gene (which would unnecessarily increase the parameter space in order to accommodate splicing events of negligible probability). *p*-values for the four regions are 0.000, 0.029, 0.001, 0.000; the overall *p*-value is 0.000. All 89 genes were similarly tested: of them, 13 were deemed not to comprise any testable regions. Of the remaining 76 genes, 20 (26%) were accepted at the 5% level to be described by the in-out model. These seem to be the genes for which the assumption of independence between exons is most applicable.

We infer that ABCB5 is most suitably described by the pairwise model. Let us suppose then that transcripts are generated for ABCB5 under the pairwise model. Reconstruction of the ASG under this model is summarized in Figure 4, with the minimal number of transcripts required to recover the ASG annotated. The size of the ASG is measured in the number of its recovered edges. The probabilities for the pairwise model are chosen using MLEs described previously. Consider Figure 4(*left*). The 10 example simulations generally follow the growth curves one would expect of sampling with replacement. In some simulations the last few edges persist in remaining undiscovered even after the generation of 100 transcripts, but by 20 transcripts the mean proportion of the ASG to have been recovered is 90.8% (Figure 4(*centre*)). What does this indicate about the probability that the 20 ESTs used to construct the ASG in the first place did in fact construct the complete ASG? If we apply our bootstrap-like method, none of the set of simulated transcript samples successfully recovers the full ASG resulting in a *p*-value of 0.

But how much can the *p*-value be trusted? To answer this we set up an experiment using the ABCB5 pairwise model as the true source for generating

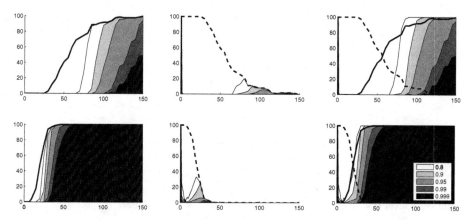

Fig. 5. Results of experiment described in text. Number of sampled transcripts (x-axis) is plotted against percentage of experiments (y-axis): percentage recovering the full ASG (*left*), percentage not recovering the full ASG (*centre*) and both (*right*). The fraction of such experiments for which the inferred ASG is accepted as the true ASG is shown for various confidence levels. The first row illustrates results for full recovery of the ASG, the second row for $\alpha = 90\%$ recovery of the full ASG.

transcripts. From this we repeatedly sampled m transcripts and computed the p-value for the ASG inferred from these m transcripts. This was repeated for various choices of m. The outcome of this experiment is illustrated in Figure 5(*top*). Both the fraction of graphs inferred from m transcripts coinciding with the full ASG, and the fraction of inferred graphs accepted at various acceptance rates are plotted. Encouragingly, it is evident from the righthand graph that there is a strong correlation between when we start to recover the full ASG and when we start to accept the inferred ASG. This indicates that our p-value does indeed capture whether the transcripts contain sufficient redundancy.

Note that the central graph, which plots acceptances of non-fully recovered inferred ASGs, separates the type II errors; any accepted graphs here are false positives. Similarly the lefthand graph, which plots acceptances of fully recovered inferred ASGs, separates the type I errors; any graph not accepted here is a false negative. As anticipated in Section 3.1, for very low m most experiments yield a high false positive rate, but in all our simulations this effect quickly dies away by $m = 3$.

For ABCB5, no acceptances are observed at the 20 transcript level, and we safely deduce that a scenario of independent random samples from a fully recovered ASG is not supported. This implies that either the ASG derived from the real 20 transcripts is a proper subset of the true ASG, or that the collection of transcripts used to infer the ASG is likely to be biased in the sense that there is little redundancy in the collection, and an emphasis rather on novel transcripts. As mentioned above, such an observation seems likely in a database with both human and biological biases. We performed a similar detailed investigation into all 56 genes with more than 5000 putative transcripts satisfying the pairwise model (data not shown) and found that in no cases was the reconstructed ASG

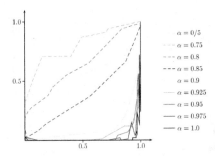

Fig. 6. Partial recovery results for nine different values of α, ranging from full ASG recovery to recovery of half the ASG. Fraction accepted as recovered to degree α at confidence level 0.95 is plotted against fraction recovered to degree α.

accepted at confidence 0.95. Thus all could reasonably be said to exhibit a bias in their transcript records. This should of course be taken with the caveats associated with ESTs and the assumption that transcript generation is assumed to be correctly described by the pairwise model, along with the fact that by choosing complex genes to begin with these results will not be indicative of the rest of the genome. But this illustration offers a novel first step towards a method for teasing out the complex relationships discussed, which are not discernible from the ASG alone.

As mentioned in Section 3.1 we cannot expect our assessment of partial ASG recovery to be as precise as our assessment of full ASG recovery. To further investigate dependence on α of the quality of the p-value computed we ran experiments similar to those plotted in Figure 5 for a range of α values with confidence level 0.95. Figure 6 plots the fraction of accepted ASGs against the fraction of inferred ASGs containing at least a fraction α of the edges in the ABCB5 ASG. Ideally we would expect a phase transition from no accepted ASGs to all ASGs being accepted around the point where 95% of the inferred ASGs contain at least α of all edges. This is indeed observed for high values of α, but for α values lower than 0.9 there is an increasing tendency toward a mere linear relationship between ASG recovery and ASG acceptance. Remembering that ASG acceptance is more likely for a false positive than for a true positive it is thus clear that our method should not be applied for low α values.

5 Discussion

In this work we have proposed a mathematical framework to consider how to predict the nature of transcript generation in alternatively splicing genes. These models can be used make inferences on questions such as the levels of independence in exon splicing and the confidence with which we can be sure that a complete ASG has been recovered. We have also considered algorithms for calculating the minimum and maximum number of informative transcripts available from an ASG. Source code, as well as the statistical tests outlined and their results, are available from `http://www.stats.ox.ac.uk/~jenkins/ASG/`. Our method for testing the coverage of an ASG by its transcripts can provide experimentalists with a way to quantify any bias in the distribution of the transcriptome. In our examples we have been restricted to existing EST data, which can be somewhat limited both in quality and quantity. Quantitative analysis of the ASG will become far more fruitful when high-throughput microarray data on alternative splicing is more readily available, from which accurate probabilities can be associated with each splicing event. An important next step will then be to begin to incorporate knowledge of tissue-specific expression of particular isoforms, which has thus far been naïvely omitted.

Unfortunately most current microarray studies focus on individual splicing events—only 12.8% of alternative splicing relationships have been detected in full-length transcripts [10], but we envisage this to improve as the need to observe whole transcripts pushes the technology in this direction. When full-length transcripts are available, one way to look more closely at the conditional probabilities inherent in an ASG would be to focus on those transcripts revealing new edges to the ASG during sampling. The resulting 'signature' histogram can be compared to the same histogram generated by transcript simulation from one of the models, i.e. assuming no exon coupling (Figure 4(*right*)). This figure and other of our tests suggest that a simple model for the distribution is Gaussian with mean between the minimum and maximum number of informative transcripts. Thus for example, strong positive correlation between exons would skew the distribution towards the minimum, compared to the distribution observed under the models. Across the 56 genes satisfying the pairwise model, the distribution of the mean number of informative transcripts reported was centred about 0.61 of the genes' ranges (i.e. between the minimum and maximum number of informative transcripts). All but 29 reported a mean in the range $(0.5, 0.7)$ and all but 7 were inside $(0.4, 0.8)$. For each gene the standard deviation in informative transcripts was less than 0.13 of the range.

Acknowledgements

Gil Ast and Richard Copley are thanked for their advice on which genes would be interesting. An anonymous reviewer is thanked for helpful comments. Thanks also to the LSI DTC at Oxford and to the EPSRC and BBSRC for its funding.

References

1. Johnson, J.M., Castle, J., Garrett-Engele, P., Kan, Z., Loerch, P.M., Armour, C.D., Santos, R., Schadt, E.E., Stoughton, R., Shoemaker, D.D.: Genome-wide survey of human alternative pre-mRNA splicing with exon junction microarrays. Science **302** (2003) 2141–2144
2. Leipzig, J., Pevzner, P., Heber, S.: The alternative splicing gallery (ASG): bridging the gap between genome and transcriptome. Nucleic Acids Research **32** (2004) 3977–3983

3. Lareau, L.F., Green, R.E., Bhatnagar, R.S., Brenner, S.E.: The evolving roles of alternative splicing. Current Opinions in Structural Biology **14** (2004) 273–282

4. Black, D.L.: Mechanisms of alternative pre-messenger RNA splicing. Annual Review of Biochemistry **72** (2003) 291–336

5. Lopez, A.J.: Alternative splicing of pre-mRNA: developmental consequences and mechanisms of regulation. Annual Review of Genetics **32** (1998) 279–305

6. Pan, Q., Shai, O., Misquitta, C., Zhang, W., Saltzman, A.L., Mohammad, N., Babak, T., Siu, H., Hughes, T.R., Morris, Q.D., Frey, B.J., Blencowe, B.J.: Revealing global regulatory features of mammalian alternative splicing using a quantitative microarray platform. Molecular Cell **16** (2004) 929–941

7. Heber, S., Alekseyev, M., Sze, S.H., Tang, H., Pevzner, P.A.: Splicing graphs and EST assembly problem. Bioinformatics **18** (2002) S181–188

8. Black, D.L.: A simple answer for a splicing conundrum. Proceedings of the National Academy of Sciences of the United States of America **102** (2005) 4927–4928

9. Ibrahim, E.C., Schaal, T.D., Hertel, K.J., Reed, R., Maniatis, T.: Serine/arginine-rich protein-dependent suppression of exon skipping by exonic splicing enhancers. Proceedings of the National Academy of Sciences of the United States of America **102** (2005) 5002–5007

10. Lee, C., Atanelov, L., Modrek, B., Xing, Y.: ASAP: the alternative splicing annotation project. Nucleic Acids Research **31** (2003) 101–105

11. Li, W.N., Reddy, S.M., Sahni, S.: On path selection in combinational logic circuits. IEEE Transactions on Computer Aided Design of Integrated Circuits and Systems **8** (1989) 56–63

12. Lee, C., Roy, M.: Analysis of alternative splicing with microarrays: successes and challenges. Genome Biology **5** (2004) 231

13. Lee, C., Wang, Q.: Bioinformatics analysis of alternative splicing. Briefings in Bioinformatics **6** (2005) 23–33

14. Castle, J., Garrett-Engele, P., Armour, C.D., Duenwald, S.J., Loerch, P.M., Meyer, M.R., Schadt, E.E., Stoughton, R., Parrish, M.L., Shoemaker, D.D., Johnson, J.M.: Optimization of oligonucleotide arrays and RNA amplification protocols for analysis of transcript structure and alternative splicing. Genome Biology **4** (2003) R66

15. Modrek, B., Lee, C.: A genomic view of alternative splicing. Nature Genetics **30** (2002) 13–19

16. Tabuchi, K., Südhof, T.C.: Structure and evolution of neurexins: insight into the mechanism of alternative splicing. Genomics **79** (2002) 849–859

17. Frank, N.Y., Margaryan, A., Huang, Y., Schatton, T., Waaga-Gasser, A.M., Gassser, M., Sayegh, M.H., Sadee, W., Frank, M.H.: ABCB5-mediated doxorubicin transport and chemoresistance in human malignant melanoma. Cancer Research **65** (2005) 4320–4333

Multiple Structure Alignment and Consensus Identification for Proteins

Jieping Ye[1], Ivaylo Ilinkin[2], Ravi Janardan[3], and Adam Isom[3]

[1] Arizona State University, Tempe, AZ 85287, USA
jieping.ye@asu.edu
www.public.asu.edu/~jye02/
[2] Rhodes College, Memphis, TN 38112, USA
ilinkin@rhodes.edu
www.rhodes.edu/MathematicsandComputerScience/FacultyandStaff/ilinkin.cfm
[3] University of Minnesota, Minneapolis, MN 55455, USA
{janardan, aisom}@cs.umn.edu
www.cs.umn.edu/~{janardan, aisom}

Abstract. An algorithm is presented to compute a multiple structure alignment for a set of proteins and to generate a consensus structure which captures common substructures present in the given proteins. The algorithm is a heuristic in that it computes an approximation to the optimal alignment that minimizes the sum of the pairwise distances between the consensus and the transformed proteins. A distinguishing feature of the algorithm is that it works directly with the coordinate representation in three dimensions with no loss of spatial information, unlike some other multiple structure alignment algorithms that operate on sets of backbone vectors translated to the origin; hence, the algorithm is able to generate true alignments. Experimental studies on several protein datasets show that the algorithm is quite competitive with a well-known algorithm called CE-MC. A web-based tool has also been developed to facilitate remote access to the algorithm over the Internet.

1 Introduction

Proteins are macromolecules that regulate all biological processes in a cellular organism [1]. The human body has about one hundred thousand different proteins that control functions as diverse as oxygen transport, blood clotting, tissue growth, immune system response, inter-cell signal transmission, and the catalysis of enzymatic reactions.

Proteins are synthesized within the cell and immediately after its creation each protein folds spontaneously into a three-dimensional (3D) configuration that is determined uniquely by its constituent amino acid sequence [14]. It is this 3D structure that ultimately determines the function of a protein, be it the catalysis of a reaction or the growth of muscle tissue or arming the body's immune system. Indeed, it is the case that where proteins are concerned "function follows form" [13].

Proteins have evolved over time through the modification and re-use of certain substructures that have proven successful [7]. It is well known [9] that during this

P. Bücher and B.M.E. Moret (Eds.): WABI 2006, LNBI 4175, pp. 115–125, 2006.
© Springer-Verlag Berlin Heidelberg 2006

process, structure is better conserved than sequence, i.e., proteins that are related through evolution tend to have similar structures even though their sequences may be quite different. Thus, the ability to identify common substructures in a set of proteins could yield valuable clues to their evolutionary history and function. This motivates the *multiple structure alignment problem* that we consider in this paper: Informally, given a collection of protein structures, we seek to align them in space, via rigid motions, such that large matching substructures are revealed. A further goal is to extract from this alignment a consensus structure that can serve as a proxy for the whole set, and could be used, for instance, as a template to perform fast searches through protein structure databases, such as the PDB, to identify structurally similar proteins.

1.1   Contributions

In this paper, we present an algorithm to compute a multiple structure alignment for a set of proteins and to generate a consensus structure. Our algorithm computes an approximation to the optimal multiple structure alignment, i.e., one that minimizes the so-called *Sum-of-Consensus distance (SC-distance)* between the consensus structure and the given set of proteins. (A more precise definition is given in Section 2.)

Our algorithm represents the input proteins and the consensus as sequences of triples of coordinates of the alpha-carbon (or C_α) atoms along the backbone. It then computes a correspondence between the coordinate triples of the C_α atoms in the different protein structures, by choosing one of the proteins as the initial consensus and applying an algorithm that is analogous to the center-star method for multiple sequence alignment [7]. Next, it derives a set of translation and rotation matrices that are optimal for the computed correspondence and uses these to align the structures in space via rigid motions and obtain the new consensus. The process is repeated until the change in SC-distance is less than a prescribed threshold. The computation of the optimal translations and rotations and the new consensus is itself an iterative process that both uses the current consensus and generates simultaneously a new one. A distinguishing feature of our algorithm is that it works directly with the coordinate representation in three dimensions with no loss of spatial information, unlike some other multiple structure alignment algorithms that operate on sets of backbone vectors translated to the origin; hence, the algorithm is able to generate true alignments.

As discussed in Section 1.2 below, there are many algorithms for multiple structure alignment. In general, it is difficult to make comparisons among them as they all operate under different sets of assumptions and problem formulations. Nonetheless, we have been able to compare our algorithm to one recent algorithm, called CE-MC [6], which also works with coordinate triples, but employs a different objective function. As discussed in Section 4, the two algorithms are very comparable in terms of the sizes of the so-called conserved regions they discover in several datasets drawn from a well-known database called HOM-STRAD [11]; however, our algorithm runs about two orders of magnitude faster than CE-MC, hence it can potentially scale to much larger datasets.

We have also incorporated our algorithm into a web-based tool to facilitate remote access and experimentation via the Internet.

Due to space constraints, we omit several details and proofs here. These can be found in the full paper [19].

1.2 Related Work

Some prior work on multiple structure alignment includes [3,4,6,10,12,20]. Orengo and Taylor [12] give algorithms for aligning pairs of proteins using a 2-level dynamic programming approach, and obtain a multiple alignment from this by aligning the pairs according to their similarity scores. Leibowitz *et al.* [10] use the technique of geometric hashing to compute a multiple alignment and core; unlike most other algorithms, theirs does not require an ordered sequence of atoms along the protein backbone. Gerstein and Levitt [4] use an iterative dynamic programming method to compute a multiple structure alignment. The CE-MC algorithm by Guda *et al.* [6], mentioned earlier, uses Monte Carlo optimization techniques to refine an initial alignment found by pairwise structure alignment using the Combinatorial Extension algorithm in [15]. Recently, Chew and Kedem [3] and Ye and Janardan [20] have shown how to compute both a multiple structure alignment and a consensus structure. In [3,20], proteins are represented as sets of unit vectors at the origin. An iterative, dynamic programming-based method is used to compute the alignment and consensus. However, a limitation of the algorithms in [3,20], is that they require the backbone vectors to be translated to the origin, hence information about the relative positions of the C_α atoms in $\Re^3$ is lost. As a result, it is not possible to generate a true alignment. By contrast, the algorithm proposed in this paper retains spatial information by representing each protein as a sequence of coordinate triples in $\Re^3$ and is able to generate a true alignment.

An important special case of multiple structure alignment is pairwise structure alignment, which involves aligning only two protein structures. Indeed, the problem arises in this paper when computing correspondences between different proteins. We use here a variant of an algorithm that we have developed recently for pairwise structure alignment [18]. Some other algorithms for pairwise alignment that may also be used (with minor modifications) include LOCK [16], DALI [8], CE [15], and the method in [2].

2   Multiple Structure Alignment: Problem Formulation

Let $\mathcal{S} = \{P_i\}_{i=1}^K$ be a set of K proteins. Protein P_i, of length L_i, consists of a chain of C_α atoms, numbered $1, 2, \cdots, L_i$, along the backbone in $\Re^3$. (As is customary [8,16], we consider only the backbone, not the amino acid residues.) We represent P_i as a sequence of *coordinate triples* $\boldsymbol{u}_j^i = (x_j^i, y_j^i, z_j^i)$, $1 \leq j \leq L_i$, on the backbone, where x_j^i, y_j^i, and z_j^i, are the coordinates of the jth C_α atom of P_i. Let $P_0 = \boldsymbol{u}_1^0, \cdots, \boldsymbol{u}_{L_0}^0$ be the *consensus structure*, of length L_0.

A *correspondence*, $\mathcal{C}$, of the K proteins $\{P_i\}_{i=1}^K$ and the consensus structure P_0 can be represented as a matrix $H = (\boldsymbol{h}_{ij})_{0 \leq i \leq K, 1 \leq j \leq L}$, for some

$L \geq \max_{0 \leq i \leq K}\{L_i\}$, where h_{ij} is either a coordinate triple belonging to the ith protein, $i \geq 0$, or a gap^1. Distances between coordinate triples are based on the squared Euclidean distance between them in $\Re^3$. The distance between a coordinate triple and a gap is called a *gap penalty*, and is denoted by ρ.

A *multiple structure alignment*, $\mathcal{M}$, of the K proteins based on the correspondence $\mathcal{C}$, can be represented as another matrix $G = (g_{ij})_{0 \leq i \leq K, 1 \leq j \leq L}$, where the ith row, for $i > 0$, is obtained via transformation (i.e., rotation and translation) of the corresponding row of H. More specifically, we combine $\{h_{ij}\}_{j=1}^{L}$ from the ith protein, i.e., the ith row of the matrix H, into a column vector $H_i \in \Re^{L \times 3}$ as below. G_i is defined similarly from the matrix G.

$$H_i = \begin{pmatrix} h_{i1} \\ \vdots \\ h_{iL} \end{pmatrix} \in \Re^{L \times 3} \quad \text{and} \quad G_i = \begin{pmatrix} g_{i1} \\ \vdots \\ g_{iL} \end{pmatrix} \in \Re^{L \times 3}, \quad \text{for} \quad i = 1, \cdots, K.$$

Then, $G_i = (H_i - T_i) \cdot R_i = (H_i - e \cdot t_i) \cdot R_i$, for $i > 0$, where $R_i \in \Re^{3 \times 3}$ is the rotation matrix, $e \in \Re^{L \times 1}$ is a vector with 1 in each entry, $t_i \in \Re^{1 \times 3}$ is a translation vector, and $T_i = e \cdot t_i \in \Re^{L \times 3}$ is the translation matrix. The transformation of a gap is chosen to remain a gap. Note that P_0 remains unchanged; that is, $G_0 = H_0$.

Under the multiple structure alignment $\mathcal{M}$, we define the *distance between the consensus structure P_0 and protein P_j* as $D^{\mathcal{M}}(P_0, P_j) = \sum_{\ell=1}^{L} d(g_{0\ell}, g_{j\ell})^2$, where $d(\cdot, \cdot)$ denotes the following distance function:

$$d(u, v) = \begin{cases} ||u - v||_2, & \text{if both } u \text{ and } v \text{ are coordinate triples.} \\ \rho, & \text{if only one of } u \text{ and } v \text{ is a coordinate triple vector.} \\ 0, & \text{if both } u \text{ and } v \text{ are gap vectors.} \end{cases}$$

The distance between P_0 and P_j can be represented compactly as $D^{\mathcal{M}}(P_0, P_j) = ||G_0 - G_j||_F^2$, where $|| \cdot ||_F$ denotes the *Frobenius norm* [5], with the additional convention that the squared difference between a coordinate triple and a gap is ρ^2. The total distance of the K proteins to the consensus structure under $\mathcal{M}$, called the *Sum-of-Consensus distance*, or *SC-distance*, is then defined as

$$SC(\mathcal{M}) = \sum_{1 \leq j \leq K} D^{\mathcal{M}}(P_0, P_j) = \sum_{1 \leq j \leq K} ||G_0 - G_j||_F^2. \tag{1}$$

Intuitively, the *SC-distance* measures how well the consensus structure represents the given set of K proteins. A similar distance function is used in [3], where each protein is represented as a set of vectors in $\Re^4$.

Note that $G_i = (H_i - T_i) \cdot R_i$, for some rotation and translation matrices R_i and T_i. We can now define our multiple structure alignment problem as follows: *Given a set $\mathcal{S} = \{P_1, \cdots, P_K\}$ of protein structures, compute a transformation (i.e., rotation and translation) for each protein, and generate a consensus structure*

---

[1] In our implementation, a gap is represented by a special symbol.

P_0, such that the resulting multiple structure alignment, $\mathcal{M}$, has minimum SC-distance, $SC(\mathcal{M})$, as defined in Equation (1).

In Section 3, we present a heuristic for this problem which approximates the global minimum of the SC-distance, $SC(\mathcal{M})$, by iterative refinement of an initial multiple structure alignment and converges to a local minimum.

3   Multiple Structure Alignment: The Algorithm

Our algorithm works iteratively, by starting with an initial multiple structure alignment and updating it incrementally with decreasing SC-distance. The algorithm finally stops at some local minimum. The expectation is that with a good starting alignment, the final alignment will be close to the optimal solution. The pseudo-code for the algorithm is given as **Algorithm MAPSCI**, which stands for **M**ultiple **A**lignment of **P**rotein **S**tructures and **C**onsensus **I**dentification. The various steps are discussed in detail below.

Algorithm MAPSCI: M̲ultiple A̲lignment of P̲rotein S̲tructures and C̲onsensus I̲dentification

1. Choose initial consensus structure P_0^0 from $\{P_i\}_{i=1}^K$. $i \leftarrow 0$. $SC^0 \leftarrow \infty$.
2. Do
3.   if $i = 0$ then compute pairwise structure alignment between P_0^i and every P_j.
4.   else use standard dynamic programming to align P_0^i with every P_j.
5. $i \leftarrow i + 1$.
6.   Compute correspondence $\mathcal{C}^i$ from the above alignments (either pairwise or dynamic programming) using center-star-like method.
7.   Compute optimal translation matrix T_j^i and optimal rotation matrix R_j^i iteratively, as in Section 3.4. Transform P_j by R_j^i and T_j^i for every j to obtain multiple structure alignment $\mathcal{M}^i$. $SC^i \leftarrow SC(\mathcal{M}^i)$.
8.   Post-process $\mathcal{M}^i$ by removing all columns consisting of only gaps.
9.   Compute new consensus structure P_0^i from $\mathcal{M}^i$ by Theorem 1.
10. Until $\left| \frac{SC^i - SC^{i-1}}{SC^{i-1}} \right| \leq \eta$.   // η is a user-specified threshold

---

3.1   Step I (Line 1): Choose the Initial Consensus Structure

There are many ways to choose the initial consensus structure P_0^0. One possibility is to choose P_0^0 as the *center protein*, so that it minimizes the sum of the minimum pairwise distances to all the other proteins. That is, P_0^0 is the $k^\star$th protein, where $k^\star = \arg\min_{1 \leq k \leq K} \sum_{i=1}^K D(P_k, P_i)$, and $D(P_k, P_i)$ denotes the distance between P_k and P_i, as computed by a pairwise structure alignment.

This choice makes sense intuitively, since it yields a consensus structure which is "not too far away" from the others; however it is expensive computationally, as it involves $\frac{K(K-1)}{2}$ pairwise alignments. A simple and less expensive choice that appears to work well is to pick P_0^0 as the *median protein*, i.e., the protein of median length. We report our experimental results for both choices in Section 4.

3.2   Step II (Lines 3 and 4): Pairwise Structure Alignment

After we determine the consensus structure P_0^0 in Step I, the $K - 1$ pairwise structure alignments between P_0^0 and $P_i \neq P_0^0$, for $i = 1, \cdots, K$, are computed using the pairwise structure alignment algorithm in [18]. (Other pairwise structure alignment algorithms, such as LOCK [16], DALI [8], CE [15], etc. could also be used instead.) After the initial step, the consensus structure is expected to be close to the proteins in the dataset. Dynamic programming on the coordinates of the C_α atoms is then applied to compute the alignment (see Line 4).

3.3   Step III (Line 6): Compute an Initial Correspondence

An initial correspondence between the K proteins is obtained by applying to the consensus structure and the pairwise alignments computed in Steps I and II a method adapted from the center-star method [7] for multiple sequence alignment. We call our extension of this method to protein structures a *center-star-like* method. There are two key differences between the two methods. First, in the center-star-like method we will be aligning not alphabet characters representing amino acids but coordinate triples derived from the protein backbones. Second, in a multiple structure alignment, the correspondence computed using the center-star-like method is only a first step; it is followed by an optimization step to compute the optimal transformation matrices. Subsequent correspondences are computed similarly.

3.4   Step IV (Lines 7 and 9): Compute Optimal Rotation and Translation Matrices and Consensus Structure

Given a correspondence $\mathcal{C}$ and a consensus structure J, we show how to find both the optimal rotation matrix R_j and translation matrix T_j for each protein P_j as well as the new consensus structure $\bar{J}$.

Assume the correspondence $\mathcal{C}$ is represented as a matrix $H = (h_{ij})$. Protein P_j in $\mathcal{C}$ can be represented as H_j consisting of $\{h_{ji}\}_{i=1}^L$, where h_{ji} represents either a coordinate triple or a gap. The objective is to find the rotation and translation matrices R_j and T_j, for $j = 1, \cdots, K$, and the consensus structure $\bar{J}$, such that sum of the pairwise alignment distances between $\bar{J}$ and each (transformed) P_j is minimum.

Mathematically, given H_j, for $j = 1, \cdots, K$, we wish to compute optimal rotation and translation matrices R_j and T_j, for each protein P_j, and the consensus structure $\bar{J}$, such that $S = \sum_{j=1}^K ||\bar{J} - (H_j - T_j) \cdot R_j||_F^2$ is minimized.

Direct minimization of S over $\bar{J}$, and the T_j's and R_j's seems difficult. Instead, we propose an iterative procedure for minimizing S. Within each iteration, the minimization of S is carried out in two stages that are interleaved: computation of the optimal $\bar{J}$ for given R_j's and T_j's; and computation of the optimal R_j's and T_j's for a given $\bar{J}$.

First, we show how to compute the consensus structure, given the rotation and translation matrices R_j's and T_j's, as stated in the following theorem:

Theorem 1. *Assume that the correspondence C is represented as a matrix $H = (h_{ij})$ and $\bar{J} = (J_1, \cdots, J_L)^T$ is the optimal consensus structure. For each column j, let I_n be the set of indices of proteins with a non-gap in the jth column and I_g be the set of indices of proteins with a gap in the jth column. Then $\bar{J}_j$, the jth position of the optimal consensus structure, is either a coordinate triple $\boldsymbol{u}_j$, where $\boldsymbol{u}_j = \frac{1}{|I_n|} \sum_{i \in I_n} \boldsymbol{h}_{ij}$, or a gap.*

Next, we show how to compute the optimal translation matrix T_i, for each i, for a given consensus structure $\bar{J}$. From the equation above for S, it is clear that the optimal T_i and T_j, for $i \neq j$ are independent of each other. Hence, we focus on the computation of T_i, for a specific i. The translation matrix T_i can be decomposed as $T_i = e \times t_i$, where $t_i \in \Re^{1 \times 3}$ is the translation vector.

As mentioned earlier, the transformation of a gap remains a gap. Hence the computation of the translation and rotation matrices is independent of the mismatches (i.e., where at least one of the two elements being compared is a gap). We can thus simplify the computation by removing all mismatches in the alignment between the consensus structure $\bar{J}$ and the ith protein P_i. Let $A \in \Re^{n \times 3}$ and $B \in \Re^{n \times 3}$ consist of the coordinate triples from the consensus structure and the ith protein, respectively, after removing the mismatches. Here n is the number of matches between the consensus structure and the ith protein (i.e., comparison of two non-gaps).

The optimal rotation matrix can be obtained by computing the Singular Value Decomposition (SVD) [5] of $A^T B$ as in [17], while it is known that the optimal translation vector is the one that matches the centroids of the coordinate triple vectors from A and B as follows:

Theorem 2. *Let A and B be defined as above. Assume that $e^T A = [0, 0, 0]$. Then for any rotation matrix R_i, the optimal translation vector t_i for minimizing $S_i = ||A - (B - T_i) \cdot R_i||_F^2 = ||A - (B - e \cdot t_i) \cdot R_i||_F^2$ is given by $t_i = \frac{1}{n} e^T B$.*

3.5   Analysis

We show in the full paper [19] that, in **Algorithm MAPSCI**, the SC-distance is non-increasing from one iteration to the next and the algorithm thus converges.

Let ℓ be the maximum length of the K proteins, m_1 the number of iterations in computing the optimal rotation and translation matrices, and m_2 the number of iterations of the loop in lines 2–10. The time complexity of the algorithm can be shown to be either $O(K^2\ell^2 + (K\ell^2 + K^2\ell m_1)m_2)$ or $O(K + (K\ell^2 + K^2\ell m_1)m_2)$, depending on the choice of the initial consensus. In practice, m_1 and m_2 tend to be small constants, so the running time is either $O(K^2\ell^2)$ or $O(K\ell^2 + K^2\ell)$.

4   Experimental Results

4.1   Software and Web Server

Algorithm MAPSCI has been implemented in C, and has been tested on several protein structure datasets, as described below. Moreover, it has been incorporated into a web-based tool (also called **MAPSCI**) for remote access over the Internet. This tool allows proteins to be specified by their PDB ids (or uploaded from local files). The results of the alignment are annotated in the JOY output format and the standard NBRF/PIR format, and can be visualized using the molecular viewer applet Chemis 3D integrated with the web tool. The software and the web-based tool can be accessed at www.geom-comp.umn.edu .

4.2   Datasets

We evaluated our algorithm using the seventeen datasets shown in Table 1. The ten proteins in Set 1 are from the *Globin* family, while the ten proteins in Set 2 are from the *Thioredoxin* family. Set 3 contains four all-alpha proteins, which are structural neighbors of the protein 1mbc (*Myoglobin*) from the DALI database. The four alpha-beta proteins in Set 4 are all structural neighbors of the protein 3trx (*Thioredoxin-Reduced Form*) from the DALI database. Sets 5–7 are from [3]: Set 5 contains sixteen proteins from the *Globin* family, Set 6 contains six all-beta proteins from the *immunoglobulin* family, and Set 7 contains five proteins that are unrelated. Sets 8–17 are from the HOMSTRAD database [11]. HOM-STRAD (HOMologous STRucture Alignment Database) is a curated database of structure-based alignments for homologous protein families. Datasets 8–17 have sequence identities ranging from 17% to 47%.

Table 1. The test datasets used for the experiments

	Dataset	Proteins (PDB ID)
1	*Globin*	1mbc, 1mba, 1dm1, 1hlm, 2lhb, 2fal, 1hbg, 1flp, 1eca, 1ash
2	*Thioredoxin*	3trx, 1aiu, 1erv, 1f9mA, 1ep7A, 1tof, 2tir, 1thx, 1quw, 1fo5A
3	*all-alpha*	1le2, 2fha, 1nfn, 1grj
4	*Alpha-beta*	1mek, 1a81, 1f37B, 1ghhA
5	*Globin*	1hlb, 1hlm, 1babA, 1babB, 1ithA, 1mba, 2hbg, 2lhb
		3sdhA, 1ash, 1flp, 1myt, 1eca, 1lh2, 2vhbA, 5mbn
6	*all-beta*	1cd8, 1ci5A, 1qa9A, 1cdb, 1neu, 1qfoA
7	*mixed*	1cnpB, 1jhgA, 1hnf, 1aa9, 1eca
8	*biotin_lipoyl*	1bdo, 1lac, 1ghk,1iyv, 1pmr, 1qjoA, 1fyc
9	*msb*	1esl, 1rdol, 1msbA, 1tn3, 2afp, 1qddA, 1ixxA, 1ixxB
10	*cyt3*	2cdv, 2cym, 1wad, 3cyr, 2cy3, 1aqe
11	*ghf7*	1celA, 1eg1A, 2ovwA, 2a39A
12	*Acetyltransf*	1bo4A, 1cmOA, 1yghA, 1qstA, 1b87A, 1cjwA
13	*HMG_bo*	1hrzA, 1nhn, 1cktA, 1hma, 2lefA
14	*intb*	1afcA, 2afgA, 2fgf, 2mib, 1i1b, 1iraX
15	*lipocalin*	1i4uA, 1bbpA, 1exsA, 1bebA, 2a2uA, 1jv4A, 1ew3A, 1bj7,
		1e5pA, 1hn2, 1dzkA, 1iiuA, 1aqb, 1epaA, 1qqsA
16	*sh3*	1awj, 1shg, 1shfA, 2src, 1qcfA, 1lckA, 1qlyA, 1bbzA,
		1griA, 1gbrA, 1cskA, 1ckaA, 1semA, 1griA, 1ycsB, 2hsp,
		1ark, 1aojA, 1pht, 1bb9
17	*TPR*	1a17, 1elwA, 1elrA, 1e96B, 1fchA, 1ihgA

4.3   Results

Our first experiment compared two choices for the initial consensus: "center" and "median". The results for the seventeen datasets are illustrated in Figure 1, where the *x*-axis denotes the different datasets and the *y*-axis denotes the SC-distance (left) and the running time (right). Throughout our experiments, we

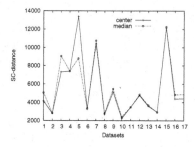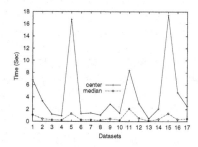

Fig. 1. Comparison of choices of initial consensus structure: "center" and "median," using the seventeen datasets. Time measurements made on a 500MHz Sun-Sparc with 512MB memory.

used $\eta = 0.1$ and $\rho = 16$. (Recall that η is a user-specified threshold for convergence—see line 10 of **Algorithm MAPSCI**; ρ is the gap penalty.)

As seen from the figure, the "center" method produced SC-distances that were slightly lower than those produced by "median" (except for Set 5). However, the computational cost of "center" is much higher than that of "median".

Our second experiment was to visualize the multiple structure alignments. An example for Set 2 is shown in Figure 2, for the two choices of the initial consensus, together with the computed final consensus. In each case, the computed consensus is seen to be visually quite similar to the input proteins. (Color versions of all the figures may be accessed at www.geom-comp.umn.edu .)

Our third experiment compared our method with CE-MC [6], using the manually-aligned HOMSTRAD database as the "gold" standard. Only conserved

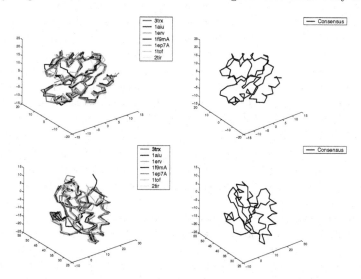

Fig. 2. Multiple structure alignment using dataset 2, with two choices for the initial consensus: "center" (top; SC-distance: 2783.47), and "median" (bottom; SC-distance: 2814.20). To avoid clutter, only the first seven proteins from the dataset are displayed.

columns (i.e., ones with no gaps) were considered for the comparison, as they are viewed to be more important biologically. We ran our algorithm and CE-MC on the ten datasets, Sets 8–17, from HOMSTRAD and reported the percentage of conserved columns from HOMSTRAD that also appeared in the alignments from these two algorithms. See Figure 3(a). Overall, our algorithm is seen to be quite competitive with CE-MC, in terms of the sizes of the conserved regions. Figure 3(b) shows the running times for both algorithms on a log scale, i.e., $\ln(1 + t)$ is plotted for each t. The running time of our algorithm is seen to be much smaller (about two orders of magnitude) than that of CE-MC, implying that it can potentially scale to much larger datasets than can CE-MC.

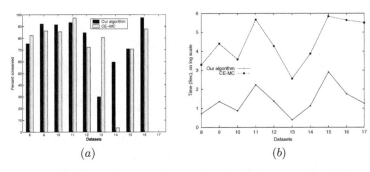

(a) (b)

Fig. 3. Comparison of our algorithm and CE-MC on the ten datasets from HOM-STRAD: the percentage conserved (graph (a)), and the computational time (graph (b)), on a log scale. Time measurements made on a 500MHz Sun-Sparc with 512MB memory.

References

1. C. Branden, and J. Tooze. Introduction to Protein Structure, Garland, 1999.
2. L.P. Chew, K. Kedem, D.P. Huttenlocher, and J. Kleinberg. Fast detection of geometric substructure in proteins. *J. Comput. Bio.*, 6:(3-4), 1999, 313–325.
3. L.P. Chew and K. Kedem. Finding the consensus shape of a protein family. Proc. *ACM Symp. Comput. Geometry SoCG'02*, 64–73, 2002.
4. M. Gerstein and M. Levitt. Using iterative dynamic programming to obtain accurate pairwise and multiple alignments of protein structures. Proc. *ISMB'96*, 59–66, 1996.
5. G.H. Golub and C.F. Van Loan. Matrix Computations, John Hopkins University Press, 3rd edition, 1996.
6. C. Guda, E.D. Scheeff, P.E. Bourne, I.N. Shindyalov. A new algorithm for the alignment of multiple protein structures using Monte Carlo optimization. Proc. *PSB'01*, 275–286, 2001.
7. D. Gusfield. Algorithms on Strings, Trees, and Sequences: Computer Science and Computational Biology, Cambridge University Press, 1997.
8. L. Holm and C. Sander. Protein Structure Comparison by Alignment of Distance Matrices. *J. Mol. Bio.*, 233, 1993, 123–138.
9. L. Holm and C. Sander. Mapping the protein universe. *Science*, 273, 1996, 595–602.

10. N. Leibowitz, Z. Fligelman, R. Nussinov, and H. Wolfson: Multiple Structural Alignment and Core Detection by Geometric Hashing. Proc. *ISMB'99*, 169–177, 1999.
11. K. Mizuguchi, C.M. Deane, T.L. Blundell, J.P. Overington. HOMSTRAD: a database of protein structure alignments for homologous families. *Prot. Sci.*, 7, 1998, 2469–2471.
12. C. Orengo and W. Taylor. SSAP: Sequential structure alignment program for protein structure comparison. *Meth. Enzymol.*, 266, 1996, 617–635.
13. G. Rose. No assembly required. *The Sciences*, 36, 1996, 26–31.
14. M. Sela, F. H. White Jr, and C. B. Anfinsen. Reductive cleavage of disulfide bridges in Ribonuclease. *Science*, 125, 1957, 691-692.
15. I.N. Shindyalov and P.E. Bourne. Protein structure alignment by incremental combinatorial extension (CE) of the optimal path. *Prot. Eng.*, 11, 1998, 739–747.
16. A.P. Singh and D.L. Brutlag. Hierarchical protein structure superposition using both secondary structure and atomic representation. Proc. *ISMB'97*, 284–293, 1997.
17. S. Umeyama. Least-square estimation of transformation parameters between two point patterns. *IEEE Trans. Pattern Anal. Mach. Intell.*, 13 (4), 1991, 376–380.
18. J. Ye, R. Janardan, and S. Liu. Pairwise protein structure alignment based on an orientation-independent backbone representation. *J. Bio. Comput. Bio.*, 2 (4), 2004, 699–717.
19. J. Ye, I. Ilinkin, R. Janardan, and A. Isom. Multiple structure alignment and consensus identification for proteins. Submitted. Available at www.geom-comp.umn.edu.
20. J. Ye, R. Janardan. Approximate multiple protein structure alignment using the Sum-of-Pairs distance. *J. Comput. Bio.*, 11 (5), 2004, 986–1000.

Procrastination Leads to Efficient Filtration for Local Multiple Alignment

Aaron E. Darling[1], Todd J. Treangen[2], Louxin Zhang[4], Carla Kuiken[5], Xavier Messeguer[2], and Nicole T. Perna[3]

[1] Department of Computer Science, University of Wisconsin, USA
darling@cs.wisc.edu
[2] Department of Computer Science, Technical University of Catalonia,
Barcelona, Spain
treangen@lsi.upc.edu
[3] Department of Animal Health and Biomedical Sciences, Genome Center,
University of Wisconsin, USA
[4] Department of Mathematics, National University of Singapore, Singapore
[5] T-10 Theoretical Biology Division, Los Alamos National Laboratory, USA

Abstract. We describe an efficient local multiple alignment filtration heuristic for identification of conserved regions in one or more DNA sequences. The method incorporates several novel ideas: (1) palindromic spaced seed patterns to match both DNA strands simultaneously, (2) seed extension (chaining) in order of decreasing multiplicity, and (3) procrastination when low multiplicity matches are encountered. The resulting local multiple alignments may have nucleotide substitutions and internal gaps as large as w characters in any occurrence of the motif. The algorithm consumes $\mathcal{O}(wN)$ memory and $\mathcal{O}(wN \log wN)$ time where N is the sequence length. We score the significance of multiple alignments using entropy-based motif scoring methods. We demonstrate the performance of our filtration method on Alu-repeat rich segments of the human genome and a large set of Hepatitis C virus genomes. The GPL implementation of our algorithm in C++ is called `procrastAligner` and is freely available from http://gel.ahabs.wisc.edu/procrastination

1   Introduction

Pairwise local sequence alignment has a long and fruitful history in computational biology and new approaches continue to be proposed [1,2,3,4]. Advanced filtration methods based on spaced-seeds have greatly improved the sensitivity, specificity, and efficiency of many local alignment methods [5,6,7,8,9]. Common applications of local alignment can range from orthology mapping [10] to genome assembly [11] to information engineering tasks such as data compression [12]. Recent advances in sequence data acquisition technology [13] provide low-cost sequencing and will continue to fuel the growth of molecular sequence databases. To cope with advances in data volume, corresponding advances in computational methods are necessary; thus we present an efficient method for local multiple alignment of DNA sequence.

P. Bücher and B.M.E. Moret (Eds.): WABI 2006, LNBI 4175, pp. 126–137, 2006.

Unlike pairwise alignment, local multiple alignment constructs a single multiple alignment for all occurrences of a motif in one or more sequences. The motif occurrences may be identical or have degeneracy in the form of mismatches and indels. As such, local multiple alignments identify the basic repeating units in one or more sequences and can serve as a basis for downstream analysis tasks such as multiple genome alignment [14,15,16,17], global alignment with repeats [18,19], or repeat classification and analysis [20]. Local multiple alignment differs from traditional pairwise methods for repeat analysis which either identify repeat families *de novo* [21] or using a database of known repeat motifs [22].

Previous work on local multiple alignment includes an Eulerian path approach proposed by Zhang and Waterman [23]. Their method uses a *de Bruijn* graph based on exactly matching k-mers as a filtration heuristic. Our method can be seen as a generalization of the *de Bruijn* filtration to arbitrary spaced seeds or seed families. However, our method employs a different approach to seed extension that can identify long, low-copy number repeats.

The local multiple alignment filtration method we present has been designed to efficiently process large amounts of sequence data. It is not designed to detect subtle motifs such as transcription factor binding sites in small, targeted sequence regions–stochastic methods are better suited for such tasks [24].

2   Overview of the Method

Our local multiple alignment filtration method begins by generating a set of candidate multi-matches using *palindromic* spaced seed patterns, listed in Table 1. The seed pattern is evaluated at every position of the input sequence, and the lexicographically-lesser of the forward and reverse complement subsequence induced by the seed pattern is hashed to identify seed matches—see Figure 1. The use of *palindromic* seed patterns offers computational savings by allowing both strands of DNA to be processed simultaneously.

Given an initial set of matching sequence regions, our algorithm then maximally extends each match to cover the entire surrounding region of sequence identity. A visual example of maximal extension is given by the black match

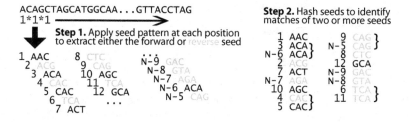

Fig. 1. Application of the palindromic seed pattern 1*1*1 to identify degenerate matching subsequences in a nucleotide sequence of length N. The lexicographically-lesser of the forward and reverse complement subsequence induced by the seed pattern is used at each sequence position.

Table 1. Palindromic spaced seeds used by `procrastAligner`. The sensitivity ranking of a seed at various levels of sequence identity is given in the columns at right. A seed with rank 1 is the most sensitive seed pattern for a given weight and percent sequence identity. The default seeds used by `procrastAligner` are listed here, while the full list of high-ranking seeds appears on the website.

Weight	Pattern	Seed Rank by Sequence Identity 65%	70%	75%	80%	85%	90%
5	11*1*11	1	1	1	1	1	1
6	1*11***11*1	1	1	1	1	1	1
7	11**1*1*1**11	1	1	1	1	1	1
8	111**1**1**111	1	1	1	1	1	1
9	111*1**1**1*111	1	1	1	1	1	1
10	111*1**1*1**1*111	1	1	1	1	1	1
11	1111**1*1*1**1111	1	1	1	1	1	2
12	1111**1*1*1*1**1111	5	3	1	1	1	1
13	1111**1**1*1*1**1*1**1111	> 10	5	1	1	1	1
14	1111**11*1*1*11**1111	2	2	1	1	1	1
15	1111*1*11**1**11*1*1111	1	1	1	1	1	1
16	1111*1*11**11**11*1*1111	2	1	1	1	1	1
18	11111**11*1*11*1*11**11111	1	1	1	1	1	1
19	1111*111**1*111*1**111*1111	5	2	1	1	1	1
20	11111*1*11**11*11**11*1*11111	> 10	> 10	3	1	1	1
21	11111*111*11*1*11*111*11111	1	1	1	3	3	2

in Figure 2. In order to extend over each region of sequence $\mathcal{O}(1)$ times, our method extends matches in order of decreasing multiplicity–we extend the highest multiplicity matches first. When a match can no longer be extended without including a gap larger than w characters, our method identifies the neighboring *subset* matches within w characters, i.e., the light gray seed in Figure 2. We then *link* each neighboring subset match to the extended match. We refer to the

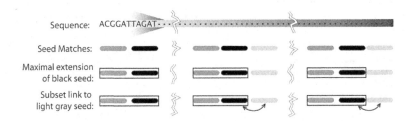

Fig. 2. Seed match extension. Three seed matches are depicted as black, gray, and light gray regions of the sequence. Black and gray have multiplicity 3, while light gray has multiplicity 2. We maximally extend the black seed to the left and right and in doing so, the black seed chains with the gray seed to the left. The light gray seed is adjacent to only two out of three components in the extended black seed. We *procrastinate* and extend the light gray seed later. We create a link between light gray and the extended black seed match.

extended match as a *superset* match. Rather than immediately extend the subset match(es), we *procrastinate* and extend the subset match later when it has the highest multiplicity of any match waiting to be extended. When extending a match with a linked superset (light gray in Figure 2), we immediately include the entire region covered by the linked superset match–obviating the need to re-examine sequence already covered by a previous match extension.

We score alignments generated by our method using the entropy equation and exact *p*-value method in [25]. Our method may produce many hundreds or thousands of local multiple alignments for a given genome sequence, thus it is important to rank them by significance. When computing column entropy, we treat gap characters as missing data.

3 Algorithm

3.1 Notation and Assumptions

Given a sequence $S = s_1, s_2, \ldots, s_N$ of length N defined over an alphabet $\{A, C, G, T\}$, our goal is to identify local multiple alignments on subsequences of S. Our filtration method first generates candidate chains of ungapped alignments, which are later scored and possibly re-aligned. Denote an ungapped alignment, or match, among subsequences in S as an object M. We assume as input a set of ungapped alignments $\mathbf{M}$. We refer the number of regions in S matched by a given match $M_i \in \mathbf{M}$ as the *multiplicity* of M_i, denoted as $|M_i|$. We refer to each matching region of M_i as a *component* of M_i. Note that $|M_i| \geq 2 \ \forall \ M \in \mathbf{M}$. We denote the left-end coordinates in S of each component of M_i as $M_i.L_1, M_i.L_2, \ldots, M_i.L_{|M_i|}$, and similarly we denote the right-end coordinates as $M_i.R_x$. When aligning DNA sequences, matches may occur on the forward or reverse complement strands. To account for this phenomenon we add an orientation value to each matching region: $M_i.O_x \in \{1, -1\}$, where 1 indicates a forward strand match and -1 for reverse.

Our algorithm has an important limitation on the matches in $\mathbf{M}$: no two matches M_i and M_j may have the same left-end coordinate, e.g. $M_i.L_x \neq M_j.L_y \ \forall \ i, j, x, y$ except for the identity case when $i = j$ and $x = y$. This constraint has been referred to by others as *consistency* and *transitivity* [26] of matches. In the present work we only require consistency and transitivity of matches longer than the seed length, e.g. seed matches may overlap.

3.2 Data Structures

Our algorithm begins with an initialization phase that creates three data structures. The first data structure is a set of *Match Records* for each match $M \in \mathbf{M}$. The *Match Record* stores M, a unique identifier for M, and two items which will be described later in Section 3.3: a set of linked match records, and a *subsuming match pointer*. The linked match records are further subdivided into four classes: a left and right *superset link*, and left and right *subset links*. The *subsuming match pointer* is initially set to a *NULL* value. Figure 3 shows a schematic

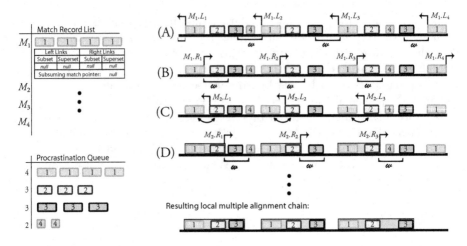

Fig. 3. The match extension process and associated data structures. **(A)** First we pop the match at the front of the procrastination queue: M_1 and begin its leftward extension. Starting with the leftmost position of M_1, we use the *Match Position Lookup Table* to enumerate every match with a left-end within some distance w. Only $M_4.L_1$ is within w of M_1, so it forms a singleton *neighborhood group* which we discard. **(B)** M_1 has no *neighborhood groups* to the left, so we begin extending M_1 to the right. We enumerate all matches within w to the right of M_1. M_2 lies to the right of 3 of 4 components of M_1 and so is not subsumed, but instead gets linked as a right-subset of M_1. We add a left-superset link from M_2 to M_1. **(C)** Once finished with M_1 we pop M_2 from the front of the procrastination queue and begin leftward extension. We find the left-superset link from M_2 to M_1, so we extend the left-end coordinates of M_2 to cover M_1 accordingly. No further leftward extension of M_2 is possible because M_1 has no left-subset links. **(D)** Beginning rightward extension on M_2 we construct a neighborhood list and find a chainable match M_3, and a subset M_4. We extend M_2 to include M_3 and mark M_4 as inconsistent and hence not extendable. Upon completion of the chaining process we have generated a list of local multiple alignments.

of the match record. We refer to the second data structure as a *Match Position Lookup Table*, or **P**. The table has N entries $p_1, p_2, \ldots, p_N$, one per character of $\mathcal{S}$. The entry for p_t stores the unique identifier of the match M_i and x for which $M_i.L_x = t$ or the *NULL* identifier if no match has t as a left-end coordinate. We call the third data structure a *Match extension procrastination queue*, or simply the *procrastination queue*. Again, we denote the multiplicity of a match M by $|M|$. The *procrastination queue* is a binary heap of matches ordered on $|M|$ with higher values of $|M|$ appearing near the top of the heap. The heap is initially populated with all $M \in \mathbf{M}$. This queue dictates the order in which matches will be considered for extension.

3.3 Extending Matches

Armed with the three aforementioned data structures, our algorithm begins the chaining process with the match at the front of the *procrastination queue*. For a match M_i that has not been subsumed, the algorithm first attempts extension to the left, then to the right. Extension in each direction is done separately in an identical manner and we arbitrarily choose to describe leftward extension first. The first step in leftward match extension for M_i is to check whether it has a left superset link. If so, we perform a *link extension* as described later. For extension of M_i without a superset link, we use the *Match Position Lookup Table* **P** to enumerate all matches within a fixed distance w of M_i. For each component $x = 1, 2, \ldots, |M_i|$ and distance $d = 1, 2, \ldots, w$ we evaluate first whether $p_{M_i.L_x-(d\cdot M_i.O_x)}$ is not *NULL*. If not then $p_{M_i.L_x-(d\cdot M_i.O_x)}$ stores an entry $\langle M_j, y \rangle$ which is a pointer to neighboring match M_j and the matching component y of M_j.

In order to consider matches on both forward and reverse strands, we must evaluate whether $M_i.O_x$ and $M_j.O_y$ are consistent with each other. We define the relative orientation of $M_i.O_x$ and $M_j.O_y$ as $o_{i,j,x,y} = M_i.O_x \cdot M_j.O_y$ which causes $o_{i,j,x,y} = 1$ if both $M_i.O_x$ and $M_j.O_y$ match the same strand and -1 otherwise. We create a tuple of the form $\langle M_j, o_{i,j,x,y}, x, d, y \rangle$ and add it to a list called the *neighborhood list*. In other words, the tuple stores (1) the unique match ID of the match with a left-end at sequence coordinate $M_i.L_x - (d \cdot M_i.O_x)$, (2) the relative orientation of $M_i.O_x$ and $M_j.O_y$, (3) the matching component x of M_i, (4) the distance d between M_i and M_j, and (5) the matching component y of M_j. If $M_j = M_i$ for a given value of d, we stop adding *neighborhood list* entries after processing that one. The *neighborhood list* is then scanned to identify groups of entries with the same match ID M_j and relative orientation $o_{i,j,x,y}$. We refer to such groups as *neighborhood groups*. Entries in the same *neighborhood group* that have identical x or y values are considered "ties" and need to be broken. Ties are resolved by discarding the entry with the larger value of d in the fourth tuple element: we prefer to chain over shorter distances. After tiebraking, each *neighborhood group* falls into one of several categories:

- **Superset:** The *neighborhood group* contains $|M_i|$ separate entries. M_j has higher multiplicity than M_i, e.g. $|M_j| > |M_i|$. We refer to M_j as a superset of M_i.
- **Chainable:** The *neighborhood group* contains $|M_i|$ separate entries. M_j and M_i have equal multiplicity, e.g. $|M_j| = |M_i|$. We can chain M_j and M_i.
- **Subset:** The *neighborhood group* contains $|M_j|$ separate entries such that $|M_j| < |M_i|$. We refer to M_j as a subset of M_i.
- **Novel Subset:** The *neighborhood group* contains r separate entries such that $r < |M_i| \wedge r < |M_j|$. We refer to the portion of M_j in the list as a *novel subset* of M_i and M_j because this combination of matching positions does not exist as a match in the initial set of matches **M**.

The algorithm considers each *neighborhood group* for chaining in the order given above: chainable, subset, and finally, novel subset. Superset groups are ignored, as any superset links would have already been created when processing the superset match.

Chainable matches. To chain match M_i with *chainable* match M_j we first update the left-end coordinates of M_i by assigning $M_i.L_x \leftarrow \min(M_i.L_x, M_j.L_y)$ for each $\langle i, j, x, y \rangle$ in the *neighborhood group* entries. Similarly, we update the right-end coordinates: $M_i.R_x \leftarrow \max(M_i.R_x, M_j.R_y)$ for each $\langle i, j, x, y \rangle$ in the group. If any of the coordinates in M_i change we make note that a *chainable* match has been chained. We then update the *Match Record* for M_j by setting its *subsuming match pointer* to M_i, indicating that M_j is now invalid and is subsumed by M_i. Any references to M_j in the *Match Position Lookup Table* and elsewhere may be lazily updated to point to M_i as they are encountered. If M_j has a left superset link, the link is inherited by M_i and any remaining neighborhood groups with *chainable* matches are ignored. *Chainable* groups are processed in order of increasing d value so that the nearest *chainable* match with a superset link will be encountered first. A special case exists when $M_i = M_j$. This occurs when M_i represents an inverted repeat within w nucleotides. We never allow M_i to chain with itself.

Subset matches. We defer subset match processing until no more chainable matches exist in the neighborhood of M_i. A subset match M_j is considered to be completely contained by M_i when for all x, y pairs in the *neighborhood group*, $M_i.L_x \leq M_j.L_y \wedge M_j.R_y \leq M_i.R_x$. When subset match M_j is completely contained by M_i, we set the *subsuming match pointer* of M_j to M_i. If the subset match is not contained we create a *link* from M_i to M_j. The subset link is a tuple of the form $\langle M_i, M_j, x_1, x_2, \ldots, x_{|M_j|} \rangle$ where the variables $x_1 \ldots x_{|M_j|}$ are the x values associated with the $y = 1 \ldots |M_j|$ from the *neighborhood list* group entries. The link is added to the left subset links of M_i and we remove any pre-existing right superset link in M_j and replace it with the new link.

Novel subset matches. A novel subset may only be formed when both M_i and M_j have already been maximally extended, otherwise we discard any novel subset matches. When a novel subset exists matches we create a new match record M_{novel} with left- and right-ends equal to the outward boundaries of M_i and M_j. Rather than extend the novel subset match immediately, we *procrastinate* and place the novel subset in the *procrastination queue*. Recall that the novel subset match contains r matching components of M_i and M_j. In constructing M_{novel}, we create links between M_{novel} and each of M_i and M_j such that M_{novel} is a left and a right subset of M_i and M_j, respectively. The links are tuples of the form outlined in the previous section on subset matches.

Occasionally a *neighborhood group* representing a novel subset match may have $M_i = M_j$. This can occur when M_i has two or more components that form a tandem or overlapping repeat. If $M_i.L_x$ has $M_i.L_y$ in its neighborhood, and $M_i.L_y$ has $M_i.L_z$ in its neighborhood, then we refer to $\{x, y, z\}$ as a tandem unit of M_i. A given tandem unit contains between one and $|M_i|$ components of M_i, and the set of tandem units forms a partition on the components of M_i. In this situation we construct a novel subset match record with one component for each tandem unit of M_i. If M_i has only a single tandem unit then we continue without creating a novel subset match record. Figure 4 illustrates how we process tandem repeats.

Fig. 4. Interplay between tandem repeats and novel subset matches. There are two initial seed matches, one black, one gray. The black match has components labelled 1-7, and the neighborhood size w is shown with respect to component 7. As we attempt leftward extension of the black match we discover the gray match in the neighborhood of components 2 and 5 of black. A subset link is created. We also discover that some components of the black match are within each others' neighborhood. We classify the black match as a tandem repeat and construct a novel subset match with one component for each of the four tandem repeat units: $\{1\}, \{2, 3, 4\}, \{5, 6\}, \{7\}$.

After the first round of chaining. If the *neighborhood list* contained one or more chainable groups we enter another round of extending M_i. The extension process repeats starting with either *link extension* or by construction of a new *neighborhood list*. When the boundaries of M_i no longer change, we classify any subset matches as either subsumed or outside of M_i and treat them accordingly. We process novel subsets. Finally, we may begin extension in the opposite (rightward) direction. The rightward extension is accomplished in a similar manner, except that the neighborhood is constructed from $M_i.R_x$ instead of $M_i.L_x$ and d ranges from $-1, -2, \ldots, -w$ and ties are broken in favor of the largest d value. Where left links were previously used, right links are now used and vice-versa.

Chaining the next match. When the first match popped from the *procrastination queue* has been maximally extended, we pop the next match from the *procrastination queue* and consider it for extension. The process repeats until the *procrastination queue* is empty. Prior to extending any match removed from the *procrastination queue*, we check the match's *subsuming match pointer*. If the match has been subsumed extension is unnecessary.

3.4   Link Extension

To be considered for leftward link extension, M_i must have a left superset link to another match, M_j. We first extend the boundaries of M_i to include the region covered by M_j and unlink M_i from M_j. Then each of the left subset links in M_j are examined in turn to identify links that M_i may use for further extension. Recall that the link from M_i to M_j is of the form $\langle M_j, M_i, x_1, \ldots, x_{|M_i|} \rangle$. Likewise, a left subset link from M_j to another match M_k is of the form $\langle M_j, M_k, z_1, \ldots, z_{|M_k|} \rangle$. To evaluate whether M_i may follow a given link in the left subsets of M_j, we take the set intersection of the x and z values for each M_k that is a left subset of M_j. We can classify the results of the set intersection as:

- **Superset:** $\{x_1, \ldots, x_{|M_i|}\} \subset \{z_1, \ldots, z_{|M_k|}\}$ Here M_k links to every component of M_j that is linked by M_i, in addition to others.
- **Chainable:** $\{x_1, \ldots, x_{|M_i|}\} = \{z_1, \ldots, z_{|M_k|}\}$ Here M_k links to the same set of components of M_j that M_i links.

- **Subset:** $\{x_1, \ldots, x_{|M_i|}\} \supset \{z_1, \ldots, z_{|M_k|}\}$ Here M_i links to every component of M_j that is linked by M_k, in addition to others.
- **Novel Subset:** $\{x_1, \ldots, x_{|M_i|}\} \cap \{z_1, \ldots, z_{|M_k|}\} \neq \emptyset$ Here M_k is neither a superset, chainable, nor subset relative to M_i, but the intersection of their components in M_j is non-empty. M_k and M_i form a novel subset.

Left subset links in M_j are processed in the order given above. Supersets are never observed, because M_k would have already unlinked itself from M_j when it was processed (as described momentarily). When M_k is a chainable match, we extend M_i to include the region covered by M_k and set the subsuming match pointer in M_k to point to M_i. We unlink M_k from M_j, and M_i inherits any left superset link that M_k may have. When M_k is a subset of M_i we unlink M_k from M_j and add it to the *deferred subset list* to be processed once M_i has been fully extended. Finally, we never create novel subset matches during link extension because M_k will never be a fully extended match.

If a chainable match was found during leftward link extension, we continue for another round of leftward extension. If not, we switch directions and begin rightward extension.

3.5   Time Complexity

A *neighborhood list* may be constructed at most w times per character of $\mathcal{S}$, and construction uses sorting by key comparison, giving $\mathcal{O}(wN \log wN)$ time and space. Similarly, we spend $\mathcal{O}(wN \log wN)$ time performing link extension. The upper bound on the total number of components in the final set of matches is $\mathcal{O}(wN)$. Thus, the overall time complexity for our filtration algorithm is $\mathcal{O}(wN \log wN)$.

4   Results

We have created a program called `procrastAligner` for Linux, Windows, and Mac OS X that implements the described algorithm. Our open-source implementation is available as C++ source code licensed under the GPL.

We compare the performance of our method in finding Alu repeats in the human genome to an Eulerian path method for local multiple alignment [23]. The focus of our algorithm is efficient filtration, thus we use a scoring metric that evaluates the filtration sensitivity and specificity of the ungapped alignment chains produced by our method. We compute sensitivity as the number of Alu elements hit by a match, out of the total number of Alu elements. We compute specificity as the ratio of match components that hit an Alu to the sum of match multiplicity for all matches that hit an Alu. Thus, we do not penalize our method for finding legitimate repeats that are not in the Alu family.

The comparison between `procrastAligner` and the Eulerian method is necessarily indirect, as each method was designed to solve different (but related) problems. The Eulerian method uses a *de Bruijn* graph for filtration, but goes

Table 2. Performance of `procrastAlign` and the Eulerian path approach on Alu repeats. Rep: total number of Alu elements; Family: number of Alu families; Alu: average Alu length in bp (S.D.); Div: average Alu divergence (S.D.); Sn: sensitivity; Sp: specificity; T: compute time; Sw: palindromic seed weight; w: max gap size. Alus were identified by RepeatMasker [22]. We report data for the fast version of the Eulerian path method as given by Table 1 of [23]. Sensitivity and specificity of `procrastAlign` was computed as described in the text.

Accession	Length	Rep	Family	Alu (bp)	Div, %	Method	Sn %	Sp %	T (s)	Sw	w
AF435921	22 Kb	28	10	261 (69)	15.0 (6.4)	Eulerian	96.3	99.4	1	-	-
						procrast	100	95.9	1	9	27
Z15025	38 Kb	52	13	245 (85)	15.7 (5.7)	Eulerian	98.6	96.7	4	-	-
						procrast	100	82.5	2	9	27
AC034110	167 Kb	87	18	261 (72)	12.2 (5.9)	Eulerian	93.5	95.2	14	-	-
						procrast	100	97.9	3	15	45
AC010145	199 Kb	118	13	277 (55)	15.0 (5.6)	Eulerian	85.2	93.7	32	-	-
						procrast	99.1	99.2	3	15	45
Hs Chr 22	1 Mbp	404	32	252 (79)	15.2 (6.1)	Eulerian	72.4	99.4	85	-	-
						procrast	98.3	97.3	20	15	45

beyond filtration to compute gapped alignments using banded dynamic programming. We report scores for a version of the Eulerian method that computes alignments only on regions identified by its *de Bruijn* filter. The results suggest that by using our filtration method, the sensitivity of the Eulerian path local multiple aligner could be significantly improved. A second important distinction is that our method reports *all* local multiple alignment chains in its allotted runtime, whereas the Eulerian method identifies only a single alignment.

We also test the ability of our method to provide accurate anchors for genome alignment. Using a manually curated alignment of 144 Hepatitis C virus genome sequences [27], we measure the anchoring sensitivity of our method as the fraction of pairwise positions aligned in the correct alignment that are also present in `procrastAligner` chains. We measure positive predictive value as the number of match component pairs that contain correctly aligned positions out of the total number of match component pairs. `procrastAligner` may generate legitimate matches in the repeat regions of a single genome. The PPV score penalizes `procrastAligner` for identifying such legitimate repeats, which subsequent genome alignment would have to disambiguate. Using a seed size of 9 and $w = 27$, `procrastAligner` has a sensitivity of 63% and PPV of 67%.

5   Discussion

We have described an efficient method for local multiple alignment filtration. The chains of ungapped alignments that our filter outputs may serve as direct input to multiple genome alignment algorithms. The test results of our prototype implementation on Alu sequences demonstrate improved sensitivity over *de Bruijn* filtration. A promising avenue of further research will be to couple our filtration method with subsequent refinement using banded dynamic programming.

The alignment scoring scheme we use can rank alignments by information content, however a biological interpretation of the score remains difficult. If a phylogeny and model of evolution for the sequences were known *a priori* then a biologically relevant scoring scheme could be used [28]. Unfortunately, the phylogenetic relationship for arbitrary local alignments is rarely known, especially among repetitive elements or gene families within a single genome and across genomes. It may be possible to use simulation and MCMC methods to score alignments where the phylogeny and model of evolution is unknown *a priori*, but doing so would be computationally prohibitive for our application.

Acknowledgements

AED was supported by NLM Training Grant 5T15LM007359-05. TJT was supported by Spanish Ministry MECD Grant TIN2004-03382 and AGAUR Training Grant FI-IQUC-2005. LZ was supported by AFT Grant 146-000-068-112.

References

1. Ma, B., Tromp, J., and Li, M.: PatternHunter: faster and more sensitive homology search. Bioinformatics **18** (2002) 440–445
2. Brudno, M., and Morgenstern, B.: Fast and sensitive alignment of large genomic sequences. Proc. IEEE CSB'02 (2002) 138–147
3. Noé, L., and Kucherov, G.: Improved hit criteria for DNA local alignment. BMC Bioinformatics **5** (2004)
4. Kahveci, T., Ljosa, V., and Singh, A.K.: Speeding up whole-genome alignment by indexing frequency vectors. Bioinformatics **20** (2004) 2122–2134
5. Choi, P, K., Zeng, F., and Zhang, L.: Good spaced seeds for homology search. Bioinformatics **20** (2004) 1053–1059
6. Li, M., Ma, B., and Zhang, L.: Superiority and complexity of the spaced seeds. Proc. SODA'06. (2006) 444–453
7. Sun, Y., and Buhler, J.: Designing multiple simultaneous seeds for DNA similarity search. J. Comput. Biol. **12** (2005) 847–861
8. Xu, J., Brown, D.G., Li, M., and Ma, B.: Optimizing multiple spaced seeds for homology search. Proc. CPM'04 (2004) 47–58
9. Flannick, J., and Batzoglou, S.: Using multiple alignments to improve seeded local alignment algorithms. Nucleic Acids Res. **33** (2005) 4563–4577
10. Li, L., Stoeckert, C.J., and Roos, D.S.: OrthoMCL: identification of ortholog groups for eukaryotic genomes. Genome Res. **13** (2003) 2178–2189
11. Jaffe, D.B., Butler, J., Gnerre, S., Mauceli, E., Lindblad-Toh, K., Mesirov, J.P., Zody, M.C., and Lander, E.S.: Whole-genome sequence assembly for mammalian genomes: Arachne 2. Genome Res. **13** (2003) 91–96
12. Ane, C., and Sanderson, M.: Missing the forest for the trees: phylogenetic compression and its implications for inferring complex evolutionary histories. Syst. Biol. **54** (2005) I311–I317

13. Margulies, M., et al. 55 other authors: Genome sequencing in microfabricated high-density picolitre reactors. Nature **437** (2005) 376–380
14. Darling, A.C.E., Mau, B., Blattner, F.R., and Perna, N.T.: Mauve: multiple alignment of conserved genomic sequence with rearrangements. Genome Res. **14(7)** (2004) 1394–403.
15. Hohl, M., Kurtz, S., and Ohlebusch, E.: Efficient multiple genome alignment. Bioinformatics **18 Suppl 1** (2002) S312–20.
16. Treangen, T., and Messeguer, X.: M-GCAT: Multiple Genome Comparison and Alignment Tool. Submitted (2006)
17. Dewey, C.N., and Pachter, L.: Evolution at the nucleotide level: the problem of multiple whole-genome alignment. Hum. Mol. Genet. **15 Suppl 1** (2006)
18. Sammeth, M., and Heringa, J.: Global multiple-sequence alignment with repeats. Proteins (2006)
19. Raphael, B., Zhi, D., Tang, H., and Pevzner, P.: A novel method for multiple alignment of sequences with repeated and shuffled elements. Genome Res. **14(11)** (2004) 2336–46.
20. Edgar, R.C., and Myers, E.W.: PILER: identification and classification of genomic repeats. Bioinformatics **21 Suppl 1** (2005)
21. Kurtz, S., Ohlebusch, E., Schleiermacher, C., Stoye, J., and Giegerich, R.: Computation and visualization of degenerate repeats in complete genomes. Proc. 8th Intell. Syst. Mol. Biol. ISMB'00 (2000) 228–38.
22. Jurka, J., Kapitonov, V.V., Pavlicek, A., Klonowski, P., Kohany, O., and Walichiewicz, J.: Repbase Update, a database of eukaryotic repetitive elements. Cytogenet. Genome Res. **110** (2005) 462–467
23. Zhang, Y., and Waterman, M.S.: An Eulerian path approach to local multiple alignment for DNA sequences. PNAS **102** (2005) 1285–90.
24. Siddharthan, R., Siggia, E.D., and van Nimwegen, E.: PhyloGibbs: a Gibbs sampling motif finder that incorporates phylogeny. PLoS Comput. Biol. **1** (2005)
25. Nagarajan, N., Jones, N., and Keich, U.: Computing the P-value of the information content from an alignment of multiple sequences. Bioinformatics **21 Suppl 1** (2005)
26. Szklarczyk, R., and Heringa, J.: Tracking repeats using significance and transitivity. Bioinformatics **20 Suppl 1** (2004) I311–I317
27. Kuiken, C., Yusim, K., Boykin, L., and Richardson, R.: The Los Alamos hepatitis C sequence database. Bioinformatics **21** (2005) 379–84
28. Prakash, A., and Tompa, M.: Statistics of local multiple alignments. Bioinformatics **21** (2005) i344–i350

Controlling Size When Aligning Multiple Genomic Sequences with Duplications

Minmei Hou[1], Piotr Berman[1], Louxin Zhang[2], and Webb Miller[1]

[1] Department of Computer Science and Engineering, Penn State, University Park, PA 16801, USA,
mhou@cse.psu.edu, berman@cse.psu.edu, webb@bx.psu.edu
[2] Department of Mathematics, National University of Singapore, Science Drive 2, Singapore 117543
matzlx@nus.edu.sg

Abstract. For a genomic region containing a tandem gene cluster, a proper set of alignments needs to align only orthologous segments, i.e., those separated by a speciation event. Otherwise, methods for finding regions under evolutionary selection will not perform properly. Conversely, the alignments should indicate every orthologous pair of genes or genomic segments. Attaining this goal in practice requires a technique for avoiding a combinatorial explosion in the number of local alignments. To better understand this process, we model it as a graph problem of finding a minimum cardinality set of cliques that contain all edges. We provide an upper bound for an important class of graphs (the problem is NP-hard and very difficult to approximate in the general case), and use the bound and computer simulations to evaluate two heuristic solutions. An implementation of one of them is evaluated on mammalian sequences from the α-globin gene cluster.

1 Introduction

The ENCODE project [22] has the goal of identifying all functional genomic segments in 1% of the human genome. As part of the project, genomic sequence data from a number of mammals are being generated for the targeted 1%, in the belief that alignment and analysis of the sequences will help predict the functional segments. Several computer programs for aligning genomic regions have been used for this purpose [15,16]. In our opinion, the current crop of alignment programs performs acceptably in many parts of the genome. However, for regions containing tandem gene clusters, more software development is necessary. The deficiency of current methods can be explained using the following long-accepted concepts.

According to standard biological jargon [6,7], two sequences are *homologs* if they are evolutionarily related, in which case they diverged at either a duplication event (and are called *paralogs*) or a speciation event (and are called *orthologs*). It is widely appreciated that a basic goal of alignment algorithms is to align sequences if and only if they are homologs. We feel that a better statement of the goal is to align precisely the orthologs. That is, we want the evolutionary

P. Bücher and B.M.E. Moret (Eds.): WABI 2006, LNBI 4175, pp. 138–149, 2006.

relationship among aligned sequences to be the same as the phylogenetic tree relating the species for those sequences. A main (and probably *the* main) use of alignments is to identify intervals within the aligned segments in which the similarity/divergence pattern differs from neutral evolution, and modern methods for detecting such intervals [21,5,23] require, for their proper functioning, that aligned rows be orthologs. In regions of the genome where no intervals have been duplicated, orthology is equivalent to homology, and existing alignment methods are effective. However, for tandem gene clusters, we know of no existing aligner that does a good job.

To represent duplications and other large-scale evolutionary rearrangements, our programs for aligning several genomic sequences produce a set of alignment "blocks", each of which is in essence a traditional alignment of segments from the given sequences or their reverse complements [2]. With duplications, the same sequence position can appear in several blocks. It is useful to note that if rows of a block are pairwise orthologous, then no two rows can be from the same species.

To build a new alignment program that obeys the two requirements

(a) any two rows of a computed block are orthologous and
(b) any pair of orthologous positions appears together in at least one block,

two hurdles had to be overcome. The first was to distinguish orthologs from paralogs, and for this there was a large literature to draw from. We provide our solution in another paper [11]. The second difficulty was that the number of possible blocks can grow exponentially with the number of sequences and duplications, which is the topic of this paper. For instance, a straightforward implementation meeting our requirements produced over 900 Mbytes of alignments when applied to intervals containing the α-globin gene clusters of 20 mammals, where the total length of the original sequences was only 3.9 Mb. We designed and implemented a space-saving strategy that decreased the amount of output to 8.7 Mb, while still fulfilling requirements (1) and (2). New ideas were required to achieve this savings, and we were led to the development of a theoretical model that turned out, in the general case, to be equivalent to a previously studied NP-complete combinatorial optimization problem, which we will call MINCLIQUECOV, namely, finding a minimum cardinality set of cliques that contains all edges of a given undirected graph. We show that the graphs we study have special properties, and they can be utilized to apply divide-conquer techniques, which would not work well with an arbitrary graph.

Here we describe our graph-theoretic model and derive a theoretical upper bound on the number of blocks that are needed to meet our requirements in an important subclass of problems. Also, using the model, we formulate two heuristic methods, and with the help of our upper bound and some computer simulations, we measure where the two methods lie in the tradeoff between computation time and output size. We also compare our solutions to an existing heuristic method for general graphs [13]. Finally, we describe the performance on the α-globin gene cluster of our alignment software that is based on the new ideas.

2   Methods

2.1   A Graph-Theoretic Model

Let $G = (V, E)$ be a graph with vertex set V and edge set E. An m-vertex complete subgraph of G is called an m-clique. A clique cover of G is a set of cliques whose edges contain every edge $e \in E$. The clique cover number, $cc(G)$, of G is defined to be the minimum number of cliques in a clique cover of G.

Assume we align genomic sequences from K species in a genomic region containing a family of tandemly duplicated genes. Suppose that each member of that gene family can be aligned to every orthologous member. In our model, a vertex represents a gene in one of the species, and there is an edge between two vertices if the genes that they represent are orthologous. Thus, we obtain a K-partite graph, called an *alignment graph*, where each part contains the nodes that represent the gene family members in a given species. A multi-alignment block with pairwise orthologous rows corresponds to a clique, and a set of multi-alignment blocks that contains every pairwise alignment (condition (2) in the Introduction) corresponds to a clique cover. Thus it would be helpful to solve MINCLIQUECOV for the alignment graph. Figure 1 gives an example in which

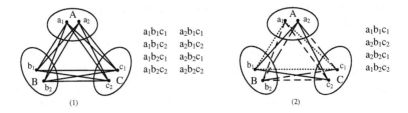

Fig. 1. A trivial example on minimum clique cover. Panel 1 shows 8 cliques to cover all edges, while panel 2 shows 4 cliques to cover all edges.

there are three species (A, B and C), each containing two members of a gene family. Each pair of genes from species is orthologous. In these example, there are eight possible alignment blocks, but four blocks are sufficient to include each orthologous pair in a block.

Unfortunately, MINCLIQUECOV is NP-hard [17]. The restriction to multi-partite graphs does not make this problem easier since a graph of n vertices is trivially an n-partite graph in which each part has only one vertex. Various techniques have been applied to solve MINCLIQUECOV and closely related problems [9,18,3]. For instance, when the degree of any vertex in G is at most 4, the problem is solvable in linear time [19].

2.2   A Special Case

In this section, we investigate the graph structure that arises under certain natural conditions, namely when all duplications have occurred after all speciation events. That is, we suppose that each gene is orthologous to every gene in a different species. Moreover, to keep things simple, we supposed that each of K species has precisely P copies of the gene. The resulting alignment graph is a complete K-partite graph: each part has P nodes, and there is an edge between any two nodes that are in two different parts. We denote such a graph by $G_{K,P}$. Note that the shape of such graph is determined for each pair of K and P. Thus MINCLIQUECOV restricted to graphs of the form $G_{K,P}$ has $O(n)$ distinct instances with n nodes, where $n = KP$ (because n can be factored into KP in less than n ways). It was shown that problems with polynomially many instances per size cannot be NP-hard (unless P=NP) [1]. But even for the case of $P = 2$, we do not know the exact solution. The technical result of this paper does not imply the problem is easy either. The purpose of this section is to derive a non-trivial upper bound on $cc(G_{K,P})$, which will help us to interpret the results of simulations that we report below.

First, though, let us mention lower bounds. For $K \geq 2$, $cc(G_{K,P}) \geq P^2$, since there are P^2 edges between any two parts of the graph, and any clique can contain at most one of those edges. Moreover, it was recently proved that $cc(G_{K,P}) \geq \log_b(KP)$, where $b = \frac{P}{(P-1)^{(P-1)/P}}$ [4]. That lower bound is approximately equal to $P(\log_P K + 1)$.

For an upper bound, it has been known for some time that $cc(G_{K,2}) = \Theta(\log_2 K)$ [9], but we seek a bound for general P. Assume $G_{K,P}$ has the following node set and edge set:

$$V = \{u(i,j): \ 0 \leq i < K \text{ and } 0 \leq j < P\}$$
$$E = \{\{u(i,j), u(k,l)\} \subset V: \ i \neq k\}$$

To present a recursive construction of small clique covers of $G_{K,P}$, we start with two simple observations.

Observation 1. *Let $U \subseteq V$. If $\mathcal{C}$ is a clique cover of G, then $\{C \cap U : \ C \in \mathcal{C}\}$ is a clique cover of $G(U)$. Thus $cc(G(U)) \leq cc(G)$.*

Observation 2. *If $\mathcal{C}_i$ is a clique cover of $< V, E_i >$ for $i = 1, 2 \ldots, k$ then $\bigcup_{i=1}^{k} \mathcal{C}_i$ is a clique cover of $< V, \bigcup_{i=1}^{k} E_i >$.*

The edges of $G_{K,P}$ can be split into two sets

$$E_0 = \{\{u(i,j), u(k,l)\} \in E: \ j = l\}$$
$$E_1 = E - E_0$$

We can cover $< V, E_0 >$ with cliques $C_j = \{u(i,j) \in V\}$, $j = 0, \ldots, P-1$. Now it remains to find a clique cover for $G_{K,P}^1 = < V, E_1 >$.

If there are $K = M \times L$ species, the edges of $G_{ML,P}^1$ can be represented as $E_2 \cup E_3$, where

$$E_2 = \{\{u(i,j), u(k,l)\} \in E_1: \ \lfloor i/L \rfloor \neq \lfloor k/L \rfloor\}$$
$$E_3 = \{\{u(i,j), u(k,l)\} \in E_1: \ i \bmod L \neq k \bmod L\}$$

To make that split more intuitive, put the ML parts of $G_{ML,P}$ into a matrix with M rows and L columns. An edge between different parts will either connect parts from different rows, or parts from different columns. Denote the set of edges connecting parts from different rows by E_2, and the set of edges connecting parts from different columns by E_3. Note that E_2 and E_3 are not necessarily disjoint.

Lemma 1. $cc(< V, E_2 >) \leq cc(G^1_{M,P})$.

Proof. Consider a clique cover $\mathcal{C}$ of $G^1_{M,P}$. Obtain $\mathcal{C}'$ by transforming each clique $C \in \mathcal{C}$ into

$$C' = \{u(i,j) \in V : u(\lfloor i/L \rfloor, j) \in C\}.$$

C' is still a clique, because if $\lfloor i/L \rfloor \neq \lfloor k/L \rfloor$ than $i \neq k$. Now an edge $e = \{u(i,j), u(k,l)\}$ of E_1 is covered by $\mathcal{C}'$ unless $\lfloor i/L \rfloor = \lfloor k/L \rfloor$, hence $e \notin E_2$. ❑

Lemma 2. $cc(< V, E_3 >) \leq cc(G^1_{L,P})$.

Proof. Similar to Lemma 1. ❑

Lemma 3. $cc(G^1_{P,P}) \leq P(P-1)$ if P is prime.

Proof. We construct a set of cliques $C_{a,b} = \{\{u(i, ai + b \bmod P) : 0 \leq i < P\}$ where $0 < a < P$ and $0 \leq b < P$. Given an edge $\{u(i,j), u(k,l)\} \in E_1$ we can find the parameters a, b of the clique that covers it by solving linear system

$$ai + b = j \bmod P$$
$$ak + b = l \bmod P$$

This system yields the following equation for a: $a(i - k) = j - l \bmod P$. Because $i \neq k$, $i - k$ has a reciprocal **mod** P, and because $j \neq l$, the a computed from this is non-zero. ❑

Theorem 1. $cc(G_{K,P}) \leq P + P(P-1)\lceil \log_P K \rceil$ if P is a prime.

Proof. By Observation 1, it suffices to prove that $cc(G_{K,P}) \leq P + aP(P-1)$ for $K = P^a$, where a is an integer. Because we can cover E_0 with P cliques, it suffices to prove that $cc(G^1_{K,P}) \leq aP(P-1)$. We can show it by induction on a. For $a = 1$ this is proven in Lemma 3. Assuming it is proven for $a - 1$, we have $K = ML$ where $M = P$ and $L = P^{a-1}$. This allows to apply Lemmas 1 and 2 to show that $cc(G^1_{K,P}) \leq (a-1)P(P-1) + P(P-1) = aP(P-1)$. ❑

When P is not a prime, we can use the above result for P', where P' is the smallest prime larger than P. For example, when $P = 6$ and $K = 2$, we have a trivial solution with 36 cliques (indeed, edges), but when $K = 3$, we can use a solution for $P' = 7$ that has 49 cliques, and this solution works for $K \leq 7$, while for $K \leq 343$ we have a solution with $7 + 2 \times 42 = 91$ cliques.

Of course, better solutions may exist. It is easy to find a solution for $G^1_{3,4}$ with 12 cliques; see Figure 2. We can apply the reasoning of Theorem 1 to show that $cc(G_{K,4}) \leq 4 + 12\lceil \log_3 K \rceil$ which, except for $K = 4, 5$, is better than $5 + 20\lceil \log_5 K \rceil$. Thus, $cc(G_{K,P})$ is at most $P(P-1)\log_P K + P$ when P is a prime number and K is in power of P. In cases that P and K do not satisfy these conditions, the values can be approximated by the nearest prime number and its power. This upper bound is conjectured to be tight, but this has not been proved.

$$\overbrace{\phantom{\{a_0,b_0,c_0\}}}^{E_0}$$

E_0

$\{a_0, b_0, c_0\}$
$\{a_1, b_1, c_1\}$
$\{a_2, b_2, c_2\}$
$\{a_3, b_3, c_3\}$

E_1

$\{a_0, b_1, c_2\}$ $\{a_1, b_0, c_3\}$ $\{a_2, b_0, c_1\}$ $\{a_3, b_0, c_2\}$
$\{a_0, b_2, c_3\}$ $\{a_1, b_2, c_0\}$ $\{a_2, b_1, c_3\}$ $\{a_3, b_1, c_0\}$
$\{a_0, b_3, c_1\}$ $\{a_1, b_3, c_2\}$ $\{a_2, b_3, c_0\}$ $\{a_3, b_2, c_1\}$

Fig. 2. Clique cover of $G_{3,4}$ with 16 cliques

2.3   Heuristic Solutions

MinCliqueCov is NP-hard, but also has no efficient approximation algorithm unless NP=P [14], so we have to use heuristics. We propose two heuristic methods for generating clique covers for complete multi-partite graphs studied in the previous section. It is then relatively straightforward to adapt these methods for the multi-alignment problem in tandem gene clusters, even in the general case of arbitrary orthology relationships; the next section discusses some of the issues that arise. However, in the idealized setting of $G_{K,P}$, with the upper bound derived above, we can evaluate how close they come to an ideal reduction in output size.

Both heuristic methods follow a divide-and-conquer strategy: Partition G into two graphs, G_1 and G_2, each having about $K/2$ parts, find clique covers CC_1 and CC_2 of G_1 and G_2 respectively, and merge them to obtain a clique cover CC of G. While these methods do not give the fewest cliques, they efficiently find a relatively small clique cover. The merge procedures can be described as follows, where "uncovered" refers to an edge not currently in a clique in CC.

Heuristics for MinCliqueCov were studied already in 1978 [13]. Recently an exact solution was also proposed [8] for the general graph. Unfortunately, our complete multi-partite graph is quite dense, and the reduction rules from [8] cannot be applied here. Thus we only compare our heuristic methods' performance with [13]'s method.

Merge I forms a new clique from the pair of cliques that maximizes the number of additional covered edges. Merge II processes cliques of CC_1 and CC_2 in a random order and forms any new clique that covers at least one new edge. Running times are dominated by the number of executions of the loops 2a and 2b, respectively, which are essentially the number of cliques generated. The lowest-level operation inside loop 2a is to examine whether an edge is covered or not; for loop 2b it is to decide whether a clique contains a certain edge or not. Both operations involve an array access and can be regarded as taking unit time. The number of unit operations inside loop 2a is $|CC_1| \cdot |CC_2| \cdot K_1 \cdot K_2$, where K_1 and K_2 are the number of partitions (species) of G_1 and G_2. The number of unit operations inside loop 2b is $|CC_1| \cdot K_1 + |CC_2| \cdot K_2$. The performance of these two methods, in terms of both actually running time and the number of cliques generated, is analyzed below.

2.4   Application in the Aligner Program

The above divide-and-conquer methods can be adapted to so-called "progressive" multiple alignment programs, which work leaves-to-root in the phylogenetic tree

Merge I (CC_1, CC_2)
1   $CC \leftarrow \phi$
2a **while** there exists an uncovered edge
3   For each pair of cliques c_i and c_j from CC_1 and CC_2 respectively
4           $u_{ij} \leftarrow$ the number of uncovered edges of $c_i \cup c_j$
5       $(c_{maxi}, c_{maxj}) \leftarrow$ the pair with maximum u_{ij}
6   insert $c_{maxi} \cup c_{maxj}$ to CC
7   Output CC

Merge II (CC_1, CC_2)
1   $CC \leftarrow \phi$
2b **while** there exists an uncovered edge (u, v) between subproblems
3       $c_1 \leftarrow$ a clique from CC_1 that contains u
4       $c_2 \leftarrow$ a clique from CC_2 that contains v
5   insert $c_1 \cup c_2$ into CC
// We still have to incorporate unused cliques from each subproblem
6   **while** there exists an unused clique c_1 from CC_1
7       $c_2 \leftarrow$ an unused clique in CC_2; if none, then any clique in CC_2
8   insert $c_1 \cup c_2$ into CC
9   **while** there exists an unused clique c_2 from CC_2
10      $c_1 \leftarrow$ an unused clique in CC_1; if none, then any clique in CC_1
11   insert $c_1 \cup c_2$ to CC
12 Output CC

for the given species. At a tree node, these aligners merge multiple alignments from the sub-trees, which is analogous to merging cliques for subgraphs G_1 and G_2, except that the split into subproblems might not be balanced. The process of merging blocks from the left and right subtree is guided by pairwise alignments between a species in the left subtree and a species in the right subtree. The set of multi-alignment blocks corresponding to the tree's root constitutes the multiple alignment of the original K species.

Thus, the use of the guide tree by the aligner can be viewed as the recursive partition of the alignment graph, and such a partition can be used both by MERGE I and MERGE II. In our current BOAST aligner, we have decided to apply MERGE II since our tests indicated that MERGE I could be about 1000 times slower (see Figure 4). While MERGE I produced fewer cliques(see Figure 3), i.e., alignment blocks, MERGE II produced a number that was acceptably small.

We need to address the following issue. The alignment graph generally has fewer edges than a complete K-partite graph, since some speciation events are preceded by duplications. Consider what we should expect if we have a perfect alignment graph, with all ortholog/paralog relationships properly diagnosed. We have two siblings T_1 and T_2 with common ancestor T and let c_i be a clique in the subgraph of T_i, $i = 1, 2$. One can see that either $c_1 \cup c_2$ is also a clique, or there are no edges between c_1 and c_2 (see [11] for the analysis of orthologous inference). In this case, when MERGE II selects a pair of cliques with union that covers at least one new edge (lines 3-4), this union is a clique. When MERGE II processes an unused clique c_1 in line 7 and 10, we first look for a clique c_2 in another sub-tree that is connected to c_1, and the union of them becomes a new clique; if none exists we simply add c_1 to the output.

Clearly, this adaptation may be incorrect if we have an imperfect alignment graph, ie., with some edges established wrongly and some correct edges missing. Consequently, it may happen that a pair of selected cliques, c_1 and c_2, has some connections, but not all. If only a minority of the possible edges between c_1 and c_2 are present, we behave as if c_1 and c_2 were not connected at all, and if the majority is present, we behave as if $c_1 \cup c_2$ was a clique. This majority criterion can have the effect of correcting some errors created by incorrect identification of orthologous pairwise alignments [11]. Note that we have to choose between two kinds of discrepancies in the output. One is that we do not include all actual orthologous relationship in a blocks. The second is that we contaminate a block with a paralogous relationship. If one kind of discrepancy is more harmful than the other, we can replace the majority criterion with some other ratio.

3 Results

3.1 Simulations

Methods Merge I and II were tested on graphs of the form $G_{K,P}$, i.e, complete K-partite graphs, where each part has P nodes. The results are plotted in Figure 3(a) and 3(c), which shows that Merge I substantially outperforms Merge II. Values of upper bound $P + P(P-1)\lceil \log_p K \rceil$ are shown in Figure 3(b). The bounds for P=4 are determined by $cc(G_{K,4}) \leq 4 + 12\lceil \log_3 K \rceil$ discussed below Theorem 1. The lower bound from [4] is lower than the trivial P^2 bound in most of our cases, so we do not show it in the figure.

Though MERGE I produces fewer cliques, it requires a longer running time. To estimate the difference in CPU requirements, we counted the numbers of previously described unit operations, so the impact of different number of cliques produced by the two methods is included. Although it is possible to improve the time efficiency of MERGE I by designing better data structures, MERGE I will always be slower than MERGE II.

The heuristic method from [13] takes even more time than Merge I, so its running time analysis is not shown here. However, we plot its clique numbers. As shown in Figure 3(d), it produces more cliques than Merge II for any instances.

3.2 α-Globin Gene Cluster

A recent study carefully identified ortholgous genes in the α-globin clusters of a number of mammals [12]. There are four types of genes in those clusters: ζ, αD, α and θ. Each species discussed below has exactly one αD-related gene, so those genes are not pertinent to this analysis. The studied species have 1 to 3 ζ-related genes, 1 to 4 α-related genes, and 0 to 3 θ-related genes (counts include pseudogenes). Table 1 shows details. Each gene copy is regarded as a node in our graph.

Our earlier TBA program [2,15,16] guarantees that every position in the reference sequence (human in this case) is in exactly one multiple alignment block,

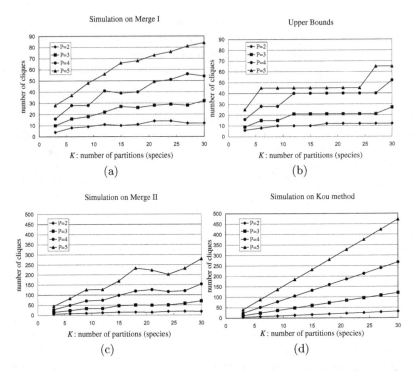

Fig. 3. Simulations of three heuristic procedures together with plots of upper bounds, with different values of K and P. The curves, top to bottom, refer to P=2,3,4,5.

and thus TBA is not able to capture all pairwise orthologous relationship in a tandem gene cluster. For example, each human α gene is aligned to only one rat α gene, despite the fact that a human α is actually orthologous to two more rat α genes, and those alignments are lost. Our new aligner, called BOAST, which implements MERGE II, captures all pairwise orthologous relationship.

We aligned sequences containing the α-globin gene clusters from 20 mammals. Each sequence is around 200K bases. Both TBA and BOAST utilize pairwise alignments computed by blastz [20]. For BOAST, the pairwise alignments are filtered by a program called TOAST [11], which retains only the putatively orthologies (i.e., deletes paralogous matches). After computation of pairwise alignments, TBA produces 7.5 Mb of alignments after 170 CPU seconds, while BOAST produces 8.7 Mb of alignments in around 112 CPU seconds.

Each aligner outputs a set of blocks, whose endpoints do not in general correspond to gene or exon boundaries. Moreover, each functional globin gene has three exons, and many alignments do not extend from one exon to the next. We estimated how many cliques were formed for each type of genes as follows. For a given gene sequence, we manually determined three positions distributed roughly evenly throughout the gene, and counted the number of times each position appeared in the multi-alignment blocks; the maximum of the three counts was used to estimate the number of times the gene copy appears in the blocks.

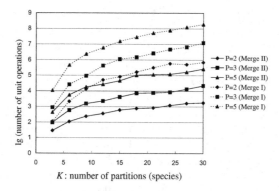

Fig. 4. Comparisons of running time on methods Merge I and II

In this example (α-globin clusters of 20 species), the clique sizes (i.e., the number of rows in a multi-alignment block) vary from 4 to 20, with most between 16 and 20. Alignment blocks containing θ-related genes have at most 17 rows since three species do not have a θ-related gene. Other blocks with less than 20 rows result from a number of factors, including inconsistent pairwise alignments, pseudogenes, or retention of a non-orthologous alignment. With many-to-many orthologous relationships, there will be combinatorial increase on the number of alignment blocks. Table 1 shows that utilizing MERGE II, we reduce the number of alignment blocks to 26 for ζ-related alignments and 58 for α-related alignment. It means that 26 and 58 multiple alignment blocks contain all pairwise orthologous relationships of a certain region for ζ-related and α-related genes respectively.

The BOAST alignments are reference-independent, which means that no sequence data from any of the species is missing from the alignment. One of our tools extracts a reference-sequence-based alignment from the BOAST alignment for any specified reference. This gives an alignment similar in size to the output of a typical reference-based multiple aligners, for example multiz [2]. Moreover, the BOAST alignments capture complete and accurate orthology information, which is currently lost by other aligners.

Table 1. Number of copies for each type of α-globin genes of 20 mammals, and number of cliques formed for each type of genes in the multiple alignment employing the heuristic Merge II. When the number of genes is given in the form x-y, x refers to genes, and y refers to genes together with pseudogenes.

gene	armadillo	baboon	cat	chimp	colobus	cow	dog	dusktit	galago	hedgehog	human	lemur	macaca	marmoset	mouse	owlm	pig	rat	rfbat	sqmonkey	#cliques
ζ-related	1	2	2	2	2	2	2	1	2	2	1-2	1	2	1	1	1	1	1	1	2	26
α-related	1	1-2	1	1-3	2-4	2	1	2	2	2-3	2-3	2	2	1-2	2-3	2	1	3	1	2	58
θ-related	0	1	0	1	1	1	1	1	1	1	1	1	1	1	1-3	1	1	3	0	1	7

4   Conclusion and Further Work

In gene clusters having a significant level of lineage-specific duplications (i.e., producing many-to-many orthology relationships), it is not practical to enumerate all possible multi-alignment blocks having pairwise orthologous rows. However, we have shown here that it is still frequently feasible to produce a set of blocks with the property that every pair of orthologs appears together in one of the blocks. The essence of the situation is captured by the problem of finding a minimum-cardinality clique cover. Both the problems of the finding the minimum number of cliques to cover all edges and the minimum number of cliques to cover all nodes (see below) are NP-complete. However, our simulations show that sizes can in practice be reduced (especially by Merge I) to be close to the upper bound. Though our alignment program has only incorporated Merge II, it still dramatically reduces the alignment size, and makes it feasible to align tandem gene clusters from many species. In the future, we hope to implement Merge I in our alignment program.

Though the clique cover problem for an arbitrary graph is NP-complete, it is open whether the problem's restriction to alignment graphs is intractable. The structure of the graphs is constrained by the phylogenetic tree for the species in question and by properties of the orthology relationship [11], and these restriction might be helpful for determining clique covers.

It also remains to investigate other criteria for aligning regions containing duplicated segments or genes. For instance, one could loosen the requirement that each orthologous pair of positions occur in two rows of the same block, and ask only that a position of one species that has an ortholog in a second species must appear in the same block as some (perhaps different orthologous) position in that second species. In essence, this can be modeled as seeking the minimum number of cliques to cover all *nodes* in the graph constructed above, which in general requires fewer cliques than the problem studied here. We think that the problem of aligning tandem gene clusters is sufficiently important that a variety of approaches should be investigated.

References

1. Berman, P.: Relationship between density and deterministic complexity of NP-complete languages. *Lecture Notes in Compute Science* **62**, 1978 63–71
2. Blanchette, M. et al.: Aligning multiple genomic sequences with the threaded block-set aligner. *Genome Research* **14**, 2004 708–715
3. Cacceta, L., P. Erdos, E.T. Ordman and N.J. Pullman: On the difference between clique numbers of a graph. *Ars Combinatoria* **19A**, 1985 97–106.
4. Cavers, M.: Clique partitions and coverings of graphs 2005 (Masters thesis, University of Waterloo)
5. Cooper, G. M., et al.: Distribution and intensity of constraint in mammalian genomic sequences. *Genome Research* **15**, 2005 901–913.
6. Fitch, W. M.: Distinguishing homologous from analogous proteins. *Syst. Zool.* **19**, 1970 99–113.

7. Fitch, W. M.: Homology, a personal view on some problems. *Trends Genet.* **16**, 2000 227–231.
8. Gramm, J., et al: Data reduction, exact, and heuristic algorithms for clique cover. *ALENEX* , 2006 86–94.
9. Gregory, D. A., and N.J. Pullman: On a clique covering problem of Orlin. *Discrete Math.* **41**, 1982 97–99.
10. Hall, M. Jr.: A problem in partition, Bull. *Amer. Math. Soc.* **47**, 1941 801–807.
11. Hou, M., et al: Aligning multiple genomic sequences that contain duplications. Manuscript.
12. Hughes, J. R., et al: Annotation of cis-regulatory elements by identification, subclassification, and functional assessment of multispecies conserved sequences. *Proc. Natl. Acad. Sci. USA* **102**. 2005 9830–9835
13. Kou, L.T., et al: Covering edges by cliques with regard to keyword conflicts and intersection graphs. *Communications of the ACM* **21(2)**. 1978 135–139
14. Lund C. and M. Yannakakis, On the hardness of approximation minimization problems. *J. Assoc. for Comput. Mach.* **41**, 1994 961–981.
15. Margulies, E. H., et al. Relationship between evolutionary constraint and genome function in 1% of the human genome. Submitted to *Nature*.
16. Margulies, E. H., et al. Annotation of the human genome through comparisons of diverse mammalian sequences. Submitted to *Genome Research*.
17. Orlin, J. Contentment in graph theory: covering graphs with cliques. *Indag. Math.* **39**, 1977 406–424.
18. Pullman, N. J., and A. Donald: Clique coverings of graphs II: complements of cliques. *Utilitas Math.* **19**, 1981 207–213.
19. Pullman, N. J.: Clique coverings of graphs IV: algorithms. *SIAM J. on Computing* **13**, 1984 57–75.
20. Schwartz, S., et al. Human-Mouse Alignments with BLASTZ. *Genome Res.* **13(1)**, 2003 103–107.
21. Siepel, A., et al. Evolutionarily conserved elements in vertebrate, insect, worm, and yeast genomes. *Genome Research* **15**, 2005 1034–1050.
22. The ENCODE Project Consortium: The ENCODE (ENCyclopedia of DNA Elements) Project. *Science* **306**, 2004 636–640.
23. Wakefield, M. J., P. Maxwell and G. A. Huttley: Vestige: maximum likelihood phylogenetic footprinting. *BMC Bioinformatics* **6**, 2005 130.

Reducing Distortion in Phylogenetic Networks

Daniel H. Huson[1], Mike A. Steel[2], and Jim Whitfield[3]

[1] Center for Bioinformatics (ZBIT), Tübingen University, Germany
huson@informatik.uni-tuebingen.de
[2] Allan Wilson Centre, University of Canterbury, Christchurch, New Zealand
m.steel@math.canterbury.ac.nz
[3] Department of Entomology, University of Illinois at Urbana-Champaign, USA
jwhitfie@life.uiuc.edu

Abstract. When multiple genes are used in a phylogenetic study, the result is often a collection of incompatible trees. Phylogenetic networks and super-networks can be employed to analyze and visualize the incompatible signals in such a data set. In many situations, it is important to have control over the amount of imcompatibility that is represented in a phylogenetic network, for example reducing noise by removing splits that do not recur among the source trees. Current algorithms for computing hybridization networks from trees are based on a combinatorial analysis of the arising set of splits, and are thus sensitive to false positive splits. Here, a filter is desirable that can identify and remove splits that are not compatible with a hybridization scenario. To address these issues, the concept of the distortion of a tree relative to a split is defined as a measure of how much the tree needs to be modified in order to accommodate the split, and some of its properties are investigated. We demonstrate the usefulness of the approach by recovering a plausible hybridization scenario for buttercups from a pair of gene trees that cannot be obtained by existing methods. In a second example, a set of seven gene trees from microgastrine braconid wasps is investigated using filtered networks. A user-friendly implementation of the method is provided as a plug-in for the program SplitsTree4.

1   Introduction

In systematics, the evolution of different species is of interest, however, phylogenetic inference is often based on the DNA or protein sequence of homologous genes and the resulting *gene trees* are usually interpreted as estimations of an underlying *species tree*. A common observation is that different genes give rise to different trees, even in the absence of tree-reconstruction errors, and this fact can usually be explained by mechanisms such as incomplete lineage sorting, duplication-and-loss, horizontal gene transfer (e.g. in bacteria) or hybridization (e.g. in plants).

Although phylogenies based on single gene analysis [32] continue to play a central role in phylogenetics, biologists interested in the evolution of specific groups of taxa often sequence and use more than one gene to infer the phylogeny

P. Bücher and B.M.E. Moret (Eds.): WABI 2006, LNBI 4175, pp. 150–161, 2006.
© Springer-Verlag Berlin Heidelberg 2006

of the taxa [23], the hope being that as more data is brought into the analysis, a better "species-signal" to "gene-noise" ratio will be obtained and that deviating signals from individual genes can be filtered out.

If the goal is simply to obtain a good estimation of the species tree and if there is evidence that a majority of the genes under study have evolved in a similar way along the same species tree, then one approach is to concatenate the alignments given for each of the genes to produce one large dataset, to which tree-building methods are then applied [23,25]. If each of the genes is long enough to contain strong phylogenetic signals for the group of taxa under investigation, then a second approach is to compute individual gene trees, to summarize them using a (usually somewhat unresolved) consensus tree and then to interpret the consensus as a representation of the well-supported parts of the species tree [30,10,26].

In both cases, the final result suppresses all incompatible signals. However, if the actual incongruencies of the individual gene trees are themselves of interest, then a representation of the data set that maintains (some of) the incompatible signals may be useful. Such a representation is given by the concept of a "split network" [1] and methods for computing such networks are presented in [8] and are implemented in the program SplitsTree4 [15].

To obtain an explicit model of reticulate evolution, reticulate networks are used [15] that explain a given set of trees in terms of hybridization, horizontal gene transfer or recombination events [13,7,19,17,18]. Current methods for determining a hybridization scenario that explains a given set of trees operate by performing a combinatorial analysis of the total set of splits of the trees to identify a hybridization network that generates the trees [22,17]. By definition, combinatorial methods are very sensitive to false positive splits, that is, splits that are incompatible to other splits in the input due to reasons such as homoplasy, tree-estimation error, incomplete lineage sorting etc.

Given a collection (or *profile*) P of k gene trees all inferred on the same set of taxa X, one approach to constructing a set of splits that summarize the set of trees, without eliminating all incompatibilities, is given by the consensus network method [2,14]. This method consists of returning all splits that occur in at least αk of the given input trees, for a given threshold $\alpha \in [0, 1]$.

A main drawback of the consensus network approach is that in practice typical data sets often consist of *partial trees*, that is, gene trees that each only mention some subset X' of the total taxon set X. Partial trees arise because the sequence data for some gene has not yet been sequenced, or because the gene is not present in the genome, for some taxon.

Given a profile of partial gene trees, the Z-closure method [16] computes a *super network* on the full taxon X that summarizes all the input trees. This approach first uses an inference rule to construct a set of splits on the full taxon set and then, as above, a network construction algorithm [8] is employed to obtain a split network. A practical weakness of this method is that it does not provide a natural parameter (such as α above) with which one can control the amount of incompatibility that is represented in the resulting network.

The goal of this paper is to develop an adjustable parameter than can be used with any super network method or consensus method to generate split networks that represent a controlled amount of incompatible signals. The approach that we take is to filter splits by the amount of "distortion" that they generate. We have implemented this approach as a plug-in `FilteredSuperNetwork` for the SplitsTree4 program [15].

This concept is particularly useful in the context of computing hybridization networks from gene trees, because it can be used to remove splits from a data set that are not compatible with a simple hybridization scenario. This is due to the fact that the distortion of a split equals the number of SPR or TBR operations required to modify a tree to accommodate the split, which will be small for incompatibilities caused by hybridization.

We illustrate this use of a distortion filter for a set of 46 *Ranunculus* (buttercup) species, represented by two gene trees, one based on a chloroplast *JSA* region, and the other based on a nuclear *ITS* region [20]. Although this dataset is known to contain examples of both allopolyploid and diploid hybridization events (Pete Lockhart, personal communication), past attempts to compute a corresponding hybridization network from the two trees have failed [17]. Here we demonstrate that a plausible hybridization network can be computed when employing an appropriate distortion filter.

A second example is given by a set of seven gene trees for 45 species of wasps [3]. Mixed-model Baysian analysis [24] of the combined data set indicates that there is little support for internal edges of the phylogeny and here we show how filtered network methods can be used to investigate whether this lack of support is due to conflict between the different gene trees, or whether it represents a lack of real coherent signal in the data.

In the following Section 2 we provide the necessary formal definitions, and then introduce the concept of distortion and explore some of its properties. Then, in Section 3, we present an algorithm for efficiently computing the distortion of a tree relative to a split. Finally, in Section 4, we illustrate the application of the algorithm to two different biological data sets.

We are grateful to the Cass Field Station of the University of Canterbury, where we developed the main ideas of this paper. D.H.H. would like to thank the DFG and the Erskine Programme for funding. J.W. would like to thank the Allan Wilson Centre for sponsoring his trip to NZ, and National Science Foundation Grant DEB 0316566 for funding the generation of the wasp data. Thanks to Pete Lockhart for providing the buttercup trees and for many useful discussions.

2   The Distortion of a Tree Relative to a Split

We mostly follow the notation of [29]. By a *partial X–tree* we mean a tree $\mathcal{T}$ together with a labeling map ϕ from some subset X' of X into the vertices of $\mathcal{T}$ so that each vertex of degree at most 2 receives at least one label. Given an X–split $\sigma = A|B$ we may regard σ as a map from X into $\{0, 1\}$ (where elements of A are sent to 0 and elements of B are sent to 1) and so, by restricting σ to X', we may view σ as a binary character for $\mathcal{T}$.

If T is a *phylogenetic tree* (that is, the only vertices of T labeled by X' are the leaves and these each receive exactly one label), then let $h(T, \sigma)$ denote the *homoplasy score* of the binary character σ on T, that is, the parsimony score of σ, minus 1.

For any X-split σ and partial X-tree T, we define the *distortion of T relative to σ* as

$$\partial(T, \sigma) := \min_{T' \in Phy(T)} h(T', \sigma),$$

where $Phy(T)$ denotes the set of *phylogenetic refinements of T*, that is, the phylogenetic trees with the same label set as T and that contain all the splits of T.

The following result provides an interpretation of the distortion as a measure of how much a tree needs to be modified in order to accommodate the split σ, see Figure 1. Recall that two commonly-used ways to transform trees are by

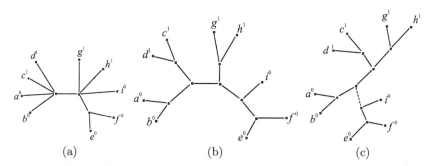

(a)   (b)   (c)

Fig. 1. (a) A tree T labeled by taxa $X = \{a, \ldots, i\}$, with superscript 0 or 1 indicating that the taxon lies in part A or B, of the split $\sigma = A \mid B = \{a, b, e, f, i\} \mid \{c, d, g, h\}$; we have $h(T, \sigma) = 3$. (b) A refinement T' of T, with $h(T', \sigma) = 1$, leading to $\partial(T, \sigma) = 1$. (c) $\partial(T, \sigma) = h(T', \sigma) = 1$ matches the transformation of T' into T'' on which σ is compatible, using one SPR move.

SPR ('subtree prune and regraft') and TBR ('tree bisection and reconnection') operations, which are explained further in [29,9]. In particular, the result explains why a filter based on distortion will be a useful tool for removing false positive splits when computing a hybridization network.

Proposition 1. *For any partial X-tree T and X-split σ, the value $\partial(T, \sigma)$ equals the smallest number of (SPR or TBR) tree rearrangement operations required to transform at least one phylogenetic refinement of T into a tree that has the split σ.*

Proof. The result follows from Theorem 5.2 of Bryant [6].   □

A tree $T' \in Phy(T)$ that minimizes $h(T', \sigma)$ is the optimal refinement of T, with respect to maximum parsimony, for the binary character that corresponds to σ, in the sense of [4]. Moreover, the value of $\partial(T, \sigma)$ is unaltered if one replaces in the definition the set $Phy(T)$ by the set of binary phylogenetic refinements of T. Notice also that if we replace the partial X-tree T by its minimal phylogenetic refinement $\mathcal{T}_p$ (i.e. the partial phylogenetic X–tree whose splits consist of the splits of $\mathcal{T}$ together with the trivial splits on the label set of $\mathcal{T}$) then we have

$$\partial(\mathcal{T}, \sigma) = \partial(\mathcal{T}_p, \sigma),$$

so it suffices to describe an algorithm for computing ∂ for partial phylogenetic trees.

The score ∂ has a dual 'max-flow' description. Let $p(\mathcal{T}, \sigma)$ denote the maximum number of vertex-disjoint paths that each connect an A–type leaf to a B–type leaf. By Menger's Theorem (see [12]) this is equal to the minimum number of vertices of $\mathcal{T}$ that need to be deleted from $\mathcal{T}$ in order to separate each A–type leaf from each B–type leaf.

Theorem 1. *For any phylogenetic X–tree and X–split, σ,*

$$\partial(\mathcal{T}, \sigma) = p(\mathcal{T}, \sigma) - 1.$$

Proof. Omitted due to space restrictions.   □

Given a collection ('profile') of partial X–trees $P = \{\mathcal{T}_1, \ldots, \mathcal{T}_k\}$ define the *distortion of P relative to σ* as follows:

$$\partial(P, \sigma) := \sum_{i=1}^{k} \partial(\mathcal{T}_i, \sigma).$$

Proposition 1 implies that $\partial(P, \sigma)$ is the minimum total number of transformations required on refinements of trees in P so that σ is a split of each resulting tree. In Section 3 we present an algorithm that efficiently computes $\partial(\mathcal{T}, \sigma)$ directly from σ and $\mathcal{T}$.

One approach to super-network construction from a profile P of partial trees would be to identify those X–splits σ for which $\partial(P, \sigma)$ is less than some (adjustable) threshold $k \geq 0$. However this problem seems in general to be intractable due to the following result.

Proposition 2. *The following problem is NP–hard. Given a profile P of partial X–trees, determine whether there exists a non-trivial X–split σ with $\partial(P, \sigma) = 0$.*

Proof. The result follows from the NP–hardness of 'Split-quartet compatibility' by [5].   □

In view of Proposition 2 an alternative approach is to use P to first construct a large set of 'feasible' X–splits, and then to use ∂ to prune this set to a more conservative subset. More concretely, we propose to first use the Z-closure algorithm to compute a set of X–splits for P and then to return all splits σ with $\partial(P, \sigma) \leq k$, for a given integer threshold $k \geq 0$.

Another option for a profile P of partial X–trees – which generalizes the consensus network approach– is, for a non-negative integer r, and real number $\alpha \in [0, 1]$ to consider the set of X–splits defined by:

$$\{\sigma : |\{\mathcal{T} \in P : \partial(\mathcal{T}, \sigma) \leq r\}| \geq \alpha|P|\}.$$

For $r = 0$ and a profile P consisting of binary phylogenetic X–trees, then using the set of all splits contained in P, this corresponds to the consensus network (with threshold α).

Proposition 2 indicates that this is a hard problem, if we do not restrict the set of splits under consideration. For partial trees, one can use the Z-closure to compute a set of candidate splits. We have implemented this approach as a plug-in for SplitsTree [15] and discuss this in detail below.

Finally, assume we are given a profile P of (non-partial) X-trees. For small values of r we can compute all possible X–splits σ with $\partial(P, \sigma) \leq r$ as follows: For each tree $T \in P$, consider all $O(\binom{n-3}{r})$ possible ways of selecting up to r vertex-disjoint edges in the tree, where $n = |X|$. By placing a change on each selected edge, each such choice of edges defines a binary character σ with distortion $\partial(T, \sigma) \leq r$. Return all splits whose total score over all trees does not exceed r.

3 Computation of the Distortion

Given a partial X–tree T and an X–split σ, the definition of $\partial(T, \sigma)$ in Section 2 does not immediately lead to an algorithm. To compute this value, we describe a modification of Sankoff's algorithm [27,28] for computing the parsimony score of a character on a tree.

In the following, we will assume that T is a phylogenetic X'–tree, with $X' \subseteq X$. However, our algorithm is easily extended to the case that T is multi-labeled (i.e., has nodes labeled by more than one taxon), and has labels on (some or all) internal vertices.

Algorithm 2 (Distortion)
Input: *A phylogenetic partial X–tree T and an X–split σ.*
Output: *The distortion $\partial(T, \sigma)$.*
Root T at the midpoint of an edge and let ρ denote the root vertex.
Initialization: For all vertices v and all $a \in \{0, 1\}$ set:

$$S_v(a) = \begin{cases} 0, & \text{if } a = 0 \text{ and } \phi^{-1}(v) \subseteq A, \text{ or } a = 1 \text{ and } \phi^{-1}(v) \subseteq B \\ \infty, & \text{if } a = 0 \text{ and } \emptyset \neq \phi^{-1}(v) \subseteq B, \text{ or } a = 1 \text{ and } \emptyset \neq \phi^{-1}(v) \subseteq A. \end{cases}$$

Compute S_ρ using the following recursion:
For $a, b \in \{0, 1\}$, and a vertex v with children $w_1, \ldots, w_k$, set

$$S_v(a) = \sum_{w_i : S_{w_i}(a) < S_{w_i}(b)+1} S_{w_i}(a) + \sum_{w_i : S_{w_i}(a) \geq S_{w_i}(b)+1} S_{w_i}(b) + \Delta,$$

where

$$\Delta = \begin{cases} 1, & \text{if there exists } w_i : S_{w_i}(a) \geq S_{w_i}(b) + 1; \\ 0, & \text{otherwise.} \end{cases}$$

The result is given by $\partial(T, \sigma) = \min\{S_\rho(0), S_\rho(1)\} - 1$.

Proposition 3. *Let T be a partial phylogenetic X–tree and $\sigma = A \mid B$ be an X-split. Algorithm 2 computes the distortion $\partial(T, \sigma)$ in linear time.*

Proof. The algorithm considers each parent-child pair of vertices exactly once, and hence the time requirement is linear.

We will prove the result by induction. First, consider the initialization step. The map $S_v(0)$ is set to 0 for every internal vertex v, and otherwise to 0 or ∞, depending on whether the label of the leaf v lies in A or B, respectively. Vice-versa for $S_v(1)$.

Now, consider a vertex v and assume by induction that we have correctly computed $S_{w_i}(a)$ for all children $W = \{w_1, w_2, \ldots, w_k\}$ of v and all $a \in \{0, 1\}$.

Define $W_0 := \{w_i \in W \mid S_{w_i}(0) < S_{w_i}(1) + 1\}$ and $W_1 := \{w_i \in W \mid S_{w_i}(0) \geq S_{w_i}(1) + 1\}$.

To compute $S_v(0)$, consider a refinement T' of T such that v has one or two out-edges (depending on whether one or both of the sets W_A and W_B are non-empty), $e_0 = (v, u_0)$ and $e_1 = (v, u_1)$, leading to one or two subtrees containing the sets W_0 and W_1, respectively. We choose state 0 and state 1 on the nodes $W_0 \cup \{u_0\}$ and $W_1 \cup \{u_1\}$, respectively, and pay a penalty of 1 for a change along edge e_1, if $W_1 \neq \emptyset$. Note that the degree of u_0 or u_1 may be 2, which we allow for purposes of the proof, as this does not alter the achievable score. We compute $S_v(1)$ in a similar manner. □

4   Implementation and Applications

We have implemented the above ideas as a new plug-in `FilteredSuperNetwork` for the program SplitsTree4 [15]. This method takes as input a profile P of (partial) X-trees and produces as output a filtered set of X–splits Σ. These splits can then be visualized as a split network using the algorithm described in [8], or used to compute a hybridization network, using the algorithm described in [17].

The method proceeds by first computing the Z-closure Σ' of all partial X–splits in P and then computing the profile score of every split $\sigma \in \Sigma'$. The user must provide two parameters. The first parameter, `maxDistortion`, determines the maximal distortion $\partial(T, \sigma)$ acceptable to consider $\sigma \in \Sigma'$ as being *supported* by the tree $T \in P$. The second parameter, `minSupportingTrees`, determines the minimum number of trees $T \in P$ that are required to support σ so that σ is present in the set of output splits Σ. Either parameter can be set by a slider that is coupled to a histogram that shows how many splits will be present in the output for any given choice of the parameter, given the current value of the other parameter.

As mentioned above, an important application of the distortion filter is as a preprocessing step in the computation of hybridization networks [17]. Given a set of gene trees that show significant incongruencies due to hybridization events, the goal here is to compute a hybridization network that "explains" the gene trees. Existing approaches perform a combinatorial analysis of the set of trees or splits to derive a network, and thus are very sensitive to false positive splits in the data set. If the underlying hybridization scenario is relatively simple, e.g.

involving only isolated events, then the distortion filter can be used to remove interfering splits.

For example, consider the set $P = \{T_1, T_2\}$ of two gene trees on 46 *Ranunculus* (buttercup) species depicted in Figure 2, based on (a) a chloroplast *JSA* region, and (b) a nuclear *ITS* region [20]. The split network representing the set Σ of all splits from either tree is shown in Figure 2(c). Although this dataset

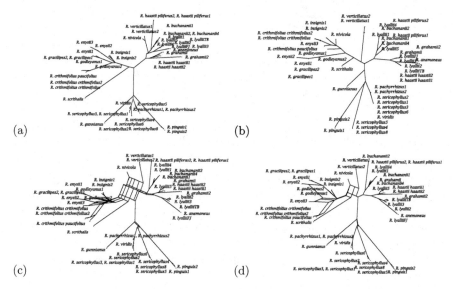

Fig. 2. Two phylogenetic trees for 46 buttercup species, obtained (a) using a nuclear ITS gene and (b) using a chloroplast JSA region [20]. (c) A split network displaying all splits contained in the two trees: (d) The split network for those splits whose distortion is at most 1 on each of the two trees.

is known to contain examples of both allopolyploid and diploid hybridization events (Pete Lockhart, personal communication), previous attempts to compute a corresponding hybridization network from the two trees have failed [17], due to interfering splits.

For this dataset, it makes sense to apply the distortion filter to obtain the set

$$\Sigma' = \{\sigma \in \Sigma \mid \partial(T, \sigma) \leq 1, \forall T \in P\},$$

as this consists of every split that is contained in one of the trees, and is also contained in the other, or in a tree that differs by one tree rearrangement from the other. Figure 2(d) shows the split network for the reduced data set Σ'.

Application of a hybridization network algorithm [17] produces the network depicted in Figure 3. The network clearly indicates that *R. nivicola* arises as a (allopolyploid) hybrid between *R. insignis*, and *R. verticallatus*. Moreover, the network indicates two further possible hybridization events, one leading to *R. enysii3* (as this involves a single lineage, probably diploid hybridization), and the other leading to *R. pinguis*.

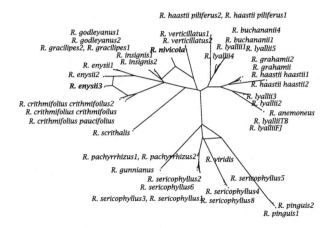

Fig. 3. The hybridization network computed from the filtered set of splits

We now discuss a second example that derives from a study of the phylogeny of microgastrine braconid wasps, a diverse and terrestrially ubiquitous group of small insects that live parasitically as immatures within the bodies of caterpillars. This insect group has been proposed to have diversified rapidly about 50 million years ago into what are now recognized as the modern genera [21,31,3]. At about this time their host insects, and the plants they live upon, were also strongly diversifying [11].

Recent work [3] presents DNA sequence data from seven genes for a set of 45 species of wasps representing a number of microgastrine genera and related subfamilies of wasps. In most cases not all species were successfully sequenced; as many as six (and as few as zero) of the species were missing from a gene tree. Mixed-model Bayesian analysis [24] of the combined seven-gene data set resolved most phylogenetic relationships at the species level (external edges) and among wasp subfamilies (deeply internal edges connecting the ingroup to outgroups), but showed short and relatively poorly-supported internal edges subtending many of the combinations of wasp genera. The internal relationships among wasp genera approximate a "star phylogeny".

It was thus of interest to investigate via filtered network methods whether this star phylogeny pattern is due to conflict between splits supported by different sets of data, or whether it represents a real lack of a coherent signal in data patterns (splits).

We consider seven unrooted, multifurcating gene trees as independently analyzed using Bayesian analysis (GTR + I + Γ substitution model for the two mtDNA genes 16S and COI, HKY + I + Γ for the nuclear genes EF1α, LW rhodopsin, *wingless*, 28S and argK). The five nuclear genes are widely believed to provide stronger phylogenetic signal for deeper relationships than the two mtDNA genes, which are more widely employed for inference of close species relationships.

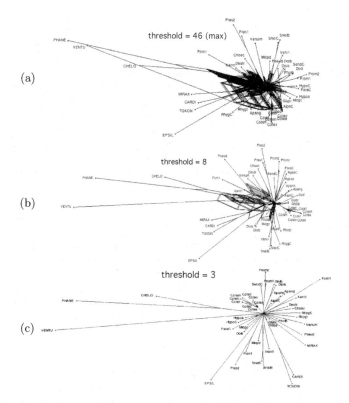

Fig. 4. Effect on Z-closure network of reducing the distortion threshold. (a) With all splits included (distortion threshold 46), (b) with distortion threshold 8, and (c) with distortion threshold 3.

This prediction seems to be borne out by the more distantly related outgroups (shown as all capitals in the taxon labels) more often being erroneously connected closely to ingroups in the mtDNA trees, along with stronger diversification among species within genera (longer edges near the periphery of the tree) than is shown with the more slowly-evolving nuclear genes.

Figure 4 shows the effect of first applying the Z-closure method for combining partial trees into a super-network (a), and then using the distortion filter with different thresholds. When all splits (threshold = 46, encompassing 213 splits) are allowed to contribute to the super-network, the result is a tangled mess, Figure 4(a). Reducing the threshold to 8, Figure 4(b), results in a clear simplification of the network (124 splits), with most of the remaining multidimensionality deriving from uncertainty in relationships between outgroups and the near star-phylogeny of ingroup generic relationships. Reducing the threshold to 3, Figure 4(c), results in only a single remaining reticulation (from 66 splits); at lower values even this uncertainty disappears.

References

1. H.-J. Bandelt and A.W.M. Dress. A canonical decomposition theory for metrics on a finite set. *Advances in Mathematics*, 92:47–105, 1992.
2. H.-J. Bandelt, P. Forster, B.C. Sykes, and M.B. Richards. Mitochondrial portraits of human population using median networks. *Genetics*, 141:743–753, 1995.
3. J.C. Banks and J.B. Whitfield. Dissecting the ancient rapid radiation of microgastrine wasp genera using additional nuclear genes. *Molecular Phylogenetics and Evolution*, in press, 2006.
4. M. Bonet, M.A. Steel, T. Warnow, and S. Yooseph. Better methods for solving parsimony and compatibility, 1998. Proc. RECOMB'98.
5. D. Bryant. Hunting for trees, building trees and comparing trees: theory and method in phylogenetic analysis. Ph.D. thesis, Dept. Mathematics, University of Canterbury, 1997.
6. D. Bryant. The splits in the neighbourhood of a tree. *Annals of Combinatorics*, 8(1):1–11, 1997.
7. S. Eddhu D. Gusfield and C. Langley. The fine structure of galls in phylogenetic networks. to appear in: INFORMS J. of Computing Special Issue on Computational Biology, 2004.
8. A.W.M. Dress and D.H. Huson. Constructing splits graphs. *IEEE/ACM Transactions in Computational Biology and Bioinformatics*, 1(3):109–115, 2004.
9. J. Felsenstein. *Inferring Phylogenies*. Sinauer Associates, Inc., 2004.
10. S. Gadagkar, M.S. Rosenberg, and S. Kumar. Inferring species phylogenies from multiple genes: concatenated sequence tree versus consensus gene tree. *J. of Experimental Zoology (Mol. Dev. Evol.)*, 304B:64–74, 2005.
11. D. Grimaldi. The co-radiations of pollinating insects and angiosperms in the Cretaceous. *Ann. Missouri Bot. Garden*, 86:373–406, 1999.
12. F. Harary. *Graph Theory*. Series in Mathematics. Addison-Wesley, Reading MA, 1969.
13. J. Hein. Reconstructing evolution of sequences subject to recombination using parsimony. *Math. Biosci.*, pages 185–200, 1990.
14. B. Holland and V. Moulton. Consensus networks: A method for visualizing incompatibilities in collections of trees. In G. Benson and R. Page, editors, *Proc. 3rd Workshop on Algorithms in Bioinformatics WABI'03*, volume 2812 of *LNBI*, pages 165–176. Springer, 2003.
15. D.H. Huson and D. Bryant. Application of phylogenetic networks in evolutionary studies. *Molecular Biology and Evolution*, 23:254–267, 2006. Software available from www.splitstree.org.
16. D.H. Huson, T. Dezulian, T. Kloepper, and M.A. Steel. Phylogenetic supernetworks from partial trees. *IEEE/ACM Transactions in Computational Biology and Bioinformatics*, 1(4):151–158, 2004.
17. D.H. Huson, T. Kloepper, P.J. Lockhart, and M.A. Steel. Reconstruction of reticulate networks from gene trees. In *Proc. 9th Int'l Conf. on Research in Computational Molecular Biology RECOMB'05*, 2005.
18. D.H. Huson and T.H. Kloepper. Computing recombination networks from binary sequences. *Bioinformatics*, 21(suppl. 2):ii159–ii165, 2005.
19. C.R. Linder and L.H. Rieseberg. Reconstructing patterns of reticulate evolution in plants. *Am. J. Bot.*, 91(10):1700–1708, 2004.
20. P.J. Lockhart, P.A. McLenachan, D. Havell, D. Glenny, D.H. Huson, and U. Jensen. Phylogeny, dispersal and radiation of New Zealand alpine buttercups: molecular evidence under split decomposition. *Ann. Missouri Bot. Garden*, 88:458–477, 2001.

21. P. Mardulyn and J.B. Whitfield. Phylogenetic signal in the COI, 16S and 28S genes for inferring relationships among genera of Microgastrinae (Hymenoptera: Braconidae); evidence of a high diversification rate in this group of parasitoids. *Molecular Phylogenetics and Evolution*, 12:282–294, 1999.

22. L. Nakhleh, T. Warnow, and C.R. Linder. Reconstructing reticulate evolution in species - theory and practice. In *Proc. 8th Int'l Conf. on Research in Computational Molecular Biology RECOMB'04*, pages 337–346, 2004.

23. A. Rokas, B.L. Williams, N. King, and S.B. Carroll. Genome-scale approaches to resolving incongruence in molecular phylogenies. *Nature*, 425:798–804, 2003.

24. F. Ronquist and J.P. Huelsenbeck. Mrbayes3: Bayesian phylogenetic inference under mixed models. *Bioinformatics*, 19:1572–1574, 2003.

25. N.A. Rosenberg. The probability of topological concordance of gene trees and species trees. *Theor. Pop. Biol.*, 61:225–247, 2002.

26. M.J. Sanderson. and A.C. Driskell. The challenge of constructing large phylogenetic trees. *Trends in Plant Sciences*, 8:374–379, 2003.

27. D. Sankoff. Minimal mutation trees of sequences. *SIAM J. of Applied Mathematics*, pages 35–42, 1975.

28. D. Sankoff and P. Rousseau. Locating the vertices of a Steiner tree in an arbitrary metric space. *Mathematical Programming*, 9:240–246, 1975.

29. C. Semple and M.A. Steel. *Phylogenetics*. Oxford University Press, 2003.

30. D.L. Swofford. When are phylogeny estimates from molecular and morphological data incongruent? In M.M. Miyamoto and J. Cracraft, editors, *Phylogenetic Analysis of DNA Sequences*, pages 295–333. Oxford University Press, Oxford UK, 1991.

31. J.B. Whitfield. Estimating the age of the polydnavirus/braconid wasp symbiosis. *Proc. of the National Academy of Sciences USA*, 99:7508–7513, 2002.

32. C.R. Woese. Bacterial evolution. *Microbiol. Rev.*, 51:221–272, 1987.

Imputing Supertrees and Supernetworks from Quartets

Barbara Hollan[1], Glenn Conner[1], Katharina T. Huber[2], and Vincent Moulton[2]

[1] Allan Wilson Centre for Molecular Ecology and Evolution
Massey University, New Zealand
[2] School of Computing Sciences, University of East Anglia,
Norwich, NR4 7TJ, UK

A contemporary and sometimes contentious problem in genome phylogeny is to reconcile the fact that an accurately reconstructed gene tree does not necessarily correspond to a species phylogeny. Thus, in practice, species phylogenies are commonly obtained by applying consensus tree/supertree methods to collections of gene trees. However, such methods can suppress true conflicts in gene trees arising from processes such as gene transfer and gene duplication/loss.

To help deal with this dilemma, Holland et al. 2004 proposed constructing consensus networks (Holland and Moulton 2003) instead of consensus trees. This requires that all genes are sequenced for all of the taxa in question, a shortcoming that was circumvented in (Huson et al. 2004) where the alternative Z-closure method for generating supernetworks as opposed to supertrees was introduced.

Here, we present a new method to generate supernetworks called Q-imputation [*Syst. Bio.*, to appear]. It works by sequentially inserting all missing taxa into a set of partial gene trees, after which a consensus network is constructed. To insert a missing taxon, a score function is used that rewards inserting the taxon into the partial gene tree in such a way that the resulting tree has as many quartet subtrees as possible in common with the other original gene trees.

Theoretical results, simulations, and studies of real data sets indicate that Q-imputation and Z-closure supernetworks have complementary strengths and weaknessess. We therefore expect that Q-imputation will provide a useful additional tool for computing supernetworks.

This work has been accepted for publication in *Systematic Biology*.

References

1. Holland B., K. T. Huber, V. Moulton, and P. Lockhart. 2004. Using consensus networks to visualize contradictory evidence for species phylogeny. *Mol. Bio. Evol.* 21:1459–1461.
2. Holland B. and V. Moulton. 2003. Consensus Networks: A Method for Visualising Incompatibilities in Collections of Trees. *Proc. 3rd Workshop on Algorithms in Bioinformatics WABI'03*, LNCS volume 2812, pp. 165–176. Springer-Verlag.
3. Huson, D., T. Dezulian, T. Klopper, and M. Steel. 2004. Phylogenetic supernetworks from partial trees. *IEEE/ACM Trans. Comp. Bio. Bioinf.* 1:151–158.

P. Bücher and B.M.E. Moret (Eds.): WABI 2006, LNBI 4175, p. 162, 2006.
© Springer-Verlag Berlin Heidelberg 2006

A Unifying View of Genome Rearrangements

Anne Bergeron[1], Julia Mixtacki[2], and Jens Stoye[3]

[1] Comparative Genomics Laboratory, Université du Québec à Montréal, Canada
bergeron.anne@uqam.ca
[2] International NRW Graduate School in Bioinformatics and Genome Research,
Universität Bielefeld, Germany
julia.mixtacki@uni-bielefeld.de
[3] Technische Fakultät, Universität Bielefeld, Germany
stoye@techfak.uni-bielefeld.de

Abstract. Genome rearrangements have been modeled by a variety of operations such as inversions, translocations, fissions, fusions, transpositions and block interchanges. The *double cut and join* operation, introduced by Yancopoulos *et al.*, allows to model all the classical operations while simplifying the algorithms. In this paper we show a simple way to apply this operation to the most general type of genomes with a mixed collection of linear and circular chromosomes. We also describe a graph structure that allows simplifying the theory and distance computation considerably, as neither capping nor concatenation of the linear chromosomes are necessary.

1   Introduction

The problem of sorting multichromosomal genomes can be stated as: Given two genomes A and B, the goal is to find a shortest sequence of rearrangement operations that transforms A into B. The length of such a shortest sequence is called the *distance* between A and B. Clearly, the solutions depend on what kind of rearrangement operations are allowed.

Given their prevalence in eukaryotic genomes [1], the usual choices of operations include translocations, fusions, fissions and inversions. However, there are some indications that transpositions should also be included in the set of operations [2], but the lack of theoretical results showing how to include transpositions in the models led to algorithms that simulate transpositions as sequences of inversions.

In [3], the authors describe a general framework in which circular and linear chromosomes can coexist throughout evolving genomes. They model inversions, translocations, fissions, fusions, transpositions and block interchanges with a single operation, called the *double cut and join* operation. This general model accounts for the genomic evidence of the coexistence of both linear and circular chromosomes or plasmids in many genomes [4,5].

In this paper, we present a simplified formalization of genomes with coexisting circular and linear chromosomes, and a formal treatment of sorting such genomes by the double cut and join operation. We introduce a very simple data structure,

P. Bücher and B.M.E. Moret (Eds.): WABI 2006, LNBI 4175, pp. 163–173, 2006.
© Springer-Verlag Berlin Heidelberg 2006

the *adjacency graph*, that is symmetric with respect to the two genomes under study and is closely related to the visual picture of the genomes themselves. We also show how the algebraic simplicity of the double cut and join operation yields efficient sorting algorithms that can be tailored to optimize the use of certain types of operations.

2   Notes on Graphs with Vertices of Degree One or Two

An essential ingredient in genome rearrangment studies are graphs where each vertex has degree one or two. Here we recall some of their properties.

Let G be a graph where each vertex has degree one or two. We call a vertex of degree one *external* and a vertex of degree two *internal*. An internal vertex connecting edges p and q is denoted by the unordered multiset $\{p, q\}$ and an external vertex incident to an edge p by the singleton set $\{p\}$.

It follows immediately from the definition of G that any connected component of G is either circular, consisting only of internal vertices, or it is linear, consisting of internal vertices bounded by two external vertices, one at each end. We denote circular components as *cycles* and linear components as *paths*. A cycle or path is *even* if it has an even number of edges, otherwise it is *odd*.

Example 1. The following graph has four vertices of degree one and six vertices of degree two. It has two cycles and two paths, one of which is even and one of which is odd.

Definition 1. *The double cut and join (DCJ) operation acts on two vertices u and v of a graph with vertices of degree one or two in one of the following three ways:*

(a) *If both $u = \{p, q\}$ and $v = \{r, s\}$ are internal vertices, these are replaced by the two vertices $\{p, r\}$ and $\{s, q\}$ or by the two vertices $\{p, s\}$ and $\{q, r\}$.*
(b) *If $u = \{p, q\}$ is internal and $v = \{r\}$ is external, these are replaced by $\{p, r\}$ and $\{q\}$ or by $\{q, r\}$ and $\{p\}$.*
(c) *If both $u = \{q\}$ and $v = \{r\}$ are external, these are replaced by $\{q, r\}$.*

In addition, as an inverse of case (c), a single internal vertex $\{q, r\}$ can be replaced by two external vertices $\{q\}$ and $\{r\}$.

Figure 1 illustrates the definition.

The DCJ operation, although defined locally on a pair of vertices, has global effects on the connected components of the graph. In order to describe these, we use a terminology essentially borrowed from biology.

First, consider Figure 2. If the two vertices are contained in two different paths and at least one of them is internal, then these paths exchange their ends, which

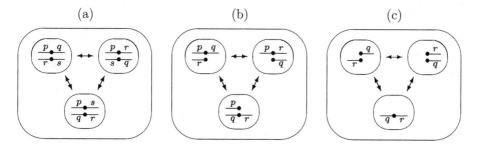

Fig. 1. Definition of the double cut and join operation. Note that the operations between the two top graphs of part (c) are the identity.

is called a *path translocation*. If both are external vertices of different paths, as in Figure 2 (c), then these paths are merged, called a *path fusion*. The inverse of a path fusion is a *path fission*.

The case shown in Figure 3, where both linear and circular components are mixed, is more intricate. If the DCJ operation acts on vertices contained in the same path and at least one of them is internal, then the intermediate part of the path is either reversed, called an *inversion*, or spliced out producing a new cycle, called an *excision*. The inverse operation of an excision is called an *integration*. If both are external vertices of the same path, as in Figure 3 (c), then a cycle is formed, called a *circularization*. Its opposite is a *linearization*.

If the vertices are contained in the same cycle, or in two different cycles, as shown in Figure 4, then we have either an *inversion*, a *cycle fusion* or a *cycle fission*.

The following lemma is an immediate consequence of the enumeration of all possible cases in Figures 2, 3 and 4:

Lemma 1. *The application of a single DCJ operation changes the number of circular or linear components by at most one.*

We will see in the next two sections how graphs with vertices of degree one or two appear in two natural ways when modeling genomes and genome rearrangements.

3 Genes, Chromosomes and Genomes

In this section we introduce our notation of genomes and how they are modeled as graphs with vertices of degree one or two.

A *gene* is an oriented sequence of DNA that starts with a *tail* and ends with a *head*. These are called the *extremities* of the gene. The tail of a gene a is denoted by a_t, and its head is denoted by a_h. In biology, the tail of a gene is often called its 3' end and the head its 5' end.

Two consecutive genes do not necessarily have the same orientation, since DNA is double stranded and the complementary strands are read by the transcription machinery in opposite direction. Thus an *adjacency* of two consecutive genes a and b, depending on their respective orientation, can be of four different types:

$$\{a_h, b_t\}, \{a_h, b_h\}, \{a_t, b_t\}, \{a_t, b_h\}.$$

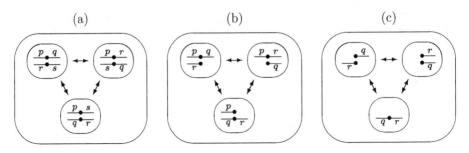

Fig. 2. The DCJ operation applied on one or two paths yields path translocations, fusions and fissions

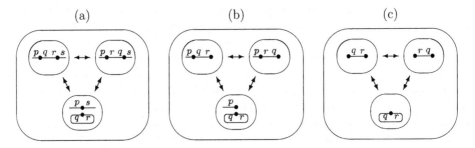

Fig. 3. The DCJ operation applied on a single path or a path and a cycle yields inversions, excisions, integrations, circularizations and linearizations

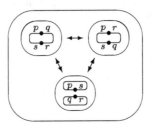

Fig. 4. The DCJ operation applied on a single cycle or on two cycles yields inversions, cycle fusions and fissions

An extremity that is not adjacent to any other gene is called a *telomere*, represented by a singleton set $\{a_h\}$ or $\{a_t\}$.

A *genome* is a set of adjacencies and telomeres such that the tail or the head of any gene appears in exactly one adjacency or telomere.

Given a genome, one reconstructs its *chromosomes* by representing the telomeres and adjacencies as vertices and then joining for each gene its tail and its head by an edge. Note that the *genome graph* obtained this way is a graph with vertices of degree one or two. The connected paths and cycles are chromosomes of the genome which are either linear or circular. Linear chromosomes are bounded by telomeres.

Chromosomes are often represented by lists of gene labels. These lists are obtained by choosing a telomere in a linear chromosome, or an arbitrary gene in a circular chromosome, and then enumerating the gene labels along the component, using positive signs to indicate genes that are read from tail to head and negative signs to indicate genes that are read from head to tail. For linear chromosomes, the enumeration stops at its other telomere, for circular chromosomes when the initial gene appears for the second time in the list. Positive signs may be omitted where convenient.

Example 2. Let

$$A = \{\{a_t\}, \{a_h, c_t\}, \{c_h, d_h\}, \{d_t\}, \{b_h, e_t\}, \{e_h, b_t\}, \{f_t\}, \{f_h, g_t\}, \{g_h\}\}$$

be a genome with seven genes $\{a, b, c, d, e, f, g\}$. The corresponding genome graph is the following:

One possible list representation of A is $\{(a, c, -d), (b, e, b), (f, g)\}$.

Since the chromosome graph is a graph with vertices of degree one or two, the double cut and join operation defined in Section 2 can be applied to these graphs. This operation is the same as defined, in different notation, by Yancopoulos *et al.* [3].

We can now formulate the problem that we consider:

The DCJ Sorting and Distance Problem. Given two genomes A and B defined on the same set of genes, find a shortest sequence of DCJ operations that transforms A into B. The length of such a sequence is called the *DCJ distance* between A and B, denoted by $d_{DCJ}(A, B)$.

Example 3. Consider the following two genomes that are defined over the set of genes $\{a, b, c, d, e, f, g\}$:

$$A = \{\{a_t\}, \{a_h, c_t\}, \{c_h, d_h\}, \{d_t\}, \{b_h, e_t\}, \{e_h, b_t\}, \{f_t\}, \{f_h, g_t\}, \{g_h\}\}$$
$$B = \{\{a_h, b_t\}, \{b_h, a_t\}, \{c_t\}, \{c_h, d_t\}, \{d_h\}, \{e_t\}, \{e_h\}, \{f_h, g_t\}, \{g_h, f_t\}\}$$

Sorting A into B can, for example, be done in the following five steps, where the affected gene extremities are underlined:

$$A = \{\{a_t\}, \{a_h, \underline{c_t}\}, \{c_h, d_h\}, \{d_t\}, \{b_h, e_t\}, \{e_h, \underline{b_t}\}, \{f_t\}, \{f_h, g_t\}, \{g_h\}\}$$
$$\{\{\underline{a_t}\}, \{a_h, b_t\}, \{c_h, d_h\}, \{d_t\}, \{b_h, \underline{e_t}\}, \{e_h, c_t\}, \{f_t\}, \{f_h, g_t\}, \{g_h\}\}$$
$$\{\{e_t\}, \{a_h, b_t\}, \{c_h, \underline{d_h}\}, \{\underline{d_t}\}, \{b_h, a_t\}, \{e_h, c_t\}, \{f_t\}, \{f_h, g_t\}, \{g_h\}\}$$
$$\{\{e_t\}, \{a_h, b_t\}, \{c_h, d_t\}, \{d_h\}, \{b_h, a_t\}, \{\underline{e_h}, \underline{c_t}\}, \{f_t\}, \{f_h, g_t\}, \{g_h\}\}$$
$$\{\{e_t\}, \{a_h, b_t\}, \{c_h, d_t\}, \{d_h\}, \{b_h, a_t\}, \{e_h\}, \{c_t\}, \{\underline{f_t}\}, \{f_h, g_t\}, \{\underline{g_h}\}\}$$
$$B = \{\{a_h, b_t\}, \{b_h, a_t\}, \{c_t\}, \{c_h, d_t\}, \{d_h\}, \{e_t\}, \{e_h\}, \{f_h, g_t\}, \{g_h, f_t\}\}$$

The DCJ distance between A and B is $d_{DCJ}(A, B) = 5$.

4   The Adjacency Graph

In order to solve the DCJ Distance Problem stated above, another graph of the type discussed in Section 2 proves to be useful, this time defined on the pair of genomes A and B.

Definition 2. *The* adjacency graph $AG(A, B)$ *is a graph whose set of vertices are the adjacencies and telomeres of A and B. For each $u \in A$ and $v \in B$ there are $|u \cap v|$ edges between u and v.*

Example 4. The adjacency graph of our two genomes

$$A = \{\{a_t\}, \{a_h, c_t\}, \{c_h, d_h\}, \{d_t\}, \{b_h, e_t\}, \{e_h, b_t\}, \{f_t\}, \{f_h, g_t\}, \{g_h\}\}$$
$$B = \{\{a_h, b_t\}, \{b_h, a_t\}, \{c_t\}, \{c_h, d_t\}, \{d_h\}, \{e_t\}, \{e_h\}, \{f_h, g_t\}, \{g_h, f_t\}\}$$

is the following:

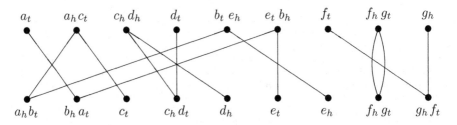

Obviously, every vertex in the adjacency graph has degree one or two, therefore it is a union of cycles and paths. Since the graph is bipartite, all cycles have even length.

The adjacency graph can easily be constructed as shown in Algorithm 1. Let N be the number of genes in genomes A and B, respectively. Then Algorithm 1 takes $O(N)$ time and uses $O(N)$ space if the genomes are stored in a data structure where, for each gene extremity, one has constant time access to the adjacency or telomere that it is contained in. For example, this can be a table with two rows of length at most $2N$ storing the adjacencies and telomeres of the genome, and another table with two rows of length N storing for each gene in which columns of the first table to find its head and its tail. For genome

---

Algorithm 1 (Construction of the adjacency graph)

1: create a vertex for each adjacency and each telomere in genomes A and B
2: **for each** adjacency $\{p, q\}$ in genome A **do**
3:     create an edge connecting $\{p, q\}$ and the vertex of genome B that contains p
4:     create an edge connecting $\{p, q\}$ and the vertex of genome B that contains q
5: **end for**
6: **for each** telomere $\{p\}$ of genome A **do**
7:     create an edge connecting $\{p\}$ and the vertex of genome B that contains p
8: **end for**

Table 1. Table storing the adjacencies and telomeres of genome A. Adjacencies have two entries, telomeres just one.

	1	2	3	4	5	6	7	8	9
first	a_t	a_h	c_h	d_t	b_h	e_h	f_t	f_h	g_h
second	–	c_t	d_h	–	e_t	b_t	–	g_t	–

Table 2. Table storing for each gene in A the location of its head and its tail in Table 1

	a	b	c	d	e	f	g
head	2	5	3	3	6	8	9
tail	1	6	2	4	5	7	8

$$A = \{\{a_t\}, \{a_h, c_t\}, \{c_h, d_h\}, \{d_t\}, \{b_h, e_t\}, \{e_h, b_t\}, \{f_t\}, \{f_h, g_t\}, \{g_h\}\}$$

from the previous example, the two tables are shown in the following.

5   Sorting by DCJ Operations

As we will see in this section, the adjacency graph allows a simple characterization of many of the properties of sorting by DCJ operations.

Lemma 2. *Let A and B be two genomes defined on the same set of N genes, then we have*

$$A = B \quad \textit{if and only if} \quad N = C + I/2$$

where C is the number of cycles and I the number of odd paths in $AG(A, B)$.

Proof. Let a be the number of adjacencies and t the number of telomeres in $A = B$, then $N = a + t/2$. The adjacency graph $AG(A, B)$ has $C = a$ cycles and $I = t$ odd paths, hence $N = a + t/2 = C + I/2$.

To show that $N = C + I/2$ implies $A = B$, assume an adjacency graph $G = AG(A, B)$ such that $N = C + I/2$. Let a be the number of adjacencies and t the number of telomeres in A, then $N = a + t/2$. Each cycle in G contains at least one adjacency of A, thus $C \leq a$. Each odd path in G contains exactly one telomere of A, thus $I \leq t$. From $C + I/2 = N = a + t/2$ it follows that $C = a$ and $I = t$. Thus all cycles have length two and all odd paths have length one, which is only possible if the genomes are equal.   □

When a DCJ operation is applied to genome A, it acts on the adjacencies and telomeres of genome A. The same DCJ operation acts *also* on the adjacency graph since the adjacencies and telomeres of genome A are vertices of this graph. Since the adjacency graph is a union of paths and cycles, all the tools and terminology of Section 2 can be used.

In Lemma 1, we showed that the number of circular and linear components can change by at most one when a DCJ operation is applied to a graph that is a union of paths and cycles. In the case of adjacency graphs we have also constraints on the possible changes in the number of odd paths:

Lemma 3. *The application of a single DCJ operation changes the number of odd paths in the adjacency graph by –2, 0, or 2.*

Algorithm 2 (Greedy sorting by DCJ)

1: **for each** adjacency $\{p, q\}$ in genome B **do**
2:     let u be the element of genome A that contains p
3:     let v be the element of genome A that contains q
4:     **if** $u \neq v$ **then**
5:         replace u and v in A by $\{p, q\}$ and $(u \setminus \{p\}) \cup (v \setminus \{q\})$
6:     **end if**
7: **end for**
8: **for each** telomere $\{p\}$ in genome B **do**
9:     let u be the element of genome A that contains p
10:    **if** u is an adjacency **then**
11:        replace u in A by $\{p\}$ and $(u \setminus \{p\})$
12:    **end if**
13: **end for**

Proof. Consider operations that are path translocations, fusions or fissions (Figure 2). Two odd paths can be either transformed into two odd paths, or into one or two paths of even length. Path(s) of even length(s) can be either transformed into path(s) of even length, or into two paths of odd length. One even and one odd path are always transformed into one even and one odd path. Finally, splitting one odd path always yields an even and an odd path.

Inversions, excisions, integrations, circularizations and linearizations (Figure 3) do not change the number of odd paths since all cycles have even length. No paths are involved in the DCJ operations of Figure 4.   □

Lemma 3 allows to derive the following lower bound for the DCJ distance:

Lemma 4. *Let A and B be two genomes defined on the same set of N genes, then we have*

$$d_{DCJ}(A, B) \geq N - (C + I/2)$$

where C is the number of cycles and I the number of odd paths in $AG(A, B)$.

Proof. Since none of the cases of the DCJ operation modifies the number of cycles and odd paths simultaneously, this follows immediately from Lemmas 1, 2 and 3.   □

The adjacency graph is also very useful when one wants to find an optimal sequence of sorting operations.

Observe that any pair of edges in the adjacency graph that connect two different vertices of genome A with an adjacency $\{p, q\}$ in genome B can be transformed by a single DCJ operation into a cycle of length two, plus the remaining structure, reduced by the two edges. This operation always increases $C + I/2$ by one since C is increased by one and we have already seen that no DCJ operation can simultaneously change C and I.

Now assume that all adjacencies of genome B are contained in cycles of length two. There might still be pairs of telomeres of B that form an adjacency in A. These adjacencies can be split into two telomeres, thus creating two odd paths of length one each, increasing I by two.

Pseudocode for this greedy sorting procedure is given in Algorithm 2. Note that the adjacency graph does not need to be constructed explicitly if the genomes are stored in the way sketched at the end of Section 4. Interestingly, the algorithm is optimal:

Theorem 1. *Let A and B be two genomes defined on the same set of N genes, then we have*

$$d_{DCJ}(A, B) = N - (C + I/2)$$

where C is the number of cycles and I the number of odd paths in $AG(A, B)$. An optimal sorting sequence can be found in $O(N)$ time by Algorithm 2.

Proof. Lemma 4 together with the fact that Algorithm 2 increments in each iteration either C by one or I by two prove the distance formula.

The linear time complexity follows from the fact that our genome representation allows to find and perform each sorting operation in constant time and the DCJ distance is never larger than N.   □

Remark 1. It is worth mentioning that our distance formula is equivalent to the result $d_{DCJ} = b - c$ given by Yancopulos *et al.* [3], where b is the number of breakpoints and c is the number of cycles of the breakpoint graph after appropriate *capping* of the linear chromosomes.

To see this, let l_A and l_B be the number of linear chromosomes in genomes A and B, respectively. Then the total number of breakpoints, as defined in [3], is $b = N + l_B + aa = N + l_A + bb$ where aa is the number of even paths that start and end in genome A and bb is the number of even paths that start and end in genome B. The number of cycles is $c = C + I + E$ where C is the number of cycles, I the number of odd paths and E the number of even paths in the adjacency graph $AG(A, B)$ as defined in this paper. Obviously $E = aa + bb$. Moreover, each linear chromosome is associated to two path ends, thus the number of linear chromosomes equals the number of paths, $l_A + l_B = I + E$. Together this implies that $2b = 2N + 2E + I$, giving $b - c = N - C - I/2$.

6   Conclusion

We have shown that, with a suitable representation, it is possible to model all rearrangement operations on the most general genome structure that mixes both circular and linear chromosomes.

The basic tools for this representation are graphs that are unions of paths and cycles. Surprisingly, this type of graph can be used for representing genomes, for computing the DCJ distance, and for suggesting rearrangement scenarios. This variety of uses suggests many interesting problems.

The first one is to investigate formal properties of graphs that are unions of paths and cycles, with respect to the DCJ operation. For example, the cyclic organization of these operations is a striking feature of Figures 2, 3 and 4 and offers new ways to classify rearrangement operations. These graphs also give a firm starting point to explore difficult rearrangement problems that involve either gene duplications [6] or missing information about the actual order of genes in a genome [7].

Last, but not the least, adding constraints on the type of allowed operations often yields equations of the form

$$d(A, B) = d_{DCJ}(A, B) + t$$

where t represents the additional cost of not resorting to DCJ operations. For example, the Hannenhalli-Pevzner distance, that allows only translocations and inversions on linear chromosomes [8], can be recast as avoiding all DCJ operations that create a circular chromosome in either genome A or B. These operations live only on Figure 2 and the upper half of Figure 3.

Another kind of restriction has recently been studied in [9], where operations are fusions and fissions between circular unsigned chromosomes, and block interchanges within a circular unsigned chromosome. The authors assign equal weight to the three operations, even if a block interchange requires two DCJ operations, and propose an $O(N^2)$ time algorithm to sort these circular genomes. Their algorithm first applies fusions to both source and target genome, until they have two genomes whose chromosomes have equal gene content. These fusions can be identified in linear time by a search of the adjacency graph. They then sort the resulting genomes by block interchanges using an $O(N^2)$ time algorithm described in [10]. This can be done in the same time complexity, but with elementary means, using a modification of our Algorithm 2 where every intermediate chromosome created by a fission is immediately re-absorbed in the next step, such that only block interchanges are performed. The modification is to search, in the newly created circular chromosomes, a pair of genes that are adjacent in the target genome, but on different chromosomes in the source genome.

References

1. Sankoff, D., Mazowita, M.: Stability of rearrangement measures in the comparison of genome sequences. In Miyano, S., Mesirov, J., Kasif, S., Istrail, S., Pevzner, P., Waterman, M., eds.: Proceedings of RECOMB 2005. Volume 3500 of LNBI., Springer Verlag (2005) 603–614
2. Bergeron, A., Stoye, J.: On the similarity of sets of permutations and its applications to genome comparison. In Warnow, T., Zhu, B., eds.: Proceedings of COCOON 2003. Volume 2697 of LNCS., Springer Verlag (2003) 68–79

3. Yancopoulos, S., Attie, O., Friedberg, R.: Efficient sorting of genomic permutations by translocation, inversion and block interchange. Bioinformatics **21** (2005) 3340–3346

4. Casjens, S., Palmer, N., van Vugt, R., Huang, W.M., Stevenson, B., Rosa, P., Lathigra, R., Sutton, G., Peterson, J., Dodson, R.J., Haft, D., Hickey, E., Gwinn, M., White, O., Fraser, C.M.: A bacterial genome in flux: The twelve linear and nine circular extrachromosomal DNAs in an infectious isolate of the Lyme disease spirochete *Borrelia burgdorferi*. Mol. Microbiol. **35** (2000) 490–516

5. Volff, J.N., Altenbuchner, J.: A new beginning with new ends: Linearisation of circular chromosomes during bacterial evolution. FEMS Microbiol. Lett. **186** (2000) 143–150

6. Zheng, C., Lenert, A., Sankoff, D.: Reversal distance for partially ordered genomes. Bioinformatics **21** (2005) i502–i508 (Proceedings of ISMB 2005).

7. El-Mabrouk, N.: Genome rearrangements with gene families. In Gascuel, O., ed.: Mathematics of Evolution and Phylogeny. Oxford University Press (2005) 291–320

8. Hannenhalli, S., Pevzner, P.A.: Transforming men into mice (polynomial algorithm for genomic distance problem). In: Proceedings of FOCS 1995, IEEE Press (1995) 581–592

9. Lu, L., Huang, Y.L., Wang, T.C., Chiu, H.T.: Analysis of circular genome rearrangement by fusions, fissions and block-interchanges. BMC Bioinformatics **7** (2006)

10. Lin, Y.C., Lu, C.L., Chang, H.Y., Tang, C.Y.: An efficient algorithm for sorting by block-interchanges and its application to the evolution of vibrio species. J. Comp. Biol. **12** (2005) 102–112

Efficient Sampling of Transpositions and Inverted Transpositions for Bayesian MCMC

István Miklós[1,3], Timothy Brooks Paige[2], and Péter Ligeti[3,4]

[1] eScience Regional Knowledge Center, Eötvös Loránd University
1117 Budapest, Pázmány Péter sétány 1/c, Hungary
miklosi@ramet.elte.hu
[2] Box 786, Amherst College, Amherst MA, 01002 USA
tbpaige@gmail.com
[3] Bioinformatics group, Alfréd Rényi Institute of Mathematics, Hungarian Academy of Sciences
1053 Budapest, Reáltanoda u. 13-15, Hungary
{miklosi, ligeti}@renyi.hu
[4] Department of Computer Science, Eötvös Loránd University
1117 Budapest, Pázmány Péter sétány 1/c, Hungary
turul@cs.elte.hu

Abstract. The evolutionary distance between two organisms can be determined by comparing the order of appearance of orthologous genes in their genomes. Above the numerous parsimony approaches that try to obtain the shortest sequence of rearrangement operations sorting one genome into the other, Bayesian Markov chain Monte Carlo methods have been introduced a few years ago. The computational time for convergence in the Markov chain is the product of the number of needed steps in the Markov chain and the computational time needed to perform one MCMC step. Therefore faster methods for making one MCMC step can reduce the mixing time of an MCMC in terms of computer running time.

We introduce two efficient algorithms for characterizing and sampling transpositions and inverted transpositions for Bayesian MCMC. The first algorithm characterizes the transpositions and inverted transpositions by the number of breakpoints the mutations change in the breakpoint graph, the second algorithm characterizes the mutations by the change in the number of cycles. Both algorithms run in $O(n)$ time, where n is the size of the genome. This is a significant improvement compared with the so far available brute force method with $O(n^3)$ running time and memory usage.

1 Introduction

The differences between the order of genes in two genomes have been used as a measurement of evolutionary distance already more than six decades ago [1]. The rediscovery of inversion distance is dated back to the eighties [2,3], and since then a large set of papers on optimization methods for genome rearrangement problems has been published. However, except the case of sorting signed permutations by inversions [4,5,6,7,8,9] or by translocations [10], only approximations

P. Bücher and B.M.E. Moret (Eds.): WABI 2006, LNBI 4175, pp. 174–185, 2006.

[11,12,13,14,15] and heuristics [16] exist. Most of the methods concerning with more types of mutations either penalize all the mutations with the same weight [14], or exclude a whole set of possible mutations due to a special choice of weights [13]. (A nice exception can be found in [17].)

Above the numerous parsimony approaches that try to obtain the shortest sequence of rearrangement operations sorting one genome into the other, Bayesian Markov chain Monte Carlo methods have been introduced a few years ago. They define different models where genomes can evolve by reversals [18,19,20], reversals and translocations [21] or reversals, transpositions and inverted transpositions [22,23]. It has been shown that transpositions and inverted transpositions could happen in unichromosomal genomes [24], therefore it is natural to incorporate such events into the Bayesian model. So far the available computer program for the model accommodating transpositions and inverted transpositions used $O(n^3)$ memory and had $O(n^4)$ running time per MCMC step [23]. Though this memory usage and running time allowed the analysis of short genomes (for example, Metazoan mithochondrial genomes), the program suffered memory problems with large genomes containing hundreds of genes.

We introduce two algorithms for characterizing and sampling transpositions and inverted transpositions. The first algorithm characterizes the mutations by the number of breakpoints they remove and samples from a distribution in which breakpoint-removing mutations are preferred. The second algorithm characterizes the mutations by the change in the number of cycles in the graph of desire and reality and samples from a distribution in which cycle-increasing mutations are preferred. Both algorithms run in $O(n)$ time where n is the length of the genome. Since linear running time algorithms for characterizing and sampling reversals have already been developed earlier [24,23], an MCMC step in the reversals, transpositions and inverted transpositions accommodating model takes only $O(n^2)$ running time (the sampling algorithm might be repeated $O(n)$ times in an MCMC step), and needs only linear memory with these algorithms.

2   Preliminaries

2.1   Mathematical Description of Genome Rearrangement

Genomes are assumed to have the same gene content, and each gene is represented in one copy in both genomes. Gene orders are described as signed permutations, numbers correspond to genes, signs represent the reading direction of genes. Since mutations are actions on the signed permutation group, transforming a genome π_1 to genome π_2 is equivalent with sorting $\pi_2^{-1}\pi_1$ to the identical permutation, and thus, we are going to talk about sorting permutations instead of transforming one into another. By following the convention, a signed permutation of length n is represented as an unsigned permutation of length $2n$, $+i$ is replaced by $2i-1, 2i$, and $-i$ is replaced by $2i, 2i-1$. This unsigned permutation is then framed to 0 and $2i+1$. To properly mimic the signed permutation case, only segments $[2i+1, 2j]$ are allowed to mutate in the unsigned representation.

The graph representation of a signed permutation is called *graph of reality and desire*, whose vertexes are the numbers from 0 to $2n + 1$, and edges are the reality and desire edges. The reality edges connect every second position in the permutation starting with 0. Mutations act on the reality edges; a reversal acts on two reality edges, while a transposition or an inverted transposition on three ones. The desire edges are arcs connecting $2i$ with $2i+1$ for each i. A desire edge is unoriented if it spans even number of points otherwise it is oriented. Since each vertex has a degree of 2, the graph of desire and reality can be unequivocally decomposed into cycles. A reality edge is a breakpoint if its cycle is longer than 2.

The identity permutation has 0 breakpoints and $n + 1$ cycles, all other mutations have more breakpoints and less cycles. Therefore the sorting of a permutation is equivalent with increasing the number of cycles to $n + 1$ or decreasing the number of breakpoints to 0. Mutations can be characterized by the number of breakpoints they remove or the change in the number of cycles. We will talk about e.g., -3-b-transpositions meaning that they remove 3 breakpoints or +1-c-inversions, which increase the number of cycles by 1.

2.2   Stochastic Modeling and Bayesian MCMC

Time-continuous Markov models have been the standard approaches for stochastic modeling of molecular evolution. Unlike the case of nucleic acid substitution models, modeling genome rearrangements is computationally demanding and no analytical solutions are known for transition probabilities. What we can calculate is the likelihood of a trajectory, which is the probability that a given sequence of mutations happened in a time span conditional on a set of parameters describing the model [22,23,24].

To sample trajectories from the posterior distribution, we apply Bayesian Markov chain Monte Carlo (MCMC) [25,26] which is a random walk on the possible trajectories, and whose stationary distribution is the posterior distribution of trajectories. The random walk is constructed in two steps. In the first step, a new trajectory is drawn from a proposal distribution, and in the second step, the discrepancy between the proposal and the target distribution is corrected by accepting the proposal with probability

$$\min\left\{1\,,\,\frac{P(X|Y)\pi(Y)}{P(Y|X)\pi(X)}\right\} \tag{1}$$

where P is the proposal distribution, π is the target one, X is the actual state of the chain, and Y is the proposal, and the chain remains in state X with the complement probability [25,27]. The proposal step replaces a part of the trajectory. The new sub-trajectory is obtained step by step, each mutation is drawn from a distribution that mimics the target distribution we would like to sample from, and the new proposal is independent from the old sub-trajectory.

The mixing of the Markov chain depends on how well the proposal distribution can mimic the target distribution. When proposing a new sub-trajectory step by step, published methods measure the departure of the actual rearrangement from the rearrangement where the sub-trajectory must arrive to, and propose mutations decreasing the measurement of the departure ('good' mutations) with high probability and propose other ones ('bad' mutations) with low probability. This philosophy seems to be essential since random mutations would reach the target rearrangement with a very small probability.

Since there are $3\binom{n+1}{3}$ transpositions and inverted transpositions and $\binom{n+1}{2}$ reversals, an algorithm that spends only constant time with each possible mutation to decide its goodness will already run in $\Omega(n^3)$ time. Therefore it is not a trivial problem how to characterize and sample mutations in less time. Below we show two algorithms characterizing and sampling transpositions and inverted transpositions in linear time, a very simple for breakpoints and a more sophisticated for cycles.

3   Characterizing and Sampling Transpositions and Inverted Transpositions

Figure 1 shows the two decision trees that the below described algorithms use to sample random mutations. At an internal node, a random decision is made only if both subtrees are non-empty. If one of the subtrees is empty, then the algorithm chooses the other subtree with probability 1. For example, in Figure 1 a), if there is no transposition or inverted transposition decreasing the number of breakpoints by 3, and there is no reversal decreasing the number of breakpoints by two, then there is no random decision at the root of the tree, the algorithm will go to the right subtree with probability 1.

3.1   Sampler Based on the Change of Breakpoints

The first algorithm characterizes the mutations with the number of breakpoints the mutation removes. The algorithm calculates in linear time the number of

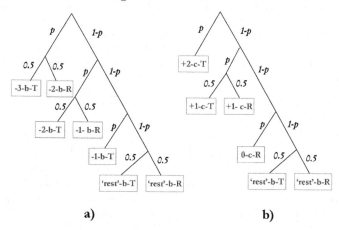

a)   b)

Fig. 1. Decision trees used by the introduced algorithms. T stands for transpositions and inverted transpositions, R stands for reversals. Numbers on the edges means probabilities, p is between 0.5 and 1. In practice, $p = 0.8$ gives a proposal distribution which is reasonably close to the target distribution, acceptance ratio is about $20 - -30\%$.

transpositions and inverted transpositions for each category in Figure 1 **a)**, namely -3-b, -2-b, -1-b and "rest" mutations, and it is able to sample from a uniform distribution for each category also in linear time.

Preprocessing. For each i, we calculate $b(i)$, which is the number of breakpoints after position i; $od(i)$, which is the number of oriented desire edges going from the left end of a reality edge to the right after position i, and $ud(i)$, which is the number of unoriented desire edges going from the left end of a reality edge to the right after position i. These numbers can be trivially calculated by traversing the permutation from right to left.

Counting the mutations. For each category and reality edge, we calculate in $O(1)$ time the number of mutations that fall in the given category and their left-most reality edge is the given edge on which they act. Since there are at most three such mutations for categories -3-b and -2-b —see Figure 2—and these cases can be checked in $O(1)$ time for each reality edge, this is the trivial part of the algorithm.

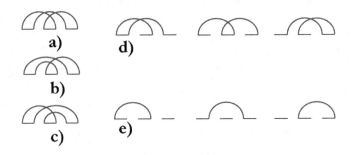

Fig. 2. Configurations on which a mutation can decrease the number of breakpoints. **a)-c):** 3-cycles on which if **a)** a transposition, **b)** an inverted transposition to the left or **c)** an inverted transposition to the right acts, the number of breakpoints decreases by 3. **d)** The three possible cases on which a transposition can decrease the number of breakpoints by two. Similarly for inverted transpositions, there are 3-3 cases derived from the 3-long cycles showed at **b)** and **c)**. **e)** The three possible situation on which a transposition can decrease the number of breakpoints by one. The empty reality edge must be a breakpoint. Similar configurations can be obtained for inverted transpositions.

The maximum number of possible -1-b-transpositions and inverted transpositions is $O(n)$ for each reality edge. These mutations fall into three categories, see Fig 2. **e)**. Having known $b(i)$, $od(i)$ and $ud(i)$ in advance, the number of mutations can be calculated in constant time for each category. For example, for the third case in Fig 2 **e)**, it is $ud(i)$ minus the possible zero, one or two mutations which are actually -2-b- or -3-b-transpositions.

The number of "rest" mutations can be easily calculated if the number of -3-b, -2-b and -1-b-mutations are subtracted from the number of all possible mutations, which is $\binom{n+1-i}{2}$ each for transpositions, inverted transpositions to the right and inverted transpositions to the left.

Sampling from each category. Since we know the number of possible mutations for each category and each leftmost reality edges, we first sample the leftmost reality edge from the properly weighted distribution. For -3-b and -2-b-mutations, the number of mutations having a fixed leftmost reality edge is constant, and the algorithm can choose a random one from this constant size set.

To sample from the $O(n)$ possible -1-b-transpositions or inverted transpositions, the algorithm first chooses one of the three possible sub-cases for the selected leftmost reality edge, and depending on the chosen sub-case, it chooses a breakpoint, an oriented or unoriented desire edge that defines the corresponding mutation.

To sample from the "rest" mutations, the algorithm first chooses a rightmost reality edge after fixing the leftmost reality edge. It calculates the number of "rest" mutations for each possible rightmost reality edge. This is the number of all possible mutations minus the number of -3-b, -2-b and -1-b mutations. The subtracted numbers can be calculated in $O(1)$ time, hence the number of "rest" mutations for each rightmost edge. After this, the algorithm chooses a rightmost edge from the properly weighted distribution, and finally, the algorithm chooses one from the $O(n)$ possible middle reality edges, given the fixed leftmost and rightmost edges.

3.2   Sampler Based on the Change of Cycles

The second algorithm characterizes the mutations by the change in the number of cycles. Though this algorithm does not tell the exact number of mutations falling into a given class, it does tell for each category and for each reality edge whether or not there exists a mutation that falls into the given category and its leftmost edge is the given one. This is enough for using the decision tree in Figure 1 **b)** and for sampling from a distribution for which the sampling probabilities can be calculated. (We would like to mention for non-experts that the ability of sampling from a distribution does not imply that sampling probabilities can be calculated, see for example [28,26,24].)

It is easy to show that cycle-increasing mutations act on one cycle. If three reality edges are in one cycle, they are in one of the eight possible configurations in Table 1. The idea of the algorithm is that for each configuration and reality edge, the algorithm decides whether or not there are other two reality edges to the right being in the given configuration with the third edge. If so, then the reality edge goes to a set from which the algorithm chooses a random leftmost reality edge. Once the algorithm has chosen the mutation type and the leftmost reality edge, it decides for each reality edge on the right hand side of the leftmost edge whether or not it can be together with a rightmost reality edge in a configuration that is good for the given mutation type. After choosing a random middle edge from the ensemble of possible middle edges, the algorithm finally chooses a random good leftmost edge. This method also takes only $O(n)$ time and memory.

Preprocessing. The algorithm works on each cycle independently. Starting with the leftmost edge of the cycle, the algorithm traverses the cycle and stores

Table 1. The possible configurations of three reality edges in a cycle and the category of mutations acting on them. Dotted arcs are not necessarily reality edges but alternating paths of reality end desire edges.

Configuration	transposition	inv. trans. to the left	inv. trans. to the right
	+2-c	+1-c	+1-c
	+1-c	+2-c	"rest"
	+1-c	"rest"	+2-c
	+1-c	"rest"	"rest"
	"rest"	+1-c	+1-c
	"rest"	+1-c	"rest"
	"rest"	"rest"	+1-c
	"rest"	"rest"	"rest"

the visiting order of reality edges, as well as the direction of the reality edges on the cycle-traversing. $\pi(i)$ tells the visit order of the reality edge in the ith position, and $pos(i)$ tells the position of the edge which was the ith in the cycle tour and $sign(i)$ tells the direction of the edge.(We will denote by plus sign the left to right direction and by minus sign the right to left direction.) These arrays can be trivially calculated in $O(n)$ time.

After this, the algorithm traverses the reality edges in reverse position order (namely, from right to left), and calculates $s\max(i) = \max_{j \geq i}\{\pi(j)|sign(j) = s\}$ both for positive and negative signs.

Existence of mutations. Each configuration in Table 1 can be traversed in six possible ways, see for example in Figure 3 how the first configuration in

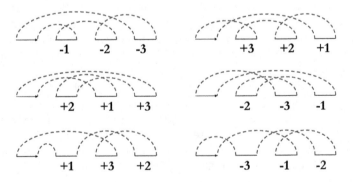

Fig. 3. The possible visiting order of three reality edges on which a transposition increases the number of cycles by two. Dotted arcs are not necessarily reality edges but alternating paths of reality and desire edges.

Table 1 can be traversed. Eight configurations times the six possible traversing gives 48 cases, and this is the 3! possible permutations of the visiting order of the three edges multiplied by the 2^3 possible signs of the three edges. Instead of configurations and traversing, we will talk about visiting permutations and signs, there is a one-to-one correspondence between them. Therefore the problem is to tell in constant time for each permutation, sign pattern and reality edge whether or not there are other two reality edges to the right being in the given permutation and sign pattern. Any sign pattern can be discussed in a general way, the three signs will be denoted by s_1, s_2 and s_3 from left to right.

Another observation is that it is enough to give algorithms for the $1, 2, 3$, the $2, 1, 3$ and $1, 3, 2$ permutations since the cycle can be traversed with starting the tour on the leftmost edge in the other direction. This will cause a change in the permutation such that 3 and 1 will be swapped, and all signs will change to the other sign. For example, in Figure 3, the cases on the right column will turn to the cases on the left column if the cycle is traversed in a reverse order.

The $1, 2, 3$ case. The $1, 2, 3$ permutation is the easy case for any signs. The algorithm traverses again the reality edges in a reverse position order, and calculates

$$s_2 \max s_3 \max(i) = \max_{j \geq i}\{\pi(j)|\pi(j) < s_3 \max(j) \ \& \ sign(j) = s_2\} \qquad (2)$$

There is a $1, 2, 3$ permutation with a good sign pattern for a position i if $sign(i) = s_1$ and $\pi(i) < s_2 \max s_3 \max(i)$.

The $2, 1, 3$ case. The algorithm runs an index i from 1 to n and is in the rightmost position j for which $\pi(j) < i$, $sign(j) = s_2$ and $s_3 \max(j) > i$. If $pos(i) < j$ and $sign(i) = s_1$, then there is a $2, 1, 3$ case with proper signs starting in position $pos(i)$, otherwise such configuration does not exist in that position. Knowing the $pos()$ and $s_3 \max()$ arrays, it is easy to jump to the proper rightmost position until $i > s_3 \max(j)$. Then the algorithm must go back to the rightmost position j for which $\pi(j) < i < s_3 \max(j)$. Directly traversing back the positions would take $O(n)$ time and such traversing back might be necessary $O(n)$ times, giving the algorithm an $O(n^2)$ running time. Therefore some preprocessing is necessary.

In the preprocessing, the algorithm marks the anchor points of the $s_3 \max$ threshold function (rectangles in Figure 4). Then for each interval between two consecutive anchor points, it traverses backward the interval, and creates the chained list of the local $s_2 \min$ anchor points (black circles in Figure 4, locality also means that it checks only points which are smaller than the right anchor $s_3 \max$ value). For the local minimum, it finds on the previous chained list the first anchor point which is smaller than the actual local minimum, traversing the chain from up to down. The actual list is then augmented with the rest of the list. With up-to-down search, each anchor point is visited only once while searching, providing the $O(n)$ running time of the preprocessing algorithm.

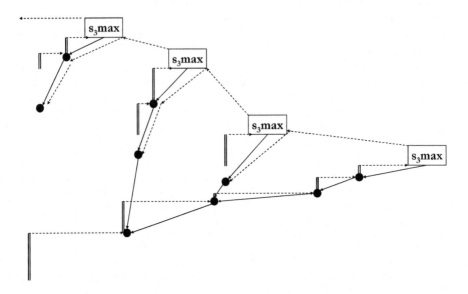

Fig. 4. Explanatory figure for the 2, 1, 3 algorithm. For details, see the text.

Increase of i is indicated with a double line in Figure 4, jumping in positions is indicated with a dashed line. While there is no j for which $\pi(j) < i < s_3 \max(j)$ and $sign(j) = s_2$, the algorithm remains in position 1 and marks all $pos(i)$ having no good 2, 1, 3 configuration. The algorithm jumps positions toward the right end of the permutation whenever a good position j appears, until $i > s_3 \max(j)$. Then it jumps to the next s_3 max anchor point to the left, and slides down on the s_2 min chained list until for the current position j, $\pi(j) < i$. Each edge of the s_2 min anchor chains is used at most once for back-traversing. To see this, suppose that the algorithm used an edge in a back-traversing, let the value of the starting s_3 max anchor point be x, and let the starting point of the edge in question be in position j, hence having value $\pi(j)$. Clearly, $\pi(j) < x$, and next time the traceback starts when $i > x$. Although the back-traversing might arrive to position j, it will stop since $i > \pi(j)$. Since the total size of the chained s_2 min anchor list is $O(n)$, the algorithm spends only $O(n)$ time with back-traversing, and hence, has only $O(n)$ running time altogether.

The 1, 3, 2 case. For this case, the preprocessing creates a double chained list of the numbers having sign s_3. It traverses the permutation in position order (namely, from left to right) and pulls out the numbers having sign s_3 from the chained list. The preprocessing remembers the neighbours of each number being pulled out, hence it will be possible to put back the numbers in reverse order.

After the preprocessing, the algorithm traverses the permutation in reverse position order (namely, from right to left), and puts back the visited s_3-signed numbers into the chained list. The algorithm remembers the maximal visited number with sign s_3 having to the right a greater number with s_2 sign, denoted by M. Whenever the algorithm arrives to an s_2-signed number, S, it updates M. To do this, it walks on the chained list of s_3-signed numbers till the last number in the chain that is smaller than S. For any s_1-signed number X, there is a $1, 3, 2$ configuration iff $X < M$.

Mutations with leftmost reality edge of position 1, and sampling the middle and rightmost edges. The above mentioned algorithms work for the reality edge in position 1, with the notation that the given permutation patterns must be compared with the configurations in Table 1.

Once we choose a rightmost edge in position i and the type of the mutation, deciding whether or not a reality edge can be in a pattern being good for the prescribed mutation is very easy, one should only check the s_3 min and s_3 max values with the possible restriction they might not be bigger or smaller than $\pi(i)$, depending on the searched permutation pattern. Similarly, once the rightmost and middle edge have been chosen, it is very easy to find the list of possible leftmost reality edges.

Weighting the reality edges. Sampling from the uniform list of possible rightmost edges might lead to a very skewed distribution where mutations on the right ends of cycles are preferred. This is because there might be significantly more mutations of a category with a leftmost reality edge at the left end of a cycle than at the right end of a cycle. Therefore some sophisticated weighting yields better distribution also in terms of acceptance ratios. This statistical issue will be discussed in another paper.

"Other" mutations. We must mention that mutations acting on more than one cycle all fall into the "other" category. Knowing whether or not there are reality edges being in other cycles, it is trivial to decide whether or not mutations acting on different cycles and having the current reality edges as leftmost edge exists is a trivial problem.

4   Discussion

We introduced two strategies for efficient sampling of transpositions and inverted transpositions. Both algorithms run in $O(n)$ time and memory, and can be used in Bayesian MCMC. With these sampling algorithms, one MCMC step can be performed in $O(n^2)$ time and in linear memory, which is a significant improvement to the so far available algorithm having $O(n^4)$ running time and $O(n^3)$ memory.

We hope we could convince the readers that designing Markov chain Monte Carlo methods in bioinformatics is not only a statistical problem but an at least as important algorithmic problem, too.

Acknowledgments

I.M. was supported by the National Office for Research and Technology at the e-Science Regional Knowledge Centre, Eötvös Loránd University. I.M. is also supported by a Bolyai postdoctoral fellowhip. I.M. and T.B.P. thank for the Rényi Institute for hosting T.B.P. in 2005 summer. P.L. would like to thank for the Rényi Institute for hosting him as a young researcher. Zsuzsanna Márton is thanked for some very useful comments.

References

1. Sturtevant, A.H., and Novitski, E.: The homologies of chromosome elements in the genus Drosophila. Genetics **26** (1941) 517–541
2. Nadau, J.H., and Taylor, B.A.: Lengths of chromosome segments conserved since divergence of man and mouse. PNAS **81** (1984) 814–818
3. Palmer, J.D., and Herbon, L.A.: Plant mitochondrial DNA evolves rapidly in structure, but slowly in sequence. J. Mol. Evol. **28** (1988) 87–97
4. Bader, D.A., Moret, B.M.E., and Yan, M.: A linear-time algorithm for computing inversion distance between signed permutations with an experimental study. J. Comp. Biol. **8(5)** (2001) 483–491
5. Bergeron, A.: A very elementary presentation of the Hannenhalli-Pevzner theory. Proc. 13th CPM'01, LNCS (2001) 106–117
6. Hannenhalli, S., and Pevzner, P.A.: Transforming Cabbage into Turnip: Polynomial Algorithm for Sorting Signed Permutations by Reversals. J. ACM **46(1)** (1999) 1–27
7. Kaplan, H., Shamir, R., and Tarjan, R.: A faster and simpler algorithm for sorting signed permutations by reversals. SIAM J. Comput. **29(3)** (1999) 880–892
8. Siepel, A.: An algorithm to find all sorting reversals. Proc. RECOMB'02 (2002) 281–290
9. Tannier, E., and Sagot, M.-F.: Sorting by reversals in subquadratic time. Proc. 15th CPM'04, LNCS (2004) 1–13.
10. Hannenhalli, S.: Polynomial algorithm for computing translocation distance between genomes. Proc. 7th CPM'96, LNCS (1996) 168–185
11. Bafna, V., and Pevzner, A.: Sorting by transpositions. SIAM J. Disc. Math. **11(2)** (1998) 224–240
12. Berman, P., Hannenhalli, S., and Karpinski, M.: 1.375-Approximation Algorithm for Sorting by Reversals. Proc. ESA'02, LNCS (2002) 200–210
13. Eriksen, N.: $(1+\varepsilon)$-approximation of sorting by reversals and transpositions. Proc. 1st WABI'01, LNCS **2149** (2001) 227–237
14. Gu, Q-P., Peng, S., and Sudborough, H.I.: A 2-Approximation Algorithm for Genome Rearrangements by Reversals and Transpositions. Theor. Comp. Sci. **210(2)** (1999) 327–339
15. Kececioglu, J.D., and Sankoff, D.: Exact and Approximation Algorithms for Sorting by Reversals, with Application to Genome Rearrangement. Algorithmica **13(1/2)** (1995) 180–210
16. Blanchette, M., Kunisawa, T., and Sankoff, D: Parametric genome rearrangement. Gene **172** (1996) GC11–GC17
17. Bader, M., and Ohlebusch, E.: Sorting by weighted reversals, transpositions and inverted transpositions. Proc. RECOMB'06, LNBI **3909** (2006) 563–577.
18. Larget, B., Simon, D.L., and Kadane, B.J.: Bayesian phylogenetic inference from animal mitochondrial genome arrangements. J. Royal Stat. Soc. B **64(4)** 681–695
19. York, T.L., Durrett, R., and Nielsen, R.: Bayesian estimation of inversions in the history of two chromosomes. J. Comp. Biol. **9** (2002) 808–818

20. Larget B, Simon DL, Kadane JB, and Sweet D.: A Bayesian analysis of metazoan mitochondrial genome arrangements Mol. Biol. Evol. **22(3)** (2005) 486–495

21. Durrett, R., Nielsen, R., and York, T.L.: Bayesian estimation of genomic distance. Genetics **166** (2004) 621–629

22. Miklós, I.: MCMC Genome Rearrangement. Bioinformatics **19** (2003) ii130–ii137

23. Miklós, I., Ittzés, P., and Hein, J.: ParIS genome rearrangement server. Bioinformatics **21(6)** (2005) 817-820.

24. Miklós, I., and Hein, J.: Genome rearrangement in mitochondria and its computational biology. Proc. 2nd RECOMB Satellite Workshop on Computational Genomics RECOMBCG'06, LNBI **3388** (2005) 85–96.

25. Metropolis, N., Rosenbluth, A.W., Rosenbluth, M.N., Teller, A.H., and Teller, E.: Equations of state calculations by fast computing machines. J. Chem. Phys. **21(6)** (1953) 1087–1091

26. Liu, J.S.: Monte Carlo strategies in scientific computing. Springer Series in Statistics, New-York. (2001)

27. Hastings, W.K.: Monte Carlo sampling methods using Markov chains and their applications. Biometrika **57(1)** (1970) 97–109

28. von Neumann, J.: Various techniques used in connection with random digits. National Bureau of Standards Applied Mathematics Series **12** (1951) 36–38.

Alignment with Non-overlapping Inversions in $O(n^3)$-Time

Augusto F. Vellozo[1], Carlos E.R. Alves[2], and Alair Pereira do Lago[3]

[1] Instituto de Matemática e Estatística da Universidade de São Paulo (IME-USP)
Rua do Matão, 1010 - Cidade Universitária CEP:05508-090 São Paulo - SP - Brasil
vellozo@ime.usp.br
[2] Universidade São Judas Tadeu (FTCE-USJT), Rua Taquari, 546, Mooca
CEP:03166-000 São Paulo - SP - Brasil
prof.carlos_r_alves@usjt.br
[3] Instituto de Matemática e Estatística da Universidade de São Paulo (IME-USP)
Rua do Matão, 1010 - Cidade Universitária CEP:05508-090 São Paulo - SP - Brasil
alair@ime.usp.br

Abstract. Alignments of sequences are widely used for biological sequence comparisons. Only biological events like mutations, insertions and deletions are usually modeled and other biological events like inversions are not automatically detected by the usual alignment algorithms.

Alignment with inversions does not have a known polynomial algorithm and a simplification to the problem that considers only non-overlapping inversions were proposed by Schöniger and Waterman [20] in 1992 as well as a corresponding $O(n^6)$ solution[1]. An improvement to an algorithm with $O(n^3 \log n)$-time complexity was announced in an extended abstract [1] and, in this present paper, we give an algorithm that solves this simplified problem in $O(n^3)$-time and $O(n^2)$-space in the more general framework of an edit graph.

Inversions have recently [4,7,13,17] been discovered to be very important in Comparative Genomics and Scherer et al. in 2005 [11] experimentally verified inversions that were found to be polymorphic in the human genome. Moreover, 10% of the 1,576 putative inversions reported overlap RefSeq genes in the human genome. We believe our new algorithms may open the possibility to more detailed studies of inversions on DNA sequences using exact optimization algorithms and we hope this may be particularly interesting if applied to regions around known rearrangements boundaries. Scherer report 29 such cases and prioritize them as candidates for biological and evolutionary studies.

1 Introduction

Alignments of sequences are widely used for biological sequence comparisons and can be associated with a set of edit operations that transform one sequence to the other. Usually, the only edit operations that are considered are the *substitution* (mutation) of one symbol by another one, the *insertion* of one symbol

---

[1] In this case, n denotes the maximal length of the two aligned sequences.

P. Bücher and B.M.E. Moret (Eds.): WABI 2006, LNBI 4175, pp. 186–196, 2006.
© Springer-Verlag Berlin Heidelberg 2006

and *deletion* of one symbol. If costs are associated with each operation, there is a classic $O(n^2)$ dynamic program that computes a set of edit operations with minimal total cost and exhibit the associated alignment, which has good quality and high likelihood for realistic costs.

Other important biological events like inversions are not automatically detected by the usual alignment algorithms and we can define a new edit operation, the *inversion* operation, which substitutes any segment by its *reverse complement* sequence. We can define a new alignment problem: given two sequences and fixed costs for each kind of edit operation, the *alignment with inversions* problem is an optimization problem that queries the minimal total cost[2] of an edit operations series that transforms one sequence to the other. Moreover, one may also be interested in the exhibition of its corresponding alignment and/or edit operations. Unfortunatley, the decision problem associated with alignment with inversions for an unlimited alphabet size is NP-hard as consequence of Jiang et al. [5].

Some simplifications of this problem have been studied and were proved to be NP-complete [3,22]. Many approximation algorithms were also proposed [6,16]. Another important simplification is the problem known as *sorting signed permutations by reversals* and polynomial algorithms were obtained in a sequence of papers [2,14,15,21]. These approaches are mainly used for the study of inversions on sequences of genes, but new comparative results given by Sherer et al. [11] show also the importance of DNA inversion studies where those methods can not be used. Moreover, Sherer et al. reported 83 inversions that are contained within a gene.

Another important approach was introduced in 1992, by Schöniger and Waterman [20]. They introduced a *simplification hypothesis*: *all regions involved in the inversions do not overlap*. This simplification is realistic for local DNA comparisons on relatively close sequences. This led to the *alignment with non-overlapping inversions* problem and they presented a simple $O(n^6)$ dynamic programming solution for this problem and also introduced a *heuristic* for it that reduced the average running-time to something between $O(n^2)$ and $O(n^4)$.

Recently, independent works [8,9,10,12] gave exact algorithms for alignments with non-overlapping inversions with $O(n^4)$-time and $O(n^2)$-space complexity. An algorithm with $O(n^3 \log n)$-time [1] was later announced. In this paper, we give an algorithm that solves this simplified problem in $O(n^3)$-time and $O(n^2)$-space.

2   Alignments with Non-overlapping Inversions

The standard alignment of two strings is called *standard alignment* in this text. This kind of alignment, when viewed as the process of transforming a string s in a string t, uses the well known string edit operations of insertion, deletion and substitution of symbols.

---

[2] In this work, we deal with the dual approach of maximization of similarity score.

The alignment of s and t is usually represented by the insertion of some *spaces* $(-)$ in certain places of each string and the matching (alignment) of each symbol or space of s with the symbol or space in the corresponding position in t. If $s[i]$ and $t[j]$ are symbols from s and t, respectively, then a pair $(s[i], t[j])$ is a *substitution* of $s[i]$ by $t[j]$ (if they are equal we say it's a *match*), $(-, t[j])$ is the *insertion* of $t[j]$ and $(s[i], -)$ is the deletion of $s[i]$. Usually, there are costs associated with each edit operation and a score is given to the alignment based on the pairs that were formed.

An extra operation is considered here: the *inversion* of a substring. A string that suffers this operation has a substring removed, reverted, complemented and inserted back in its original place. For example, the inversion of the string ACCATGC gives GCATGGT.

When evaluating an alignment with inversions, there is a cost associated with the inversion operation. Besides that, insertions, substitutions and deletions may be applied in an inverted substring, incurring in additional costs.

In this paper we consider only *non-overlapping* inversions. This means that when aligning two strings we may consider multiple inversions in s, but any symbol of s may be involved in at most one inverted substring. When dealing with non-overlapping inversions, the order in which the inversions are performed is unimportant.

In the following sections, $\bar{s}$ is the inverted string s while $\overline{s[a..b]}$ is the inverted substring os s that starts in position a and ends in position b. These positions are taken from s, not $\bar{s}$, as would be the case in $\bar{s}[a..b]$ (notice the extension of the bar in each case).

3   Edit Graph

Let s and t be two sequences of lengths n and m respectively.

Definition 3.1 (Edit Graph of s and t). *Consider* $V = \{(i, j) | 0 \le i \le n, 0 \le j \le m\}$ *and* $E = E_H \cup E_D \cup E_V$, *such that,*

- $E_H = \{e_H^{i,j} = ((i, j-1), (i, j)) | 0 \le i \le n, 0 < j \le m\}$ *is the set of horizontal edges that end on vertex* (i, j),
- $E_D = \{e_D^{i,j} = ((i-1, j-1), (i, j)) | 0 < i \le n, 0 < j \le m\}$ *is the set of diagonal edges that end on vertex* (i, j),
- $E_V = \{e_V^{i,j} = ((i-1, j), (i, j)) | 0 < i \le n, 0 \le j \le m\}$ *is the set of vertical edges that end on vertex* (i, j).

Consider the function $\omega : E \longrightarrow \mathbb{R} \cup \{-\infty\}$, *that associates each edge* $e \in E$ *with weight* $\omega(e)$. *The directed graph* $G = (V, E, \omega)$ *is the edit graph of s and t.*

In this work, the weight of edge $e_V^{i,j}$ is the score of the deletion of letter $s[i]$ when $s[1..i-1]$ is aligned with $t[1..j]$, the weight of edge $e_H^{i,j}$ is the score of the insertion of letter $t[j]$ when $s[1..i]$ is aligned with $t[1..j-1]$ and the weight of edge $e_D^{i,j}$ is the score of the substitution of letter $s[i]$ by letter $t[j]$ when

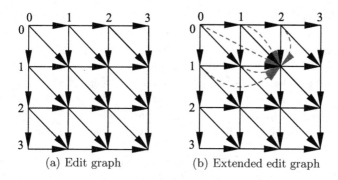

(a) Edit graph   (b) Extended edit graph

Fig. 1. Examples of edit graph and extended edit graph. Edge weights are not shown, and the only extended edges shown are those that arrive at $(1, 2)$.

$s[1..i-1]$ is aligned with $t[1..j-1]$. These weights are usually defined by a function $\phi : \Sigma \cup \{-\} \times \Sigma \cup \{-\} \longrightarrow \mathbb{R} \cup \{-\infty\}, - \notin \Sigma$, such that $\omega(e_V^{i,j}) = \phi(s[i], -)$, $\omega(e_H^{i,j}) = \phi(-, t[j])$ and $\omega(e_D^{i,j}) = \phi(s[i], t[j])$, where Σ is the set of symbols used in the sequences.

Therefore, there is a one-to-one relation between paths in G and standard alignments of s against t. In others words, one path from $(0, 0)$ to (i, j) in G corresponds to one and only one standard alignment of $s[1..i]$ against $t[1..j]$. The score of an alignment without inversions is the total weight of its corresponding path in G.

We say that a path p from $u = (i, j)$ to $v = (i', j')$ is optimal if there is no other path from u to v with total weight greater than the weight of p. We denote $w_u^v = w_{i,j}^{i',j'}$ to be the weight of this optimal path path p. If there is no such a path from u to v, we denote $w_u^v = -\infty$.

Notice that the score of an optimal standard alignment of s against t is the weight of an optimal path from $(0, 0)$ to (i, j) in G.

Definition 3.2 (Extended edit graph of s and t). *Consider E_H, E_D, E_V and V as described in the definition of edit graph of s and t. Consider $E = E_H \cup E_D \cup E_V \cup E_X$ where $E_X = \bigcup_{i=0}^n \bigcup_{j=0}^m E_X^{i,j}$ and $E_X^{i,j}$ is the set of extended edges that end on vertex (i, j), that is*

$$E_X^{i,j} = \{e_{i',j'}^{i,j} = ((i', j'), (i, j)) \mid 0 \le i' \le i \le n, 0 \le j' \le j \le m \ e \ (i', j') \ne (i, j)\}.$$

The directed graph $G = (V, E, \omega)$ is the extended edit graph of s and t and the weight function ω is defined like in the edit graph, but extended to assign weights to the extended edges.

In this paper, the extended edges represent optimal standard alignments of substrings of t against inverted substrings of s.

Let G be an extended edit graph of s and t. The graph obtained by removing the extended edges from G is an edit graph of s and t. Like in edit graphs, an optimal path in an extended edit graph is a path with maximal weight.

4   The Algorithm

Let $s = s[1..n]$ and $t = t[1..m]$ be the sequences to be aligned.

Let $\overline{G} = (V, \overline{E}, \overline{w})$ be the edit graph of $\overline{s}$ and t. This graph is used to evaluate the alignments of substrings of t and inverted substrings of s. In $\overline{G}$, the weights $\overline{w}(e_H^{i,j})$, $\overline{w}(e_D^{i,j})$ and $\overline{w}(e_V^{i,j})$ correspond, respectively, to the scores of insertion of $t[j]$, substitution of $\overline{s}[i] = \overline{s[n+1-i]}$ by $t[j]$ and deletion of $\overline{s}[i] = \overline{s[n+1-i]}$.

Let $G = (V, E, w)$ be the extended edit graph of s and t, such that

$$
\begin{aligned}
w(e_H^{i,j}) &= \text{ score of insertion of } t[j], \\
w(e_V^{i,j}) &= \text{ score of deletion of } s[i], \\
w(e_D^{i,j}) &= \text{ score of substitution of } s[i] \text{ by } t[j], \\
w(e_{i',j'}^{i,j}) &= \overline{w}_{(n-i,j')}^{(n-i',j)} + w_{inv},
\end{aligned}
$$

where w_{inv} is a penalty value for inversions and $\overline{w}_{(n-i,j')}^{(n-i',j)}$ is the weight of an optimal path from $(n-i, j')$ to $(n-i', j)$ in $\overline{G}$. In others words $\overline{w}_{(n-i,j')}^{(n-i',j)}$ is the score of the standard alignment of $\overline{s[i'+1..i]}$ against $t[j'+1..j]$.

Since there is a one to one relation between paths in G and alignments with non-overlapping inversions of s against t, the weight of an optimal path from $(0,0)$ to (n,m) in G is the score of an optimal alignment with non-overlapping inversions of s against t.

The following definitions help us to understand how the weight of an optimal path from $(0,0)$ to (n,m) in G is obtained through Algorithm 1.

Definition 4.1 (Matrix B). $B[i,j] = w_{0,0}^{i,j}$ *is the weight of an optimal path from* $(0,0)$ *to* (i,j) *on* G, $0 \le i \le n$ *and* $0 \le j \le m$.

In others words $B[i,j]$ is the score of an optimal alignment with non-overlapping inversions of $s[1..i]$ against $t[1..j]$.

Definition 4.2 (Matrix $Out_{i'}^i$). *Given* i' *and* i *such that* $0 \le i' \le i \le n$ *we define the matrix* $Out_{i'}^i[1..m, 1..m]$ *of* G *as*

$$
Out_{i'}^i[j', j] = \begin{cases} B[i', j'] + w_{i',j'}^{i,j}, & \text{if } 0 \le j' \le j \le m, \\ -\infty & \text{if } 0 \le j < j' \le m, \end{cases}
$$

The element $Out_{i'}^i[j', j]$ stores the optimal alignment score of $s[1..i]$ against $t[1..j]$ such that $\overline{s[i'+1..i]}$ is aligned with $t[j'+1..j]$.

Definition 4.3 ($hDif_{i'}^{i,j}$ vector). *Let* G *be an edit graph. Given* i' *and the vertex* (i,j) *of* G *such that* $0 \le i' \le i$, *we define* $hDif_{i'}^{i,j}$ *of* G *by the vector of size* j *such that* $hDif_{i'}^{i,j}[j'] = w_{i',j'}^{i,j} - w_{i',j'}^{i,j-1}$, $0 \le j' < j$.

The vector $hDif_{i'}^{i,j}$ has an important property that is used by our algorithm: it is nondecreasing.

Lemma 4.4 *The vector $hDif_{i'}^{i,j}$ of an edit graph G is nondecreasing.*

Proof. Let (i', j_1), (i', j_2), (i, j_3) and (i, j_4) be vertices of G, such that $0 \leq j_1 < j_2 \leq j_3 < j_4 \leq m$. There is at least one common vertex v that belongs to the paths from (i', j_2) to (i, j_3) and from (i', j_1) to (i, j_4), as one can see at Figure 2. To simplify, we define: $a = w_{i',j_1}^{i,j_3}$, $b = w_{i',j_2}^{i,j_4}$, $c = w_{(i',j_1)}^{v}$, $d = w_v^{(i,j_4)}$, $e = w_{(i',j_2)}^{v}$ and $f = w_v^{(i,j_3)}$. As a and b are the optimal path scores then $a \geq c + f$ and $b \geq e + d$. Adding the two previous inequalities we have $a + b \geq c + f + e + d \Rightarrow b - (e + f) \geq (c + d) - a$. Consider $j_3 = j_4 - 1$. Therefore $w_{i',j_2}^{i,j_4} - w_{i',j_2}^{i,j_4-1} \geq w_{i',j_1}^{i,j_4} - w_{i',j_1}^{i,j_4-1} \Rightarrow hDif_{i'}^{i,j_4}[j_2] \geq hDif_{i'}^{i,j_4}[j_1]$. ∎

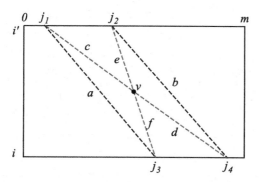

Fig. 2. Illustration of the proof of Lemma 4.4

The number of times that $hDif_{i'}^{i,j}[j']$ increases when we sweep through $hDif_{i'}^{i,j}$ from $j' = 0$ to $j - 1$ is called $\psi H_{i'}^{i,j}$.

Usually, the adopted score system has integer values: r for rewarding a match, q for a mismatch and E for a gap. Usually $2E \leq q < r$. Using the edit graph notation, the weights of the edges can be defined as $\omega(e_D^{i,j}) = r$ if $s[i] = t[j]$, $\omega(e_D^{i,j}) = q$ if $s[i] \neq t[j]$ and $\omega(e_H^{i,j}) = \omega(e_V^{i,j}) = E \ \forall(i, j)$. In these cases $\psi H_{i'}^{i,j} \leq r - 2E$, so $\psi H_{i'}^{i,j}$ is limited by a constant. For instance, if the score system is the LCS (Longest Common Subsequence), $r = 1$ and $q = E = 0$, then $\psi H_{i'}^{i,j} \leq 1$. The Figure 3 shows a case where $\psi H_{i'}^{i,j} \leq 3$.

In this text, we consider $\psi H_{i'}^{i,j}$ limited by a constant.

We store the values of j' where occur each increment of $hDif_{i'}^{i,j}$ in a matrix called $BLH_{i'}^{i}$.

Definition 4.5 ($BLH_{i'}^{i}$ matrix). *Given i' and i such that $0 \leq i' \leq i \leq n$, we define the column j, $0 \leq j \leq m$, of $BLH_{i'}^{i}$ as a vector of size $\psi H_{i'}^{i,j}$ such that $BLH_{i'}^{i}[\alpha, j]$ is the α-th j' where $hDif_{i'}^{i,j}[j'] \neq hDif_{i'}^{i,j}[j' - 1]$, for j' from 1 to $j - 1$.*

Weights of optimal paths from $(0,j')$ to (n,j)

	j=0	j=1	j=2	j=3	j=4	j=5	j=6	j=7	j=8	j=9
j'=0	-4	-2	-2	-2	-3	-2	0	-1	-2	-3
j'=1	-	-4	-2	-2	-3	-1	1	0	-1	-2
j'=2	-	-	-4	-4	-2	0	2	1	0	-1
j'=3	-	-	-	-4	-2	0	2	1	0	0
j'=4	-	-	-	-	-4	-2	0	-1	-2	-1
j'=5	-	-	-	-	-	-4	-2	-2	-2	0
j'=6	-	-	-	-	-	-	-4	-2	-2	0
j'=7	-	-	-	-	-	-	-	-4	-4	-2
j'=8	-	-	-	-	-	-	-	-	-4	-2
j'=9	-	-	-	-	-	-	-	-	-	-4

hDif

	j=0	j=1	j=2	j=3	j=4	j=5	j=6	j=7	j=8	j=9
j'=0	-	2	0	0	-1	1	2	-1	-1	-1
j'=1	-	-	2	0	-1	2	2	-1	-1	-1
j'=2	-	-	-	0	2	2	2	-1	-1	-1
j'=3	-	-	-	-	2	2	2	-1	-1	0
j'=4	-	-	-	-	-	2	2	-1	-1	1
j'=5	-	-	-	-	-	-	2	0	0	2
j'=6	-	-	-	-	-	-	-	2	0	2
j'=7	-	-	-	-	-	-	-	-	0	2
j'=8	-	-	-	-	-	-	-	-	-	2

BLH

	j=0	j=1	j=2	j=3	j=4	j=5	j=6	j=7	j=8	j=9
1	-	-	1	-	2	1	-	5	5	3
2	-	-	-	-	-	-	-	6	-	4
3	-	-	-	-	-	-	-	-	-	5

Fig. 3. In this example we used the sequences $s = AATG$ and $t = TTCATGACG$ to build an edit graph G. All vertical and horizontal edges of G have weight -1, the weight $\omega(e_D^{i,j}) = -1$ if $s[i] \neq t[j]$ and $\omega(e_D^{i,j}) = 1$ if $s[i] = t[j]$.

---

algorithm 1. Algorithm $O(n^3)$ that builds matrix B

$\text{BIMN3}(s, t)$

```
1    for i from 0 to |s| do
2            ▷ Get the optimal path ended with non-extended edges
3            if i = 0 then
4                    B[0, 0] ← 0
5            else    B[i, 0] ← B[i − 1, 0] + ω(e_V^{i,j})
6            for j from 1 to |t| do
7                    if i = 0 then
8                            B[0, j] ← B[0, j − 1] + ω(e_H^{i,j})
9                    else    aux ← max(B[i, j − 1] + ω(e_H^{i,j}), B[i − 1, j] + ω(e_V^{i,j}))
10                           B[i, j] ← max(aux, B[i − 1, j − 1] + ω(e_D^{i,j}))
11           ▷ Get the optimal path ended with extended edges
12           for i' from i downto 0 do
13                   BLH ← buildBlh(Ḡ, BLH, i')
14                   maxOut_{i'}^i ← getMaxOut(BLH, B, i')
15                   for j from 0 to |t| do
16                           B[i, j] ← max(B[i, j], maxOut_{i'}^i[j] + ω_{inv})
17   return B
```

The elements of matrix $BLH_{i'}^i$ are called *borderline points* in [18]. Figure 3 shows an example of $hDif$ and BLH.

Algorithm 1 builds matrix B and Figure 4 shows its execution.

The function $buildBlh(\overline{G}, BLH, i')$ builds the $BLH_{i'}^i$ matrix. It was developed based on the algorithm described in section 6 of [19] and runs in $O(m)$

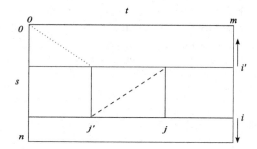

Fig. 4. Execution of Algorithm 1. The dotted line is a path from $(0,0)$ to (i',j') in G. The dashed line represents an alignment of $\overline{s[i'+1..i]} \times t[j'+1..j]$.

time. Remembering that each column of a borderline matrix has $O(1)$ elements, the function $buildBlh(\overline{G}, BLH, i')$ builds each column of $BLH_{i'}^{i}$ based on the respective column of matrix $BLH_{i'+1}^{i}$ in constant time.

The function $getMaxOut(BLH, B, i')$ returns a vector with the maximum value of each column of $Out_{i'}^{i}$ in $O(m)$ time and was developed based on the algorithm described in subsection 6.2 of [18]. The linear time complexity of this function is attained through a procedure that sweeps through $BLH_{i'}^{i}$ and line i' of matrix B, both with $O(m)$ data.

Using these functions one can see that Algorithm 1 is correct and runs in $O(n^2m)$ time ($O(n^3)$ time, if $m = O(n)$).

5   Experiments

We implemented Algorithm 1 in Java. We worked with two sequences pair of different lengths, 867 and 95.319 bp (base pairs) of human and chimpanzee. These sequences are cited in [11]. The human/chimp sequences were downloaded from the University of California at Santa Cruz website (http://genome.ucsc.edu/). The sequences were taken from the November 2003 chimpanzee (panTro1) genome assembly and the May 2004 (hg17) human genome assembly[3].

The shortest pair is formed by human genome chr7:95119414-95120280 and chimpanzee genome chr6:96726524-96727390. The alignment obtained by the algorithm shows 98,6% of total identities and an inversion involving chr7-95119717-95119979 of human and chr6:96726825-96727087 of chimpanzee.

The longest pair is formed by human genome chr7:80523522-80618840 and chimpanzee genome chr6:81751455-81846825. To cope with sequences of this length faster we broke the sequences into fragments of 100 pairs each.

The fragments were submitted to a standard alignment procedure, such that each fragment from the human genome was aligned against every fragment of the chimpanzee genome twice: inverted and not inverted. Our algorithm was used considering the sequences like sequences of fragments instead of sequences

---

[3] http://genome.ucsc.edu/cgi-bin/hgTrackUi?hgsid=59218717&g=netPanTro1

of base pairs. A match between two fragments occurs when their alignment has a score greater than a threshold. The alignment obtained by the algorithm shows 94,8% of total matches and an inversion involving chr7-80553522-80588821 of human and chr6:81781455-81816854 of chimpanzee. One can see this inversion at Figure 5 for fragment size 1000 for better resolution.

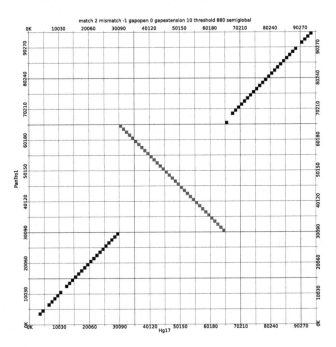

Fig. 5. Alignment of human genome chr7:80523522-80618840 and chimpanzee genome chr6:81751455-81846825. Fragment size is 1000 for better visualization.

We also tested the algorithm on simulated data for random DNA sequences with length in average 700. Each pair of sequences differ from each other by a number of indels ranging from 5% to 10%, mismatches ranging from 5% to 15%, and number of non-overlapping inversions ranging from 1 to 15. We obtained consistent results and detected all the inversions as one would expect.

We also implemented in Java the $O(n^3 \log n)$ algorithm described in [1], the $O(n^4)$ algorithm described in [9] and the sparse algorithm described in [10] that has complexity $O(r^2 \log^2 r)$, where $r = O(n^2)$ is the number of matches between symbols in one sequence against symbols in the other sequence. The tests showed that Algorithm 1 is, as it is expected, always faster than the algorithm $O(n^3 \log n)$, which is in turn always faster than the algorithm $O(n^4)$. If the sequences to be aligned were DNA sequences then Algorithm 1 was faster than sparse algorithm, but if the sequences to be aligned were sequences of DNA fragments, where the number of matches is small, then the sparse algorithm was faster than the Algorithm 1.

6   Conclusion

In this paper we described a new algorithm that solves the alignment with non-overlapping inversions problem in $O(n^3)$-time and $O(n^2)$-space. We hope that this speed up opens the possibility to studies of inversions on DNA sequences by an exact optimization algorithm. Algorithms that are applied to the study of inversions of sequences of genes cannot be applied in theses cases, since they do not allow repetitions of symbols, nor insertions, nor deletions.

Our algorithm may be particularly interesting when applied to regions around known rearrangement boundaries, since many biologists suppose that inversions at DNA level are very probable in these cases.

Many studies have been done with inversions in DNA sequences.

Acknoledgements

This work was supported by Proj. Pronex-FAPESP/CNPq proc. 2003/09925-5.

References

1. Carlos E. R. Alves, Alair Pereira do Lago, and Augusto F. Vellozo. Alignment with non-overlapping inversions in $O(n^3 \log n)$-time. In *Proceedings of GRACO2005*, volume 19 of *Electron. Notes Discrete Math.*, pages 365–371 (electronic), Amsterdam, 2005. Elsevier.
2. David A. Bader, Bernard M. E. Moret, and Mi Yan. A linear-time algorithm for computing inversion distance between signed permutations with an experimental study. *Journal of Computational Biology*, 8(5):483–491, 2001.
3. Alberto Caprara. Sorting permutations by reversals and Eulerian cycle decompositions. *SIAM J. Discrete Math.*, 12(1):91–110 (electronic), 1999.
4. Cerdeño-Tárraga, Patrick, Crossman, Blakely, Abratt, Lennard, Poxton, Duerden, Harris, Quail, Barron, Clark, Corton, Doggett, Holden, Larke, Line, Lord, Norbertczak, Ormond, Price, Rabbinowitsch, Woodward, Barrell, and Parkhill. Extensive DNA inversions in the B. fragilis genome control variable gene expression. *Science*, 307(5714):1463–1465, Mar 2005.
5. Xin Chen, Jie Zheng, Zheng Fu, Peng Nan, Yang Zhong, Stefano Lonardi, and Tao Jiang. Assignment of orthologous genes via genome rearrangement. *IEEE/ACM Trans. Comput. Biol. Bioinformatics*, 2(4):302–315, 2005.
6. David A. Christie. A 3/2-approximation algorithm for sorting by reversals. In *Proceedings of the Ninth Annual ACM-SIAM Symposium on Discrete Algorithms (San Francisco, CA, 1998)*, pages 244–252, New York, 1998. ACM.
7. Cáceres, Ranz, Barbadilla, Long, and Ruiz. Generation of a widespread Drosophila inversion by a transposable element. *Science*, 285(5426):415–418, Jul 1999.
8. A. P. do Lago, C. A. Kulikowski, E. Linton, J. Messing, and I. Muchnik. Comparative genomics: simultaneous identification of conserved regions and their rearrangements through global optimization. In *The Second University of Sao Paulo/Rutgers University Biotechnology Conference*, Rutgers University Inn and Conference Center, New Brunswick, NJ, August 2001.

9. Alair Pereira do Lago, Ilya Muchnik, and Casimir Kulikowski. An $O(n^4)$ algorithm for alignment with non-overlapping inversions. In *Second Brazilian Workshop on Bioinformatics, WOB 2003*, Macaé, RJ, Brazil, 2003. http://www.ime.usp.br/~alair/wob03.pdf.

10. Alair Pereira do Lago, Ilya Muchnik, and Casimir Kulikowski. A sparse dynamic programming algorithm for alignment with non-overlapping inversions. *Theor. Inform. Appl.*, 39(1):175–189, 2005.

11. Feuk, MacDonald, Tang, Carson, Li, Rao, Khaja, and Scherer. Discovery of human inversion polymorphisms by comparative analysis of human and chimpanzee DNA sequence assemblies. *PLoS Genet*, 1(4):e56, Oct 2005.

12. Yong Gao, Junfeng Wu, Robert Niewiadomskil, Yang Wang, Zhi-Zhong Chen, and Guohui Lin. A space efficient algorithm for sequence alignment with inversions. In *Computing and Combinatorics, 9th Annual International Conference, COCOON 2003*, volume 2697 of *Lecture Notes in Computer Science*, pages 57–67. Springer-Verlag, 2003.

13. Graham and Olmstead. Evolutionary significance of an unusual chloroplast DNA inversion found in two basal angiosperm lineages. *Curr Genet*, 37(3):183–188, Mar 2000.

14. Sridhar Hannenhalli and Pavel A. Pevzner. Transforming cabbage into turnip: polynomial algorithm for sorting signed permutations by reversals. In *ACM Symposium on Theory of Computing*, pages 178–189. Association for Computing Machinery, 1995.

15. Sridhar Hannenhalli and Pavel A. Pevzner. Transforming cabbage into turnip: polynomial algorithm for sorting signed permutations by reversals. *J. ACM*, 46(1):1–27, 1999.

16. J. Kececioglu and D. Sankoff. Exact and approximation algorithms for sorting by reversals, with application to genome rearrangement. *Algorithmica*, 13(1-2): 180–210, 1995.

17. Kuwahara, Yamashita, Hirakawa, Nakayama, Toh, Okada, Kuhara, Hattori, Hayashi, and Ohnishi. Genomic analysis of Bacteroides fragilis reveals extensive DNA inversions regulating cell surface adaptation. *Proceedings of the National Academy of Sciences U S A*, 101(41):14919–14924, Oct 2004.

18. Gad M. Landau and Michal Ziv-Ukelson. On the common substring alignment problem. *J. Algorithms*, 41(2):338–359, 2001.

19. Jeanette P. Schmidt. All highest scoring paths in weighted grid graphs and their application to finding all approximate repeats in strings. *SIAM J. Comput.*, 27(4):972–992 (electronic), 1998.

20. M. Schöniger and M. S. Waterman. A local algorithm for DNA sequence alignment with inversions. *Bulletin of Mathematical Biology*, 54(4):521–536, Jul 1992.

21. Eric Tannier and Marie-France Sagot. Sorting by reversals in subquadratic time. In *Combinatorial pattern matching*, volume 3109, pages 1–13, 2004. CPM 2004.

22. R. Wagner. On the complexity of the extended string-to-string correction problem. In *Seventh ACM Symposium on the Theory of Computation*. Association for Computing Machinery, 1975.

Accelerating Motif Discovery: Motif Matching on Parallel Hardware

Geir Kjetil Sandve[1], Magnar Nedland[2], Øyvind Bø Syrstad[1],
Lars Andreas Eidsheim[1], Osman Abul[3], and Finn Drabløs[3]

[1] Department of Computer and Information Science,
Norwegian University of Science and Technology, Trondheim, Norway
`sandve@idi.ntnu.no`, {`syrstad, eidsheim`}`@stud.ntnu.no`
[2] Interagon A.S.,
Trondheim, Norway
`magnar.nedland@interagon.com`
[3] Department of Cancer Research and Molecular Medicine,
Norwegian University of Science and Technology, Trondheim, Norway
{`osman.abul, finn.drablos`}`@ntnu.no`

Abstract. Discovery of motifs in biological sequences is an important problem, and several computational methods have been developed to date. One of the main limitations of the established motif discovery methods is that the running time is prohibitive for very large data sets, such as upstream regions of large sets of cell-cycle regulated genes. Parallel versions have been developed for some of these methods, but this requires supercomputers or large computer clusters. Here, we propose and define an abstract module PAMM (Parallel Acceleration of Motif Matching) with motif matching on parallel hardware in mind. As a proof-of-concept, we provide a concrete implementation of our approach called MAMA. The implementation is based on the MEME algorithm, and uses an implementation of PAMM based on specialized hardware to accelerate motif matching. Running MAMA on a standard PC with specialized hardware on a single PCI-card compares favorably to running parallel MEME on a cluster of 12 computers.

1 Introduction

Computational discovery of motifs in biological sequences has many important applications, the best known being discovery of transcription factor binding sites (TFBS) in DNA and active sites in proteins. More than a hundred methods have been developed for this problem, all with different strengths and characteristics. Methods that use probabilistic motifs (typically PWMs) are often favored because of their high expressibility. One of the best known and most widely used methods is MEME [1]. MEME is a flexible tool that uses Expectation Maximization (EM) to discover motifs as position weight matrices (PWMs) in both proteins and DNA.

One of the main limitations of current PWM-based motif discovery methods is that the running time is prohibitive for large datasets such as upstream regions

P. Bücher and B.M.E. Moret (Eds.): WABI 2006, LNBI 4175, pp. 197–206, 2006.
© Springer-Verlag Berlin Heidelberg 2006

of large sets of cell-cycle regulated genes. Parallel versions have been developed for some methods, for instance the paraMEME [2] version of MEME, but this typically requires supercomputers or computer clusters. Specialized hardware, such as Field Programmable Gate Arrays (FPGAs), may be a very viable alternative to this. FPGAs have previously been used in bioinformatics for instance to accelerate homology search [3], multiple sequence alignment [4] and phylogeny inference [5].

In this paper, we propose and define an abstract module PAMM (Parallel Acceleration of Motif Matching). Proposing the PAMM module serves two purposes. Firstly, it introduces acceleration of motif matching by parallel hardware to the motif discovery field. Secondly, PAMM serves as an interface between the development of modules for parallel matching of motifs and the development of algorithms that can make use of parallel motif matching.

As a first implementation of our methodology, we propose a method MAMA (Massively parallel Acceleration of the Meme Algorithm) that accelerates MEME by the use of an existing pattern matching hardware called the Pattern Matching Chip (PMC) [6]. The PMC can match a subset of regular expressions with massive parallelization[1]. Since this chip was not intended for weighted pattern matching, some transformations are needed when representing and matching motifs. Nonetheless, with these transformations in place we achieve very efficient matching of PWMs against sequences. Running MAMA on a standard PC with specialized hardware on a single PCI-card compares favorably to running paraMEME on a cluster of 12 computers.

2   Parallel Acceleration of Motif Matching

An ever increasing number of computing platforms offer capabilities for parallel execution of programs. Specialized hardware exists to relieve the main CPU of specific tasks, and FPGAs allow the creation of modules for application specific hardware acceleration. To allow the field of motif discovery to realize the full potential of modern computing hardware, the algorithms need to take advantage of this.

Here we propose and define an abstract module PAMM that can be used for accelerating motif discovery by matching motifs against sequences in parallel. The purpose of PAMM is to serve as an interface between development of modules for parallel matching of motifs and the development of algorithms that can make use of parallel motif matching. An overview of the PAMM module is presented in Figure 1. The input to PAMM is a set of motifs M and a set of sequences S, while the output depends on the requirements of the algorithm in question. Each motif is represented as a matrix. As the figure shows, there are two main parts in the PAMM module; a motif matcher and a post processing unit. The motif matcher calculates the match scores for each motif, while the post processing unit refines the results.

---

[1] More information at http://www.interagon.com

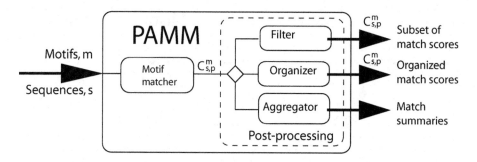

Fig. 1. The structure of the PAMM module

2.1   Motif Matching

The core of a PAMM implementation is a motif matcher that determines match scores $c_{s,p}^m$ for each motif m when aligned at each position p in each sequence s. As the number of motifs and sequences that can be processed in parallel will be limited in any practical implementation of the module, the algorithm must partition the inputs accordingly.

As a standard set-up we propose that a limited number of motifs are first loaded into the PAMM, and that sequence data are then streamed through. The motif matcher will continually calculate match scores for each motif against the sequences. When all motifs have been matched against the complete sequence data, a new set of motifs can be loaded into the module and matched against the sequences. As this means that the same sequences will typically be streamed through the PAMM many times, practical implementations could have an option to store a limited amount of sequence data in local memory to further accelerate matching and reduce bandwidth usage. This set-up is illustrated in Figure 2(a).

An alternative set-up could be to first load a limited amount of sequence data into the PAMM, and then stream motifs through the module. This could be an effective solution for cases with relatively short sequence data and large number of motifs. This setup is illustrated in Figure 2(b).

2.2   Post-processing of Match Scores

The number of results from the motif matcher is $|M| * |S|$, where M is the set of motifs and S is the set of all sequence data. This potentially large amount of results must somehow be processed by the system. By incorporating post processing, the number of results returned from a PAMM implementation can be reduced substantially. This reduces result processing in the algorithm module, as well as bandwidth requirements in the case where the PAMM and algorithm modules reside on different (sub)systems.

We envision three main branches of post processing for PAMM implementations; organizing, filtering, or aggregating (or a combination of these).

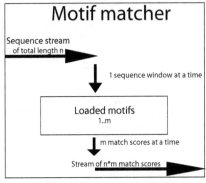

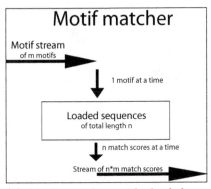

(a) Motif matcher with loaded motifs and sequences streamed through

(b) Motif matcher with loaded sequences and motifs streamed through

Fig. 2. Two possible set-ups of the motif matcher

An organizing post processor organizes the results in a way that facilitates efficient further processing of results outside the PAMM module. It could for instance return the match scores sorted by value. Although this does not decrease bandwidth usage, it may allow the CPU to process the results more efficiently.

A filtering post processor filters out uninteresting match scores to save processing time outside the PAMM module. It could for instance make the PAMM return only match scores above a threshold given for each motif. Although this discards some information, our own experiments (not presented here) show that the normalized match scores typically follow a distribution where most sequence offsets have a negligible likelihood of being motif locations. In combination with an organizing post processor, the k highest match scores could be returned, or all scores at most l lower than the highest match score.

An aggregating post processor is tailored to a specific motif discovery algorithm and may be particularly (computationally) effective. If the PAMM is to be used in connection with stochastic optimization methods like Gibbs sampling, it can be set to return one sequence offset per sequence, with offsets chosen randomly based on the normalized probabilities of motif occurrences. Alternatively, if the PAMM is used in connection with EM methods, a new motif may be constructed from the match scores directly in hardware (maximization step of EM). This new motif would represent a weighted average of every window in the sequences, with windows weighted by the match score of a previous motif.

2.3 Motif Representations

The representation of a motif in PAMM is as a motif matrix $m \in M$ with element values $m_{i,x}$, where i is motif position and x is a symbol from the alphabet, i.e., $x \in \{A, C, G, T\}$. The element values represent individual scores for each symbol x from the alphabet at each position i in the motif. The motif is aligned against sequences as a sliding window. For a given alignment at position p in sequence

s, the score of motif position i is $m_{i,x}$, where x is the symbol at position $p + i$ in sequence s. The match score $c^m_{s,p}$ of the motif is the sum of scores at each motif position. This motif representation maps directly to PWMs (log-likelihood or log-odds) that are often used for motif discovery.

In addition to PWMs, strings allowing mismatches [7,8] (a consensus string allowing a certain Hamming distance to an occurrence) and IUPAC strings [9,10] (strings of characters and character classes) are commonly used models in motif discovery. Both of these can be represented by a motif matrix. For a motif matrix representing a mismatch string, elements $m_{i,x}$ corresponding to the consensus symbol at a position have value 1, and all other matrix elements are 0. Matrix scores $c >= n - h$ corresponds to a hit for the mismatch expression, where n is motif length and h is allowed number of mismatches. This is shown in Figure 3(a). For a motif matrix representing an IUPAC string, elements $m_{i,x}$ corresponding to symbols in the character class at a position are valued 1, and all other matrix elements are 0. Matrix scores $c = n$ corresponds to a hit for the IUPAC expression. This is shown in Figure 3(b)

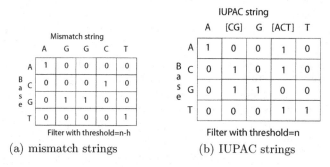

(a) mismatch strings   (b) IUPAC strings

Fig. 3. Matrix representation of discrete motif models

Other and more complex motif models could also be represented with such a matrix (variants of Markov models and bayesian trees have for instance been used in motif discovery). This will typically require a larger motif matrix and some preprocessing of the sequence data. Such preprocessing could be done by additional hardware modules within the PAMM. The generality of the matrix representation makes it suitable as a standard motif representation for the PAMM module.

3   Practical Implementation

This section describes a motif discovery algorithm that uses a PAMM implementation to accelerate motif matching. To explore the potential of PAMM in motif discovery, we have used available hardware (PMC) to implement a PAMM module.

We have analyzed the running time of the MEME algorithm and developed a motif discovery algorithm MAMA based on MEME that uses the PAMM implementation for motif matching in the performance-critial parts. As this is a first implementation and a proof-of-concept, we have only made adjustments to the MEME algorithm that make it run faster while not altering which motifs are discovered.

3.1   Motif Discovery Using the PAMM Module

MEME is a a motif discovery algorithm based on Expectation Maximization (EM) that match motifs against sequences in the expectation step. Profiling of the MEME implementation showed that matching initial motifs (starting points) against sequences consumed most of the total running time. We have therefore made the necessary adjustments to allow parallel acceleration of this first iteration of MEME.

MEME Running Time. EM was first used for motif discovery by Lawrence *et al.* [11]. As EM is easily trapped in local minima, they used several random starting points (initial PWMs) for EM. This was improved in the MEME algorithm of Bailey and Elkan [1], which use every substring of a given length in the data set as starting point. More specifically, for every substring a PWM is constructed with a fixed weight to the elements in the matrix corresponding to symbols in the substring, and another, lower fixed weight to the other elements. As this typically amounts to very many starting points, they run EM for one iteration from each starting point, and then only continue with those PWMs that seem most promising.

Inspection of the MEME implementation[2] shows that specialized code is used for this first iteration, using dynamic programming to exploit overlap between starting points. PWMs generated from each substring in the data set are first matched against the sequences (expectation step). For each PWM, the sequence offsets are then sorted by match score and the k highest scoring offsets used to generate a PWM candidate for the next iteration (maximization step). Finally, the significance values for all candidate PWMs are computed, and the most significant ones kept and refined (iterated until convergence).

MEME tries a very large number of starting points in the first iteration, and only continues with a few most promising motifs. Our profiling showed that the first iteration amounted to around 97% of total running time in our tests, using data sets supplied with MEME, the TCM model, and otherwise default parameters. Although this number might vary for different test cases and parameter settings, it shows that the first iteration is the bottleneck when it comes to running time of the algorithm. Furthermore, matching motifs against sequences and sorting offset scores dominate the running time.

Exploration of Starting Points. As the first iteration dominates the running time of MEME, we have focused on accelerating this part. More specifically, we have used the PAMM module to match PWMs and sort offset scores in the first iteration, and left the remaining parts of MEME unaltered.

---

[2] Version 3.5.0, downloaded from http://meme.nbcr.net/downloads/

Exploration of starting points differs a bit from all other iterations in MEME. First, all matrix elements of starting point PWMs has one of two values: a fixed high value for elements corresponding to the symbol of the substring it is based on, and a fixed low value for every other element. Thus, all sequence windows at a given Hamming distance from the substring a PWM is based on will get the same PWM score. Ranking of sequence offsets based on PWM score will therefore in the first iteration be equal to ranking of sequences windows based on Hamming distance. Secondly, in a general EM iteration each sequence window is used in the maximization step (weighted by the expectation values). When maximizing the PWMs in the first iteration, however, only the sequence windows corresponding to the top k expectation values are used.

These properties are exploited in MAMA by using a PAMM implementation that represents motifs efficiently and returns sequence offsets sorted by match score. The motif discovery algorithm thus only needs to consider the first k sequence offsets returned by the PAMM implementation.

3.2   Implementation of the PAMM Module

We have implemented PAMM using available hardware for parallel pattern matching. This hardware, The Pattern Matching Chip (PMC) [6], is a multiple instruction single data (MISD) parallel hardware on a PCI card. One PCI-card can match up to one thousand simple patterns against 100 MB of sequences per second, and it is quite straightforward to set up searches. Because of its efficiency and ease of use, we have used the PMC for this first implementation of the PAMM module. The PMC implementation covers both motif matching and organization of match scores.

Motif Matching. As the PMC only supports binary matching of patterns, and integer summation, the PWM match scores need to be discretized. The discretization is based on the fact that the log-likelihood for any base pair in any location is in the interval $\left[log(\frac{\beta}{n+4\beta}), log(\frac{n+\beta}{n+4\beta})\right]$, where β is the pseudo-count and n is the number of motif sites, given as parameters to MEME. Instead of using a fixed granulation of the interval, we define a granulation parameterized with ϵ. Then, each value $m_{i,x}$ in the PWM m is represented by a number $c_{i,x} = \lfloor \frac{log(m_{i,x}) - log(\frac{\beta}{n+4\beta})}{\epsilon} \rfloor$ of processing elements (PEs) in the specialized hardware. The number of PEs matching a symbol of the alphabet at a given position is thus proportional to the log-likelihood value of that symbol at that position. When the PWM is aligned with a sequence window, the sum of PE match scores at a motif position then corresponds to the *score* at that position. Note that since only one of the four nucleotides can match at a position, the other three do not contribute to the *score*. Furthermore, as PWM log-likelihood is the the sum of log-likelihoods for each position, the total PWM score is given by the sum of *scores* of all positions.

Two optimizations are worth mentioning. First, if the minimum score $c_i = min_x(c_{i,x})$ at a given position i is higher than zero, we may subtract c_i from each score value at that position, and then add c_i to the score after the search. Secondly, if $c = max_{i,x}(c_{i,x})$ is the maximum score value of the motif, and more score values are close to c than are close to zero, we then use transformed score values $c'_{i,x} = c - c_{i,x}$ and compute total PWM score as: $c \cdot I - \sum_i \sum_x c'_{i,x}$, where i runs over all I positions of m. Both optimizations give equivalent results to the basic method while using less PEs on the PMC, thus allowing more matrices to be matched simultaneously.

The discretization method considered above can be used generally for matching arbitrary PWMs against sequences. The approximation accuracy clearly depends on the granulation parameter ϵ. As discussed in section 3.1, the PWMs are regular in the first iteration of MEME. Motif matching can then be done with degenerate use of discretization, thereby avoiding approximation problems. To ensure that MAMA gives the same results as MEME, we have therefore only used hardware-acceleration in the first iteration, and used a standard software solution for motif matching in the remaining iterations. Since the running time of MEME is strongly dominated by the first iteration, we still achieve significant speed-ups.

Organizing Match Scores. As the PMC provides massive parallelity, we are able to calculate expectation values for many PWMs in parallel. We also use this parallelity to scan each PWM against the sequences several times with different hit thresholds. By searching with several thresholds in parallel, we can make the PMC return sequence offsets sorted by decreasing match score. This corresponds to a PAMM organizing module for post-processing of match scores, and avoids CPU-intensive sorting of offsets after the expectation step.

4   Results

We have compared the performance of our hardware accelerated version MAMA with the CPU based version of MEME on data sets of different sizes. On all test referred to here we have used the TCM model of MEME, which is the most general model and presented as the main model in the original MEME article [1]. We ran our tests with the following hardware configuration:

- MAMA: 2.8 Ghz Pentium4 PC with 1 GB memory and the specialized hardware on a single PCI card.
- MEME: 2.8 Ghz Pentium4 PC with 1 GB memory.
- ParaMEME: a cluster of 12 computers, each 3.4 Ghz Pentium4 PC with 1 GB memory.

We evaluated the performance of MAMA on the largest data set (mini-drosoph) supplied with MEME and on 5 data sets of human promoter regions, consisting of from 100 to 1600 sequences of 5000 base pair length from cell cycle regulated genes (J.P.Diaz, in preparation). Data sets, sizes and running times are given in Table 1 for both MEME, paraMEME and MAMA. We see that MAMA gives a

Table 1. Results for MEME, paraMEME and MAMA on 6 data sets

Data set	Size (Mbp)	Running time (hours)		
		MEME	paraMEME	MAMA
mini-drosoph	0.5	2.6	0.19	0.27
hs_100	0.5	2.7	0.20	0.23
hs_200	1	11	0.87	0.50
hs_400	2	104	3.6	1.7
hs_800	4	X^3	15	6.4
hs_1600	8	X^3	64	13

significant speed-up compared to MEME on all datasets, and that the speed-up increases with data set size. On the 1 Mbp (Million base pairs) data set, MAMA is more than twenty times as fast as MEME, and on the 8 Mbp data set it is even four times as fast as paraMEME on the 12-computer cluster. For all data sets, standard MEME and the hardware-accelerated version MAMA discovers the same motifs.

5 Discussion and Conclusion

We have proposed an abstract module PAMM for parallel hardware-acceleration of motif discovery. This module could be used for acceleration of many different motif discovery methods. The acceleration could be especially large if postprocessing of match scores is tailored to a specific algorithm.

As an exemplification and proof-of-concept we have developed a version of the MEME algorithm called MAMA that uses available hardware to implement a PAMM module. As shown in section 4, MAMA achieves a speed-up of more than a factor of 10 as compared to MEME on a single CPU. Our working implementation thus shows that the PAMM module indeed has a potential.

Furthermore, our work shows examples of both problematic issues and potential rewards in connection with hardware acceleration of algorithms within bioinformatics. Since we have implemented weighted motif matching on hardware that was not specifically built for that purpose, we had to do some transformations of the problem. The issues and solutions with regards to discretization and parallelization are relevant for many algorithmic solutions involving specialized hardware.

A natural continuation of the work presented in this paper is to develop a FPGA-based implementation of PAMM. Such a solution would be more readily available for practical use and further refinement by the scientific community. It could potentially also give even higher speed-ups. On the other hand, such a solution presumes a solution of representing PWMs on FPGA that is both efficient and flexible. We have ongoing work in this direction that shows promising results.

---

[3] Not tested due to excessive running times.

References

1. Bailey, T.L., Elkan, C.: Fitting a mixture model by expectation maximization to discover motifs in biopolymers. Proc. Conf. Intell. Syst. Mol. Biol. ISMB'94 (1994) 28–36
2. Grundy, W.N., Bailey, T.L., Elkan, C.P.: ParaMEME: a parallel implementation and a web interface for a DNA and protein motif discovery tool. Comput. Appl. Biosci. **12** (1996) 303–310
3. Yamaguchi, Y., Miyajima, Y., Maruyama, T., Konagaya, A.: High speed homology search using run-time reconfiguration. LNCS. Volume 2438. (2002) 281–291
4. Oliver, T., Schmidt, B., Nathan, D., Clemens, R., Maskell, D.: Using reconfigurable hardware to accelerate multiple sequence alignment with ClustalW. Bioinformatics **21**(16) (2005) 3431–3432
5. Mak, T.S.T., Lam, K.P.: Embedded computation of maximum-likelihood phylogeny inference using platform FPGA. In Proc. Comput. Systems Bioinformatics Conf. CSB'04, IEEE. (2004) 512–514
6. Halaas, A., Svingen, B., Nedland, M., Sætrom, P., Snøve Jr., O., Birkeland, O.R.: A recursive MISD architecture for pattern matching. IEEE Trans. Very Large Scale Integr. Syst. **12**(7) (2004) 727–734
7. Marsan, L., Sagot, M.F.: Extracting structured motifs using a suffix tree-algorithms and application to promoter consensus identification. In: Proc. 4th Int'l Conf. Comput. Mol. Bio. RECOMB'00, ACM Press (2000) 210–219
8. Blanchette, M., Tompa, M.: Discovery of regulatory elements by a computational method for phylogenetic footprinting. Genome Res. **12**(5) (2002) 739–748
9. Sinha, S., Tompa, M.: YMF: A program for discovery of novel transcription factor binding sites by statistical overrepresentation. Nucleic Acids Res. **31**(13) (2003) 3586–3588
10. Bortoluzzi, S., Coppe, A., Bisognin, A., Pizzi, C., Danieli, G.: A multistep bioinformatic approach detects putative regulatory elements in gene promoters. BMC Bioinformatics **6**(1) (2005) 121
11. Lawrence, C.E., Reilly, A.A.: An expectation maximization (EM) algorithm for the identification and characterization of common sites in unaligned biopolymer sequences. Proteins **7**(1) (1990) 41–51

Segmenting Motifs in Protein-Protein Interface Surfaces[*]

Jeff M. Phillips[1], Johannes Rudolph[2], and Pankaj K. Agarwal[1]

[1] Department of Computer Science, Duke University
[2] Department of Biochemistry, Duke University

Abstract. Protein-protein interactions form the basis for many inter-cellular events. In this paper we develop a tool for understanding the structure of these interactions. Specifically, we define a method for identifying a set of structural motifs on protein-protein interface surfaces. These motifs are secondary structures, akin to α-helices and β-sheets in protein structure; they describe how multiple residues form knob-into-hole features across the interface. These motifs are generated entirely from geometric properties and are easily annotated with additional biological data. We point to the use of these motifs in analyzing hotspot residues.

1 Introduction

Interactions between proteins govern many intercellular events, yet are poorly understood. These interactions lie at the heart of cell division and cell growth, which in turn dictate the pattern of health versus disease. A better understanding of protein-protein interactions will enhance our understanding of biological processes and how we can manipulate them for the benefit of human health.

Protein-protein interfaces. The protein-protein interface defines the essential region of a protein-protein interaction. Most attempts [10,12,20] at defining this interface include all atoms from one protein within some distance cutoff $(4-5\text{Å})$ from atoms of the other protein. This approach does not provide independent structural information and makes it difficult to identify features or subregions of the interface. To follow standard notation, we refer to the interface between two protein chains (say A and D) in a protein complex (say 1brs) as 1brsAD.

A more recent approach to defining the protein-protein interface [3] constructs a surface equidistant to both chains of a complex using the Voronoi diagram. This otherwise infinite structure is bounded using topological techniques. Each atom that contributes to the interface is associated with at least one polygon on the interface surface, and each polygon is associated with two atoms, one from

---

[*] Research supported by NSF grant CCR-00-86013 and NIH GM061822. J.M.P. is also supported by an NSF GRF and a JB Duke Fellowship; and P.K.A. is also supported by NSF under grants EIA-98-70724, EIA-01-31905, and CCR-02-04118, and by a grant from the U.S.–Israel Binational Science Foundation.

P. Bücher and B.M.E. Moret (Eds.): WABI 2006, LNBI 4175, pp. 207–218, 2006.

each side. The freely available MAPS: Protein Docking Interfaces software [2] has been developed to visualize protein interfaces and their relevant properties. The MAPS software displays the interface surface embedded between the protein structures as well as a simplified flattened view. The flattened interface surface can be colored to show physico-chemical properties such as residue type, atom type, distance to closest atom, or electrostatics, as associated with each polygon. A merged view allows both sides of the interface to be viewed simultaneously.

Features of surfaces. Proteins have evolved extensive shape complementarity to their interacting proteins. Because protein surfaces are not flat, this leads to knob-into-hole structures at the protein-protein interface, as noted by Connolly [6]. These knobs consist of a set of atoms, often from a single residue, that protrude from one protein into the other protein of the complex. This interaction may be difficult to capture by examination of the structure of the protein complex. However, these knob-into-hole features can be readily identified visually in nearly all protein-protein complexes as bumps on the interface surface. Yet, their biological significance is not well understood, in part because there exists no standard definition or automated method of indentification. Our goal is to automatically segment these regions on the interface surface in a consistent way. We call these knob-in-hole features motifs.

Feature extraction is a common problem in computer vision [5,13,19], where features points are used to provide a correspondence between two surfaces. However, these techniques only identify specific points as opposed to interesting regions. For closed curves and surfaces, such as protein surfaces, interesting regions corresponding to knobs and holes can be defined using a variety techniques like elevation [1,18]. However, these approaches do not generalize to surfaces with boundary, such as the interface surface, and thus the bumps at the interaction site corresponding knobs and holes could not be identified. On a surface with boundary, the problem of defining regions of significance is quite challenging. There is no clear measure of depth, the overall geometry can obscure local features of interest, and local features can be excessively fragmented. As such, it is hard to quantify bumps or pockets, much less segment them.

Our contribution. We adapt standard notions of discrete curvature on a polygonal surface to be less local. Using this globally-aware definition we grow regions with large curvature to segment regions of interest on the interface surface. We then integrate these motifs with the MAPS software to incorporate structural information into this convenient visualization tool. Finally we demonstrate some interesting properties of these motifs.

2   Discrete Curvature and Watershed Procedures on Polygonal Surfaces

The protein-protein interface surface as defined in [3] is the piecewise-linear surface, that is everywhere equidistant from the closest atom in each of the two proteins. We represent this structure as a set of triangles that are glued together with edges and vertices denoted $\Sigma = \{T, E, V\}$, respectively. All interface surfaces we deal with separate two proteins[1], and thus Σ is an orientable piecewise-linear 2-manifold with boundary in $\mathbb{R}^3$. We can thus arbitrarily choose one of these proteins and let the normal direction of every triangle $t \in T$ point towards this protein.

Vertices, edges, and triangles are *simplices*. A vertex is a 0-simplex, an edge is a 1-simplex, and a triangle is a 2-simplex. The *star* of a simplex σ, denoted $\mathrm{St}(\sigma)$, is the set of simplices incident to σ. More generally, if Υ is a set of simplices, then $\mathrm{St}(\Upsilon)$ is the set of simplices incident to any simplex in Υ. The *link* of a set of simplices Υ, denoted $\mathrm{Lk}(\Upsilon)$, is the boundary of the closure of $\mathrm{St}(\Upsilon)$. All interior vertices have the same number of incident edges and incident triangles while boundary vertices have one fewer incident triangles than they do incident edges. All vertices have at least 3 incident edges. For each triangle $t \in \mathrm{St}(v) \cap T$, let $\alpha_{v,t}$ be the angle between the two edges of t incident to v, as shown in Figure 1.

2.1   Discrete Curvature

To formalize the notion of how much the interface surface is locally bending, we appeal to the idea of curvature. Any point p on a smooth surface can be assigned a value of Guassian curvature and mean curvature [7]. We extend these notions to vertices on a polygonal surface. The standard [15] definition for *discrete Gaussian curvature* of a vertex v is

$$K(v) = 2\pi - \sum_{t \in \mathrm{St}(v) \cap T} \alpha_{v,t}.$$

Intuitively, $K(v)$ describes the angle defect, or how far the angles of the surface surrounding v are from those of a flat surface.

There is no standard definition for the discrete mean curvature of a vertex; it is only defined on edges. We present a new definition for discrete mean curvature, H, on vertices. For a triangle $t \in T$, let $\mathbf{n}_t$ denote the unit vector in the direction normal to triangle t. We define the normal direction of a vertex v to be the weighted mean of the normal directions of all of its incident triangles, as follows

$$\mathbf{n}_v = \left(\sum_{t \in \mathrm{St}(v) \cap T} \alpha_{v,t} \mathbf{n}_t \right) \Big/ \left(\sum_{t \in \mathrm{St}(v) \cap T} \alpha_{v,t} \right). \tag{1}$$

For a triangle t and a vertex v incident to t, we define the face vector $\mathbf{f}_{v,t}$ to be the unit vector from v to the mid point of the edge opposite v on t. Now we define *discrete mean curvature* of v as

---

[1] An interface surface can be defined to separate more than two proteins in a larger complex. In this case the surface is not an orientable 2-manifold because an edge can have three adjacent triangles. However, we can always divide this larger surface into orientable 2-manifold components.

$$H(v) = \left(\sum_{t \in \mathrm{St}(v) \cap T} \alpha_{v,t}(\mathbf{n}_v \cdot \mathbf{f}_{v,t}) \right) \Big/ \left(\sum_{t \in \mathrm{St}(v) \cap T} \alpha_{v,t} \right), \tag{2}$$

where $\cdot$ indicates a dot product (see Figure 1). Intuitively, $H(v)$ denotes the

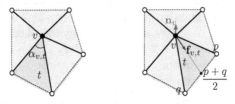

Fig. 1. Star of vertex v to illustrate the calculation of K (left) and H (right)

average deviation from $\mathbf{n}_v$ of the normals of all triangles incident to v, weighted by their incident angles. For a vertex v, if $K(v)$ is positive, then the sign of $H(v)$ determines if the surface surrounding v is curving towards the normal direction, or away from the normal direction. If a polygonal surface is a piecewise-linear approximation of a smooth surface, then the area of triangles, instead of angles can be used as weights in (1) and (2). Since, an interface surface does not approximate a smooth surface, we use the angles of the incident triangles as weights. As the surface is subdivided, the curvature of the surface does not change, and neither do the incident angles.

2.2　Watershed Procedures

Let $h : \Sigma \to \mathbb{R}$ be a function defined on a surface Σ. If Σ is $\mathbb{R}^2$, we can imagine h representing the height of a terrain. Suppose we start pouring water on the terrain and monitor the structure of the flooded regions. The connectivity of the flooded regions changes at the critical points of the terrain: minima, maxima, and saddle points. Lakes are created at minima, islands disappear at maxima, and lakes merge or islands form at saddle points. At saddle points, instead of always merging lakes or creating islands, a dam can be built to define a boundary between two regions. This approach, known as the watershed algorithm [17], can be extended to height functions on 2-manifolds and the dam locations can be used to segment Σ.

However, there is no clear notion of a height function on an arbitrary 2-manifold. Using K or H as the height does not segment Σ well because curvature is a local property. Hence, we modify the above watershed algorithm in two ways. First we run the watershed algorithm from two sides of Σ simultaneously using K, but use the sign of H to determine on which of the two sides a flooded component lies. We create dams when two components from opposite sides meet,

and we merge components from the same side. We face another technical problem
if we use K as the height function. Consider the motif (in purple) on the interface
surface 1atnAD shown in Figure 2. The value of $K(v_1)$ is large, but $K(v_2)$ is
small. A watershed algorithm directly using the function K would not identify
the motif shown; rather it would just segment the tip of the motif. By altering
the value of $K(v)$ in a careful manner, the algorithm lets components grow in
large steps. How we redefine the values of $K(v)$ will be explained in more detail
in Section 3.1.

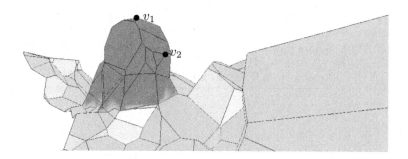

Fig. 2. One motif on interface surface 1atnAD. $K(v_1)$ is large, but $K(v_2)$ is not.

3   Algorithm for Finding Motifs

Each motif we construct is a component of the subset of the interface surface
visited by the modified watershed algorithm. A motif is represented by the set
of vertices in its interior. We let M denote both a motif and its set of vertices. A
sign $\varsigma(M) \in \{-, +\}$ is assigned to each motif M when it is created. The *rim* of
M, denoted by $\mathrm{rim}(M)$, is the set of vertices in $\mathrm{Lk}(M)$. If there are two motifs
M and M', with $\varsigma(M) \neq \varsigma(M')$, and there is a vertex $r \in \mathrm{rim}(M) \cap \mathrm{rim}(M')$,
then r is called a *dam* vertex. A dam vertex never becomes an interior vertex of
a motif.

Algorithm 3.1 outlines our algorithm. We set a threshold τ. At each step the
algorithm chooses a vertex v with $K(v) \geq \tau$. It makes v a dam vertex, adds v to
a motif, creates a new motif with v as its interior vertex, or merges two motifs
that share v on their rims. It then recomputes the value of $K(\cdot)$ on the "affected"
rim vertices. The worst-case running time of the algorithm is $O(|V|^2)$. However
the algorithm took 1 to 75 seconds on the 143 interface surfaces we tested; these
surfaces consisted of between 943 and 9096 simplices.

Figure 3 displays the interface 1brsAD from two angles. The motifs are colored
in two shades, representing either a knob of chain A protruding into a hole in
chain D or vice-versa. We only show motifs larger than 20Å^2 to avoid clutter.
There are three large motifs (in front on the left view and on the left in the right
view) that form significant knob-into-hole structures. The four smaller motifs
mirror this knob-into-hole pattern, albeit less dramatically.

Algorithm 3.1. FIND-MOTIFS(Σ, τ)

1:   U: set of interior vertices of Σ.
2:   **while** $(\max_{v \in U} K(v) \geq \tau)$ **do**
3:      Let $v = \arg\max_{v \in U} K(v)$; $U := U \setminus \{v\}$.
4:      **if** $(v \in \mathrm{rim}(M) \cap \mathrm{rim}(M') \wedge \varsigma(M) \neq \varsigma(M'))$ **then**
5:         Mark v as a dam vertex.
6:      **else**
7:         **if** (v is not a rim vertex) **then**
8:            Create a new motif $M = \{v\}$; $\varsigma(M) = \mathrm{sign}(H(v))$.
9:         **else**
10:           Merge all motifs whose rim contains v into a single motif M.
11:           Add v to M.
12:         Compute $\mathrm{rim}(M)$.
13:         **for all** $r \in \mathrm{rim}(M)$ **do**
14:           Recompute $K(r)$ as described in Section 3.1.

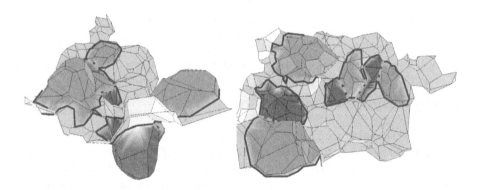

Fig. 3. Interface surface and motifs for 1brsAD from two views. Darkly shaded motifs and lightly shaded motifs have opposite signs of H, meaning they are protruding in different directions. The rims are the edges in bold, except between the dam vertices where they have light dashes.

3.1 Computing the Curvature on the Rim

To compute the curvature for vertices on the rim we implicitly remove the vertices of the motif and treat this part of the surface as if Saran wrap was placed over the removed motif. The idea is that, if an entire motif is chopped off then the surface around it should appear flat; however, if part of the motif is chopped off then the vertices on the rim are now at the tip of a motif and thus should still have large curvature. Actually reconstructing this Saran wrap surface is difficult, and in general can not be done without adding new vertices. However, we can conservatively approximate the star of any vertex in the rim without constructing the full Saran wrap surface. For a vertex $r \in \mathrm{rim}(M)$ we call $\mathrm{St}(r) \setminus \mathrm{St}(M)$—the part of the star of r outside of M—the *upper star* of r and we denote it $\mathrm{St}_+(r)$.

Similarly, the *lower star* of r is denoted $\mathrm{St}_-(r) = \mathrm{St}(r) \cap \mathrm{St}(M)$. We compute the upper normal of r,

$$\mathbf{n}_r^+ = \left(\sum_{t \in \mathrm{St}_+(r) \cap T} \alpha_{r,t} \mathbf{n}_t \right) \Big/ \left(\sum_{t \in \mathrm{St}_+(r) \cap T} \alpha_{r,t} \right).$$

If $\varsigma(M)$ is negative (the surface is curving away from $\mathbf{n}_r^+$) we define h_r as the highest vertex in $\mathrm{rim}(M) \setminus \{r\}$, in the direction of $\mathbf{n}_r^+$. If $\varsigma(M)$ is positive (the surface is curving towards $\mathbf{n}_r^+$), we define h_r as the lowest vertex in $\mathrm{rim}(M) \setminus \{r\}$. We then construct a new star of r by using $\mathrm{St}_+(r)$ and replacing $\mathrm{St}_-(r)$ with the two faces and the edge connecting h_r to the part of $\mathrm{St}_+(r)$ that is incident to M, as in Figure 4. The choice of h_r from $\mathrm{rim}(M) \setminus \{r\}$ causes the new surface around r to be as flat as possible. Using the reconstructed star of r, $K(r)$ can be computed as before.

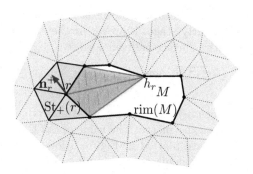

Fig. 4. Recreation of $\mathrm{St}(r)$ on $\mathrm{rim}(M)$. $\mathrm{St}_+(r)$ is colored lighter. The Saran wrap surface is colored darker.

3.2   Choosing the Threshold τ

Algorithm 3.1 can be run incrementally by gradually lowering τ. At certain values of τ a new vertex v is handled in Step 3. Let τ_v denote the threshold τ at which v first becomes part of a motif M. Often the value $K(r)$ for a vertex $r \in \mathrm{rim}(M)$ increases when reevaluated on the rim and sometimes becomes greater than τ_v. In this case $\tau_r = \tau_v$ and r is immediately handled before τ is lowered. This process often continues until a large motif has been created. As a result, motifs usually grow in large spurts at particular thresholds, as seen in Figure 5. Eventually, as τ becomes quite small, U becomes empty. This means flat parts of the surface have become part of motifs, often concatenating several large motifs. Choosing an appropriate value of τ in Algorithm 3.1 will halt the algorithm before this later stage happens.

Although any threshold between 0.3 and 0.4 radians seems to work reasonably well for most surfaces, we can do better by choosing τ for each surface individually. For instance, surfaces bent around a protein often require larger thresholds, while relatively flat surfaces often require smaller ones. Instead of trying to quantify this bend, we find the threshold that optimizes a score function over the motifs.

As a vertex v is added to a motif M it is given a weight

$$w(v) = \text{Area}(\text{St}_+(v)) \cdot \tau_v$$

which is equal to τ_v times the area that M is incremented. The value τ_v is never increasing during the algorithm, so vertices added earlier are given more weight.

Motifs are also given a penalty for merging. The penalty $p(M)$ is initially set to 0. When M merges with another smaller motif, M', $p(M)$ is incremented by $\text{Area}(M')$. When a smaller motif is merged with a larger motif, it disappears so its score is essentially set to 0. Each motif M is given a score

$$s_\tau(M) = \sum_{v \in M} w(v) - p(M)$$

for its state at a threshold τ. Thus larger motifs are given more weight, especially if they are created at an early stage. However, if they merge, their scores are heavily penalized. An entire surface is given a score

$$S_\tau(\Sigma) = \text{sign}(\xi_\tau(\Sigma))\sqrt{|\xi_\tau(\Sigma)|}, \text{ where } \xi_\tau(\Sigma) = \sum_{M \in \Sigma} \text{sign}(s_\tau(M)) \cdot s_\tau(M)^2.$$

As τ decreases, $S_\tau(\Sigma)$ usually increases to an optimal value and then decreases as the motifs grow too large and merge. The value of $S_\tau(\Sigma)$ as well as the values $s_\tau(M)$ for all motifs in Σ are plotted in Figure 5 for the interface surface for 1nmbN{L,H}. The score is calculated by running Algorithm 3.1 for τ set to .50 through .25 at .01 intervals. The threshold which returns the largest score $S_\tau(\Sigma)$ is used, which in this case is $\tau = 0.34$. The interface surface and motifs for 1nmbN{L,H} are also shown for different values of τ. Only motifs with surface area larger than 20Å^2 are displayed to avoid clutter.

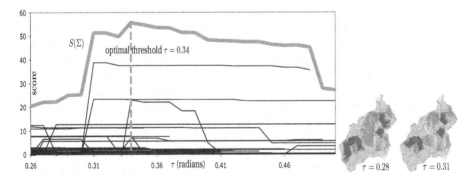

Fig. 5. Plot of $S(\Sigma)$ and $s(M)$ for all motifs on interface surface for 1nmbN{L,H}. (top) Interface surface of 1nmbN{L,H} at $\tau = \{0.27, 0.31, 0.34, 0.37, 0.40\}$. (bottom)

4   Applications

The reproducible and automated identification of motifs on the interface surface
provides a novel tool for biochemists to characterize protein-protein interactions.
We present here two preliminary forays into possible applications. Just as the
designation of α-helices and β-sheets has helped enhance the understanding of
the structure of a protein beyond what could be extracted from the atomic
coordinates and their ordering on the backbone alone, we envision interface
motifs having the same contributions for understanding the structure of the
protein-protein interface.

4.1   Visualizing the Motifs on the Interface Surface

To readily visualize these structures we have integrated the motifs into the MAPS
site [2]. The MAPS software contains a database of over 150 protein-protein interfaces. It displays both a flattened version of the interface and a 3D version
embedded within the protein structures, side-by-side. By flattening the interface,
the 3D structure is removed so that the physical and chemical properties from
one or both sides can be more easily visualized. Mapping the motifs into this flattened view reintroduces structural information and allows for facile comparisons
between knobs and holes and physical and chemical properties.

Figure 6 shows a snapshot of the MAPS software displaying the interface
surface for 1brsAD. On the left is the standard 3-dimensional view embedded in
the structures of the protein. In the middle the flattened interface is colored by its
motifs as seen from chain D. The two shades of motifs correspond to either chain
A protruding into chain D or vice-versa. The thick black lines show the outline
of the surface patches corresponding to different residues. Note how the large

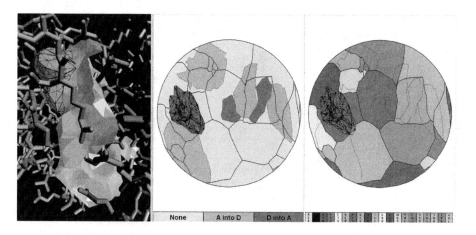

Fig. 6. Snapshot of MAPS software: Shows interface surface 1brsAD with protein complex drawn as a stick model (left), motifs on flattened interface (middle), and residue
type on flattened interface. Motif corresponding to Asp35 is highlighted in all views.

lightly shaded motifs are at the junction of many residue patches wherein the hole in chain D, made by the knob in chain A, is composed of multiple residues. On the other hand, the darkly shaded motifs are generally at the center of a single residue patch wherein the knob in chain D consists of only one residue. This pattern is reversed if the other side of the interface is shown. On the right the flattened interface is colored by its amino acid composition as seen from chain D. A single motif is highlighted in all views. Note the large overlap of the highlighted motif with Asp35, one of the key hotspot residues in chain D of the 1brsAD complex.

4.2   Hotspot Residues and Motifs

Hotspot residues for a protein complex are those few residues (5-10%) at a protein interface whose mutation leads to a significant reduction in the binding energy ($\Delta\Delta G \geq 2$ kcal/mol) [4]. The identification of hotspots is not necessarily a matter of trivial inspection as essentially any type of residue can be a hotspot. Thus various computational methods have been developed to predict hotspots, relying either on traditional force fields [9,14] or simpler physical energy functions parameterized on experimental data [8,11], with reasonable to good results.

Intuitively their exists a potential relationship between motifs and hotspot residues. In particular, one might expect prominent knobs associated with a single residue to be likely hotspot residues. For illustration we consider the complex 1brsAD. For the 14 residues on the interface characterized by mutation [16] we

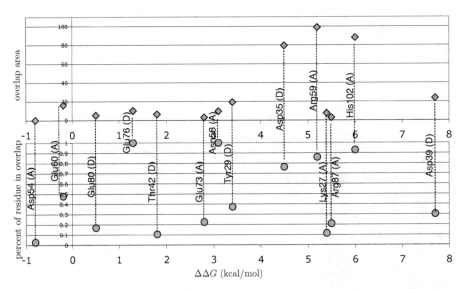

Fig. 7. Plots comparing hotspot residues to motifs on the interface surface 1brsAD. Top chart plots the area of the overlap between the hotspot residue and the motif best covering it on the interface surface versus $\Delta\Delta G$ of that residue. Bottom chart plots the percent of the residue area which is in the overlap region versus the $\Delta\Delta G$ of the residue.

compare each residue to the motif which has the largest area of overlap on the interface surface. That is, for each residue we compute the area of overlap and the percent of its area that is in this overlap region. In Figure 7 we plot these values versus the $\Delta\Delta G$s of the mutated residues. Three residues, including Asp35, are clearly identified as hotspots whereas three hotspot residues with $\Delta\Delta G > 4$ kcal/mol are not identified by this method. Satisfyingly, none of the non-hotspot residues are incorrectly labeled as such.

Future inquires into this relationship will incorporate a larger dataset of hotspot residues and the chemical properties of the potential hotspot residues.

5 Future Work

In addition to a possible correlation with hotspot residues, we envision that our structural annotation tool could aid other biological applications. For instance, statistical analysis may reveal preferred knob residues whose identification may be useful for prediction of protein docking. Moreover, overlaying motifs from different interfaces may lead to classifications by size or shape that can be useful for describing recurring interfacial motifs. Also, by observing higher order relationships of motif arrangements on the protein interface (tertiary structures), we envision new ways of comparing and classifying protein-protein complexes.

Acknowledgments

We thank Tammy Bailey and Herbert Edelsbrunner for observations and suggestions and Jeffrey Headd for help integrating the motifs into the MAPS software.

References

1. P.K. Agarwal, H. Edelsbrunner, J. Harer, and Y. Wang. Extreme elevation on 2-manifold. *Proc. 20th ACM Symp. on Computational Geometry SoCG'04*, 2004.
2. Y.-E.A. Ban, P.L. Brown, H. Edelsbrunner, J.J. Headd, and J. Rudolph. MAPS: Protein docking interfaces. http://biogeometry.cs.duke.edu/research/docking/index.html, May 2006.
3. Y.-E.A. Ban, H. Edelsbrunner, and J. Rudolph. Interface Surface for Protein-Protein Complexes. *in press, J. ACM*.
4. A.A. Bogan and K.S. Thorn. Anatomy of hot spots in protein interfaces. *J. of Molecular Biology*, 280:1–9, 1998.
5. C.S. Chua and R. Jarvis. Point signatures: A new representation for 3d object recognition. *Int'l J. of Computer Vision*, 25(1):63–85, 1997.
6. M.L. Connolly. Shape complementarity at the hemoglobin $\alpha_1\beta_1$ subunit interface. *Biopolymers*, 25(7):1229–1247, 1986.
7. M.P. do Čarmo. *Differential Geometry of Curves and Surfaces*. Prentice-Hall, Upper Saddle River, NJ, 1976.
8. R. Guerois, J.E. Nielsen, and L. Serrano. Predicting changes in the stability of proteins and protein complexes: A study of more than 1000 mutations. *J. of Molecular Biology*, 320:369–387, 2002.

9. S. Huo, I. Massova, and P.A. Kollman. Computational alanine scanning of the 1:1 human growth hormone-receptor complex. *J. of Computational Chemistry*, 23:15–27, 2002.

10. S. Jones and J.M. Thornton. Analysis of protein-protein interaction sites using surface patches. *J. of Molecular Biology*, 272:121–132, 1997.

11. T. Kortemme and D. Baker. A simple physical model for binding energy hot spots in protein-protein complexes. *Proc. National Academy of Science USA*, 99:14116–14121, 2002.

12. L. Lo Conte, C. Chothia, and J. Janin. The atomic structure of protein-protein recognition sites. *J. of Molecular Biology*, 285:2177–2198, 1999.

13. D.G. Lowe. Object recognition and local scale-invariant features. *Proc. 7th IEEE Int'l Conf. on Computer Vision ICPV'99*, 2:1150–1157, 1999.

14. I. Massova and P.A. Kollman. Computational alanine scanning to probe protein-protein interactions: A novel approach to evaluate binding free energies. *J. of the American Chemical Society*, 121:8133–8143, 1999.

15. M. Meyer, M. Desbrun, P. Schröder, and A.H. Barr. Discrete differential-geometry operators for triangulated 2-manifolds. *Visualization and Mathematics III*, 2003.

16. G. Schreiber and A.R. Fersht. Energetics of protein-protein interactions: Analysis of the barnase-barstar interface by single mutations and double mutation cycles. *J. of Molecular Biology*, 248:478–486, 1995.

17. L. Vincent and P. Soille. Watersheds in digital spaces: An efficient algorithm based on immersion simulations. *IEEE Transactions on Pattern Analysis and Machine Intelligence*, 13(6):583–598, 1991.

18. Y. Wang, P.K. Agarwal, P. Brown, H. Edelsbrunner, and J. Rudolph. Coarse and reliable geometric alignment for protein docking. *10th Pacific Symp. on Biocomputing PSB'05*, pages 64–75, 2005.

19. Y. Wang, B.S. Peterson, and L.H. Staib. 3d brain surface matching based on geodesics and local geometry. *Computer Vision and Image Understanding*, 89:252–271, 2003.

20. D. Xu, C.-J. Tsai, and R. Nussinov. Hydrogen bonds and salt bridges across protein-protein interfaces. *Protein Engineering*, 10(9):999–1012, 1997.

Protein Side-Chain Placement Through MAP Estimation and Problem-Size Reduction

Eun-Jong Hong and Tomás Lozano-Pérez

Computer Science and Artificial Intelligence Lab, MIT,
Cambridge, MA 02139, USA
{eunjong, tlp}@mit.edu

Abstract. We present an exact method for the global minimum energy conformation (GMEC) search of protein side-chains. Our method consists of a branch-and-bound (B&B) framework and a new subproblem-pruning scheme. The pruning scheme consists of upper/lower-bounding methods and problem-size reduction techniques. We explore a way of using the tree-reweighted max-product algorithm for computing lower-bounds of the GMEC energy. The problem-size reduction techniques are necessary when the size of the subproblem is too large to rely on more accurate yet expensive bounding methods. The experimental results show our pruning scheme is effective and our B&B method exactly solves protein sequence design cases that are very hard to solve with the dead-end elimination.

1 Introduction

A computational approach to the protein structure prediction problem is to solve the "inverse folding problem": to find a sequence or conformation that will fold to the target structure [1]. In this approach, the search of the minimum energy conformation is an important computational challenge. Two major applications where finding the minimum energy conformation is useful and necessary are the conformation modeling (homology modeling) problem [2] and the sequence design problem [3]. In finding the minimum energy conformation, the problem is discretized and simplified by computing the interaction energies only for some finite number of fixed side-chain conformations of each residue type [4]. These conformations are chosen by their statistical significance and are called rotamers. With the rotamer model, the energy function of a protein sequence folded into a specific template structure can be described in terms of [5]: (1) $E_{template}$ – the self-energy of a backbone template, (2) $E(i_r)$ – the interaction energy between the backbone and rotamer conformation r at ith position, (3) $E(i_r j_s)$ – the interaction energy between rotamer conformation r at position i and rotamer conformation s at position j, $i \neq j$. Then, the energy of a protein sequence in a specific template structure and conformation $C = \{i_r\}$ is written as $\mathcal{E}(C) = E_{template} + \sum_i E(i_r) + \sum_i \sum_{j>i} E(i_r j_s)$. Note that $E_{template}$ is constant by definition, and therefore can be ignored when minimizing $\mathcal{E}(C)$. A conformation that minimizes $\mathcal{E}(C)$ is often called the global minimum energy

P. Bücher and B.M.E. Moret (Eds.): WABI 2006, LNBI 4175, pp. 219–230, 2006.
© Springer-Verlag Berlin Heidelberg 2006

conformation (GMEC). In this work, we call the problem of finding the GMEC for a given set of rotamers and energy terms as the "GMEC problem".

The GMEC problem is a strongly NP-hard optimization problem. Despite the theoretical hardness, one finds that many instances of the GMEC problem are easily solved by the exact method of dead-end elimination (DEE) [5]. Popularly used elimination procedures such as Goldstein's conditions [6] combined with splitting [7], the magic bullet heuristic [8], and unification [6] are often able to reduce the problem size dramatically, while demanding only reasonable computational resources. However, we still find sequence design cases where DEE requires impractical amount of time and space. Other than DEE, there exist various exact approaches for the GMEC problem. Gordon and Mayo [9] used a variant of the branch-and-bound (B&B) method. Althaus et al. [10], Eriksson et al. [11], and Kingsford et al. [12] present integer linear programming (ILP) approaches. Leaver-Fay et al. [13] and Xu [14] describe methods based on tree-decomposition. Xie and Sahinidis [15] describes several residue-reduction and rotamer-reduction techniques. Each approach has advantages depending on the characteristics of the data, but most of them have not attempted to solve hard protein design cases, where there exist interactions between all possible pairs of positions and a large number of similar rotamers are allowed for each position. There also exist approximate approaches such as Yanover and Weiss [16] who used belief-propagation methods to solve side-chain placement problems.

In this work, we present an alternative exact solution method for the GMEC problem. Figure 1 illustrates the method. Our method consists of a B&B framework and a new subproblem-pruning scheme. The pruning scheme consists of upper/lower-bounding methods and problem-size reduction techniques. The basis for our upper/lower-bounding method is approximate *maximum-a-priori* (MAP) estimation. Particularly, we explore a way of using the tree-reweighted max-product algorithm (TRMP) [17]. The problem-size reduction techniques are necessary when TRMP can only compute weak bounds but the size of the subproblem is too large to rely on more accurate yet expensive bounding methods. Through an iterative use of several reduction techniques, we can obtain a problem of reasonable size that can be effectively lower-bounded. Such reduction techniques guarantee that the given subproblem can be pruned against an upper-bound U if the reduced subproblem can be pruned against U. On the other hand, if we are lucky, a subproblem can be also quickly solved using DEE only. The experimental results show that the running time of our pruning scheme is comparable to linear programming (LP) but our method is more effective in pruning subproblems than LP. We also find our B&B method exactly solves sequence design cases that are very hard to solve with DEE.

2   GMEC Problem as MAP Estimation

Probabilistic inference problems [18] involve a vector of random variables $\mathbf{x} = (x_1, x_2, \ldots, x_n)$ characterized by a probability distribution $p(\mathbf{x})$. In this work, the GMEC problem is formulated as a MAP estimation problem that asks to find the

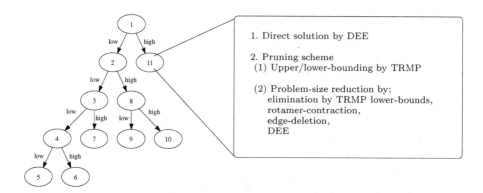

Fig. 1. An overview of the exact method for the GMEC problem. The method consists of a branch-and-bound framework and a pruning scheme, which in turn is composed of bounding by TRMP and a collection of problem-size reduction techniques. Labels on branches are related to the splitting scheme, and the numbers marked on the nodes correspond to the order by which the nodes are visited.

maximum a posteriori (MAP) assignment $\mathbf{x}^*$ such that $\mathbf{x}^* = \arg\max_{\mathbf{x} \in \mathcal{X}} p(\mathbf{x})$, where $\mathcal{X}$ is the sample space for $\mathbf{x}$. In the GMEC problem, we number the sequence positions by $i = 1, \ldots, n$, and associate with each position i a discrete random variable x_i that ranges over R_i, a set of allowed rotamers at position i. Then, we can define a probability distribution $p(\mathbf{x})$ over $\mathcal{X} = R_1 \times \ldots \times R_n$ as

$$p(\mathbf{x}) = \exp\{-e(\mathbf{x})\}/Z, \tag{1}$$

for a normalization constant Z and $e(\mathbf{x}) = \sum_{i=1}^{n} e_i(x_i) + \sum_{i=1}^{n-1} \sum_{j=i+1}^{n} e_{ij}(x_i, x_j)$, where $e_i(r) = E(i_r)$ for $r \in R_i$, and $e_{ij}(r, s) = E(i_r j_s)$ for $(r, s) \in R_i \times R_j$. Therefore, the GMEC problem for minimizing $e(\mathbf{x})$ is equivalent to the MAP estimation problem for $p(\mathbf{x})$. A probability distribution over a random vector can be related to a graphical model [18]. An undirected graphical model $\mathcal{G} = (\mathcal{V}, \mathcal{E})$ consists of a set of vertices $\mathcal{V}$ for random variables and a set of edges $\mathcal{E}$ connecting some pairs of vertices. In the MAP estimation equivalent of our GMEC problem, the graphical model is generally a complete graph with n vertices.

Wainwright et al. [17] presents an algorithm called tree-reweighted max-product algorithm that can find a MAP assignment for loopy graphical models. The basic idea of the tree-reweighted max-product algorithm is to use a set of spanning-trees $\mathcal{T}$ such that every vertex and edge of $\mathcal{G}$ are covered by some $T \in \mathcal{T}$. Kolmogorov noted [19] that we may define $\mathcal{T}$ as a set of (not necessarily spanning) trees that cover the graph. In what follows, we will use a variant of Wainwright et al.'s algorithm that lets us use an arbitrary tree cover, and call it TRMP without presenting the details of the algorithm. Although TRMP is not guaranteed to always find the optimal solution, it can be used as an upper-bounding tool for the GMEC problem in the same way that the conventional max-product algorithm is used as an upper-bounding tool on loopy graphs [16].

In addition, it can also provide useful lower-bounds for the GMEC problem, which will be explained in Section 4.1.

3   General Pair-Flags

We use general pair-flags to constrain the conformation space $\mathcal{X}$. For example, if the pair-flag for (i_r, j_s) is set, all conformations in $\mathcal{Z} = \{\mathbf{x} \in \mathcal{X} | (x_i, x_j) = (r, s)\}$ are excluded from the search space, i.e. the GMEC problem is solved over $\mathcal{X} \setminus \mathcal{Z}$. However, unlike in DEE, this does not generally imply $\min_{\{\mathbf{x} | (x_i, x_j) = (r, s)\}} e(\mathbf{x}) > \min_{\mathbf{x}} e(\mathbf{x})$. We will denote the set of pair-flags for the given GMEC problem as $\tilde{P}$ (possibly empty) and define pair-flag functions from $\tilde{P}$ as $\tilde{g}_{ij}(r, s, \tilde{P}) = 1$ if $(i_r, j_s) \in \tilde{P}$, and 0 otherwise. We also let $\tilde{g}(\mathbf{x}, \tilde{P}) = \sum_{i, j \in \mathcal{V}, i \neq j} \tilde{g}_{ij}(x_i, x_j, \tilde{P})$.
By defining $P(\{e\}, U) \overset{def}{=} \{(i_r, j_s) \mid \min_{\{\mathbf{x} | (x_i, x_j) = (r, s)\}} e(\mathbf{x}) > U\}$, we have the following lemma regarding minimization under pair-flag constraints:

Lemma 1. *For any $\tilde{P}$ and $\tilde{P}'$ such that $\tilde{P} \subset \tilde{P}'$ and $\tilde{P}' \setminus \tilde{P} \subset P(\{e\}, U)$, $\min_{\{\mathbf{x} | \tilde{g}(\mathbf{x}, \tilde{P}') = 0\}} e(\mathbf{x})$ is either infeasible or greater than U if and only if $\min_{\{\mathbf{x} | \tilde{g}(\mathbf{x}, \tilde{P}) = 0\}} e(\mathbf{x})$ is either infeasible or greater than U.*

The implication of Lemma 1 is that given a subproblem $\min_{\{\mathbf{x} | \tilde{g}(\mathbf{x}, \tilde{P}) = 0\}} e(\mathbf{x})$ in the B&B-tree, the subproblem can be pruned if and only if the modified subproblem $\min_{\{\mathbf{x} | \tilde{g}(\mathbf{x}, \tilde{P}') = 0\}} e(\mathbf{x})$ can be pruned. In addition, we can also show $\min_{\{\mathbf{x} | \tilde{g}(\mathbf{x}, \tilde{P}') = 0\}} e(\mathbf{x}) = \min_{\{\mathbf{x} | \tilde{g}(\mathbf{x}, \tilde{P}) = 0\}} e(\mathbf{x})$ if $\min_{\{\mathbf{x} | \tilde{g}(\mathbf{x}, \tilde{P}) = 0\}} e(\mathbf{x}) \leq U$.
In what follows, when we need to mention pair-flag information, we will implicitly assume we have some $\tilde{P}$, and use the notation $\tilde{g}(\mathbf{x})$ instead of $\tilde{g}(\mathbf{x}, \tilde{P})$ where specifying $\tilde{P}$ is not particularly necessary. The following condition on pair-flags can be maintained without loss of generality and will be used in Section 4:

Condition 1. *For all $r \in R_i$ and $i \in \mathcal{V}$, there exists $s \in R_j$ for each $j \in \mathcal{V}, j \neq i$ such that $(i_r, j_s) \notin \tilde{P}$.*

4   Problem-Size Reduction Techniques

4.1   Elimination by TRMP Lower-Bounds

We can exploit the properties of TRMP in computing a lower-bound of the minimum conformation energy for some given set of conformations. If such a lower-bound is greater than U, we can eliminate corresponding conformations from the problem while conserving the inequality relation between $\min_{\{\mathbf{x} | \tilde{g}(\mathbf{x}) = 0\}} e(\mathbf{x})$ and U. In addition, if $\min_{\{\mathbf{x} | \tilde{g}(\mathbf{x}) = 0\}} e(\mathbf{x}) \leq U$, the elimination does not change the optimal value. In this section, we first review the key properties of Wainwright et al.'s algorithm – ρ-reparameterization and tree-consistency of pseudo-max-marginals, before presenting how to compute the lower-bounds. Note that TRMP shares these properties.

Single max-marginals μ_i [17] are defined as the maximum of $p(\mathbf{x})$ when one of the variable x_i is fixed, i.e. $\mu_i(x_i) = \kappa_i \max_{\{\mathbf{x}'|x_i'=x_i\}} p(\mathbf{x}')$. Similarly, pairwise max-marginals μ_{ij} are defined as $\mu_{ij}(x_i, x_j) = \kappa_{ij} \max_{\{\mathbf{x}'|(x_i',x_j')=(x_i,x_j)\}} p(\mathbf{x}')$. Note that κ_i and κ_{ij} are constants that can vary depending on i or j. It is known that any tree-distribution $p(\mathbf{x})$ can be factored in terms of its max-marginals as $p(\mathbf{x}) \propto \prod_{i \in V} \mu_i(x_i) \prod_{(i,j) \in \mathcal{E}} \frac{\mu_{ij}(x_i,x_j)}{\mu_i(x_i)\mu_j(x_j)}$. Max-marginals for tree-distributions can be exactly computed by the conventional max-product algorithm. Wainwright et al. [17] use the notion of pseudo-max-marginals. By construction, pseudo-max-marginals $\nu = \{\nu_i, \nu_{ij}\}$ from the tree-reweighted max-product algorithm satisfy ρ-reparameterization, i.e.

$$
p(\mathbf{x}) \propto \prod_{T \in \mathcal{T}} \left[\prod_{i \in V(T)} \nu_i(x_i) \prod_{(i,j) \in \mathcal{E}(T)} \frac{\nu_{ij}(x_i, x_j)}{\nu_i(x_i)\nu_j(x_j)} \right]^{\rho(T)}, \tag{2}
$$

where $\rho(T) = \frac{|\{T \in \mathcal{T}\}|}{|\mathcal{T}|}$. A tree-distribution $p^T(\mathbf{x}; \nu)$ for given pseudo-max-marginals can be defined as

$$
p^T(\mathbf{x}; \nu) \overset{def}{=} \prod_{i \in V(T)} \nu_i(x_i) \prod_{(i,j) \in \mathcal{E}(T)} \frac{\nu_{ij}(x_i, x_j)}{\nu_i(x_i)\nu_j(x_j)}.
$$

Then, we have $p(\mathbf{x}) \propto \prod_{T \in \mathcal{T}} \{p^T(\mathbf{x}; \nu)\}^{\rho(T)}$ from (2). On the other hand, pseudo-max-marginals ν^* at convergence of the tree-reweighted max-product algorithm satisfy tree-consistency condition with respect to every spanning tree $T \in \mathcal{T}$. More precisely, ν^* is tree-consistent with respect to a spanning tree T if it satisfies $\nu_i^*(x_i) \propto \max_{x_j \in R_j} \nu_{ij}^*(x_i, x_j)$ for all $x_i \in R_i$ and $(i,j) \in \mathcal{E}(T)$.

In what follows, we assume ν is in a normal form [19], i.e. $\max_{r \in R_i} \nu_i(r) = 1$ for all $i \in V$, and $\max_{(r,s) \in R_i \times R_j} \nu_{ij}(r,s) = 1$ for all $(i,j) \in \mathcal{E}$. Then, since ν always satisfies ρ-reparameterization, rearranging the terms of (2) gives, for some constant $\nu_c > 0$,

$$
p(\mathbf{x}) = \nu_c \prod_{i \in V} \nu_i(x_i)^{\rho_i} \prod_{(i,j) \in \mathcal{E}} \left(\frac{\nu_{ij}(x_i, x_j)}{\nu_i(x_i)\nu_j(x_j)} \right)^{\rho_{ij}},
$$

where $\rho_{ij} = \frac{|\{T \in \mathcal{T} \mid (i,j) \in \mathcal{E}(T)\}|}{|\mathcal{T}|}$ and $\rho_i = \frac{|\{T \in \mathcal{T} \mid i \in V(T)\}|}{|\mathcal{T}|}$.

The following lemmas show how we can compute lower-bounds for some sets of conformations. For example, Lemma 2 combined with (1) can provide rotamer lower-bounds i.e. a lower-bound of $\min_{\{\mathbf{x}|x_\zeta=r\}} e(\mathbf{x})$ for each $r \in R_\zeta$ and $\zeta \in V$:

Lemma 2. *When ν satisfies the tree-consistency condition, we have, for all $r \in R_\zeta$, $\zeta \in V$, $\max_{\{\mathbf{x}|x_\zeta=r\}} p(\mathbf{x}) \leq \nu_c \nu_\zeta(r)^{\rho_\zeta}$.*

For rotamer-pair lower-bounds, i.e. to lower-bound $\min_{\{\mathbf{x}|(x_\zeta,x_\eta)=(r,s)\}} e(\mathbf{x})$, we use $\max_{\{\mathbf{x}|(x_\zeta,x_\eta)=(r,s)\}} p(\mathbf{x}) \leq \nu_c \prod_{T \in \mathcal{T}} \left[\max_{\{\mathbf{x}|(x_\zeta,x_\eta)=(r,s)\}} p^T(\mathbf{x}) \right]^{\rho(T)}$, where $\max_{\{\mathbf{x}|(x_\zeta,x_\eta)=(r,s)\}} p^T(\mathbf{x})$ for each T can be easily solved using Lemma 3 when we let $\mathcal{T} = \mathcal{S}$, a set of stars:

Lemma 3. *When ν satisfies the tree-consistency condition, the following inequalities hold:*

1. if $\zeta, \eta \notin \mathcal{V}(T)$, then $\max_{\{\mathbf{x}|(x_\zeta, x_\eta)=(r,s)\}} p^T(\mathbf{x}) = 1$.
2. if $\zeta \in \mathcal{V}(T)$ and $\eta \notin \mathcal{V}(T)$, then $\max_{\{\mathbf{x}|(x_\zeta, x_\eta)=(r,s)\}} p^T(\mathbf{x}) = \nu_\zeta(r)$.
3. if $(\zeta, \eta) \in \mathcal{E}(T)$, then $\max_{\{\mathbf{x}|(x_\zeta, x_\eta)=(r,s)\}} p^T(\mathbf{x}) = \nu_{\zeta\eta}(r,s)$.
4. if $\zeta, \eta \in \mathcal{V}(T)$ and $(\zeta, \eta) \notin \mathcal{E}(T)$ for a star T (let ξ be the center of T), then
$$\max_{\{\mathbf{x}|(x_\zeta, x_\eta)=(r,s)\}} p^T(\mathbf{x}) = \max_{x_\xi \in R_\xi} \frac{\nu_{\xi\zeta}(x_\xi, r)\nu_{\xi\eta}(x_\xi, s)}{\nu_\xi(x_\xi)}.$$

If we use pair-flags, we may improve rotamer lower-bounds by the inequality $\max_{\{\mathbf{x}|x_\zeta=r, \tilde{g}(\mathbf{x})=0\}} p(\mathbf{x}) \leq \nu_c \prod_{T \in \mathcal{T}} \left[\max_{\{\mathbf{x}|x_\zeta=r, \tilde{g}(\mathbf{x})=0\}} p^T(\mathbf{x}) \right]^{\rho(T)} \leq \nu_c \nu_\zeta(r)^{\rho_\zeta}$, which holds for tree-consistent ν. Let n_{rot} be the average number of rotamers per position. If we use a naive search, it takes $O(n_{rot}^2 n)$ comparison operations to exactly solve $\max_{\{\mathbf{x}|x_\zeta=r, \tilde{g}(\mathbf{x})=0\}} p^T(\mathbf{x})$. Therefore, computing an improved lower-bound for a rotamer takes $O(n_{rot}^2 n^2)$ since $|\mathcal{T}| = O(n)$.

4.2   Rotamer-Contraction

The idea of rotamer contraction is to reduce the number of rotamers at one selected position by first clustering similar rotamers of the position and replacing all rotamers in each cluster with one rotamer-aggregate. Let ζ be the position whose rotamers we partition into a number of clusters $C_1, \ldots, C_l$, $l < |R_\zeta|$. Then, we contract all rotamers $r \in C_k$ as one rotamer-aggregate c_k. The contracted GMEC problem has a new conformation space $\mathcal{X}^{rc}$, which is same as $\mathcal{X}$ except that R_ζ is replaced by $\{c_1, \ldots, c_l\}$. Then, we define a new energy function $e^{rc}(\mathbf{x})$ over $\mathcal{X}^{rc}$ and the set of pair-flags $\tilde{P}^{rc}$ so that the optimal value of the contracted problem $\min_{\{\mathbf{x} \in \mathcal{X}^{rc}|\tilde{g}(\mathbf{x}, \tilde{P}^{rc})=0\}} e^{rc}(\mathbf{x})$ is a lower-bound of $\min_{\{\mathbf{x} \in \mathcal{X}|\tilde{g}(\mathbf{x}, \tilde{P})=0\}} e(\mathbf{x})$. One way of choosing $e^{rc}(\mathbf{x})$ for a given clustering is given by *contract-rotamers* in Algorithm 1. We use notation $e^{rc}(\mathbf{x}, \tilde{P})$ to indicate the function is also defined by $\tilde{P}$. A lower-bounding technique similar to rotamer-contraction is used by Koster et al. [20] for the frequency assignment problem. We have the following lemma on *contract-rotamers*:

Lemma 4. *For any given clustering of rotamers of $\zeta \in \mathcal{V}$, if $\{\mathbf{x} \in \mathcal{X}|\tilde{g}(\mathbf{x}, \tilde{P}) = 0\} \neq \phi$, then $\min_{\{\mathbf{x} \in \mathcal{X}|\tilde{g}(\mathbf{x}, \tilde{P})=0\}} e(\mathbf{x}) \geq \min_{\{\mathbf{x} \in \mathcal{X}^{rc}|\tilde{g}(\mathbf{x}, \tilde{P}^{rc})=0\}} e^{rc}(\mathbf{x}, \tilde{P})$.*

In rotamer-contraction, how we cluster rotamers of position ζ determines the quality of resulting lower-bounds. Our approach is a greedy scheme that keeps placing rotamers in a cluster as long as the decrease in the optimal value is less than or equal to a specified amount. However, it is hard to exactly know the decrease $\min_{\{\mathbf{x} \in \mathcal{X}|\tilde{g}(\mathbf{x}, \tilde{P})=0\}} e(\mathbf{x}) - \min_{\{\mathbf{x} \in \mathcal{X}^{rc}|\tilde{g}(\mathbf{x}, \tilde{P}^{rc})=0\}} e^{rc}(\mathbf{x}, \tilde{P})$. In addition, it is generally not feasible to bound the decrease since rotamer-contraction may even turn an infeasible subproblem into a feasible one. We instead upper-bound $\Delta OPT^{rc} \stackrel{def}{=} \min_{\mathbf{x} \in \mathcal{X}} e(\mathbf{x}) - \min_{\{\mathbf{x} \in \mathcal{X}^{rc}|\tilde{g}(\mathbf{x}, \tilde{P}^{rc})=0\}} e^{rc}(\mathbf{x}, \tilde{P})$. Let $U^{rc}_{\Delta OPT}(\tilde{P}) = \max_{k=1,\ldots,l} \min_{r \in C_k} \sum_{j \in \Gamma(\zeta)} \max_{\{s \in R_j|(\zeta_{c_k}, j_s) \notin \tilde{P}^{rc}\}} \{ e_{\zeta j}(r,s) + \frac{e_\zeta(r)}{|\Gamma(\zeta)|} - e^{rc}_{\zeta j}(c_k, s, \tilde{P}) \}$. Then, we have the following lemma:

Lemma 5. *For any given clustering of rotamers of $\zeta \in \mathcal{V}$, we have $\Delta OPT^{rc} \leq U^{rc}_{\Delta OPT}(\tilde{P}) \leq U^{rc}_{\Delta OPT}(\phi)$*

Algorithm 1: contract-rotamers

Data: ζ, $C_1, \ldots, C_l$, $\mathcal{X}$, $\{e\}$, $\tilde{P}$
Result: $\mathcal{X}^{rc}, \{e^{rc}\}, \tilde{P}^{rc}$
begin
   $\mathcal{X}^{rc}$ is same with $\mathcal{X}$ except R_ζ is replaced with $\{c_1, \ldots, c_l\}$
   $\tilde{P}^{rc} \leftarrow \tilde{P} \backslash \{(\zeta_r, j_s), j \in \mathcal{V}, j \neq \zeta\}$
   foreach C_k, $k = 1, \ldots, l$ **do**
      foreach $s \in R_j$, $j \in \mathcal{V}, j \neq \zeta$ **do**
         $e^{rc}_{\zeta j}(c_k, s, \tilde{P}) \leftarrow \min_{\{r \in C_k, (\zeta_r, j_s) \notin \tilde{P}\}} e_{\zeta j}(r, s) + \frac{e_\zeta(r)}{|\Gamma(\zeta)|}$,
         if $(\zeta_r, j_s) \in \tilde{P}$ for all $r \in C_k$ **then** $\tilde{P}^{rc} \leftarrow \tilde{P}^{rc} \cup (\zeta_{c_k}, j_s)$
      $e^{rc}_\zeta(c_k) \leftarrow 0$.
   define $e^{rc}(\mathbf{x})$ same as $e(\mathbf{x})$ for other terms
end

Note that $U^{rc}_{\Delta OPT}(\tilde{P})$ has a finite value due to Condition 1. Lemma 4 and Lemma 5 suggests rotamer-contraction may benefit from the use of pair-flags by smaller decrease in the optimal value, and better upper-bounding of the decrease. We include a rotamer in a cluster if $U^{rc}_{\Delta OPT}(\tilde{P})$ from the inclusion is less than some constant Δ^{rc}. When $\min_{\mathbf{x}} e(\mathbf{x}) > U$, Δ^{rc} can be allowed to be at most $\min_{\mathbf{x}} e(\mathbf{x}) - U$ or some fraction of it. Since we do not know the exact value of $\min_{\mathbf{x}} e(\mathbf{x})$, Δ^{rc} is heuristically set as a fraction of the difference between an upper-bound of $\min_{\mathbf{x}} e(\mathbf{x})$ and U. Both upper-bounds are obtained by TRMP.

4.3   Edge Deletion

In edge deletion, we first identify a pair of positions (ζ, η) such that the deviation in $e_{\zeta\eta}(r, s)$ for all $(r, s) \in R_\zeta \times R_\eta$ is small, then set all the pairwise energies of (ζ, η) to the minimum of the pairwise energies. That is, the new energy function $e^{ed}(\mathbf{x})$ will be defined to be the same as $e(\mathbf{x})$ except $e^{ed}_{\zeta\eta}(r, s) = \min_{\{(r,s) \in R_\zeta \times R_\eta | (\zeta_r, \eta_s) \notin \tilde{P}\}} e_{\zeta\eta}(r, s)$, for all $(r, s) \in R_\zeta \times R_\eta$. Since $e^{ed}_{\zeta\eta}(x_\zeta, x_\eta)$ is constant, we can ignore the interaction of (ζ, η) and replace $\mathcal{E}$ by $\mathcal{E} \backslash (\zeta, \eta)$. The same idea is explored by Xie and Sahinidis [15] as an approximation procedure. Some advantages of doing edge-deletion are: (1) when the graph becomes sparse, we may use direct solution techniques such as dynamic programming. (2) Empirically, being able to cover the graph with fewer trees is favorable for obtaining tighter lower-bounds from TRMP. (3) Rotamer-contraction may obtain fewer clusters for the same Δ^{rc}. The pair-flags are kept intact through edge-deletion even for the edge being deleted. Then, it is straightforward to obtain similar properties for edge-deletion as Lemma 4 and 5.

5   Branch-and-Bound Framework

We split a subproblem by dividing rotamers of a position into two groups by their rotamer lower-bounds. If the conformation space of the current subproblem

F^i is defined by rotamer sets $\{R_i\}$ and we decide to split it into $F^{i,low}$ (low-subproblem) and $F^{i,high}$ (high-subproblem), we can define the conformation space for each with $\{R_i^{low}\}$ and $\{R_i^{high}\}$, respectively, where R_i^{low} and R_i^{high} are defined the same as R_i except $R_\zeta^{low} \cup R_\zeta^{high} = R_\zeta$, $|R_\zeta^{low}| \approx |R_\zeta^{high}|$, and $LB(\zeta_r) \leq LB(\zeta_s)$ for all $r \in R_\zeta^{low}$, $s \in R_\zeta^{high}$ ($LB(\zeta_r)$ is a rotamer lower-bound for ζ_r). The goal of such a splitting scheme is to make the optimal value of $F^{i,low}$ likely to be less than that of $F^{i,high}$. We prefer a splitting position ζ whose difference between maximum and minimum rotamer lower-bounds is large. Subproblems are selected by a mix of what are called "best-first" and "depth-first" strategies: (1) follow the depth-first strategy, (2) always dive into $F^{i,low}$ first when the current subproblem F^i is split. The goal is first to find a good upper-bound by following depth-first through the low-subproblems from the first series of splittings, then to prune the remaining subproblems using the upper-bound. Figure 1 shows an example B&B-tree that can result from our splitting scheme and subproblem-selection strategy, where optimal solution from node 5 is supposed to provide a near-optimal upper-bound.

6   Experimental Results

In our numerical experiments, a Linux workstation with a 2.2 GHz Intel Xeon processor and 3.5 GBytes of memory was used. Table 1 shows 12 protein design cases used in the experiments. DEE on each case was performed with the following options: Goldstein's singles elimination, splitting with split flags ($s = 1$), Goldstein's pair elimination with one magic bullet, and unification allowing maximum 6,000 rotamers per position. E-9 was finished in 4.8 hours but none of others were solved within 48 hours.

We first show an example use of TRMP lower-bounds in eliminating rotamers or rotamer-pairs of subproblems from E-10. In the following, we use the

Table 1. Test cases facts. All cases are from the antigen-antibody model system. Each case repacks the antigen protein or the antibody. Each column represents (1) case name, (2) number of positions, (3) maximum number of rotamers offered at a position, (4) number of total rotamers, (5) $\sum_{i=1}^{n} \log_{10} |R_i|$, (6) case composition (with m: # positions allowed to mutate, n: # positions only wild-types are allowed, and w: # water molecules to be oriented at the interface). R uses the standard rotamer library, and E multiplies each of χ_1 and χ_2 by a factor of 3 by adding $\pm 10°$. E-1 were offered only hydrophobic residues while others were offered both hydrophobic and polar residues. All energies were calculated using the CHARMM package and the parameter set 'param22'.

Case	n	max $|R_i|$	$\sum |R_i|$	$\log_{10} conf$	Composition	Case	n	max $|R_i|$	$\sum |R_i|$	$\log_{10} conf$	Composition
R-1	34	125	1422	30.0	34 m	E-5	24	1344	9585	49.6	24 m
R-2	30	133	1350	40.2	30 m	E-6	36	1984	8543	59.1	4 m, 32 n
E-1	19	617	3675	38.1	19 m	E-7	10	2075	5201	21.9	5 m, 3 n, 2 w
E-2	23	1370	9939	52.3	23 m	E-8	10	1915	5437	20.7	4 m, 4 n, 2 w
E-3	23	1320	8332	49.1	23 m	E-9	15	2091	5700	25.1	3 m, 6 n, 6 w
E-4	15	1361	7467	33.9	15 m	E-10	23	1949	9837	42.5	7 m, 7 n, 9 w

Table 2. TRMP lower-bounding results for subproblems of E-10. The meaning of each column is, in order: (1) subproblem, (2) number of rotamers, (3) number of rotamer-pairs, (3) median rotamer lower-bound(lb) when not using pair-flags, (4) number of rotamers such that $lb > U$, (5) median rotamer lower-bound when using pair-flags, (5) number of rotamers such that $lb > U$, (6) median rotamer-pair lower-bound, (7) number of rotamer-pairs such that $lb > U$. The value of U is -325.038. $+\infty$ implies the lower-bounding problem turned out to be infeasible due to pair-flags.

Subprob.	#rots	#rot-pairs	Rot-lb's w/o pair-flags		Rot-lb's w/ pair-flags		Rot-pair lb's	
			med. lb	#rots $lb > U$	med. lb	#rots $lb > U$	med. lb	#pairs $lb > ub$
2-high	3,345	4,769,691	-332.831	1,301	$+\infty$	2,220	-312.544	4,071,145
3-high	3,022	3,879,787	-315.025	1,134	$+\infty$	2,141	-313.614	3,328,424
4-high	2,665	3,036,834	-315.689	1,380	$+\infty$	2,273	-313.276	2,799,272
5-high	2,281	2,299,981	-336.173	81	-336.173	920	-323.780	1,292,520
6-high	2,171	2,071,431	-343.019	0	-343.019	200	-330.363	590,576
7-high	1,964	1,702,980	-342.556	8	-342.556	215	-329.554	508,750
8-high	1,848	1,499,857	-344.865	0	-344.636	42	-335.640	218,324
9-high	1,669	1,223,065	-337.791	0	-337.791	289	-329.037	384,812

notation "i-high" to denote the high-subproblem at depth i spawned from the first depth-first dive along the low-subproblems (root node is at depth 1). For example, in Figure 1, node 11 is 2-high and node 6 is 5-high. Table 2 shows lower-bounding results for subproblems at depth 2 to 11. In 2-high, simple rotamer lower-bounds were able to eliminate 39% of rotamers. However, we obtain even more elimination when we use rotamer lower-bounds computed using pair-flags. This is due to massive flagging of rotamer-pairs by rotamer-pair lower-bounds. Large elimination obtained for subproblems at small depth are due to our splitting scheme of dividing rotamers by their lower-bounds.

To evaluate our pruning scheme, we compared it (call it PbyR: *prune-by-reduction*) against linear programming (LP). We used subproblems of various sizes generated while solving the design cases of Table 1 with our B&B method. We used the LP formulation given by Wainwright et al. [17] and solved it with a C++ procedure using CPLEX 8.0 library. In PbyR, we alternated rotamer-contraction and edge-deletion at every iteration. At every 8th reduction, we applied DEE to see if we could solve the reduced problem or only to flag more rotamer/rotamer-pairs. (Note that we adapted DEE to make it compatible with general pair-flags.) We computed TRMP lower-bounds at every 24th reduction and flagged rotamers/rotamer-pairs. We allowed at most 300 reductions until we find a lower-bound greater than U or exactly solve the reduced problem. Figure 2 shows the result for the 156 subproblems remaining after excluding the subproblems that could be solved quickly by DEE alone. The bounding times of the two methods are comparable although LP is slightly faster in small to medium-sized subproblems. However, Figure 2 (b) shows that the bounds from PbyR are greater than LP bounds except for the one data point below the $y = x$ line. Note that a PbyR bound for a subproblem is not generally a lower-bound of the subproblem's optimal value since rotamer/rotamer-pair elimination by TRMP lower-bounds can also increase the optimal value. However, a PbyR bound is

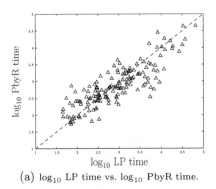

(a) $\log_{10}$ LP time vs. $\log_{10}$ PbyR time.

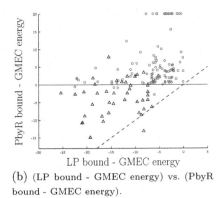

(b) (LP bound - GMEC energy) vs. (PbyR bound - GMEC energy).

Fig. 2. Comparison of LP and PbyR in pruning subproblems from B&B. In (b), a circle represents the PbyR bound was computed using less than 50 reductions. Also in (b), points such that PbyR bound - GMEC energy ≥ 20 were all clamped at 20.

greater than U only if the original subproblem's optimal value is greater than U. Therefore, if we had U equal to the GMEC energy for each design case, we could immediately prune the subproblems corresponding to the data points over the horizontal solid line in Figure 2 (b). There was no such case with LP among the tested subproblems. Figure 2 (b) suggests that performing reductions more than 50 times often resulted in lower-bounds that were useless for pruning.

Finally, we used the B&B method of Figure 1 to solve each design case. The branch-and-bound method was implemented in C++ using the PICO-library [21] as a sequential B&B framework. At each node of the B&B method, we first eliminated rotamers using DEE with the same set of options mentioned earlier. When singles-elimination condition of DEE fails to eliminate any rotamer, we let TRMP lower-bounds eliminate more rotamers. Then, we used the reduction techniques iteratively in the same mix as we used for comparison test against LP, but limited the number of reductions to be at most four times the depth of the node in the B&B-tree. When branching was necessary, the subproblem located at the end of the first dive usually had $\sum_{i=1}^{n} \log_{10} |R_i| \leq 13$ and was exactly solved by DEE. Table 3 shows the result. We were able to solve six cases at the root node without branching. Considering DEE couldn't finish five of them

Table 3. Solving the design cases using our B&B method. Each column represents (1) case name, (2) number of branches, (3) number of branches from the first depth-first dive along the low-subproblems, (4) total solution time.

Case	# Br.	#F.D.Br.	Time (h)	Case	# Br.	#F.D.Br.	Time (h)	Case	# Br.	#F.D.Br.	Time (h)
R-1	0	0	1.1	E-3	0	0	6.4	E-7	15	12	6.9
R-2	14	14	2.7	E-4	0	0	4.2	E-8	17	12	13.7
E-1	28	28	9.7	E-5	8	8	27.2	E-9	0	0	3.3
E-2	0	0	6.2	E-6	0	0	5.6	E-10	202	35	139.1

for 48 hours, rotamer/rotamer-pair elimination using TRMP lower-bounds enormously reduced the solution time. All cases were also solved efficiently except E-10 where the upper-bounds (from TRMP) of the subproblems were often very close to the GMEC energy. However, in all cases, the number of total branching is only moderately larger than that from the first dive. In all cases where branching was necessary, the upper-bound obtained at the end of the first dive was equal to the GMEC energy, confirming that our branching scheme and subproblem-selection strategy meets expectations.

7 Conclusion

In this work, we presented an exact solution method for the GMEC problem. Our branch-and-bound method using the suggested pruning scheme was able to solve hard sequence design cases that DEE couldn't solve within practical resources levels. There is certainly a decision-making flavor in using our proposed pruning scheme since a trade-off between the amount of pruning effort and the quality of the final bound should be considered in deciding when to stop the pruning attempt and to split the subproblem. Therefore, future work may include a systematic allocation of pruning effort throughout the B&B-tree for faster solution.

Acknowledgment. The authors would like to thank Bruce Tidor for suggesting the problem and for helpful advice. Shaun Lippow, Alessandro Senes, and Michael Altman gave freely of the test cases, and the DEE code.

References

1. Drexler, K.E.: Molecular engineering: an approach to the development of general capabilities for molecular manipulation. Proc. National Academy of Sciences USA **78** (1981) 5275–5278
2. Vasquez, M.: Modeling sidechain conformation. Current Opinion in Structural Biology **6** (1996) 217–221
3. Hellinga, H.W., Richards, F.M.: Optimal sequence selection in proteins of known structure by simulated evolution. Proc. National Academy of Sciences USA **91** (1994) 5803–5807
4. Janin, J., Wodak, S., Levitt, M., Maigret, B.: Conformation of amino-acid side-chains in proteins. J. of Molecular Biology **125** (1978) 357–386
5. Desmet, J., De Maeyer, M., Hazes, B., Lasters, I.: The dead-end elimination theorem and its use in protein side-chain positioning. Nature **356** (1992) 539–542
6. Goldstein, R.F.: Efficient rotamer elimination applied to protein side-chains and related spin glasses. Biophysical J. **66** (1994) 1335–1340
7. Pierce, N.A., Spriet, J.A., Desmet, J., Mayo, S.L.: Conformational splitting: a more powerful criterion for dead-end elimination. J. of Computational Chemistry **21** (2000) 999–1009
8. Gordon, D.B., Mayo, S.L.: Radical performance enhancements for combinatorial optimization algorithms based on the dead-end elimination theorem. J. of Computational Chemistry **13** (1998) 1505–1514

9. Gordon, D.B., Mayo, S.L.: Branch-and-terminate: a combinatorial optimization algorithm for protein design. Structure with Folding and Design **7** (1999) 1089–1098

10. Althaus, E., Kohlbacher, O., Lenhof, H.P., Müller, P.: A combinatorial approach to protein docking with flexible side-chains. J. of Computational Biology **9** (2002) 597–612

11. Eriksson, O., Zhou, Y., Elofsson, A.: Side chain-positioning as an integer programming problem. In: Proc./ 1st Workshop on Algorithms in Bioinformatics WABI'01. Volume 2149 of LNCS., Springer (2001) 128–141

12. Kingsford, C., Chazelle, B., Singh, M.: Solving and analyzing side-chain positioning problems using linear and integer programming. Bioinformatics **21** (2005) 1028–1036

13. Leaver-Fay, A., Kuhlman, B., Snoeyink, J.: An adaptive dynamic programming algorithm for the side-chain placement problem. In: Proc. Pacific Symp. on Biocomputing PSB'05, Singapore, World Scientific (2005) 16–27

14. Xu, J.: Rapid protein side-chain packing via tree decomposition. In: Proc. Conf. on Research in Mol. Comput. Biol. RECOMB'05. Volume 3500 of LNCS., Springer (2005) 423–439

15. Xie, W., Sahinidis, N.V.: Residue-rotamer-reduction algorithm for the protein side-chain conformation problem. Bioinformatics **22** (2006) 188–194

16. Yanover, C., Weiss, Y.: Approximate inference and protein-folding. In: Proc. of Neural Information Processing Systems. (2002)

17. Wainwright, M.J., Jaakola, T.S., Willsky, A.S.: Map estimation via agreement on (hyper)trees: Message-passing and linear programming approaches. Technical Report UCB/CSD-3-1269, Computer Science Division (EECS), UC Berkeley (2003)

18. Jordan, M.I.: Graphical models. Statistical Science (Special Issue on Bayesian Statistics) **19** (2004) 140–155

19. Kolmogorov, V.: Convergence tree-reweighted message passing for energy minimization. Technical Report MSR-TR-2005-38, Microsoft Research (2005)

20. Koster, A.M., van Hoesel, S.P., Kolen, A.W.: Lower bounds for minimum interference frequency assignment problems. Technical Report RM 99/026, Maastricht University (1999)

21. Eckstein, J., Phillips, C.A., Hart, W.E.: Pico: an object oriented framework form parallel branch and bound. Technical report, RUTCOR (2001)

On the Complexity of the Crossing Contact Map Pattern Matching Problem

Shuai Cheng Li and Ming Li

David R. Cheriton School of Computer Science
University of Waterloo
Waterloo ON N2L 3G1 Canada
{scli, mli}@cs.uwaterloo.ca

Abstract. Contact maps are concepts that are often used to represent structural information in molecular biology. The contact map pattern matching (CMPM) problem is to decide if a contact map (called the *pattern*) is a substructure of another contact map (called the *target*). In general, the problem is NP-hard, but when there are restrictions on the form of the pattern, the problem can, in some case, be solved in polynomial time. In particular, a polynomial time algorithm has been proposed [1] for the case when the patterns are so-called *crossing contact maps*. In this paper we show that the problem is actually NP-hard, and show a flaw in the proposed polynomial-time algorithm. Through the same method, we also show that a related problem, namely, the *2-interval patten matching problem with $\{<, \between\}$-structured patterns and disjoint interval ground set*, is NP-hard.

1 Introduction

Contact-maps are graph theoretic concepts that are often used in protein structure analysis [2]. The contact map matching problem (CMPM) is to decide if a contact map, called the *pattern*, is a substructure of another contact map, called the *target*. This problem is NP-hard in the most general case, but some cases with restrictions on the form of patterns to be matched have been shown to be solvable in polynomial time. In this paper we consider the case where the pattern is a *crossing contact map*. It is not previously known if there exists a polynomial-time algorithm in this case [1].

This problem is closely related to an open problem known as the *2-interval pattern matching problem*, which came about from the study of interactions of ribonucleic acids. The complexity of this problem was first investigated by Vialette [3], and then followed by Blin *et. al.* [4]. Crochemore *et. al.* proposed approximation algorithms on optimization versions of the problem [5]. The problem of whether the 2-interval pattern matching problem has a polynomial-time algorithm with *disjoint interval ground sets* and $\{<, \between\}$-*structured patterns*, was left unanswered in these works.

Gramm [1] proposed a polynomial-time algorithm that would solve the above two problems. Regrettably, we noticed a flaw in an assumption made by the algorithm. We show in this paper that the two problems are actually NP-hard.

P. Bücher and B.M.E. Moret (Eds.): WABI 2006, LNBI 4175, pp. 231–241, 2006.

This paper is organized as follows. In Section 2 we give the definitions and problem statements. In Section 3, we give a polynomial reduction from the Clique Problem to the two problems, thus showing them to be NP-hard. In Section 4, we give a concrete counterexample that the algorithm in [1] would produce an incorrect solution.

2   Problem Definition and Previous Results

We follow the notations from [1]. A *contact map* consists of a pair $(\mathcal{S}, \mathcal{A})$, where $\mathcal{S}$ is a set of integers, and $\mathcal{A}$ consists of a set of ordered pairs, which is: $\mathcal{A} = \{(s_l, s_r) | s_l, s_r \in \mathcal{S}, s_l < s_r\}$. A pair (s_l, s_r) is referred to as an *arc*. We denote $L((s_l, s_r)) = s_l$ and $R((s_l, s_r)) = s_r$.

The contact map pattern matching (CMPM) problem is: Given two contact maps $\mathcal{CM}(\mathcal{S}_p, \mathcal{A}_p)$ (called the *pattern*) and $\mathcal{CM}(\mathcal{S}, \mathcal{A})$ (called the *target*) where $|\mathcal{S}_p| \leq |\mathcal{S}|$, find a subset $\mathcal{S}'$ of $\mathcal{S}$ with $|\mathcal{S}'| = |\mathcal{S}_p|$, such that there is a one-one mapping $\mathcal{M}$ from the elements of $\mathcal{S}$ to the elements of $\mathcal{S}'$ that satisfies the following two conditions:

- if $s_1, s_2 \in \mathcal{S}_p$ and $s_1 < s_2$, then $\mathcal{M}(s_1), \mathcal{M}(s_2) \in \mathcal{S}'$ and $\mathcal{M}(s_1) < \mathcal{M}(s_2)$.
- if $(s_1, s_2) \in \mathcal{A}_p$, then $(\mathcal{M}(s_1), \mathcal{M}(s_2)) \in \mathcal{A}$.

If such a mapping exists, we say that $\mathcal{CM}(\mathcal{S}_p, \mathcal{A}_p)$ *occurs* in $\mathcal{CM}(\mathcal{S}, \mathcal{A})$. In general, the CMPM problem is NP-hard [2,3]. However, some cases with restrictions on the patterns have been shown to be solvable in polynomial time. To state these restrictions, we first define the following three types of relations between any two given arcs $a = (s_l, s_r)$, $a' = (s_l', s_r')$:

- $a < a'$ (*a is less than a'*) iff $s_r < s_l'$
- $a \sqsubset a'$ (*a is nested in a'*) iff $s_l' < s_l < s_r < s_r'$
- $a \between a'$ (*a crosses a'*) iff $s_l < s_l' < s_r < s_r'$

A contact map $(\mathcal{S}, \mathcal{A})$ is called a *crossing contact map* (CCM) iff $\forall a, a' \in \mathcal{A}$, $a \neq a'$, one of the relations: $a < a'$, $a' < a$, $a \between a'$ or $a' \between a$ is satisfied. A CCM is also called a $\{<, \between\}$-*structured* contact map. Other types of contact maps can be defined similarly, such as $\{<\}$, or $\{\sqsubset, \between\}$-*structured* contact maps. The CMPM problem with $\{<\}$, $\{\sqsubset\}$, $\{\between\}$, or $\{<, \sqsubset\}$-structured patterns can be solved in polynomial time, but is NP-hard for the $\{\sqsubset, \between\}$ and $\{<, \sqsubset, \between\}$-structured patterns [3]. In this paper, we are interested in the remaining case of when the patterns are CCMs. The following formally states the problem:

CROSSING CONTACT-MAP PATTERN MATCHING (CCMPM)PROBLEM [1]

Input: Contact maps $\mathcal{CM}(\mathcal{S}_p, \mathcal{A}_p)$ and $\mathcal{CM}(\mathcal{S}, \mathcal{A})$ with $\mathcal{CM}(\mathcal{S}_p, \mathcal{A}_p)$ as a CCM

Output: Does $\mathcal{CM}(\mathcal{S}_p, \mathcal{A}_p)$ occur in $\mathcal{CM}(\mathcal{S}, \mathcal{A})$?

---

3   Hardness Results

We use the Clique Problem, a well known NP-hard problem, for reduction in this paper. Let an instance of the Clique Problem be given by a graph $\mathcal{G}(\mathcal{V}, \mathcal{E})$ and by a positive integer ℓ. For notation simplicity, define $n = |\mathcal{V}|$ and without loss of generality, assume $\mathcal{V} = \{1, \ldots, n\}$.

In the following, we will first define some terms to facilitate the presentation of the reduction. We will then construct (1) a target map $\mathcal{CM}(\mathcal{S}_\mathcal{G}, \mathcal{A}_\mathcal{G})$, and (2) a pattern $\mathcal{CM}(\mathcal{S}_{n,\ell}, \mathcal{A}_{n,\ell})$ with parameters ℓ and n, from a given graph $\mathcal{G}(\mathcal{V}, \mathcal{E})$. We will then analyze the reduction and show its correctness.

3.1   Additional Notation and Definitions

A set $\mathcal{A}$ of k distinct arcs where $\forall a, a' \in \mathcal{A}$, either $a \between a'$ or $a' \between a$, is called a *k-arc crossing cluster*. Given two disjoint sets of arcs $\mathcal{A}_1, \mathcal{A}_2$, we say $\mathcal{A}_1$ *crosses* $\mathcal{A}_2$, or $\mathcal{A}_2$ *is crossed by* $\mathcal{A}_1$ (written $\mathcal{A}_1 \between \mathcal{A}_2$), just in case either (1) $\forall a_1 \in \mathcal{A}_1$, $\forall a_2 \in \mathcal{A}_2$, $a_1 \between a_2$, or (2) if one of $\mathcal{A}_1$ or $\mathcal{A}_2$ is an empty set. $\mathcal{A}_1 < \mathcal{A}_2$ ($\mathcal{A}_1$ is *less than* $\mathcal{A}_2$, or $\mathcal{A}_2$ is *greater than* $\mathcal{A}_1$), and $\mathcal{A}_1 \sqsubset \mathcal{A}_2$ ($\mathcal{A}_1$ is *nested* in $\mathcal{A}_2$) can be defined similarly. We also say an arc a *crosses* a set of arcs $\mathcal{A}$ to mean $\{a\} \between \mathcal{A}$ (the cases for $\sqsubset$ and $<$ can be defined similarly).

For any three sets of arcs $\mathcal{A}_1$, $\mathcal{A}_2$ and $\mathcal{A}_3$, we say that

- $\mathcal{A}_3$ is *from* $\mathcal{A}_1$ *to* $\mathcal{A}_2$ iff $\mathcal{A}_1 < \mathcal{A}_2$ and $\mathcal{A}_1 \between \mathcal{A}_3$, $\mathcal{A}_3 \between \mathcal{A}_2$, and
- $\mathcal{A}_3$ is *anchored* by $\mathcal{A}_1$ and $\mathcal{A}_2$ iff $\mathcal{A}_1 < \mathcal{A}_3$ and $\mathcal{A}_2 \between \mathcal{A}_3$.

Given two point sets $\mathcal{S}_1$ and $\mathcal{S}_2$, we write $\mathcal{S}_1 < \mathcal{S}_2$ iff $\forall s_1 \in \mathcal{S}_1$ and $\forall s_2 \in \mathcal{S}_2$, $s_1 < s_2$. For a arc set $\mathcal{A}$, we denote $L(\mathcal{A}) = \bigcup_{a \in \mathcal{A}} \{L(a)\}$, and $R(\mathcal{A}) = \bigcup_{a \in \mathcal{A}} \{R(a)\}$.

The subscript '$*$' is a special symbol which matches every defined subscript. That is, $A_{*,j}$ refers to the set $\{A_{ij} | A_{i,j}$ is defined$\}$, and $A_{*,*}$ refers to the set of all $A_{i,j}$ that has been defined.

If $\mathcal{CM}(\mathcal{S}_{n,\ell}, \mathcal{A}_{n,\ell})$ occurs in $\mathcal{CM}(\mathcal{S}_\mathcal{G}, \mathcal{A}_\mathcal{G})$, there exists a one-one mapping $\mathcal{M}$ between elements in $\mathcal{A}_{n,\ell}$ and some elements in $\mathcal{A}_\mathcal{G}$. Here, we extend the definition of the mapping to any set $\mathcal{A}'_p \subseteq \mathcal{A}_{n,\ell}$, such that $\mathcal{M}(\mathcal{A}'_p) = \bigcup_{a \in \mathcal{A}'_p} \{\mathcal{M}(a)\}$.

3.2   Target Contact Map Construction

In this Section, we construct a target contact map $\mathcal{CM}(\mathcal{S}_\mathcal{G}, \mathcal{A}_\mathcal{G})$ from a given graph $\mathcal{G}(\mathcal{V}, \mathcal{E})$. We first build some large crossing clusters, and then we construct the arcs which connect these clusters.

Large Crossing Clusters. Firstly, we construct $2n + 2$ crossing clusters, which are H, Z_u ($1 \leq u \leq n$), T and V_u ($1 \leq u \leq n$). H is a $28n^4$-arc crossing cluster, Z_u is a $5n^3$-arc crossing cluster, T is a $9n^4$-arc crossing cluster and V_u is a $5n^3$-arc crossing cluster. Let $Z = \bigcup_{u=1}^{n} Z_u$ and $V = \bigcup_{i=1}^{n} V_u$. Furthermore we define the following order for these large clusters:

$$H < Z_1 < \ldots < Z_n < T < V_1 < \ldots < V_n$$

Arcs from H to Z_u. There is a 2-arc crossing cluster from $\mathbf{H}$ to $\mathbf{Z}_u$ for each u, $1 \le u \le n$. Denote the two arcs as $A_{u,1}$, and $A_{u,2}$, $A_{u,1} \between A_{u,2}$. Let $A_u = \{A_{u,1}, A_{u,2}\}$. Furthermore, we define the following orders:

$$H \between A_u, A_u \between Z_u \qquad\qquad 1 \le u \le n \qquad\qquad (1)$$

$$A_{u_1} \sqsubset A_{u_2}, \qquad\qquad 1 \le u_1 < u_2 \le n \qquad\qquad (2)$$

Equation 1 ensures that A_u is from H to Z_u. Equation 2 forces that at most one pair of arcs in $A_{*,*}$ can be included in a CCM.

Let $A = \bigcup_{u=1}^{n} A_u$, it is clear that $|A| = 2n$.

Arcs from Z_u to Z_v. There are two types of arcs from Z_u to Z_v ($1 \le u < v \le n$): $E_{u,v}$ and $C_{u,v}$. $E_{u,v}$ consists of u crossing clusters, denoted $E_{u,v,w}$, $1 \le w \le u$. Each cluster $E_{u,v,w}$ contains 3 arcs, respectively $E_{u,v,w,1}$, $E_{u,v,w,2}$ and $E_{u,v,w,3}$ with $E_{u,v,w,1} \between E_{u,v,w,2}$, $E_{u,v,w,1} \between E_{u,v,w,3}$ and $E_{u,v,w,2} \between E_{u,v,w,3}$. Each $C_{u,v}$ is a single arc. We now define orders among the arcs $E_{*,*,*,*}$ and $C_{*,*}$ which are needed for our proof. Diagrams of these orders are depicted in the Appendix.

Firstly, we ensure that $E_{u,*,*,*}$ and $C_{u,*}$ are crossed by Z_u, while $E_{*,v,*,*}$ and $C_{*,v}$ crosses Z_v:

$$Z_u \between E_{u,*,*,*}, Z_u \between C_{u,*}, \qquad\qquad 1 \le u \le n-1 \qquad\qquad (3)$$

$$E_{*,v,*,*} \between Z_v, C_{*,v} \between Z_v, \qquad\qquad 2 \le v \le n \qquad\qquad (4)$$

Secondly, we define the orders among the arcs which are crossing Z_v ($2 \le v \le n$):

$$R(E_{*,v,1,*}) < R(E_{*,v,2,*}) < \ldots < R(E_{*,v,v-1,*}) < R(C_{*,v}) \qquad\qquad (5)$$

$$R(E_{*,v,w,1}) < R(E_{*,v,w,2}) < R(E_{*,v,w,3}), 1 \le w < v \qquad\qquad (6)$$

$$E_{v-1,v,w,i} \sqsubset E_{v-2,v,w,i} \sqsubset \ldots \sqsubset E_{w,v,w,i}, 1 \le w < v, 1 \le i \le 3 \qquad\qquad (7)$$

$$C_{v-1,v} \sqsubset C_{v-2,v} \sqsubset \ldots \sqsubset C_{1,v} \qquad\qquad (8)$$

$$E_{*,v,*,*} \sqsubset A_v, C_{*,v} \sqsubset A_v \qquad\qquad (9)$$

Equation 5 ensures that for the arcs crossing Z_v, the right endpoints are ordered according to the 3rd subscripts. Also the right endpoints for $C_{*,v}$ should be greater than the right endpoints of $E_{*,v,*,*}$. Furthermore, Equation 6 orders (the right endpoints) of $E_{*,v,w,*}$ according to the 4th subscripts for any given v and w, and then Equation 7 orders them by their first subscripts. Equation 8 defines the order between the arcs of $C_{*,*}$, and at most one arc in $C_{*,v}$ can be selected for a CCM. Equation 9 defines the relations between the arcs of $C_{*,v}$, $E_{*,v,*,*}$ and $A_{v,*}$. If $A_{v,*}$ is selected for a CCM, then none of the arcs in $E_{*,v,*,*}$ and $C_{*,v}$ can be used.

Thirdly, for the arcs which are crossed by Z_u ($1 \le u \le n-1$), we introduce the orders as below:

$$E_{u,u+1,w,*} \sqsubset E_{u,u+2,w,*} \sqsubset \ldots \sqsubset E_{u,n,w,*}, \qquad\qquad 1 \le w \le u \qquad\qquad (10)$$

$$C_{u,u+1} \sqsubset C_{u,u+2} \sqsubset \ldots \sqsubset C_{u,n} \qquad\qquad (11)$$

Equation 10 ensures that for any given u and w, at most one 3-arc crossing cluster can be chosen for a CCM, namely $E_{u,v,w,*}$ for some v. Similarly Equation 11 ensures that for a given u, at most one arc in $C_{u,*}$ appears in a CCM.

Lastly, we define the orders between those arcs which are crossed by Z_z and are crossing Z_z ($1 \leq z \leq n$):

$$E_{*,z,w,1} < E_{z,*,w,*}, E_{*,z,w,2} \lozenge E_{z,*,w,*}, \qquad\qquad 1 \leq w < z < n \qquad (12)$$

$$A_{z,1} < E_{z,*,z,*}, A_{z,2} \lozenge E_{z,*,z,*}, \qquad\qquad 1 \leq z < n \qquad (13)$$

$$A_{z,2} < C_{z,*}, \qquad\qquad 1 \leq z < n \qquad (14)$$

Equation 12 ensures that the arcs from Z_u for a given w is anchored by $E_{*,z,w,1}$ and $E_{*,z,w,2}$. Notice that for $w = z$, the set $E_{*,z,z,*}$ is not defined. The arcs $E_{z,*,z,*}$ is anchored by arcs $A_{z,1}$ and $A_{z,2}$ (by Equation 13). Combining with Equation 3, Equation 14 ensures that arc $C_{z,*}$ is anchored by arc $A_{z,2}$ and arc set Z_z.

Denote $C = C_{*,*}$, we know that $|C| = 1/2(n^2 - n)$. Let $E = E_{*,*,*,*}$ and we have $|E| = 1/2(n^3 - n)$.

Arcs from Z_u to T. Arcs from Z_u to T are denoted as F_u. F_u consists of u 2-arc crossing clusters, and the clusters are denoted as $F_{u,w}$, $1 \leq w \leq u$. $F_{u,w}$ contains two arcs: $F_{u,w,1}$ and $F_{u,w,2}$, where $F_{u,w,1} \lozenge F_{u,w,2}$. Firstly we ensure that F_u is from Z_u to T ($1 \leq u \leq n$):

$$Z_u \lozenge F_{u,*,*}, F_{u,*,*} \lozenge T.$$

Furthermore, we define the following orders:

$$E_{*,u,w,1} < F_{u,w,*}, E_{*,u,w,2} \lozenge F_{u,w,*}, \qquad\qquad 1 \leq w < u \leq n \qquad (15)$$

$$A_{u,1} < F_{u,u,*}, A_{u,2} \lozenge F_{u,u,*}, \qquad\qquad 1 \leq u \leq n \qquad (16)$$

$$E_{u,*,w,*} \sqsubset F_{u,w,*}, \qquad\qquad 1 \leq w \leq u < n \qquad (17)$$

$$R(F_{*,1,*}) < R(F_{*,2,*}) < \ldots < R(F_{*,n,*}) \qquad\qquad\qquad (18)$$

$$R(F_{*,w,1}) < R(F_{*,w,2}), \qquad\qquad 1 \leq w \leq n \qquad (19)$$

$$F_{n,w,i} \sqsubset F_{n-1,w,i} \sqsubset \ldots \sqsubset F_{w,w,i}, \qquad\qquad 1 \leq w \leq n, 1 \leq i \leq 2 \qquad (20)$$

Equation 15 and 16 ensures that $F_{u,w,*}$ are anchored by $E_{*,u,w,1}$ and $E_{*,u,w,2}$ or by $A_{u,1}$ and $A_{u,2}$ respectively. Equation 17 ensures that if some arcs of $F_{u,w,*}$ appears in a CCM, then none of the arcs of $E_{u,*,w,*}$ can appear in a CCM. The right endpoints of $F_{*,*,*}$ are ordered according to their 2nd subscripts by Equation 18, and then by the 3rd subscript (by Equation19). Furthermore, Equation 20 ensures that only one arc is possible for a CCM in the set $F_{*,w,i}$ for given w and i.

Let $F = F_{*,*,*}$. Note that $|F| = n^2 + n$.

Arcs from T to V_v and from V_u to V_v. Two types of arcs $I_{u,v}$ and $P_{u,v}$ are defined. $I_{u,v}$ and $P_{u,v}$ are induced from the edges of $\mathcal{G}(\mathcal{V}, \mathcal{E})$. $I_{u,v}$ can be either a 3-arc crossing cluster or an empty set. If $I_{u,v} \neq \emptyset$, we denote the three arcs in it as $I_{u,v,1}$, $I_{u,v,2}$ and $I_{u,v,3}$, with $I_{u,v,1} \lozenge I_{u,v,2}$, $I_{u,v,1} \lozenge I_{u,v,3}$ and $I_{u,v,2} \lozenge I_{u,v,3}$. $P_{u,v}$ contains $(n-v)$ crossing clusters, each cluster $P_{u,v,w}$ $(v < w \leq n)$ is empty or has 2 crossing arcs. If $P_{u,v,w} \neq \emptyset$, we denote the two arcs as $P_{u,v,w,1}$ and $P_{u,v,w,2}$, $P_{u,v,w,1} \lozenge P_{u,v,w,2}$.

The arcs from T to V_v are in two sets: $P_{0,v,w}$ and $I_{0,v}$. $P_{0,v,w}$ is a 2-arc crossing cluster and $I_{0,v}$ is a 3-arc crossing cluster. They are all nonempty sets.

The edge information of $\mathcal{G}(\mathcal{V}, \mathcal{E})$ is used to construct the arcs from V_u to V_v. For the case $I_{u,v}$, $1 \leq u < v \leq n$, if $(u,v) \notin \mathcal{E}_\mathcal{G}$ $I_{u,v} = \emptyset$; otherwise $(u,v) \in \mathcal{E}_\mathcal{G}$, $I_{u,v}$ is a 3-arc crossing cluster.

For the case $P_{u,v,w}$, $1 \leq u < v < w \leq n$, if $(u,w) \notin \mathcal{E}_\mathcal{G}$, we have $P_{u,v,w} = \emptyset$; otherwise $(u,w) \in \mathcal{E}_\mathcal{G}$, we have $P_{u,v,w}$ as a 2-arc crossing cluster.

Firstly we ensures that $I_{0,*,*}$ and $P_{0,*,*,*}$ are crossed by T; $I_{u,*,*}$ and $P_{u,*,*,*}$ are crossed by V_u $(1 \leq u \leq n-1)$, and $I_{*,v}$ and $P_{*,v,*,*}$ are crossing V_v:

$$T \lozenge I_{0,*,*}, \qquad\qquad\qquad T \lozenge P_{0,*,*,*} \qquad (21)$$

$$V_u \lozenge I_{u,*,*}, 1 \leq u < n \qquad V_u \lozenge P_{u,*,*,*}, 1 \leq u < n-1 \qquad (22)$$

$$I_{*,v,*} \lozenge V_v, 1 \leq v \leq n \qquad P_{*,v,*,*} \lozenge V_v, 1 \leq v < n \qquad (23)$$

For the arcs which are crossing V_v, we define the orders:

$$R(I_{*,v,*}) < R(P_{*,v,v+1,*}) \qquad\qquad 1 \leq v \leq n-1 \quad (24)$$

$$R(P_{*,v,v+1,*}) < R(P_{*,v,v+2,*}) < \ldots < R(P_{*,v,n,*}), \qquad 1 \leq v \leq n-1 \quad (25)$$

$$R(P_{*,v,w,1}) < R(P_{*,v,w,2}), \qquad\qquad 1 \leq v < w \leq n \quad (26)$$

$$P_{v-1,v,w,i} \sqsubset P_{v-2,v,w,i} \sqsubset \ldots \sqsubset P_{0,v,w,i}, 1 \leq v < w \leq n, \qquad 1 \leq i \leq 2 \quad (27)$$

$$I_{v-1,v,*} \sqsubset I_{v-2,v,*} \sqsubset \ldots \sqsubset I_{0,v,*}, \qquad\qquad 1 \leq v \leq n \quad (28)$$

Equation 25 ensures that for a given v, the right endpoints of $P_{*,v,*,*}$ are sorted according to the third subscript. Then Equation 26 ensures that for any given v and w, the right endpoints for $P_{*,v,w,*}$ are sorted according to the forth subscript. Further more, for any given v, w and i, Equation 27 ensures that at most one arc in $P_{*,v,w,i}$ can be selected for a CCM.

Next we introduce the orders for the arcs which are crossed by T, and the arcs which are crossed by V_u:

$$P_{u,u+1,w,*} \sqsubset P_{u,u+2,w,*} \sqsubset \ldots \sqsubset P_{u,n,w,*}, \qquad 0 \leq u, u+1 < w \leq n \quad (29)$$

$$I_{u,w,*} \sqsubset P_{u,*,w,*}, \qquad\qquad 0 \leq u, u+1 < w \leq n \quad (30)$$

Equation 29 and 30 ensures that either (1) one 2-arc crossing cluster $P_{u,v,w,*}$ can be selected for a CCM, or (2) the 3-arc crossing cluster $I_{u,w,*}$ is selected for a CCM, or (3) none of them are selected.

Furthermore, for the arcs which are crossed by T, we define the orders:

$$F_{*,w,1} < I_{0,w}, F_{*,w,2} \lozenge I_{0,w}, \qquad\qquad 1 \leq w \leq \ell \quad (31)$$

$$F_{*,w,1} < P_{0,*,w,*}, F_{*,w,2} \lozenge P_{0,*,w,*}, \qquad 2 \leq w \leq \ell \quad (32)$$

Equation 31 ensures that $I_{0,w}$ is anchored by $F_{*,w,1}$ and $F_{*,w,2}$, and Equation 32 ensured that $P_{0,*,w,*}$ is anchored by $F_{*,w,1}$ and $F_{*,w,2}$.

Lastly, we define orders between those arcs are crossed by V_z and the arcs which crosses V_z:

$$P_{*,z,w,1} < I_{z,w}, P_{*,z,w,2} \lozenge I_{z,w}, \qquad\qquad 1 \le z, z+1 \le w \le n \qquad (33)$$

$$P_{*,z,w,1} < P_{z,*,w,*}, P_{*,z,w,2} \lozenge P_{z,*,w,*}, \qquad 1 \le z, z+1 < w \le n \qquad (34)$$

Equation 33 ensures that $I_{z,w}$ is anchored by $P_{*,z,w,1}$ and $P_{*,z,w,2}$, and Equation 34 ensured that $P_{z,*,w,*}$ is anchored by $P_{*,z,w,1}$ and $P_{*,z,w,2}$.

Let $P = P_{*,*,*,*}$ and $I = I_{*,*}$, it is not difficult to show that $|P| \le 1/3(n^3 - n)$ $\cdot |I| \le 3/2(n^2 + n)$.

Let $\mathcal{A_G} = H \cup A \cup C \cup Z \cup E \cup F \cup I \cup P \cup V$, and $S_{\mathcal{G}}$ to be those endpoints of the arcs in $\mathcal{A_G}$. The target contact map $\mathcal{CM}(S_{\mathcal{G}}, \mathcal{A_G})$ is fully specified. The following results can be shown for $\mathcal{CM}(S_{\mathcal{G}}, \mathcal{A_G})$:

Lemma 1. *(i) An arc $a \in E$ crosses no more than $9n^3$ arcs.*
(ii) An arc $a \in F$ crosses no more than $17n^4$ arcs.
(iii) An arc $a \in I$ crosses no more than $9n^3$.
(iv) $|\mathcal{A_G} - H| < |H|$.

Proof. We know that $|A| + |C| + |E| + |F| \le 2n + 1/2(n^2 - n) + 1/2(n^3 - n) + (n^2 + n) \le 4n^3$. The only possible arcs an arc $a \in E$ can cross are from A, C, E, F, and Z_u for some u with $1 \le u \le n$.

Since except A, C, E, F, an arc $a \in F$ may cross some arcs in P and I, and T as well, we have $|P| + |I| < 4n^3$ and $|T| = 9n^4$. For an arc $a \in I$, it only crosses those arcs from P, I, and one V_u for some u with $1 \le u \le n$. It is easy to verify that $|\mathcal{A_G} - H| < |H|$. $\qquad\square$

3.3 Pattern Construction

Large Crossing Clusters. Similar to the target case, firstly, we construct $2\ell + 2$ crossing clusters, which are H', Z'_u $(1 \le u \le \ell)$, T' and V'_u $(1 \le u \le \ell)$. H' is a $28n^4$-arc crossing cluster. Z'_u is a $5n^3$-arcs crossing cluster. T' is a $9n^4$-arc crossing cluster. V'_u $(1 \le u \le \ell)$ is a $5n^3$-arc crossing cluster. We also denote $Z' = \bigcup_{u=1}^{\ell} Z'_u$ and $V' = \bigcup_{i=1}^{\ell} V'_u$. Furthermore we define the following order for these large clusters:

$$H' < Z'_1 < \ldots < Z'_\ell < T' < V'_1 < \ldots < V'_\ell$$

Arcs from H' to Z'_1. There is a 2-arc crossing cluster from H' to Z'_1, and is denoted as A'. The two arcs are denoted as A'_1 and A'_2, $A'_1 \lozenge A'_2$. Furthermore, A' is from H' to Z'_1:

$$H' \lozenge A', A' \lozenge Z'_1 \qquad (35)$$

Arcs from Z'_u to Z'_{u+1}. There are two types of arcs from Z'_u to Z'_{u+1}: E'_u and C'_u. C'_u is a single arc. E'_u contains u 3-arc crossing clusters, these clusters are denoted as $E'_{u,w}$, $1 \le w \le u$. For each cluster $E'_{u,w}$, the three arcs of it are denoted as $E'_{u,w,1}$, $E'_{u,w,2}$ and $E'_{u,w,3}$ with $E'_{u,w,1} \lozenge E'_{u,w,2}$, $E'_{u,w,1} \lozenge E'_{u,w,3}$ and $E'_{u,w,2} \lozenge E'_{u,w,3}$

Firstly, we ensure that $E'_{u,*,*}$ and C'_u are from Z'_u and to Z'_{u+1}

$$Z'_u \lozenge E'_{u,*,*}, \qquad\qquad E'_{u,*,*} \lozenge Z'_{u+1}, 1 \leq u \leq \ell - 1 \qquad (36)$$

$$Z'_u \lozenge C'_u, \qquad\qquad C'_u \lozenge Z'_{u+1}, 1 \leq u \leq \ell - 1 \qquad (37)$$

Furthermore, we define the following orders:

$$A'_1 < E'_{1,*,*}, A'_2 \lozenge E'_{1,*,*} \qquad\qquad (38)$$

$$E'_{u,w_1,*} \lozenge E'_{u,w_2,*}, \qquad\qquad 1 \leq w_1 < w_2 \leq u \leq \ell - 1 \qquad (39)$$

$$E'_{u,w,1} < E'_{u+1,w,*}, E'_{u,w,2} \lozenge E'_{u+1,w,*}, \qquad 1 \leq w \leq u < \ell - 1 \qquad (40)$$

$$E'_{u,*,*} \lozenge C'_u, \qquad\qquad 1 \leq u \leq \ell - 1 \qquad (41)$$

$$C'_{u-1} < E'_{u,u,*} \qquad\qquad 2 \leq u \leq \ell - 1 \qquad (42)$$

$E'_{1,*,*}$ (a 3-arc crossing cluster) is anchored by A'_1 and A'_2 (Equation 38). Equation 39 ensures that arcs in $E_{u,*,*}$ forms a crossing cluster. Furthermore, Equations 40 ensures that the 3-arc crossing cluster $E'_{u+1,w,*}$ is anchored by $E'_{u,w,1}$ and $E'_{u,w,2}$. Equation 41 means that the crossing cluster $E'_{u,*,*}$ crosses the arc C'_u. Combining the information from Equation 36 and Equation 42, the arc set $E'_{u,u,*}$ is anchored by C'_{u-1} and Z'_u.

Let $C' = C'_*$ and $E' = E'_{*,*,*}$.

Arcs from Z'_ℓ to T'. The arcs from Z'_ℓ to T' are denoted as F'. F' has ℓ crossing clusters, each crossing cluster contains 2 arcs. The crossing clusters are denoted as F'_w ($1 \leq w \leq \ell$), the two arcs in F'_w are denoted as $F'_{w,1}$ and $F'_{w,2}$, $F'_{w,1} \lozenge F'_{w,2}$. Furthermore, we have the following orders:

$$V'_\ell \lozenge F'_{*,*}, F'_{*,*} \lozenge T' \qquad\qquad (43)$$

$$E'_{\ell-1,w,1} < F'_{w,*}, E'_{\ell-1,w,2} \lozenge F'_{w,*}, \qquad 1 \leq w \leq \ell - 1 \qquad (44)$$

$$C'_{\ell-1} < F'_\ell \qquad\qquad (45)$$

$$F'_{w_1,*} \lozenge F'_{w_2,*}, \qquad\qquad 1 \leq w_1 < w_2 \leq \ell \qquad (46)$$

Equation 43 ensures that $F'_{*,*}$ is from V'_ℓ to T'. $F'_{w,*}$ is anchored by $E'_{\ell-1,w,1}$ and $E'_{\ell-1,w,2}$ (Equation 44) and F'_ℓ is anchored by $C'_{\ell-1}$ and V'_ℓ (Equation 45). Furthermore, arcs in $F'_{*,*}$ forms a crossing cluster by Equation 46.

Arcs from T' to V'_1 and from V'_u to V'_{u+1}. There are two types of arcs: I'_u, ($0 \leq u < \ell$), and P'_u. I'_u ($0 \leq u \leq \ell - 1$) is a 3-arc crossing cluster: the three arcs being $I'_{u,1}$, $I'_{u,2}$ and $I'_{u,3}$, where $I'_{u,1} \lozenge I'_{u,2}$, $I'_{u,1} \lozenge I'_{u,3}$ and $I'_{u,2} \lozenge I'_{u,3}$. P'_u contains $(\ell - u - 1)$ ($0 \leq u \leq n - 2$) 2-arc crossing clusters, each cluster is denoted as $P'_{u,w}$ ($u + 1 < w \leq \ell$). Denote the 2 arcs of $P'_{u,w}$ as $P'_{u,w,1}$ and $P'_{u,w,2}$, $P'_{u,w,1} \lozenge P'_{u,w,2}$.

Firstly we ensure that I'_0 and $P'_{0,*,*}$ are crossed by T'; I_u and $P_{u,*,*}$ are crossed by V'_u $(1 \le u \le \ell - 1)$, and I'_u and $P'_{u,*,*}$ crosses V'_{u+1}:

$$T' \between I'_0, T' \between P'_{0,*,*} \tag{47}$$

$$V'_u \between I'_u, \qquad\qquad 1 \le u < \ell \tag{48}$$

$$V'_u \between P'_{u,*,*}, \qquad\qquad 1 \le u < \ell - 1 \tag{49}$$

$$I'_{u,*} \between V'_{u+1}, \qquad\qquad 0 \le u < \ell \tag{50}$$

$$P'_{u,*,*} \between V'_{u+1}, \qquad\qquad 0 \le u < \ell - 1 \tag{51}$$

Furthermore, arcs in I'_u and $P'_{u,*}$ forms a crossing cluster:

$$I'_u \between P'_{u,*}, \qquad\qquad 0 \le u < \ell - 1 \tag{52}$$

$$P'_{u,w_1} \between P'_{u,w_2}, \qquad\qquad 0 \le u, u+1 < w_1 < w_2 \le \ell \tag{53}$$

Also we introduce the following orders:

$$F'_{1,1} < I'_0, F'_{1,2} \between I'_0 \tag{54}$$

$$F'_{w,1} < P'_{0,w}, F'_{w,2} \between P'_{0,w}, \qquad\qquad 2 \le w \le \ell \tag{55}$$

$$P'_{u,u+2,1} < I'_{u+1}, P'_{u,u+2,2} \between I'_{u+1}, \qquad\qquad 1 \le u < \ell - 1 \tag{56}$$

$$P'_{u,w,1} < P'_{u+1,w,*}, P'_{u,w,2} \between< P'_{u+1,w,*}, \qquad\qquad 1 \le u, u+2 < w \le \ell \tag{57}$$

Equation 54 ensures that I'_0 is anchored by $F'_{1,1}$ and $F'_{1,2}$ and Equation 55 ensures that $P'_{0,w}$ is anchored by $F'_{w,1}$ and $F'_{w,2}$. I'_{u+1} is anchored by $P'_{u,u+2,1}$ and $P'_{u,u+2,2}$, and $P'_{u+1,w,*}$ is anchored by $P'_{u,w,1}$ and $P'_{u,w,2}$.

Let $P' = P'_{*,*,*}$ and $I' = I'_*$.

$\mathcal{A}'_{n,\ell} = H' \cup A' \cup C' \cup Z' \cup E' \cup F' \cup I' \cup P' \cup V'$ and $\mathcal{S}'_{n,\ell}$ are the endpoints of those arcs in $\mathcal{A}'_{n,\ell}$. It is not difficult to verify the following result by the constructions.

Lemma 2. $\mathcal{CM}(\mathcal{S}_{n,\ell}, \mathcal{A}_{n,\ell})$ is a $\{<, \between\}$-structured contact map, and $\mathcal{CM}(\mathcal{S}_\mathcal{G}, \mathcal{A}_\mathcal{G})$ is a $\{<, \sqsubset, \between\}$-structured contact map.

3.4   Correctness

According to the construction, we can prove the following results, proofs are omitted.

Lemma 3. If $\mathcal{CM}(\mathcal{S}_{n,\ell}, \mathcal{A}_{n,\ell})$ occurs in $\mathcal{CM}(\mathcal{S}_\mathcal{G}, \mathcal{A}_\mathcal{G})$, then $\forall \mathcal{M}, \mathcal{M}(H') = H$, $\mathcal{M}(A') = A_{u_1,*}$ for some u_1, with $1 \le u_1 \le n$.

Lemma 4. If $\mathcal{CM}(\mathcal{S}_{n,\ell}, \mathcal{A}_{n,\ell})$ occurs in $\mathcal{CM}(\mathcal{S}_\mathcal{G}, \mathcal{A}_\mathcal{G})$, then $\forall \mathcal{M}, \mathcal{M}(E'_{1,1,*}) = E_{u_1,u_2,u_1,*}$ and $\mathcal{M}(C'_1) = C_{u_1,u_2}$ for some u_1, u_2 with $1 \le u_1 < u_2 \le n$.

Lemma 5. If $\mathcal{CM}(\mathcal{S}_{n,\ell}, \mathcal{A}_{n,\ell})$ occurs in $\mathcal{CM}(\mathcal{S}_\mathcal{G}, \mathcal{A}_\mathcal{G})$ and $\mathcal{M}(E'_{1,1,*}) = E_{u_1,u_2,u_1,*}$, then $\mathcal{M}(E'_{2,v,*}) = E_{u_2,u_3,u_v,*}$ and $\mathcal{M}(C'_v) = C_{u_v,u_{v+1}}$ $(v = 1,2)$ for some u_3 with $u_2 < u_3 \le n$.

Lemma 6. *If* $CM(S_{n,\ell}, A_{n,\ell})$ *occurs in* $CM(S_G, A_G)$ *and* $M(E'_{k,v_1,*}) = E_{u_k,u_{k+1},u_{v_1},*}$ *and* $M(C'_{v_1}) = C_{u_{v_1},u_{v_1}+1}$, *($v_1 = 1, \ldots, k$) for* $u_1 < \ldots < u_{k+1} \leq n$, *then* $M(E'_{k+1,v_2,*}) = E_{u_{k+1},u_{k+2},u_{v_2},*}$ *and* $M(C'_{v_2}) = C_{u_{v_2},u_{v_2}+1}$ *($v_2 = 1, \ldots, k+1$) for some* u_{k+2} *with* $u_{k+1} < u_{k+2} \leq n$.

By induction and Lemma 3-6, we have the following results:

Lemma 7. *If* $CM(S_{n,\ell}, A_{n,\ell})$ *occurs in* $CM(S_G, A_G)$, *then* $\forall M$, $M(E'_{\ell-1,v,*}) = E_{u_{\ell-1},u_\ell,u_v,*}$ *and* $M(C'_{\ell-1}) = C_{u_{\ell-1},u_\ell}$ *with* $v = 1, \ldots, \ell-1$ *and for some* $u_1, \ldots, u_\ell$, $1 \leq u_1 < \ldots < u_\ell \leq n$.

Lemma 8. *If* $CM(S_{n,\ell}, A_{n,\ell})$ *occurs in* $CM(S_G, A_G)$, *if* $M(E'_{\ell-1,v,*}) = E_{u_{\ell-1},u_\ell,u_v,*}$ *and* $M(C'_{\ell-1}) = C_{u_{\ell-1},u_\ell}$ *for* $u_1, \ldots, u_\ell$ *with* $1 \leq u_1 < \ldots < u_\ell \leq n$, *then* $M(F'_{v,*}) = F_{u_\ell,u_v,*}$ *($1 \leq v \leq \ell$).*

Lemma 9. *If* $CM(S_{n,\ell}, A_{n,\ell})$ *occurs in* $CM(S_G, A_G)$, *then* $\forall M$, $M(I'_{0,*}) = I_{0,u_1,*}$, *and* $M(P'_{0,v,*}) = P_{0,u_1,u_v,*}$, *($2 \leq v \leq \ell$) for some* $u_1, \ldots, u_\ell$ *with* $1 \leq u_1 < \ldots < u_\ell \leq n$.

Lemma 10. *If* $CM(S_{n,\ell}, A_{n,\ell})$ *occurs in* $CM(S_G, A_G)$, *then* $\forall M$, $M(I'_{w,*}) = I_{u_w,u_w+1,*}$, *and* $M(P'_{w,v,*}) = P_{u_w,u_w+1,u_v,*}$ *($1 \leq w < \ell$, $1 \leq w+1 < v \leq \ell$) for some* $u_1, \ldots, u_\ell$ *with* $1 \leq u_1 < \ldots < u_\ell \leq n$.

Then by the construction of $CM(S_G, A_G)$, we have:

Lemma 11. *If* $CM(S_{n,\ell}, A_{n,\ell})$ *occurs in* $CM(S_G, A_G)$, G *has a size* ℓ *clique.*

Finally, the following Theorem can be shown:

Theorem 1. $CM(V_{\ell,n}, A_{\ell,n})$ *occurs in* (CMV_G, A_G) *if and only if* G *contains a clique with size* ℓ, *and hence the CCMPM problem is NP-hard.*

It may be noticed that we have shown a stronger result where the problem is NP-hard even for the case that the target is a $\{<, \sqsubset, \emptyset\}$-structured contact map (in general, arcs in target can share endpoints). It is not difficult to perform the same reduction for the 2-interval pattern matching problem with disjoint interval ground set, and with $\{<, \emptyset\}$-structured pattern. Due to lack of space, we omit the formal argument here.

The maximum contact-map overlap (CMO) problem with $\{<, \emptyset\}$ structured patterns is to find a maximized common CCM between two given contact maps. The complexity of this problem was an open question [1]. We now show that the problem is NP-hard using Theorem 1.

Theorem 2. *The CMO problem is NP-hard.*

Proof. Given a CCMPM problem instance: $CM(S_p, A_p)$ and $CM(S, A)$. Find the maximized common CCM $CM(S'_p, A'_p)$ between $CM(S_p, A_p)$ and $CM(S, A)$, and then verify if $CM(S'_p, A'_p)$ is identical to $CM(S_p, A_p)$.

Clearly this reduction is polynomial. Thus the Theorem holds. □

4   Counterexample for the Algorithm in [1]

In this section, we present a counterexample for the algorithm in [1]. The example is displayed in Figure 1. The arcs are labeled with letters instead of numbers for the ease of illustration. The pattern is a CCM with 24 arcs, while the target contains 42 arcs, and is $\{<, \sqsubset, \between\}$-structured. The arcs are labeled in the way that we intend to map an arc of a pattern to an arc of the target which is labeled with the same letter in a different case.It can be verified that the pattern does not occur in the target, but the algorithm in [1] produces a 'yes' answer.

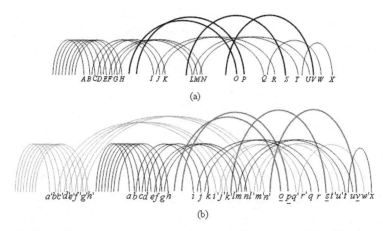

Fig. 1. An Example Demonstrating the Flaw of the Algorithm (a) A $\{\between, <\}$-Structured CM as the Pattern (b) The Target CM

References

1. Gramm, J.: A polynomial-time algorithm for the matching of crossing contact-map patterns. IEEE/ACM Trans. Comput. Biol. Bioinformatics **1** (2004) 171–180
2. Goldman, D., Papadimitriou, C., Istrail, S.: Algorithmic aspects of protein structure similarity. In: Proc. 40th Symp. on Foundations of Computer Science FOCS'99, IEEE Computer Society (1999) 512
3. Vialette, S.: On the computational complexity of 2-interval pattern matching problems. Theor. Comput. Sci. **312** (2004) 223–249
4. Blin, G., Fertin, G., Vialette, S.: New results for the 2-interval problem. In: Proc. 15th Symp. Combinatorial Pattern Matching CPM'04. Volume 3109 of LNCS, Springer-Verlag (2004) 311–322
5. Crochemore, M., Hermelin, D., Landau, G., S.Vialette: Approximating the 2-interval pattern problem. In: Proc. 13th European Symp. Algorithms ESA'05. Volume 3669 of LNCS, Springer Verlag (2005) 426–437

A Fuzzy Dynamic Programming Approach to Predict RNA Secondary Structure

Dandan Song and Zhidong Deng

Department of Computer Science and Technology, National Laboratory of
Information Science and Technology, Tsinghua University, Beijing 100084, China
sdd00@mails.tsinghua.edu.cn, michael@tsinghua.edu.cn

Abstract. Due to the recent discovery of many RNAs with great diversity of functions, there is a resurgence of research in using RNA primary sequences to predict their secondary structures, due to the discovery of many new RNAs with a great diversity of functions. Among the proposed computational approaches, the well-known traditional approaches such as the Nussinov approach and the Zuker approach are essentially based on deterministic dynamic programming, whereas the stochastic context-free grammar (SCFG), the Bayesian estimation, and the partition function approaches are based on stochastic dynamic programming. In addition, heuristic approaches like artificial neural network and genetic algorithm have also been presented to address this challenging problem. But the prediction accuracy of these approaches is still far from perfect. Here based on the fuzzy sets theory, we propose a fuzzy dynamic programming approach to predict RNA secondary structure, which takes advantage of the fuzzy sets theory to reduce parameter sensitivity and import qualitative prior knowledge through fuzzy goal distribution. Based on the experiments performed on a dataset of tRNA sequences, it is shown that the prediction accuracy of our proposed approach is significantly improved compared with the BJK grammar model of the SCFG approach.

1   Introduction

Not just a passive carrier of genetic information, RNA molecules are found to play many important roles in cell, such as regulatory and catalytic. Therefore, a complete understanding of their functions is worth exploring. Similar to proteins, the functions of RNA molecules are mainly determined by their structures. Since experimental techniques such as X-ray crystallography and nuclear magnetic resonance (NMR) to obtain the structure data usually require a great deal of time and money, the gap between the exponentially exploding number of nucleic acid sequences and the slowly accumulating number of structures data is expanding.

In recent years, many novel computational RNA structure analysis and prediction approaches have been proposed. Early in 1978, Nussinov proposed a maximal base-paring approach, which initiated a conversion of RNA secondary structure prediction into an optimal decision problem and used a dynamic programming approach to directly solve it [1]. This research work is undoubtedly

P. Bücher and B.M.E. Moret (Eds.): WABI 2006, LNBI 4175, pp. 242–251, 2006.

of great significance although the resulting prediction accuracy is poor due to its simplicity. Then Zuker developed a minimum free energy(MFE) approach based on dynamic programming algorithm [2]. The earlier version of the Zuker approach has been improved and implemented by several commonly-used packages as Mfold [3] and RNAfold(Vienna RNA Package)[4]. Apparently, either the Nussinov approach or the Zuker approach and its improvements are based on deterministic dynamic programming algorithm.

Stochastic approaches are also applied to RNA secondary structure prediction problem, e.g., stochastic context-free grammars (SCFG) [5], Bayesian statistical [6], and partition function [7], all of which are based on stochastic dynamic programming algorithm. Among these approaches, the SCFG approach is preferred due to its simple and suitable description of RNA secondary structure. In addition, heuristic methods based on artificial neural network [8] and genetic algorithm (GA) [9] have also been proposed.

However, all these promising methods still have several limitations. For example, alternative suboptimal foldings are neglected by these approaches, but they might actually pinpoint conserved regions of the RNA secondary structure. Additionally, changes in the scoring parameters often lead to drastic alterations of results.

Fuzzy set theory, originated by Zadeh [10] and dealing with a different kind of uncertainty, is well suited for incorporating human experiences, due to the fact that it can express the imprecision of meaning that may result from the use of natural language as we define a model.

Dynamic programming [11] is one of the earliest methodologies to which fuzzy sets theory has been applied, which directly results in the fuzzy dynamic programming (FDP). In essence, the FDP is a recast of dynamic programming, and has attracted wide attention in many fields during the last decades[12].

In this paper, we propose a FDP approach to predict RNA secondary structure, which takes advantage of fuzzy sets theory to reduce the parameter sensitivity, and imports qualitative prior knowledge by adding fuzzy goal distribution to the prediction. Compared to the BJK grammar model of the SCFG approach, which has been shown to have the best synthetical performance among nine SCFG grammar models [13], our experiments performed on a dataset of tRNA sequences show that the average prediction accuracy is increased by 5.34% (sensitivity) and 4.05% (specificity) respectively.

2   Method

2.1   Fuzzy Modeling

Set Up Structure. Referring to the BJK grammar of the SCFG approach [14], we define a fuzzy BJK (FBJK) structure model. In contrast to the original BJK, however, the proposed FBJK model has several unique features. By incorporating the idea of the fuzzy sets theory, it is more suited to incorporate qualitative subjective knowledge, which is beneficial to improve the prediction accuracy.

– State space:
  In the FBJK structure model, three state subspaces are used: L, S, and F. Each state subspace is composed of all allowed possible base pairs of the given sequence, which is indicated by (i, j) $(i(j) = 1, \cdots, l, j \geq i)$, where i and j denote the bases and l denotes the sequence length. As a result, each state subspace forms an upper triangle matrix, see Figure 1.

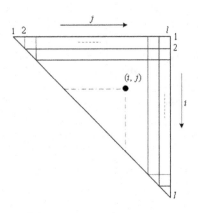

Fig. 1. One layer of state space

– Fuzzy sets of observable characters:
  The observable characters contain four bases $\{A, C, G, U\}$, and depending on whether they are base-paired or base-unpaired we define two fuzzy sets of observable characters $Pair$ and $Single$. In this case, $Pair$ is composed of all possible base-pairs and can be expressed by a 4×4 membership function matrix as Eqs. (1). $Single$ includes the four single bases and can be represented by an 1×4 membership function vector as Eqs. (2).

$$\mu_{Pair} = \begin{bmatrix} \mu_{AA} & \mu_{AC} & \mu_{AG} & \mu_{AU} \\ \mu_{CA} & \mu_{CC} & \mu_{CG} & \mu_{CU} \\ \mu_{GA} & \mu_{GC} & \mu_{GG} & \mu_{GU} \\ \mu_{UA} & \mu_{UC} & \mu_{UG} & \mu_{UU} \end{bmatrix} \tag{1}$$

$$\mu_{Single} = \begin{bmatrix} \mu_A & \mu_C & \mu_G & \mu_U \end{bmatrix}. \tag{2}$$

– Admissible fuzzy decision set:

$$\mathcal{U}_t \in \mathbf{U} = \{U_1, U_2, U_3, U_4, U_5, U_6\}.$$

where each fuzzy decision U_i corresponds to a state transition rule, which is given by Eqs.(3)-(8). For each of the three state subspaces, i.e., L, S, and F, all the fuzzy decisions are classified as three admissible fuzzy decision subsets, that is, $\mathbf{U}_L, \mathbf{U}_S, \mathbf{U}_F \subseteq \mathbf{U}$.

$$U_1 \in \mathbf{U}_L : If \quad \mathcal{X}_t \ is \ L(i,j) \quad and \quad \mathcal{U}_t \ is \ U_1,$$
$$then \quad \mathcal{X}_{t+1} \ is \ F(i+1, j-1) \ and \ Pair(i,j); \tag{3}$$

$$U_2 \in \mathbf{U}_L : If \quad \mathcal{X}_t \ is \ L(i,i) \quad and \quad \mathcal{U}_t \ is \ U_2,$$
$$then \quad \mathcal{X}_{t+1} \ is \ Single(i); \tag{4}$$

$$U_3 \in \mathbf{U}_S : If \quad \mathcal{X}_t \ is \ S(i,j) \quad and \quad \mathcal{U}_t \ is \ U_3,$$
$$then \quad \mathcal{X}_{t+1} \ is \ L(i,k) \ and \ S(k+1, j) \ (i \leq k < j); \tag{5}$$

$$U_4 \in \mathbf{U}_S : If \quad \mathcal{X}_t \ is \ S(i,j) \quad and \quad \mathcal{U}_t \ is \ U_4,$$
$$then \quad \mathcal{X}_{t+1} \ is \ L(i,j); \tag{6}$$

$$U_5 \in \mathbf{U}_F : If \quad \mathcal{X}_t \ is \ F(i,j) \quad and \quad \mathcal{U}_t \ is \ U_5,$$
$$then \quad \mathcal{X}_{t+1} \ is \ L(i,k) \ and \ S(k+1, j) \ (i \leq k < j); \tag{7}$$

$$U_6 \in \mathbf{U}_F : If \quad \mathcal{X}_t \ is \ F(i,j) \quad and \quad \mathcal{U}_t \ is \ U_6,$$
$$then \quad \mathcal{X}_{t+1} \ is \ F(i+1, j-1) \ and \ Pair(i,j). \tag{8}$$

- Optimal fuzzy policy:
  The fuzzy policy is a subsequence of fuzzy decisions at the current and past states. Accordingly, the fuzzy policy that has the biggest membership degree is defined as the optimal fuzzy policy at stage k. So we have

$$\mu(\mathcal{U}_0^*, \mathcal{U}_1^*, \cdots, \mathcal{U}_{k-1}^*) = \max_{\mathcal{U}_0, \mathcal{U}_1, \cdots, \mathcal{U}_{k-1}} \mu(\mathcal{U}_0, \mathcal{U}_1, \cdots, \mathcal{U}_{k-1}), \tag{9}$$

  where k denotes the number of stages. For the RNA secondary structure prediction problem, k refers to the average length of RNA subsequences.
- Fuzzy goal:
  As a fuzzy set, the universe of fuzzy goal is specified to be the set of the bases in the given sequence, while its linguistic variables are specified to be "3'-end of hairpins", which are conservative features of RNA secondary structures as Figure 2.1 shows. More specifically, using the state transition rules mentioned above, "3'-end of hairpin" denotes the base coordinate i when $S(i,i)$ (i.e., $S(i,j)$ satisfies $i = j$) is transferred to $L(i,i)$ using the fuzzy decision U_4.
- Termination state set:
  We use the FDP with an implicit termination time, whose current state X_t is starting from the initiate fuzzy state and transferring according to the fuzzy state transition rule of the current decision. When the current state attains for the first time the termination state set, the process is terminated. Thus, in our FBJK structure, the termination state set is specified to be the fuzzy set of observable characters $Single$.

Estimate Parameters. The training samples datasets are used to estimate parameters: fuzzy decision fitness and fuzzy goal distribution. The main idea of our estimation method is fuzzy statistics, which is also called 'polling' in

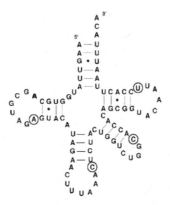

Fig. 2. Take a tRNA's secondary structure as an example. The bases marked by black circle indicate the 3'-end points of hairpins, which are taken as linguistic variables of the fuzzy goal.

[15]. Compared with iterated learning methods, this method can greatly reduce computational complexity and time. While there is sacrifice in accuracy, it can basically satisfy the present requirement.

- Fuzzy decision fitness:
  Here, each fuzzy decision for all fuzzy state is given the same fitness, which is time and state independent $\mu_C(U_i)$. As transition rules currently used are unambiguous, for each training sequence with its secondary structure annotation, the fuzzy policy is unique. After occurrence number of each decision is counted, the fitness of fuzzy decisions is determined from these counts using a Laplace (plus-one) prior idea as equation (10) follows.

$$\mu_C(U_i^r) = \frac{n_i^r + 1}{\sum_j n_j^r + n^r}, \tag{10}$$

  where n_i^r is the occurrence number of the fuzzy decision U_i^r and n^r is number of fuzzy decisions in the state subspace r. In our FBJK structure, $n^r = 2$ for $r \in \{L, S, F\}$.
- Membership of the elements in the fuzzy sets of observable characters:
  The equations in the following are extensions of Laplace (plus-one) prior idea to compute the membership of elements in *Pair* and *Single*.

$$\mu_{Pair}(X, Y) = \frac{n_{X,Y} + 1}{\sum_{X,Y} n_{X,Y} + 16} \quad X(Y) \in \{A, C, G, U\}, \tag{11}$$

$$\mu_{Single}(X) = \frac{n_X + 1}{\sum_X n_X + 4} \quad X \in \{A, C, G, U\}. \tag{12}$$

– Fuzzy goal distribution:
Typical membership functions are used to describe the distribution of element membership in fuzzy goal. Here peak-Γ membership function is selected, whose formula is as equation (13).

$$\mu(x) = \begin{cases} e^{k(x-a)} & x \leq a, k > 0; \\ e^{-k(x-a)} & x > a, k > 0. \end{cases} \quad (13)$$

where x is the relative coordinates of the base. By firstly storing the relative coordinates of the end points of hairpins (i/n) in annotated training sequences, their mean E and covariance V values are calculated, which are then used to determine the k and a parameters as follows.

$$a = E, \quad k = (\frac{4}{V})^{1/3}. \quad (14)$$

To avoid membership being zero, the plus-one prior distribution is added to the above membership functions of fuzzy goal subsets and the membership functions are slightly adjusted.

On the other hand, when the structure profile of homologous RNA sequences is known, multiple fuzzy goal subsets can be defined according to different hairpins. The membership function of each fuzzy goal subset can be estimated respectively. Then the set of fuzzy goal is the union of these fuzzy goal subsets, in which membership of each base is its biggest membership in these fuzzy goal subsets.

$$\mu_{A \cup B}(x) = \max(\mu_A(x), \mu_B(x)). \quad (15)$$

2.2 Fuzzy Inference

Having root with deterministic and stochastic dynamic programming approaches, fuzzy dynamic programming computes optimal value for each substage by the filling process and then attains the optimal fuzzy policy as well as secondary structure result by the tracing back process.

The optimal fuzzy policy is defined as

$$\mu(\mathcal{U}_0^*, \cdots, \mathcal{U}_{K-1}^* | X_0) = \max_{\mathcal{U}_0, \cdots, \mathcal{U}_{K-1}} (\mu_C(\mathcal{U}_0) \circ \cdots \circ \mu_C(\mathcal{U}_{K-1}) \circ \mu_{G^K}(X_K)), \quad (16)$$

where the states satisfy fuzzy state transition rules:

$$If \; \mathcal{X}_t \; is \; X_k \; and \; \mathcal{U}_t \; is \; U_k, \; then \; \mathcal{X}_{t+1} \; is \; X_{k+1} \quad (k = 0, 1, \cdots, K-1). \quad (17)$$

where K is the total number of fuzzy stages. For fuzzy dynamic programming with an implicit termination time, K can't be specified in advance but varies with paths. Thus Kth stage in the above equation denotes the termination time while $K - 1$th is the stage before termination and so on. Meanwhile, $\mu_{G^K}(x_K)$ is the membership of the state x_k in the fuzzy goal and the fuzzy operator $\circ$ is defined as algebraic product as the form $x \circ y = xy$.

Filling Process. The filling process of FDP is iterative, which starts from the subsequences with length 1 (i.e. the diagonal of the up-triangular matrix in Fig. 1), iterately expands the average length of the subsequences (to the up-right direction of the matrix) until the whole sequence is computed and the terminating condition is matched (reaches initiate state $S(1,l)$). In each iteration, the optimal policy of the subsequence decision process is computed.

$$\mu_{G_{K-v}}(X_{K-v}) = \max_{U_{K-v}} \left(\mu_C(U_{K-v}) \circ \mu_{G_{K-v+1}}(X_{K-v+1})\right) (v = 1, \cdots, K). \quad (18)$$

More specifically, for now used FBJK structure and the definition of fuzzy goal, the corresponding filling iteration formulations are symbolized in following equations.

First, for state $L(i,i)$:

$$\mu(L(i,i)) = \mu_C(U_2) \circ \mu_{Single}(B(i)); \quad (19)$$

Second, for state $S(i,i)$:

$$\mu(S(i,i)) = \mu_C(U_4) \circ \mu(L(i,i)) \circ \mu_G(B(i)); \quad (20)$$

Except for the above special formulas, other iterations are presented as follows:

$$\mu(L(i,j)) = \mu_C(U_1) \circ \mu(L(i+1,j-1)) \circ \mu_{Pair}(B(i), B(j));$$
$$_{i<j}$$

$$\mu(S(i,j)) = \max_{i<j} \begin{cases} \mu_C(U_3) \circ \max_{i<k<j}(\mu(L(i,k)) \circ \mu(S(k+1,j)) \\ \mu_C(U_4) \circ \mu(L(i,j)) \end{cases} ; \quad (21)$$

$$\mu(F(i,j)) = \max_{i\le j} \begin{cases} \mu_C(U_5) \circ \max_{i<k<j}(\mu(L(i,k)) \circ \mu(S(k+1,j))) \\ \mu_C(U_6) \circ \mu(F(i+1,j-1)) \circ \mu_{Pair}(B(i), B(j)) \end{cases} .$$

where $B(i)$ refers to the ith base in the sequence.

Tracing Back Process. After the filling process has determined the membership of optimal policy, the tracing back process is used to retrieve the actual optimal policy and its optimal path. It is actually a transformation process of the current state starting from the initial state and using the fuzzy state transition rules until attaining the termination state set. The optimal path is composed of these states. This process is usually completed by employing the stack structure.

Meanwhile, the optimal secondary structure is attained simply as the base pairs uniquely determine the secondary structure: bases i and j are paired in the optimal secondary structure only if $Pair(i,j)$ is on the optimal path.

3   Experimental Results

3.1   Dataset Preparation

In experiments, the tRNA datasets are used, in which 843 tRNA sequences with annotated secondary structure are taken from the EMBL databank [16], including various series such as virus, archaea, eubacteria, cyanelle, cytoplasm and mitochondria. We construct three training sample datasets. The first one is named MT100, in which 100 tRNA sequences are randomly selected from the mitochondria data; while the second one is MT10CY10, where 10 tRNA sequences are randomly selected from the cytoplasm data and 10 sequences from the mitochondria data. The Rand tRNA dataset is composed of 569 randomly selected tRNA from all the series.

3.2   Comparison with the BJK Grammar of the SCFG Approach

The paper of [13] gives a systematical analysis of nine different SCFG models, which concludes that the prediction accuracy of Knudsen/Hein's BJK grammar is only slightly lower than its extension $G6^S$ grammar with a first order Markov chain. While including stacking parameters makes the $G6^S$ grammar much more complex, in synthetic sense, the BJK grammar can be treated as the best SCFG grammar. Using the same training and testing samples datasets, we compare our results with the BJK grammar model of the SCFG approach.

The parameter minimum length of hairpin (HLEN) of the two approaches are both set to be 3 to keep them identical. Since the typical structure of tRNA consists of four hairpins, when performing on the tRNA datasets, four fuzzy goal subsets are used to describe them respectively and then merge into the set of fuzzy goal in our approach.

We use *sensitivity* and *specificity* parameters to evaluate the RNA secondary structure prediction accuracy, which are common measures of the accuracy of prediction methods. Table 1 and 2 express the prediction accuracy of our Fuzzy Dynamic Programming (FDP) approach compared with the BJK grammar of the SCFG approach (BJK). The average *sensitivity* of these experiments is 81.48% for the BJK grammar of the SCFG approach and 86.83% for our FDP approach, while the average *specificity* is 78.62% and 82.68% respectively. Our method outperforms the BJK grammar of the SCFG approach significantly.

Table 1. The FDP's predicted Sensitivity(%) result compared with SCFG's BJK grammar with the same training (columns) and testing (rows) datasets. For instance, the first number 82.57 refers to the prediction accuracy of the BJK grammar using the MT10CY10 dataset to train parameters and the ARCHAE dataset to test.

Dataset	MT10CY10		MT100		Rand tRNA	
	BJK	FDP	BJK	FDP	BJK	FDP
ARCHAE	82.57	88.73	80.98	84.79	83.26	85.77
CY	81.06	90.58	81.68	86.30	83.01	88.69
CYANELCHLORO	83.39	93.03	83.34	88.79	85.71	90.18
EUBACT	90.30	92.30	87.56	91.16	91.19	93.06
VIRUS	80.45	84.16	78.40	84.36	80.45	84.36
MT	77.43	88.00	78.42	85.75	73.47	82.59
PARTIII	77.11	84.21	77.39	79.12	73.98	77.45

Table 2. FDP's predicted Specificity(%) result compared with SCFG's BJK grammar

Dataset	MT10CY10		MT100		Rand tRNA	
	BJK	FDP	BJK	FDP	BJK	FDP
ARCHAE	75.01	80.34	73.99	77.07	77.99	78.86
CY	76.65	85.22	77.23	81.34	82.65	85.12
CYANELCHLORO	79.08	87.48	78.44	82.65	83.75	85.31
EUBACT	84.21	86.67	81.72	84.95	86.84	87.46
VIRUS	78.20	78.05	73.98	79.00	79.80	80.87
MT	75.82	84.62	76.77	83.43	78.28	84.28
PARTIII	75.74	81.79	76.88	78.63	78.04	83.06

4   Discussion

The fuzzy structure model presented in this paper is similar to the BJK grammar of the SCFG approach, with the fuzzy transition rules analogous to the BJK's production rules. But in our approach, we take the RNA secondary structure prediction process as a fuzzy inference system and uses the fuzzy dynamic programming approach to compute the optimal fuzzy policy. By taking advantage of fuzzy theory, conservative information of RNA secondary structure can be naturally and easily imported into the prediction process to get better performance than the BJK grammar of the SCFG approach, which is the one with the best synthetical performance among nine SCFG models in [13].

What must be emphasized is that, in our approach different definitions of state spaces, fuzzy decisions and fuzzy state transition rules can be developed according to realistic problems, not constrained to the description of this paper.

5   Conclusion and Future Work

This paper develops a novel Fuzzy Dynamic Programming Approach to improve the accuracy of RNA secondary structure prediction. The structure and its parameters, as well as the filling and tracing back process are described systematically. The test of our approach performed on tRNA dataset provided better outcomes than the reference BJK grammar of the SCFG approach. What remains to be improved?

- The membership functions of fuzzy decisions and fuzzy goals are both calculated by fuzzy statistical methods. This can be developed to use learning algorithms like the forward-backward algorithm of the SCFG approach. The prediction accuracy is expected to be improved, while the computational complexity is supposed to rise.
- Besides the relative coordinates of 3'-end bases of hairpin, more typical substructures such as the stem features can be taken as the conservative information to improve the prediction.
- Our approach can be modified to include processing of pseudo-knots.

Acknowledgements

This work was supported in part by the National Science Foundation (Grant No. 60321002) and the Teaching and Research Award Program for Outstanding Young Teachers in Higher Education Institutions of MOE (TRAPOYT), China.

References

1. Nussinov, Pieczenik, Griggs, Kleitman: Algorithms for loop matchings. SIAM J. on Applied Mathematics **35** (1978) 68–82
2. Zuker, M., Stiegler, P.: Optimal computer folding of large RNA sequences using thermodynamics and auxiliary information. Nucleic Acids Research **9**(1) (1981) 133–148
3. Zuker, M.: Mfold web server for nucleic acid folding and hybridization prediction. Nucleic Acids Research **31**(13) (2003) 3406–3415
4. Hofacker, I.L., Fontana, W., Stadler, P.F., Bonhoeffer, L.S., Tacker, M., Schuster, P.: Fast folding and comparison of RNA secondary structures. Monatsh. Chem. **125** (1994) 167–188
5. Eddy, S.R., Durbin, R.: RNA sequence analysis using covariance models. Nucleic Acids Research
6. Ding, Y., Lawrence, C.E.: A Bayesian statistical algorithm for RNA secondary structure prediction. Computers & Chemistry **23**(3-4) (1999) 387–400
7. Mccaskill, J.S.: The equilibrium partition function and base pair binding probabilities for RNA secondary structure. Biopolymers **29**(6-7) (1990) 1105–1119
8. Steeg, E.W.: Neural network algorithms for RNA secondary structure prediction. Technical report, University of Toronto Computer Science Dept., Toronto Canada (1990)
9. Hu, Y.J.: Gprm: a genetic programming approach to finding common RNA secondary structure elements. Nucleic Acids Research **31**(13) (2003) 3446–3449
10. Zadeh, L.A.: Fuzzy sets. Information and Control **8**(3) (1965) 338–353
11. Bellman, R.E., Zadeh, L.A.: Decision making in a fuzzy environment. Management Science **17** (1970) 141–164
12. Kacprzyk, J., Esogbue, A.O.: Fuzzy dynamic programming: main developments and applications. Fuzzy Sets Syst. **81**(1) (1996) 31–45
13. Dowell, R.D., Eddy, S.R.: Evaluation of several lightweight stochastic context-free grammars for RNA secondary structure prediction. BMC Bioinformatics **5** (2004) 71–99
14. Knudsen, B., Hein, J.: Rna secondary structure prediction using stochastic context-free grammars and evolutionary history. Bioinformatics **15**(6) (1999) 446–454
15. Bilgiç, T., Türkşen, I.B.: Measurement of membership functions: theoretical and empirical work. In Dubois, H.P.D., Zimmermann, H.J., eds.: International Handbook of Fuzzy Sets and Possibility Theory. Kluwer Academic, Norwell, MA (1999)
16. Steinberg, S., Misch, A., Sprinzl, M.: Compilation of tRNA sequences and sequences of tRNA genes. Nucleic Acids Research **21**(13) (1993) 3011–3015

Landscape Analysis for Protein-Folding Simulation in the H-P Model*

Kathleen Steinhöfel[1], Alexandros Skaliotis[1], and Andreas A. Albrecht[2]

[1] King's College London, Department of Computer Science
Strand, London WC2R 2LS, UK
[2] University of Hertfordshire, School of Computer Science
Hatfield, Herts AL10 9AB, UK

Abstract. The hydrophobic-hydrophilic (H-P) model for protein folding was introduced by Dill et al. [7]. A problem instance consists of a sequence of amino acids, each labeled as either hydrophobic (H) or hydrophilic (P). The sequence must be placed on a 2D or 3D grid without overlapping, so that adjacent amino acids in the sequence remain adjacent in the grid. The goal is to minimize the energy, which in the simplest variation corresponds to maximizing the number of adjacent hydrophobic pairs. The protein folding problem in the H-P model is NP-hard in both 2D and 3D. Recently, Fu and Wang [10] proved an $exp(O(n^{1-1/d}) \cdot \ln n)$ algorithm for d-dimensional protein folding simulation in the HP-model. Our preliminary results on stochastic search applied to protein folding utilize complete move sets proposed by Lesh et al. [15] and Blazewicz et al. [4]. We obtain that after $(m/\delta)^{O(\Gamma)}$ Markov chain transitions, the probability to be in a minimum energy conformation is at least $1 - \delta$, where m is the maximum neighbourhood size and Γ is the maximum value of the minimum escape height from local minima of the underlying energy landscape. We note that the time bound depends on the specific instance. Based on [10] we conjecture $\Gamma \leq n^{1-1/d}$. We analyse $\Gamma \leq \sqrt{n}$ experimentally on selected benchmark problems [15,21] for the 2D case.

1 Introduction

A great variety of models has been developed for protein folding simulations, with different levels of detail (for a concise discussion, cf. [20]). In the present paper, we focus on *minimal models* [11], and we distinguish roughly between lattice models [7] and off-lattice models [8,17]. For a discussion of energy functions and justifications for the use of simplified (approximated) energy functions we refer the reader to [20]. One of the most popular models of protein folding is the hydrophobic-hydrophilic (H-P) model [7]. In the H-P model, proteins are modelled as chains whose vertices are marked either H (hydrophobic) or P (hydrophilic); the resulting chain is embedded into some lattice. H nodes are considered to attract each other while P nodes are neutral. An optimal embedding is one that maximizes the number of H-H contacts. The rationale for this

---

* Research partially supported by EPSRC Grant No. EP/D062012/1.

P. Bücher and B.M.E. Moret (Eds.): WABI 2006, LNBI 4175, pp. 252–261, 2006.

objective is that hydrophobic interactions contribute a significant portion of the total energy function. Unlike more sophisticated models of protein folding, the main goal of the H-P model is to explore broad qualitative questions about protein folding such as whether the dominant interactions are local or global with respect to the chain [11].

Lattice models of protein folding have provided valuable insights into the general complexity of protein structure prediction problems: Protein structure prediction has been shown to be NP-hard for a variety of lattice models [3,11,16]. The intractability results are complemented by performance guaranteed approximation algorithms that run in linear time [11,13]. Since protein structure prediction is NP-hard, (local) search-based algorithms are a natural choice to tackle the problem, especially in lattice models; cf. literature in [11]. Lesh et al. [15] and Blazewicz et al. [4] proposed complete neighbourhood move sets for local search in 2D and 3D grids, respectively, and performed computational experiments on benchmark problems for protein folding in the H-P model. Recently, Fu and Wang [10] proved an $exp(O(n^{1-1/d}) \cdot \ln n)$ algorithm for d-dimensional protein folding simulation in the HP-model. It is interesting to note that this time bound almost exactly mirrors the folding time approximation $exp(\lambda \cdot n^{2/3} \pm \chi \cdot n^{1/2}/2)$ by Finkelstein and Badretdinov [9][1].

The present paper reports our preliminary results on stochastic search applied to protein folding in the H-P model. We utilize the complete move sets proposed in [15] and [4]. We obtain that after $(m/\delta)^{O(\Gamma)}$ Markov chain transitions, the probability to be in a minimum energy conformation is at least $1 - \delta$, where m is the maximum neighbourhood size of individual conformations, and Γ is the maximum value of the minimum escape height from local minima of the underlying energy landscape. Thus, the run-time estimation is *problem-specific*. To be competitive with the Fu/Wang run-time bound, we need to show $\Gamma \leq n^{1-1/d}$. Future research will focus on proven upper bounds of Γ in the context of complete move sets for the H-P model. In the present paper, we analyse the conjecture $\Gamma \leq \sqrt{n}$ experimentally on selected benchmark problems (taken from [15,21]) for the 2D case.

2   Preliminaries

Our stochastic local search procedure for protein folding is based on simulated annealing [6,14], where the underlying Markov chain is of inhomogeneous type [5,12]. For simplicity of presentation, we focus on the 2D rectangular grid H-P model only.

Anfinsen's thermodynamic hypothesis [2] motivates the attempt to predict protein folding by solving certain optimization problems, but there are two main difficulties with this approach: The precise definition of the energy function that has to be minimised, and the extremely difficult optimization problems arising from the energy functions commonly used in folding simulations [11,17]. In the

---

[1] The authors are grateful to one anonymous referee for drawing our attention to [9].

2D rectangular grid H-P model, one can define the minimization problem as
follows:

$$\min_{\alpha} E(S, \alpha) \quad \text{for} \quad E(S, \alpha) := \xi \cdot HH_c(S, \alpha), \tag{1}$$

where where S is a sequence of amino acids containing n elements; $S_i = 1$, if
amino acid on the ith position in the sequence is hydrophobic; $S_i = 0$, if amino
acid on the ith position is polar; α is a vector of $(n - 2)$ grid angles defined
by consecutive triples of amino acids in the sequence; HH_c is a function that
counts the number of neighbours between amino acids that are not neighbours
in the sequence, but they are neighbours on the grid (they are topological neighbours); finally, $\xi < 0$ is a constant lower than zero that defines an influence ratio
of hydrophobic contacts on the value of conformational free energy. The distances between neighbouring grid nodes is assumed to be equal to 1. We identify
sequences α with conformations of the protein sequence S, and a valid conformation α of the chain S lies along a non-self-intersecting path of the rectangular
grid such that adjacent vertices of the chain S occupy adjacent locations. Thus,
we define the set of conformations (for each S specifically) by

$$\mathcal{F}_S := \{ \alpha \text{ is a valid conformation for } S \}. \tag{2}$$

Since $\mathcal{F} := \mathcal{F}_S$ is defined for a specific S, we denote the objective function by

$$\mathcal{Z}(\alpha) := \xi \cdot HH_c(S, \alpha). \tag{3}$$

The neighbourhood relation of our stochastic local search procedure is determined by the set of *pull moves* introduced in [15] for 2D protein folding
simulations in the H-P model (and, basically, extended to the 3D case in [4]).
For details of the definition of the set of pull moves we refer the reader to [15].

Theorem 1. [15] *The set of pull moves is local, reversible, and complete within*
$\mathcal{F}$, *i.e., any* $\beta \in \mathcal{F}$ *can be reached from any* $\alpha \in \mathcal{F}$ *by executing pull moves only.*

The set of neighbours of α that can be reached by a single pull move is denoted
by $\mathcal{N}_\alpha$, where additionally α is included since the search process can remain in
the same configuration. Furthermore, we set

$$N_\alpha := |\mathcal{N}_\alpha|; \tag{4}$$

$$\mathcal{F}_{\min} := \{ \alpha : \alpha \in \mathcal{F} \text{ and } \mathcal{Z}(\alpha) = \min_{\alpha'} E(S, \alpha') \}. \tag{5}$$

In simulated annealing-based search, the transitions between neighbouring elements are depending on the objective function $\mathcal{Z}$. Given a pair of protein conformations $[\alpha, \alpha']$, we denote by $G[\alpha, \alpha']$ the probability of generating α' from
α, and by $A[\alpha, \alpha']$ we denote the probability of accepting α' once it has been
generated from α. As in most applications of simulated annealing, we take a
uniform generation probability:

$$G[\alpha, \alpha'] := \begin{cases} \dfrac{1}{N_\alpha}, & \text{if } \alpha' \in \mathcal{N}_\alpha; \\ 0, & \text{otherwise.} \end{cases} \tag{6}$$

The acceptance probabilities $A[\alpha, \alpha']$ are derived from the underlying analogy to thermodynamic systems:

$$A[\alpha, \alpha'] := \begin{cases} 1, \text{ if } \mathcal{Z}(\alpha') - \mathcal{Z}(\alpha) \leq 0; \\ e^{-\frac{\mathcal{Z}(\alpha') - \mathcal{Z}(\alpha)}{t}}, \text{ otherwise,} \end{cases} \tag{7}$$

where t is a control parameter having the interpretation of a *temperature* in annealing processes. The probability of performing the transition between α and α' is defined by

$$\mathbf{Pr}\{\alpha \to \alpha'\} = \begin{cases} G[\alpha, \alpha'] \cdot A[\alpha, \alpha'], \text{ if } \alpha' \neq \alpha; \\ 1 - \sum_{\alpha' \neq \alpha} G[\alpha, \alpha'] \cdot A[\alpha, \alpha'], \text{ otherwise.} \end{cases} \tag{8}$$

By definition, the probability $\mathbf{Pr}\{\alpha \to \alpha'\}$ depends on the control parameter t. Let $\mathbf{a}_\alpha(k)$ denote the probability of being in conformation α after k transition steps. The probability $\mathbf{a}_\alpha(k)$ is calculated in accordance with

$$\mathbf{a}_\alpha(k) := \sum_{\beta \in \mathcal{F}} \mathbf{a}_\beta(k-1) \cdot \mathbf{Pr}\{\beta \to \alpha\}. \tag{9}$$

The recursive application of (9) defines a Markov chain of probabilities $\mathbf{a}_\alpha(k)$, where $\alpha \in \mathcal{F}$ and $k = 1, 2, \dots$. If the parameter $t = t(k)$ is a constant t, the chain is said to be a *homogeneous* Markov chain; otherwise, if $t(k)$ is lowered at any step, the sequence of probability vectors $\mathbf{a}(k)$ is an *inhomogeneous* Markov chain.

In the present paper we are focusing on a special type of inhomogeneous Markov chains where the value $t(k)$ changes in accordance with

$$t(k) = \frac{\Gamma}{\ln(k+2)}, \quad k = 0, 1, \dots. \tag{10}$$

The choice of $t(k)$ is motivated by Hajek's Theorem on logarithmic cooling schedules for inhomogeneous Markov chains [12]. To explain Hajek's result, we first need to introduce some parameters characterising local minima of the objective function:

Definition 1. *A conformation $\alpha' \in \mathcal{F}$ is said to be reachable at height h from $\alpha \in \mathcal{F}$, if $\exists \alpha_0, \alpha_1, \dots, \alpha_r \in \mathcal{F}$ with $\alpha_0 = \alpha \wedge \alpha_r = \alpha'$ such that $G[\alpha_u, \alpha_{u+1}] > 0, u = 0, 1, \dots, (r-1)$, and $\mathcal{Z}(\alpha_u) \leq h$ for all $u = 0, 1, \dots, r$.*

We use the notation $H(\alpha \Rightarrow \alpha') \leq h$ for this property. The conformation α is a *local minimum*, if $\alpha \in \mathcal{F} \backslash \mathcal{F}_{\min}$ and $\mathcal{Z}(\alpha') \geq \mathcal{Z}(\alpha)$ for all $\alpha' \in \mathcal{N}_\alpha \backslash \{\alpha\}$.

Definition 2. *Let $\lambda_{\min}$ denote a local minimum, then $D(\lambda_{\min})$ denotes the smallest h such that there exists $\lambda' \in \mathcal{F}$ with $\mathcal{Z}(\lambda') < \mathcal{Z}(\lambda_{\min})$ that is reachable at height $\mathcal{Z}(\lambda_{\min}) + h$.*

The following convergence property has been proved by B. Hajek:

Theorem 2. [12] *For $t(k)$ from (10), the asymptotic convergence $\sum_{\alpha \in \mathcal{F}_{min}} \mathbf{a}_\alpha(k)$ $\underset{k \to \infty}{\longrightarrow} 1$ of the algorithm defined by (3), ..., (9) is guaranteed if and only if*

1. $\forall \alpha, \alpha' \in \mathcal{F} \, \exists \alpha_0, \alpha_1, \ldots, \alpha_r \in \mathcal{F}$ *such that* $\alpha_0 = \alpha \wedge \alpha_r = \alpha'$
   and $G[\alpha_u, \alpha_{u+1}] > 0$ *for* $u = 0, 1, \ldots, (r-1)$;
2. $\forall h : H(\alpha \Rightarrow \alpha') \leq h \iff H(\alpha' \Rightarrow \alpha) \leq h$;
3. $\Gamma \geq \max_{\lambda_{min}} D(\lambda_{min})$.

From Theorem 1 and the definition of $\mathcal{N}_\alpha$ we immediately conclude that the conditions (i) and (ii) are valid for $\mathcal{F}$. Thus, together with Theorem 2 we obtain:

Corollary 1. *If $\Gamma \geq \max_{\lambda_{min}} D(\lambda_{min})$, the algorithm defined by (3), ..., (10) and the pull move set from [15] tends to minimum energy conformations in the H-P model.*

3   Run-Time Estimates of Simulations

In this section, we outline a run-time estimation for finding optimum conformations with a certain confidence $\delta' = 1 - \delta > 0$. The run-time estimation is an extension of the convergence analysis from [1] to a more complicated objective function, and it relates the run-time to the landscape parameter Γ (cf. (10)), to the confidence parameter $\delta' = 1 - \delta$, and to the maximum size m of individual neighbourhood sets.

For any $\alpha \in \mathcal{F}$ we introduce the following parameters:

$$s(\alpha) := |\{\, \alpha' : \alpha' \in \mathcal{N}_\alpha \wedge \mathcal{Z}(\alpha') > \mathcal{Z}(\alpha)\}|, \tag{11}$$

$$r(\alpha) := |\{\, \alpha' : \alpha' \in \mathcal{N}_\alpha \wedge \alpha' \neq \alpha \wedge \mathcal{Z}(\alpha') \leq \mathcal{Z}(\alpha)\}|. \tag{12}$$

Thus, from the definition of $\mathcal{N}_\alpha$ and (4) we have

$$s(\alpha) + r(\alpha) = N_\alpha - 1. \tag{13}$$

We observe that for $\mathcal{Z}(\alpha') > \mathcal{Z}(\alpha)$ the acceptance probability (7) can be rewritten as

$$e^{-(\mathcal{Z}(\alpha')-\mathcal{Z}(\alpha))/t(k)} = \frac{1}{(k+2)^{(\mathcal{Z}(\alpha')-\mathcal{Z}(\alpha))/\Gamma}}, \quad k \geq 0. \tag{14}$$

To simplify notation, we use $\gamma := \gamma(\alpha', \alpha) := (\mathcal{Z}(\alpha') - \mathcal{Z}(\alpha))/\Gamma$, in most cases not indicating the dependence on (α', α).

In (9), we separate the probabilities according to whether or not α' equals α, and the probability to remain in α is substituted by the defining equation from (8). Thus, we obtain:

$$\mathbf{a}_\alpha(k) = \sum_{\alpha' \in \mathcal{N}_\alpha} \mathbf{a}_{\alpha'}(k-1) \cdot \mathbf{Pr}\{\alpha' \to \alpha\}$$

$$= \mathbf{a}_\alpha(k-1) \cdot \left(1 - \sum_{\alpha' \neq \alpha} \mathbf{Pr}\{\alpha \to \alpha'\}\right) + \sum_{\alpha' \neq \alpha} \mathbf{a}_{\alpha'}(k-1) \cdot \mathbf{Pr}\{\alpha' \to \alpha\}.$$

The value of $\mathbf{a}_\alpha(k)$ is now expressed by using structural parameters as defined in (11) and (12):

Lemma 1. *The value of* $\mathbf{a}_\alpha(k)$ *can be calculated from probabilities of the previous step by*

$$\mathbf{a}_\alpha(k) = \left(\frac{s(\alpha)+1}{N_\alpha} - \frac{1}{N_\alpha} \cdot \sum_{i=1}^{s(\alpha)} \frac{1}{(k+1)^\gamma}\right) \cdot \mathbf{a}_\alpha(k-1) + \sum_{i=1}^{s(\alpha)} \frac{\mathbf{a}_{\alpha_i}(k-1)}{N_{\alpha_i}} +$$

$$+ \sum_{j=1}^{r(\alpha)} \frac{\mathbf{a}_{\alpha_j}(k-1)}{N_{\alpha_j}} \cdot \frac{1}{(k+1)^\gamma}. \tag{15}$$

The backwards expansion from Lemma 1 will be used as the main relation reducing $\mathbf{a}_\alpha(k)$ to probabilities from previous steps. The elements of the conformation space are distinguished by their minimum distance to $\mathcal{F}_{\min}$: Given $\alpha \in \mathcal{F}$, we consider a shortest path of length $\mathrm{dist}(\alpha)$ with respect to neighbourhood transitions from α to $\mathcal{F}_{\min}$. We introduce a partition of $\mathcal{F}$ in accordance with $\mathrm{dist}(\alpha)$:

$$\alpha \in M_i \Longleftrightarrow \mathrm{dist}(\alpha) = i \geq 0, \quad \text{and} \quad \mathcal{M}_{d_m} = \bigcup_{i=0}^{d_m} M_i, \tag{16}$$

where $M_0 := \mathcal{F}_{\min}$ and d_m is the maximum distance. From the proof of Theorem 1 in [15] we conclude

$$d_m \leq n^{O(1)}. \tag{17}$$

Since we want to analyze the convergence to elements from $M_0 = \mathcal{F}_{\min}$, we have to show that the value

$$\sum_{\alpha \notin M_0} \mathbf{a}_\alpha(k) \tag{18}$$

becomes small as k increases. We assume $k \geq d_m$ and we are going backwards from step k: At the same backwards transition from k to $(k-1)$, the neighbours of α are generating terms containing $\mathbf{a}_\alpha(k-1)$ as a factor in the same way as $\mathbf{a}_\alpha(k)$ generates terms with factors $\mathbf{a}_{\alpha_i}(k-1)$ and $\mathbf{a}_{\alpha_j}(k-1)$, see Lemma 1. If we now consider the entire sum $\sum_{\alpha \notin M_0} \mathbf{a}_\alpha(k)$, the terms corresponding to a particular $\mathbf{a}_\alpha(k-1)$ can be collected together to form a single expression. Firstly, we consider $\alpha \in M_i$, $i \geq 2$. In this case, α does not have neighbours from M_0, i.e., the expansion from Lemma 1 appears for all neighbours of α in the reduction of $\sum_{\alpha \notin M_0} \mathbf{a}_\alpha(k)$ to step $(k-1)$. Therefore, in the expansion of $\sum_{\alpha \notin M_0} \mathbf{a}_\alpha(k)$, the following arithmetic term is generated when the particular α is from M_1:

$$\left(1 - \frac{r(\alpha)}{N_\alpha}\right) \cdot \mathbf{a}_\alpha(k-1). \tag{19}$$

We introduce the following abbreviations:

$$\varphi(\alpha, v) := \frac{1}{N_\alpha} \cdot \sum_{i=1}^{s(\alpha)} \frac{1}{(k+2-v)^{\gamma_i}} \quad \text{and} \quad D_\alpha(k-v) := \frac{s(\alpha)+1}{N_\alpha} - \varphi(\alpha, v). \tag{20}$$

Now, the backwards expansion can be summarised to

Lemma 2. *A single step of the expansion of $\sum_{\alpha \notin M_0} \mathbf{a}_\alpha(k)$ results in*

$$\sum_{\alpha \notin M_0} \mathbf{a}_\alpha(k) = \sum_{\alpha \notin M_0} \mathbf{a}_\alpha(k-1) - \sum_{\alpha \in M_1} \frac{r(\alpha)}{N_\alpha} \cdot \mathbf{a}_\alpha(k-1) + \sum_{\alpha' \in M_0} \varphi(\alpha', 1) \cdot \mathbf{a}_{\alpha'}(k-1). \quad (21)$$

The diminishing factor $(1 - r(\alpha)/N_\alpha)$ is generated by definition for all elements of M_1. At subsequent reduction steps, the factor is "transmitted" successively to all probabilities from higher distance levels M_i because any element of M_i has at least one neighbour from M_{i-1}. We denote

$$\sum_{\alpha \notin M_0} \mathbf{a}_\alpha(k) = \sum_{\alpha \notin M_0} \mu(\alpha, v) \cdot \mathbf{a}_\alpha(k - v) + \sum_{\alpha' \in M_0} \mu(\alpha', v) \cdot \mathbf{a}_{\alpha'}(k - v), \quad (22)$$

i.e., the coefficients $\mu(\tilde{\alpha}, v)$ are the factors at probabilities after v steps of a backwards expansion of $\sum_{\alpha \notin M_0} \mathbf{a}_\alpha(k)$. Starting from step $(k - 1)$, the probabilities $\mathbf{a}_{\alpha'}(k - v)$, $\alpha' \in M_0$, from (22) are expanded in the same way as the probabilities for all other $\alpha \notin M_0$. Taking into account (20), we obtain the following parameterized representation for $\mu(\tilde{\alpha}, v)$:

Lemma 3. *The following recurrent relation is valid for the coefficients $\mu(\tilde{\alpha}, v)$:*

$$\mu(\tilde{\alpha}, v) = \mu(\tilde{\alpha}, v-1) \cdot D_{\tilde{\alpha}}(k-v) + \sum_{\alpha'' < \tilde{\alpha}} \frac{\mu(\alpha'', v-1)}{N_{\tilde{\alpha}}} + \sum_{\alpha' > \tilde{\alpha}} \frac{\mu(\alpha', v-1)}{N_{\tilde{\alpha}}} \cdot \frac{1}{(k+2-v)^\gamma}. \quad (23)$$

We take advantage of the fact that for conformations α different from local and global minima the factor $D_\alpha(k - v)$, which is associated with the probability to remain in α, is smaller than $(1 - 1/(m + 1))$ for $m := \max_\alpha N_\alpha$, i.e. there is an upper bound independent of $(k - v)$; see (20). Let MIN denote the set of all global and local minima. We set $\widehat{\mathcal{M}} := \{\alpha : r(\alpha) \geq 1\} = \mathcal{F} \backslash \text{MIN}$ and consider $\mathbf{a}_\alpha(k)$ defined by (8) and (9) when all probabilities on the right hand side are recursively substituted in the same way, where we break up the paths of the expansion that lead from some α to α' with $\mathcal{Z}(\alpha) > \mathcal{Z}(\alpha')$. Such transitions generate a factor $(k + 2 - u)^{-\gamma}$, which is then used as the crucial type of factors in the upper bound of $\mathbf{a}_\alpha(k)$. By analysing this type of expansions, we obtain:

Lemma 4. *If $k > 2 \cdot (m + 1)^2 \cdot \ln(k + 2)^{\max \gamma}$ for the maximum size m of neighbourhoods, then*

$$\sum_{\alpha \in \widehat{\mathcal{M}}} \mathbf{a}_\alpha(k) < O\left(\frac{(m + 1)^3}{(k - 2 \cdot (m + 1)^2 \cdot \ln(k + 2)^{\max \gamma})^{\min \gamma}}\right). \quad (24)$$

By $\mathcal{M}^{\text{lm}} \subset \text{MIN}$ we denote the set of all local minima, and $\mathcal{A}$ stands for the RHS of (24). If $\alpha \in \mathcal{M}^{\text{lm}}$, we represent $\mu(\alpha, v)$ by $\mu(\alpha, v) = 1 - \nu(\alpha, v)$ and by straightforward calculations we obtain

$$\sum_{\alpha \notin M_0} \mathbf{a}_\alpha(k) - \sum_{\alpha \notin M_0} \mathbf{a}_\alpha(k') < \mathcal{A} + \sum_{\alpha \in \mathcal{M}^{\text{lm}}} \nu(\alpha, v') \cdot \mathbf{a}_\alpha(k).$$

Thus, it remains to analyse $\nu(\alpha, v')$, $v' \geq d_m + v$, for local minima:

Lemma 5. *If $\alpha \in \mathcal{M}^{\mathrm{lm}}$, then*

$$\nu(\alpha, v') < O\left(\frac{(m+1)}{(k+2-v')^{\min \gamma}}\right). \tag{25}$$

From (25) and Lemma 5 we obtain the main result:

Theorem 3. *If $\Gamma \geq \max_{\lambda_{\min}} D(\lambda_{\min})$ for $\mathcal{F}$ from (2) and $0 < \delta < 1$, then*

$$k \geq \left(\frac{(m+1)^3}{\delta}\right)^{O(\Gamma)} implies \sum_{\alpha' \in \mathcal{F}_{\min}} \mathbf{a}_{\alpha'}(k) \geq 1 - \delta. \tag{26}$$

4   Landscape Analysis on Selected Benchmarks

As mentioned in Section 1 already, the run-time estimation (26) from Theorem 3 is problem-specific, i.e. depends on the parameter Γ of the landscape induced by an individual protein sequence. For a problem-independent upper bound we conjecture $\Gamma \leq n^{1-1/d}$, which complies with the result from [10]. However, for individual protein sequences one can proceed as follows: Given a sequence α, the parameter Γ is estimated in a pre-processing step (landscape analysis), where the maximum increase of the objective function is monitored in-between two successive improvements of the best value obtained so far. This approach usually overestimates Γ significantly. Therefore, we are searching for a suitable constant c such that $\Gamma' = G_{\mathrm{monit}}/c$ comes closer to Γ, where G_{monit} is the maximum of the monitored increases of the objective function in-between two successive total improvements of the objective function. This estimation Γ' is then taken (together with the length of α and a choice of δ for the confidence $1 - \delta$) as the setting for the (slightly simplified) run-time estimation according to (26). In our computational experiments on 2D benchmark problems we indeed obtain optimum solutions for smaller values of Γ than $\sqrt{n}$.

The stochastic local search procedure as described in Section 2 was implemented and we analysed the following 2D benchmark problems (cf. [15,21]):

Table 1. Selected 2D benchmark problems from [15,21]

name/n	structure	$z_{\min}$
S36	3P2H2P2H5P7H2P2H4P2H2PH2P	-14
S60	2P3HP8H3P10HPH3P12H4P6HP2HPHP	-35
S64	12HPHPH2P2H2P2H2PH2P2H2P2H2PH2P2H2P2H	
	2PHPHP12H	-42
S85	4H4P12H6P12H3P12H3P12H3PH2P2H2P2H2PHPH	-53
S100	6PHP2H5P3HP5HP2H4P2H2P2HP5HP10HP2HP7H	
	11P7H2PHP3H6PHP2H	-48

Table 2. Results for selected 2D benchmarks; $1 - \delta = 0.51$

name/n	$\sqrt{n}$	G_{monit}	Γ'	$(n/\delta)^{\Gamma'}$	T_{max}
S36	6.00	9.25	3.00	$\approx 4.0 \times 10^5$	29,341
S60	≈ 7.74	14.00	3.87	$\approx 1.2 \times 10^8$	30,319
S64	8.00	18.00	4.00	$\approx 2.9 \times 10^8$	259,223
S85	≈ 9.20	21.75	4.60	$\approx 2.0 \times 10^{10}$	13,740,964
S100	10.00	21.50	5.00	$\approx 3.5 \times 10^{11}$	57,195,268

Unfortunately, information about the exact number of ground states is not provided; the ground states are equally treated. In [15], three states are reported for S85, two states for S100.

Following the experimental part of [1], we use $(m/\delta)^{\Gamma'}$ as a simplified version of (26), where Γ' is $\approx \sqrt{n}/2$. We compare Γ' to G_{monit}/c, i.e. apart from trying to approximate the real Γ by Γ', we also try to relate Γ' to G_{monit}.

In Table 2 we report results where $\mathcal{Z}_{\min}$ was achieved for all five benchmark problems from Table 1. By $T_{\max}$ we denote the average number of transitions necessary to achieve $\mathcal{Z}_{\min}$ calculated from four successive runs for the same benchmark problem. The same applies to G_{monit}, which is the average from the four runs executed for each of the five benchmark problems. Although by definition Γ has to be an integer value, we allowed rational values for Γ'. The simplified version of (26) was calculated for $m = n$ and $\delta = 0.49$, i.e. for a confidence of 51%. As already mentioned, the value of Γ' was chosen $\approx \sqrt{n}/2$, which was used in (10) for the implementation.

As can be seen, the simplified version of (26) still overestimates the number of transitions sufficient to achieve $\mathcal{Z}_{\min}$ for the selected benchmark problems, which is at least partly due to the setting $m = n$. To incorporate improved upper bounds of m will be subject of future research. Based on the data from Table 2, the constant c in $\Gamma' = G_{\text{monit}}/c$ ranges from 3.08 to 4.73. Overall, the results encourage us to attempt a formal proof of the conjecture $\Gamma \leq \sqrt{n}$.

5   Concluding Remarks

We analyzed the run-time of protein folding simulations in the H-P model, if the underlying algorithm is based on the pull move set and logarithmic simulated annealing. We obtained that the probability to be in a minimum energy conformation is at least $1 - \delta$ after $(m/\delta)^{\kappa \cdot \Gamma}$ Markov chain transitions, where $m <$ sequence length n, κ is a small constant, and Γ is a crucial parameter of the landscape induced by the energy measure, the pull move set, and the individual sequence that has to be folded. Future research will be directed towards tight upper bounds of Γ in terms of the sequence length n, improved upper bounds of the maximum neighbourhood size m, on computational experiments on benchmark problems for the 3D case, and on landscape properties related to Levinthal's paradox [18], i.e. if there are "shallow" sub-landscapes with small Γ that imply fast folding.

References

1. Albrecht, A.A.: A stopping criterion for logarithmic simulated annealing. Computing (2006); in press.
2. Anfinsen, C.B.: Principles that govern the folding of protein chains. Science **181** (1973) 223–230.
3. Berger, B., Leighton, T.: Protein folding in the hydrophobic-hydrophilic (HP) model is NP-complete. J. Comput. Biol. **5** (1998) 27–40.
4. Blazewicz, J., Lukasiak, P., Milostan, M.: Application of tabu search strategy for finding low energy structure of protein. Artif. Intell. Med. **35** (2005) 135–145.
5. Catoni, O.: Rough large deviation estimates for simulated annealing: applications to exponential schedules. Ann. Probab. **20** (1992) 1109–1146.
6. Černy, V.: A thermodynamical approach to the travelling salesman problem: an efficient simulation algorithm. J. Optim. Theory Appl. **45** (1985) 41–51.
7. Dill, K.A., Bromberg, S., Yue, K., Fiebig, K.M., Yee, D.P., Thomas, P.D., Chan, H.S.: Principles of protein folding – A perspective from simple exact models. Protein Sci. **4** (1995) 561–602.
8. Eastwood, M.P., Hardin, C., Luthey-Schulten, Z., Wolynes, P.G.: Evaluating protein structure-prediction schemes using energy landscape theory. IBM J. Res. Dev. **45** (2001) 475–497.
9. Finkelstein, A.V., Badretdinov A.Y.: Rate of protein folding near the point of thermodynamic equilibrium between the coil and the most stable chain fold. Folding & Design **2** (1997) 115–121.
10. Fu, B., Wang, W.: A $2^{O(n^{1-1/d} \cdot \log n)}$ time algorithm for d-dimensional protein folding in the HP-model. Proc. ICALP'04, pp. 630–644, LNCS 3142, 2004.
11. Greenberg, H.J., Hart, W.E., Lancia, G.: Opportunities for combinatorial optimization in computational biology. INFORMS J. Comput. **16** (2004) 211–231.
12. Hajek, B.: Cooling schedules for optimal annealing. Mathem. Oper. Res. **13** (1988) 311–329.
13. Heun, V.: Approximate protein folding in the HP side chain model on extended cubic lattices. Discrete Appl. Math. **127** (2003) 163–177.
14. Kirkpatrick, S., Gelatt Jr., C.D., Vecchi, M.P.: Optimization by simulated annealing. Science **220** (1983) 671–680.
15. Lesh, N., Mitzenmacher, M., Whitesides, S.: A complete and effective move set for simplified protein folding. Proc. RECOMB'03, pp. 188–195, 2003.
16. Nayak, A., Sinclair, A., Zwick, U.: Spatial codes and the hardness of string folding problems. J. Comput. Biol. **6** (1999) 13–36.
17. Neumaier, A.: Molecular modeling of proteins and mathematical prediction of protein structure. SIAM Rev. **39** (1997) 407–460.
18. Ngo, J.M., Marks, J., Karplus, M.: Computational complexity, protein structure prediction, and the Levinthal paradox. In: K. Merz Jr., S. LeGrand (eds.), The Protein Folding Problem and Tertiary Structure Prediction, pp. 433–506, Birkhäuser, Boston, 1994.
19. Pardalos, P.M., Liu, X., Xue, G.: Protein conformation of a lattice model using tabu search. J. Global Optim. **11** (1997) 55–68.
20. Straub, J.E.: Protein folding and optimization algorithms. The Encyclopedia of Computational Chemistry, vol. 3, pp. 2184–2191, Wiley & Sons, 1998.
21. Unger, R., Moult, J.: Genetic algorithms for protein folding simulations. J. Mol. Biol. **231** (1993) 75–81.

Rapid *ab initio* RNA Folding Including Pseudoknots Via Graph Tree Decomposition

Jizhen Zhao[1], Russell L. Malmberg[2], and Liming Cai[1]

[1] Department of Computer Science, University of Georgia, Athens, GA 30602, USA
{jizhen, cai}@cs.uga.edu *
[2] Department of Plant Biology, University of Georgia, Athens, GA 30602, USA
russell@plantbio.uga.edu

Abstract. The prediction of RNA secondary structure including pseudoknots remains a challenge due to the intractable computation of the sequence conformation from intriguing nucleotide interactions. Optimal algorithms often assume a restricted class for the predicted RNA structures and yet still require a high-degree polynomial time complexity, which is too expensive to use. Heuristic methods may yield time-efficient algorithms but they do not guarantee optimality of the predicted structure. This paper introduces a new and efficient algorithm for the prediction of RNA structure with pseudoknots for which the structure is not restricted. Novel prediction techniques are developed based on graph tree decomposition. In particular, stem overlapping relationships are defined with a graph, in which a specialized maximum independent set (IS) corresponds to the desired optimal structure. Such a graph is tree decomposable; dynamic programming over a tree decomposition of the graph leads to an efficient algorithm. The new algorithm is evaluated on a large number of RNA sequence sets taken from diverse resources. It demonstrates overall sensitivity and specificity that outperforms or is comparable with those of previous optimal and heuristic algorithms yet it requires significantly less time than other optimal algorithms.

1 Introduction

The secondary structure of an RNA molecule is formed due to short or long distance pairings between nucleotides in the sequence. Base pair regions either single, nested or parallel are called *stem-loops*; base pair regions crossing each other are called *pseudoknots* [23]. Pseudoknots are important structures in RNA molecules and often play important functional roles [12] such as catalysis, RNA splicing, transcription regulation. Knowing the secondary structures of RNA molecules is critical for determining their three dimensional structures and understanding their functions. Automated prediction of RNA secondary structure is thus in demand since it is expensive and time consuming to experimentally determine the structure.

---

* To whom correspondence should be addressed.

P. Bücher and B.M.E. Moret (Eds.): WABI 2006, LNBI 4175, pp. 262–273, 2006.

It is computationally challenging to predict RNA secondary structure including pseudoknots. In particular, the problem of predicting RNA pseudoknots with the minimum free energy is provably NP-hard [13]. Practical approaches to cope with this computational challenge are either to restrict the class of pseudoknots under consideration or to employ heuristics in the algorithms. Optimal algorithms for restricted pseudoknot classes are usually thermodynamics-based, extended from Zuker's algorithm for the prediction of pseudoknot-free structures [25]. In such algorithms, the predicted optimal structure of a single RNA sequence is the one with the global minimum free energy based on a set of experimentally determined parameters. Among these algorithms, PKNOTS [17] can handle the widest classes of pseudoknots. However, its time complexity $O(n^6)$ makes it infeasible to fold RNA sequences of a moderate length. The computation efficiency may be improved at the cost of further restricting the structure of pseudoknots [16], but still with a time complexity $O(n^5)$ or $O(n^4)$. Most such algorithms produce only the optimal solution, while suboptimal ones that may reveal the true structure are often ignored.

On the other hand, computationally efficient heuristic methods have also been explored to allow unrestricted pseudoknot structures. Iterated loop matching (ILM) [19] is one such method. It finds the most stable stem, adds it to the candidate secondary structure and then masks off the bases forming the stem and iterates on the left sequence segments until no other stable stem can be found. One structure is reported at the end. Another algorithm, HotKnots [16], does the prediction in a slightly different way. It keeps multiple candidate structures rather than only one and builds each of them in a similar but more elaborate way. These methods can usually be fast, yet they often do not provide an optimality guarantee for the predicted structure or a quality measure on the predicted structure with respect to the optimal structure. Other heuristic methods based on genetic algorithms and Monte Carlo simulation usually do not address the optimality issue either [1,5].

In this paper, we introduce a novel approach for the optimal prediction of RNA pseudoknots for which the structure is not restricted. Our method is based on a simplified thermodynamic model without accounting for loop energies [15,19]. In this method, stable stems are selected from an RNA sequence as vertices of a graph; vertices are connected with edges if corresponding stems conflict (i.e., overlap) in their positions in the sequence. The optimal structure of an RNA sequence corresponds to a collection of non-conflicting stable stems, which can be found by seeking the maximum weighted independent set (WIS) from the graph. We observe that stable stems can be so selected that the resulting graph is of a moderately small tree width t. Based on a tree decomposition of the graph, a dynamic programming algorithm for WIS of the worst-case time complexity $O(1.44^t n)$ is obtained, where n is the number of vertices in the graph, at most quadratic in the length of the RNA sequence. This is an efficient prediction algorithm parameterized on the tree width t, which is usually small.

We implemented our algorithm TdFOLD and evaluated its performance on various RNA sequence sets from different sources. The test results showed high efficiency and high accuracy for our algorithm. TdFOLD was tested against PKNOTS, ILM and HotKnots on a set of 50 tRNA's, a set of 50 small RNA sequences containing pseudoknots with length ranging from 23 to 113, and a set of 11 large RNA's with length range from 210 to 412. The results showed that overall, in terms of the sensitivity and specificity of the prediction, TdFOLD outperforms the optimal algorithm PKNOTS and the heuristic algorithms ILM and HotKnots. In time efficiency, it outperforms PKNOTS and HotKnots, and is comparable with ILM. Our algorithm will also output suboptimal structures without spending much more time than reporting the optimal structure.

Graph theoretic methods have previously been explored for RNA structure prediction [23]. Our method is different from the previous ones in two respects. Our graphs constructed from the RNA sequence contain vertices describing stems instead of nucleotides; making the stem to be the smallest structural unit can greatly simplify the complexity of the problem. More importantly, our graph algorithm takes advantage of the tree decomposition technique on the formulated graphs. In fact, it has been demonstrated that the RNA secondary structure can be profiled with a conformational graph of small tree width [21]. The underlying graph constructed for the *ab initio* structure prediction is essentially an augmentation of the conformational graph in which additional vertices and edges are added only for the overlapping stems, thus inheriting the tree decomposability which makes the algorithm efficient.

2   Methods and Algorithm

Given an RNA sequence, our algorithm first builds a pool of stable stems, then finds a number of secondary structures with (near) minimum total stem energies by a tree decomposition based procedure for a graph formed by the stable stems. These predicted secondary structures are then reordered by counting the stem and loop energies together.

2.1   Problem Formulation

A (canonical) base pair is either a Watson-Crick pair (*A-U* or *C-G*) or for wobble pair *G-U*. A *stem* is a set of stacked nucleotide base pairs on an RNA sequence s. In general a stem S can be associated with four positions (i^l, j^l, i^r, j^r), where $i^l < j^l < i^r < j^r$, on the sequence s such that (a) $(s[i^l], s[j^r])$ and $(s[j^l], s[i^r])$ are two canonical base pairs; and (b) for any two base pairs $(s[x], s[y]), (s[z], s[w])$ in the stem S, either $i^l \leq x < z \leq j^l$ and $i^r \leq w < y \leq j^r$, or $i^l \leq z < x \leq j^l$ and $i^r \leq y < w \leq j^r$. Region $s[i^l..j^l]$ is the *left region* of the stem and $s[i^r..j^r]$ is the *right region* of the stem. Stem S is *stable* if the formation of its base pairs allows the thermodynamic energy $\Delta(S)$ of the stem to be below a predefined threshold parameter $E < 0$. Figure 1(a) shows all the stable stems in Ec_Pk_4 with $E = -5$ kcal/mol, the fourth pseudoknot in E.coli tmRNA [24], and their corresponding free energy values.

A *stem graph* $G = (V, E)$ can be defined for the RNA sequence s, where each vertex in V uniquely represents a stable stem on s, and E contains an edge

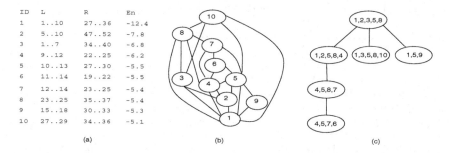

ID	L	R	En
1	1..10	27..36	-12.4
2	5..10	47..52	-7.8
3	1..7	34..40	-6.8
4	9..12	22..25	-6.2
5	10..13	27..30	-5.5
6	11..14	19..22	-5.5
7	12..14	23..25	-5.4
8	23..25	35..37	-5.4
9	15..18	30..33	-5.3
10	27..29	34..36	-5.1

(a)   (b)   (c)

Fig. 1. (a) Ten stable stems in Ec_Pk_4, the fourth pseudoknot in E.coli tmRNA molecule, including their left and right regions, and thermodynamic energies; (b) stem graph for Ec_Pk_4; and (c) a tree decomposition of the stem graph with tree width 4

between two vertices if and only if the corresponding two stems (a, b, c, d) and (x, y, z, w) conflict in their positions, i.e., one or both of the regions $s[a..b]$ and $s[c..d]$ overlap with at least one of the regions $s[x..y]$ and $s[z..w]$. Figure 1(b) shows the stem graph for Ec_Pk_4 constructed according to the stable stems given in Figure 1(a). The stem graph is a weighted graph, with a weight on every vertex. Usually, the weight of a vertex can simply be the absolute value of the thermodynamic energy $\Delta(S)$ of the stem S corresponding to the vertex. The weight may also be adjusted by scaling it (non-)linearly according to the length of the corresponding stem or the distance between the left and right regions of the stem. The problem of predicting the optimal structure of the RNA then corresponds to finding a collection of non-conflicting stems from its stem graph which achieves the maximum total weight. This is exactly the same as the graph theoretic problem: finding the maximum WIS in the stem graph. Note the weight for an IS representing a secondary structure is based on the total energies of the stems only (similar models were previously adopted by both primitive method [15] and more elaborate one [19]).

2.2   Identifying Stable Stems

For our purpose, stable stems are defined according to a set of parameters. In particular, a stem contains at least P base pairs; the loop length in between the left and right region of the stem is at least L; the thermodynamic energy is at most E. Bulges within a stem are allowed, for which the stem essentially becomes a set of substems separated by the bulges. In addition, parameter T limits the minimum substem length, and parameter B limits the maximum bulge length. The thermodynamic energy $\Delta(S)$ of stem S is calculated by taking into account both the stacking energies and the destabilizing energies caused by bulges. A procedure similar to the one used in [11] is employed to identify all the stable stems. The stable stem pool can be extended by introducing maximal substems that can resolve the conflicts and meet the requirements defined by the above parameters for each pair of overlapped stems in the pool.

2.3   Tree Decomposition Based Algorithm

Definition [18] A *tree decomposition* of graph $G = (V, E)$ is a pair (T, X) if it satisfies:

1. $T = (I, F)$ is a tree with node set I and edge set F,
2. $X = \{X_i : i \in I, X_i \subseteq V\}$, $\bigcup_i X_i = V$ and $\forall u \in V$, $\exists i \in I$ such that $u \in X_i$,
3. $\forall (u, v) \in E$, $\exists i \in I$ such that $u, v \in X_i$,
4. $\forall i, j, k \in I$, if k is on the path that connects i and j in tree T, $X_i \cap X_j \subseteq X_k$

The *width* of a tree decomposition (T, X) is $\max_{i \in I} |X_i| - 1$. The *tree width* of the graph G is the minimum tree width over all possible tree decomposition of G. If T is restricted to be a path, we refer (T, X) as a *path decomposition* and the best width over all of the path decompositions as the *path width* of G. The tree decomposition is rooted in the deep graph minor theorems by Robertson and Seymour [18]. It provides a topological view on a graph and the tree width measures how much the graph is "tree-like". Figure 1(c) shows a tree decomposition for the stem graph given in Figure 1(b).

Many computationally intractable graph problems can be easily solved on graphs of small tree width. In particular, a large number of such graph problems, while intractable on general graphs, can be solved in linear time, given a tree decomposition of tree width $\leq t$, for a fixed t. Maximum WIS is one such problem [3]; it has time complexity $O(2^t n)$. For the RNA stem graphs, we observe that vertices contained in every node of a tree decomposition can be partitioned into a small collection of maximal cliques, thus the factor 2^t can be further reduced. For example, in Figure 1(c), node $\{1, 2, 5, 3, 8\}$ contains two cliques $\{1, 2, 5\}$, and $\{3, 8\}$ (also see Figure 1(b)). In general, let $C_1, \ldots, C_q$, where $\sum_{i=1}^q |C_i| = t$, be the maximal cliques contained in a node, for some small q, then the number of valid partial ISs for the vertices in the node is at most $\prod_{i=1}^q |C_i| \leq (t/q)^q$. While the right term may reach the worst case extreme $e^{t/e} \approx 2^{0.53t}$ when $t/q = e$, the base of natural logarithm, in reality, the worst case may never occur because q usually is small. In the above example, the factor is reduced to $3 \times 2 = 6$ in contrast to the number $2^5 = 32$.

Algorithm details. Now we describe the tree decomposition based dynamic programming algorithm that finds the maximum WIS from the stem graph $G = (V, E)$. It assumes a binary tree decomposition (T, X), where $X = \cup X_{i=1}^m$, for the stem graph, where $m = O(|V|)$, $|X_i| = t$, for $i = 1, \ldots, m$. We only discuss the process for achieving the optimal solution. The technical details for getting suboptimal solutions are similar.

The algorithm constructs one dynamic programming table m_i for every tree node $X_i = \{v_1, \ldots, v_t\}$. Table m_i records all possible partial ISs in the subgraph induced by the set of all the vertices in the subtree rooted at i of the tree decomposition. There are t columns in the table m_i, one for each vertex in the corresponding tree node X_i. Rows are the combinations of these vertices; a vertex is selected if and only if the corresponding column takes value 1. There are additional three columns V, S, Opt in the table.

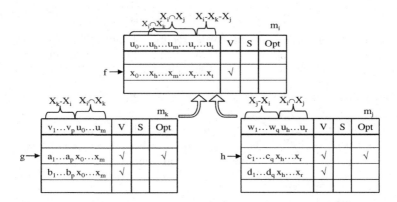

Fig. 2. Dynamic programming table construction over tree decomposition. Table m_i is computed also based on the computed tables m_k and m_j. Row $f = (x_0, \ldots, x_h, \ldots, x_m, \ldots, x_r, \ldots, x_t)$ in table m_i is computed from row g in table m_k and row h of table m_j, Row g is the optimal for columns $X_k - X_i$ given the value $(x_0, \ldots, x_m)$ for columns $X_k \cap X_i$. Similarly, row h is the optimal for columns $X_j - X_i$ given the value $(x_h, \ldots, x_r)$ for columns $X_j \cap X_i$.

These tables are constructed in a bottom-up fashion, from leaves to the apex of the tree decomposition (see Figure 2). Each row of a table is a combination of the vertices in the corresponding node. Column V is set 1 if the row represents a valid IS. For a leaf node, S is 0 if the row is not a valid IS; otherwise S is the corresponding weight of the set. For an internal node i that has two children j and k whose tables m_j, m_k have been computed, for each row in table m_i, column S is computed as $S = w_1 + w_2 + w_3 - w_4$, where

- w_1 is the weight of the row in table m_j with the same combination in the columns corresponding to the vertices in $X_j \cap X_i$ that has column $Opt = 1$;
- w_2 is the weight of the row in table m_k with the same combination in the columns corresponding to the vertices in $X_k \cap X_i$ that has column $Opt = 1$;
- w_3 is the weight of the IS formed by the choices in columns corresponding to the vertices in $X_i - X_j - X_k$; and
- w_4 is the weight of the IS formed by the same combination in the columns corresponding to the vertices in $X_i \cap X_j \cap X_k$.

Column Opt is set 1 if and only if the row represents a valid IS and S in this row is optimal among all the rows with different choices in the columns corresponding to the vertices in $X_i - X_p$ given the chosen values same as this row in the columns corresponding to the vertices in $X_i \cap X_p$, where node p is the parent of node i.

As mentioned earlier, the enumeration of the combinations of the graph vertices in tree node X_i is along a number of maximal cliques. In general, a greedy algorithm is used to partition set X_i into a collection of cliques. Consider the sequence as a straight line and the left (right) region of a stem as an interval.

Let all the left regions of the stable stems included in the tree node form an interval graph. Choose an interval (left region) with the right end at the left most position among all of the intervals, record all the intervals overlap with this interval as a clique and remove them, recursively call on the interval graph left until it is empty. A linear time in t is enough for this procedure.

Tree decomposition of stem graph. Finding the optimal tree decomposition is NP-hard [2], we use a simple, fast heuristic algorithm to produce a tree decomposition for the given stem graph. This algorithm is based on a heuristic method for greedy fill-in [?]. The method will produce a tree decomposition with small tree width but not necessary the optimal one.

Reordering suboptimal structures. The list of candidate structures, including the optimal and the suboptimal ones, are reordered based on a more sophisticated energy model. In particular, we recalculate the free energy for each of the candidate structures using a procedure implemented in [16] according to the energy model in [20,14] combined with the one in [6], which take the stem stabilizing energies, loop destabilizing, and pseudoknot energies into account.

3   Evaluation Results

3.1   Data Sets and Experiment Details

We used three sets of RNA sequences to evalute the algorithm (see Table 1). The first set is 50 tRNAs with lengths ranging from 71 to 79 (with the average 75). The second set is 50 small RNA sequences or sequence segments with pseudoknot structures of lengths ranging from 23 to 113 (with the average 53). The third set is 11 large RNA sequences of lengths ranging from 210 to 412 (with the average 344).

We compared the performance of our algorithm TdFOLD and that of algorithms PKNOTS [17], ILM [19], and HotKnots [16]. We ran all these algorithms on the tRNAs and the set of small pseudoknot RNAs, and run all but PKNOTS on the set of large RNAs. We evaluated both accuracy and efficiency of these algorithms. The accuracy is measured in both sensitivity and specificity. Let RP be the number of base pairs in the real structure, TP (true positive) be the number of correctly predicted base pairs and FP (false positive) be the number of predicted base pairs that do not exist as real structures. We define SE (sensitivity) as TP/RP, and SP (specificity) as $TP/(TP + FP)$. The perfect prediction should yield 1 for both sensitivity and specificity values.

For tRNA, we turned off the pseudoknot option for PKNOTS since we already know they are pseudoknot free. For TdFOLD, parameters were set to default values and the number of output solutions was set to 40 for tRNAs and small pseudoknotted RNAs. The parameters were adjusted for each of the large sequences. The experiments were run on a PC with 2.8 GHz Intel(R) Pentium 4 processor and 1-GB RAM, running RedHat Enterprise Linux version 4 AS.

Table 1. Test sets: sequence IDs with their reference citations

Set one: tRNA[22]		
GA0001 GA1262 GA2492 GA3755 GA4966 GC2866 GD1723 GD5199 GE2095 GE4739 GF1407 GF4687 GG0841 GG2136 GG3917 GH0128 GH4536 GI1748 GI4502 GK1078 GK4537 GM0313 GM2284 GM4471 GM5945 GN2837 GP1341 GP3879 GP5312 GQ2684 GR0044 GR0793 GR1516 GR2309 GR3541 GR4508 GR4705 GR4740 GR5278 GT0109 GT1418 GT4178 GT5273 GV0579 GV1734 GV4391 GV5554 GW1796 GW5332 GY4135		
Set two: small RNAs		
Sequence type	Sequence IDs	
aptamers	NGF-L6 [24]	
antizyme ribosomal frame shifting site	Rr_ODCanti [24]	
HIV-1-RT ligand RNA	HIVRT32, HIVRT322, HIVRT33 [16]	
hepatitis virus ri-bozyme	HDV, HDV_anti [16]	
mRNA	Bt-PrP, Ec_alpha, Ec_S15, Hs-PrP, T4_gene32 [24]	
rRNA	Sc_18S-PKE21-7 [24]	
ribozymes	HDV-It_ag [24]	
ribozymes	satRPV, Tt-LSU-P3P7, Bp_PK2 [24]	
tmRNA	Lp_PK1, Ec_PK1, Ec_PK4 [24]	
telomerase RNA	T.the_telo [24]	
viral tRNA like	OYMV, APLV, CGMMV, SBWMV1, BSMVbeta, CGMMV_PKbulge, ORSV-S1, AMV3 [24]	
viral 3'UTR	TMV-L_UPD-PK3, STMV_UPD1-PK3, BVQ3_UPD-PKb, BSBV1_ , PSLVbeta_UPD-PK1, PSLVbeta_UPD-PK3, BSBV3, UPD-PKc, SBWMV1_UPD-PKb [24]	
viral ribosomal RNA shifting signals	EIAV, PLRV-S [24]; minimal IBV, MMTV, MMTV-vpk, pKA-A, BWYV, SRV-1, T2_gene32[9]	
viral RNA	PSIV_IRES [24]; TYMV, TMV.L, TMV.R [16]	
Set three: large RNAs		
Sequence type	Sequence IDs	
RNaseP RNA	A.ferr, A.laid (pseudoknot free), A.tum, B.anth, B.halo,	
	CPB147, D.desu, EM14b-9, E.ther, T.rose [4]	
telomerase RNA	telo.human [5]	

3.2   Testing Results

Table 2 summarize the testing results for different programs on the three RNA data sets. It shows that TdFOLD has sensitivity 0.81 and specificity 0.75 on average for the tRNA prediction, which are slightly better than PKNOTS and significantly better than ILM and HotKnots. For the small pseudoknotted RNAs, TdFOLD has average sensitivity 0.76, which is less than PKNOTS but greater than ILM and HotKnots. On the other hand, TdFOLD has average specificity 0.79, which outperforms all the others. TdFOLD is slightly better in overall accuracy than PKNOTS, which reports the optimal structure according to its sophisticated energy model. This suggests that considering the stems as prediction units can filter some noise. For the large RNA's, TdFOLD maintains the same sensitivity (0.54) as HotKnots, which is slightly better than ILM. TdFOLD has the highest specificity on average.

Table 2. Summary of testing results on tRNAs, smale and large RNAs, where SE: sensitivity, SP: specificity, T: time (in seconds, if not otherwise noted)

		TdFOLD			HotKnots			ILM			PKNOTS		
		SE	SP	T	SE	SP	T	SE	SP	T	SE	SP	T
tRNA	min	0.33	0.29	0.26	0.33	0.25	0.57	0.33	0.25	0.01	0	0	0.11
	max	1.00	1.00	1.37	1.00	1.00	8.32	1.00	1.00	0.15	1.00	1.00	0.24
	average	0.81	0.75	0.54	0.72	0.66	3.33	0.75	0.61	0.03	0.78	0.73	0.41
small	min	0	0	0.04	0	0	0.05	0	0.25	0.001	0	0	0.27
	max	1.00	1.00	0.57	1.00	1.00	57.0	1.00	1.00	0.05	1.00	1.00	>1hr
	average	0.76	0.79	0.36	0.69	0.72	5.84	0.73	0.69	0.03	0.78	0.73	1066
large	min	0.18	0.17	0.46	0.24	0.18	157	0.38	0.25	0.71			
	max	0.86	0.73	14.5	0.68	0.63	29710	0.77	0.82	1.49			
	average	0.54	0.53	3.97	0.54	0.49	4456	0.51	0.44	0.97			

Efficiency comparisons are also given in Table 2 on each data set, respectively. For tRNA's, the average running time of 0.54 seconds for TdFOLD is slower than the average 0.03 of ILM and the average 0.41 of PKNOTS but faster than the average 3.33 of HotKnots. This is not a surprise because we turned the pseudoknot option off for PKNOTS. For small pseudoknotted RNA's, TdFOLD is slower than ILM (0.36 vs. 0.03 seconds), while much faster than HotKnots and PKNOTS (5.84 and 1066 seconds). For large RNA sequences, it is comparable (slightly slower) than ILM (3.97 vs. 0.97 seconds) while much faster than HotKnots (4456 seconds) on average. In general, the speed of TdFOLD is comparable to ILM and much faster than PKNOTS and HotKnots.

According to Table 2, all of the programs could predict some sequences (different for each program) totally wrong (zero sensitivity and/or specificity). This reveals that the available thermodynamic parameters for RNA secondary structures may not be optimal for all RNA classes. Thus it is hard to guarantee that the structure with the minimum free energy is the true structure. This makes the output of a list of low energy suboptimal structures a valuable feature of a structure prediction algorithm. The prediction results of TdFOLD for 23 tRNAs and 19 short pseudoknotted RNAs are improved by considering the top five structures, rather than only the top one among the 40 output predictions for each sequence. By "improved" we mean that there is at least one suboptimal prediction with both the sensitivity and specificity better than (or the same as) those of the optimal prediction. If there is more than one prediction improved over the top one, we choose the best among all the improved. For example, the average sensitivity and specificity are improved to 0.91 and 0.85 for tested tRNAs, 0.81 and 0.85 for tested short pseudoknotted RNAs.

4 Discussion and Conclusion

When related structurally homologous sequences are available, the accuracy of RNA structure prediction can usually be improved through the use of comparative analysis. A fully automated comparative analysis process exists [8,7] for consensus structure prediction of pseudoknot free RNAs, which iterates between the following two steps: (a) build an optimal (or nearly optimal) structure model given the current multiple alignment; and (b) build a multiple alignment given the current structure model. Nevertheless, for RNA pseudoknots, both algorithms for step (a) and (b) can be computationally intensive; the implementation remains a computational challenge.

The tree decomposable model and tree decomposition based techniques make it possible to implement efficiently the automated comparative analysis process. Based on an earlier work of ours, pseudoknots can be profiled with the conformational graph model [21] of small tree width; the efficient optimal structure-sequence alignment developed is ideal for step (b). In addition, the algorithm introduced in this paper can be employed for step (a), to construct a structure model for multiple RNAs. As it was done for pseudoknot-free RNAs, the mutual information content $M_{i,j}$ can be computed for every pair of aligned columns i, j, which is defined as the relative entropy

$$M_{i,j} = \sum_{x_i, y_j \in \{A,C,G,U\}} f(x_i, y_j) \log \frac{f(x_i, y_i)}{f(x_i) f(y_j)}$$

where $f(x_i, y_j)$ is the frequency for nucleotides x_i, y_j to occur in pair in these two columns i, j, and $f(x_i)$ and $f(y_j)$ are for independent occurrences. The multiple alignment can be regarded as a "generic sequence" consisting of columns as "nucleotides". The pairwise interactions between columns result in a conformation structure of the "generic sequence", yielding a consensus structure for the multiple sequences. Therefore, we can use our structure prediction algorithm TdFOLD to predict the structure of the "generic sequence" using the mutual information content $M_{i,j}$ as "pairing energy" between columns i and j.

In conclusion, in this paper, we presented a tree decomposition based fast RNA folding algorithm, which is efficient, accurate, not limited to any specific class of pseudoknots, and can report a list of suboptimal structures. Combined with an efficient structure-sequence alignment algorithm we developed earlier [21], it also can be used to implement an automated comparative RNA structure analysis process that can infer the pseudoknot consensus structure from a set of unaligned RNA sequences.

Acknowledgment

This work was supported in part by the NIH BISTI grant No: R01GM072080-01A1.

References

1. J. Abrahams, M. van den Berg, E. van Batenburg, and C. Pleij. Prediction of RNA secondary structure, including pseudoknotting, by computer simulation. *Nucleic Acids Res.*, 18:3035–3044, 1990.

2. H. L. Bodlaender. Classes of graphs with bounded tree-width. *Tech. Rep. RUU-CS-86-22, Dept. of Computer Science, Utrecht University, the Netherlands*, 1986.

3. H. L. Bodlaender. Dynamic programming algorithms on graphs with bounded tree-width. In *Proc. 15th Int'l Colloquium on Automata, Languages and Programming ICALP'87*, pages 105–119. Springer Verlag, Lecture Notes in Computer Science, vol. 317, 1987.

4. J. Brown. The ribonuclease p database. *Nucleic Acids Res.*, 27:314, 1999.

5. J.-H. Chen, S.-Y. Le, and J.V. Maize. Prediction of common secondary structures of RNAs: a genetic algorithm approach. *Nucleic Acids Research*, 28(4):991–999, 2000.

6. R. Dirks and N. Pierce. A partition function algorithm for nucleic acid secondary structure including pseudoknots. *J. Comput. Chem.*, 24:16641677, 2003.

7. R. Durbin, S.R. Eddy, A. Krogh, and G.J. Mitchison. *Biological Sequence Analysis: Probabilistic Models of Proteins and Nucleic Acids*. Cambridge University Press, 1998.

8. S.R. Eddy and R. Durbin. RNA sequence analysis using covariance models. *Nucleic Acids Research*, 22:2079–2088, 1994.

9. D. Giedroc, C. Theimer, and P. Nixon. Structure, stability and function of RNA pseudoknots involved in stimulating ribosomal frame shifting. *J. of Molecular Biology*, 298:167–185, 2000.

10. I.V. Hicks, A.M. C.A. Koster, and E. Kolotoglu. Branch and tree decomposition techniques for discrete optimization. In *Tutorials in Operations Research: INFORMS – New Orleans 2005*. 2005.

11. Y. Ji, X. Xu, and G.D. Stormo. A graph theoretical approach for predicting common RNA secondary structure motifs including pseudoknots in unaligned sequences. *Bioinformatics*, 20(10):1591–1602, 2004.

12. A. Ke, K. Zhou, F. Ding, J.H. Cate, and J.A. Doudna. A conformational switch controls hepatitis delta virus ribozyme catalysis. *Nature*, 429:201205, 2004.

13. R.B. Lyngso and C.N.S. Pedersen. RNA pseudoknot prediction in energy-based models. *J. of Computational Biology*, 7(3-4):409–427, 2000.

14. D.H. Mathews, J. Sabina, M. Zuker, and C.N.S. Pederson. Expanded sequence dependence of the thermodynamic parameters improves prediction of RNA secondary structure. *J. Mol. Biol.*, 288:911–940, 1999.

15. R. Nussinov, G. Pieczenik, J. Griggs, and D. Kleitman. Algorithms for loop matchings. *SIAM J. Applied Mathematics*, 35:68–82, 1978.

16. J. Ren, B. Rastegart, A. Condon, and H.H. Hoos. HotKnots: Heuristic prediction of RNA secondary structures including pseudoknots. *RNA*, 11:1194–1504, 2005.

17. E. Rivas and S.R. Eddy. A dynamic programming algorithm for RNA structure prediction including pseudoknots. *J. Molecular Biology*, 285:2053–2068, 1999.

18. N. Robertson and P.D. Seymour. Graph minors ii. algorithmic aspects of tree width. *J. Algorithms*, 7:309–322, 1986.

19. J. Ruan, G.D. Stormo, and W. Zhang. An iterated loop matching approach to the prediction of RNA secondary structures with pseudoknots. *Bioinformatics*, 20(1):58–66, 2004.

20. M.J. Serra, D.H. Turner, and S.M. Freier. Predicting thermodynamic properties of RNA. *Meth. Enzymol.*, 259:243–261, 1995.

21. Y. Song, C. Liu, R. L. Malmberg, F. Pan, and L. Cai. Tree decomposition based fast search of RNA structures including pseudoknots in genomes. In *Proc. Comput. System Bioinformatics Conf. CSB'05*, pages 223–234. IEEE Computer Society, 2005.

22. M. Sprinzl, C. Horn, M. Brown, A. Ioudovitch, and S. Steinberg. Compilation of tRNA sequences and sequences of tRNA genes. *Nucleic Acids Res.*, 26:148–153, 1998.

23. J. Tabaska, R. Cary, H. Gabow, and G. Stormo. An RNA folding method capable of identifying pseudoknots and base triples. *Bioinformatics*, 14(8):691–699, 1998.

24. F. van Batenburg, A. Gultyaev, C. Pleij, J. Ng, and J. Oliehoek. Pseudobase: a database with RNA pseudoknots. *Nucleic Acids Res.*, 28:201–204, 2000.

25. M. Zuker and P. Stiegler. Optimal computer folding of large RNA sequences using thermodynamics and auxiliary information. *Nucleic Acids Res.*, 9(1):133–148, 1981.

Flux-Based *vs.* Topology-Based Similarity of Metabolic Genes

Oleg Rokhlenko[1], Tomer Shlomi[2,*], Roded Sharan[2,**],
Eytan Ruppin[2], and Ron Y. Pinter[1]

[1] Dept. of Computer Science, Technion–IIT, Haifa 32000, Israel
{olegro, pinter}@cs.technion.ac.il
[2] School of Computer Science, Tel-Aviv University, Tel-Aviv 69978, Israel
{shlomito, roded, ruppin}@tau.ac.il

Abstract. We present an effectively computable measure of functional gene similarity that is based on metabolic gene activity across a variety of growth media. We applied this measure to 750 genes comprising the metabolic network of the budding yeast. Comparing the *in silico* computed functional similarities to those obtained by using experimental expression data, we show that our computational method captures similarities beyond those that are obtained by the topological analysis of metabolic networks, thus revealing—at least in part—dynamic characteristics of gene function. We also suggest that network centrality partially explains functional centrality (*i.e.* the number of functionally highly similar genes) by reporting a significant correlation between the two. Finally, we find that functional similarities between topologically distant genes occur between genes with different GO annotations.

1   Introduction

The study of biological networks has attracted considerable attention in recent years, including the construction of mathematical models to elucidate both cell activity as well as genes' function and expression. Much of the work to date has attempted to establish measures for the similarity (or distance) between genes that are based on the topological properties of metabolic networks. Even though recent analyses have provided valuable insights regarding this issue [1,2], topological characteristics alone (as devised by *e.g.* Kharchenko *et al.* [3], Chen and Vitkup [4]) offer only a static description of the properties of interest. On the other hand, accurate prediction of dynamic cell activity using kinetic models requires detailed information on the rates of enzyme activity which is rarely available; moreover, such analysis is usually limited to small-scale networks.

Fortunately, for metabolic networks, the use of stochiometry and other sources of information provides an added value over the topology of the underlying structure. Specifically, constraint-based stochiometric models have emerged as a key

---

* Supported in part by the Tauber Fund.
** Supported by an Alon Fellowship.

P. Bücher and B.M.E. Moret (Eds.): WABI 2006, LNBI 4175, pp. 274–285, 2006.

method for studying such networks permitting large-scale analysis thereof. They use genome-scale networks to predict steady-state metabolic activity, regardless of specific enzyme kinetics. In these models, stoichiometric, thermodynamic, flux capacity and possibly other constraints affect the space of possible flux distributions attainable by a metabolic network.

In this paper we devise an effectively computable functional similarity measure between genes that is based on their metabolic activity. Such a measure would allow us to perform large scale *in silico* experiments and predict functional relations that can then be validated by experimental methods. Specifically, we suggest a method for determining similarities in gene activities that is based on Flux Balance Analysis (FBA). We first suggest a knockout-based measure but find it to be only moderately correlated with experimental data (of gene co-expression, see below). We then employ a measure of *metabolic genes co-activity* (MGCA), which tells how similar gene functions are in terms of the correlation between their corresponding flux activity vectors across a large variety of growth media. This latter measure, already used in a more limited scope by [5], is significantly better than the former measure in terms of correlation with experimental data.

Our evaluation of the suggested measures is based on testing their correlations with experimental data on similarity in gene expression, to assess their veracity. The basic relation between metabolic fluxes and gene expression was already studied and established previously both computationally (showing only a moderate correlation) as well as experimentally. Recall that the metabolic state of an organism is controlled via transcriptional regulation which adjusts gene expression levels according to metabolic demands [6]. Previous studies have shown that the expression patterns of enzyme coding genes are correlated with the flux patterns predicted by FBA: Schuster *et al.* [7,8] and Famili *et al.* [5] have shown that genes, associated with fluxes which are predicted to change together when shifting from one medium to another (*e.g.* in diauxic shift), are co-expressed under these conditions; Reed and Palsson [9] have shown that the genes associated with fluxes that are correlated within the solution space also exhibit moderate levels of correlation in their expression. Recently, Bilu *et al.* [10] proposed a more direct relation between expression and flux where the range of possible optimal flux values for a given reaction reflects evolutionary constraints on the expression levels of its associated enzymes; specifically, they have shown that the regulation of reactions which have an optimal fixed value is under strong selection to maintain their flux at the precise levels needed, while the regulation of reactions which may have a broad range of optimal values is under weaker selection.

In this work we extend upon these previous studies to look into ways of building upon the reported correlation between fluxes and expression, to construct efficient measures of functional similarity among metabolic genes. To this end, in contrast with the previous studies, we examine the relation between fluxes and expression while concomitantly controlling for correlations caused solely by the network's topology.

Our comparison focuses on 750 metabolic genes of the yeast *Saccharomyces cerevisiae*. We find the correlation between MGCA and co-expression to be statistically significant. Furthermore, it remains so even after cancelling the effect of the underlying (static) network topology. These results support the notion that our measure indeed captures the true functional similarity between metabolic genes.

2   Preliminaries—Modeling Metabolism and Flux Balance Analysis

Flux Balance Analysis (FBA) [11,12] is a particular constraint-based method which assumes that the network is regulated to maximize or minimize a certain cellular function, which is usually taken to be the organism's growth rate. FBA has been demonstrated to be a very useful technique for the analysis of metabolic capabilities of cellular systems [13,14]. It involves carrying out a steady state analysis, using the stoichiometric matrix (as defined below) for the system in question. The system is assumed to be optimized with respect to functions such as maximization of biomass production or minimization of nutrient utilization; it is solved accordingly to obtain a steady state flux distribution, which is then used to interpret the metabolic capabilities of the system.

In FBA, the constraints imposed by stoichiometry in a chemical network at steady state are analogous to Kirchoff's Second Law for the flow of currents in electric circuits [15], namely—for each of the M metabolites in a network the net sum of all production and consumption fluxes, weighted by their stoichiometric coefficients, is zero:

$$\sum_{j=1}^{N} S_{ij} v_j = 0, \quad i = 1, \ldots, M \tag{1}$$

Here, S_{ij} is the element of the stoichiometric matrix S corresponding to the stoichiometric coefficient of metabolite i in reaction j. The flux v_j is the rate of reaction j at steady state, and is the j-th component of an N-dimensional flux vector v, where N is the total number of fluxes. In addition to internal fluxes, which are associated with chemical reactions, v includes exchange fluxes that account for metabolite transport through the membrane. The steady-state approximation is generally valid because of the fast equilibration of metabolite concentrations (seconds) with respect to the time scale of genetic regulation (minutes) [16].

Additional constraints, including those pertaining to the availability of nutrients or to the maximal fluxes that can be supported by enzymatic pathways, can be introduced as the following inequalities:

$$\alpha_j \leq v_j \leq \beta_j \tag{2}$$

For example, for a substrate uptake flux v_j, one can set α_j and β_j to be equal to the corresponding measured or imposed values. Eq. 2 can also be used to distinguish reversible and irreversible reactions, where $\alpha_j = 0$ for the latter.

All flux vectors that satisfy the constraints mentioned above define a feasible space, Φ. For an underdetermined system, as is typically the case in FBA models of cellular metabolic networks [13], Φ is a convex set in the N-dimensional space of fluxes [17]. Due to the linear nature of Φ, it is possible to use linear programming [18] to characterize the points in Φ that maximize or minimize a given linear objective function. A natural choice for an objective function in metabolic models of prokaryotes and simple eukaryotes is biomass production [13,14], as it is reasonable to hypothesize that unicellular organisms have evolved towards maximal growth performance. This process is formalized by introducing a growth flux that transforms a linear combination of fundamental metabolic precursors into biomass.

The maximization of biomass production is implemented by defining an additional flux v_{gro} associated with cell growth. For this flux, the stoichiometric factors of the reactants are the experimentally known proportions c_i of metabolite precursors X_i contributing to biomass production [13]:

$$c_1 X_1 + c_2 X_2 + \ldots + c_M X_M \overset{v_{gro}}{\to} Biomass \qquad (3)$$

The search for the flux vector maximizing v_{gro} under the constraints of Eqs. 1 and 2 is solved using the Simplex algorithm.

The theoretical basis of FBA is supported by several experiments. These include empirical validation of growth yield and flux predictions [13,14], measurements of uptake rates around the optimum under various conditions [19], and results from large-scale gene deletion experiments [20].

For the stoichiometric analysis of the metabolic network of *S. cerevisiae*, we have used the reconstruction by Duarte, Herrgard, and Palsson [21]. The nodes of this network correspond to metabolic genes, and the edges correspond to the connections established by metabolic reactions. Two metabolic genes are connected if the corresponding enzymes share a common metabolite among their substrates or products. The list of metabolic reactions, and the 1060 (metabolites) by 1149 (fluxes) stoichiometric matrix (available at `http://gcrg.ucsd.edu`) were compiled using data from public databases and the literature. The 1149 reactions are associated with 750 genes. As in previous FBA formulations, we use inequalities (Eq. 2) to limit nutrient uptake and to implement reactions' irreversibility. In addition to the 1149 internal reactions, we added to the model 116 uptake/excretion reactions, for each of the metabolites listed as "extracellular" in the basic model.

3   Similarity Measures for Metabolic Genes

In the context of the aforementioned motivation, we suggest two techniques for obtaining the distance between metabolic genes: a knockout-functional (KF) scheme and a growth-functional (GF) scheme. The biological plausibility of the obtained distance measures is validated by correlating them with the corresponding similarity measure obtained by expression data.

3.1 Knockout-Functional Scheme

Cellular response to a gene knockout involves rerouting of metabolic flux through alternative pathways and the utilization of isoenzymes [22,23]. We hypothesize that similar metabolic responses to gene knockouts may provide evidence for similar metabolic functionality between genes. Based on this hypothesis, we define the KF similarity measure between gene pairs as the similarity in the metabolic response following their knockout.

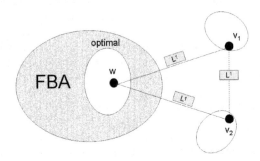

Fig. 1. Schematic illustration of the proposed flux similarity model. w stands for the optimal flux distributions on the wild-type metabolic network, v_1 stands for the optimal flux distribution on the metabolic networks with the first flux knocked-out, and v_2 stands for the optimal flux distribution on the metabolic networks with the second flux knocked-out.

Predicting the metabolic response for gene knockouts is a more difficult task than predicting the metabolic state of wild-type strains. Gene deletion is commonly modeled by constraining the flux through the reactions associated with a given gene to zero, and applying FBA [13]. However, it turns out that the metabolic state of the knocked-out strain is not necessarily optimal in terms of growth rate, and thus in many cases FBA's predictions are inaccurate. Instead, it was hypothesized that the cell adapts to gene knockouts by minimizing the change in its metabolic state. Specifically, the Minimization of Metabolic Adjustment (MOMA) approach searches for a metabolic state for a knocked-out strain with minimal distance, under the L2 norm, from the flux distribution of the wild-type strain [22]. Recently, a new method called Regulatory On-Off Minimization (ROOM) was suggested to predict metabolic states following gene knockouts, and was shown to provide better predictions of knockout phenotypes [23]. ROOM aims to minimize the number of regulatory changes required for the adaptation by minimizing the number of significant flux changes between the metabolic states of the wild-type and knocked-out states (*i.e.* using the norm L0).

A naive method for measuring the distance between the metabolic responses of two gene knockouts would be to simulate the knockout of each of them individually using ROOM, and then compute the distance between the obtained flux distributions. However, in many cases ROOM (like FBA and MOMA) provides multiple possible metabolic states for the knocked-out strain rather than a single solution. In these cases, it is not clear how to define the similarity measure between two genes.

To overcome this problem we define the KF similarity measure as the minimal distance between the optimal ROOM solutions for the two genes[1]. This is achieved by formulating a single optimization problem to find two ROOM solutions with minimal distance between them. The schematic illustration of our model is presented in Figure 1.

Notably this formulation depends on the choice of a wild-type and thus we repeat our analysis for several different wild-types. Furthermore, since ROOM requires Mixed Integer Linear Programming (MILP) optimization which is NP-hard, we use a relaxed version of ROOM and, in addition, we use the L1 norm instead of L0. The use of the L1 norm is similar to a variant of ROOM, called ROOM-LP, that was shown to provide similar predictions to ROOM [23]. The L1 norm was also used by Kuepfer *et. al.* [24] for a similar purpose of knockout prediction. The distance between the two flux distributions of the knocked-out strains is also minimized using the L1 norm.

The optimization problem is formulated as a LP problem as follows:

$$\min \|v_1 - v_2\|_{L1}$$
$$s.t.$$
$$S \cdot v_1 = 0; \quad v_{min} \leq v_1 \leq v_{max}; \quad v_1[ko1] = 0, \, ko1 \in A_1;$$
$$S \cdot v_2 = 0; \quad v_{min} \leq v_2 \leq v_{max}; \quad v_2[ko2] = 0, \, ko2 \in A_2;$$
$$\|w - v_1\|_{L1} = l_1; \quad \|w - v_2\|_{L1} = l_2;$$

where w is the wild-type flux distribution, A_1 and A_2 are sets of reactions associated with the deleted genes, and l_i $(i = 1, 2)$ are the optimal solutions of a single optimization problem:

$$\min \|v - w\|_{L1}$$
$$s.t.$$
$$S \cdot v = 0; \quad v_{min} \leq v \leq v_{max}; \quad v_{ko1} = 0, \, ko1 \in A;$$

Solving the above optimization problem we receive a measure of similarity between fluxes.

3.2 Growth-Functional Scheme

We hypothesize that the regulation of reactions that are active (different than zero) together across certain media and passive (equal to zero) together across others should be similar. In order to evaluate our hypothesis, we follow and extend the approach of [10], computing genes' activities across 100 randomly generated growth media.

To pursue this possibility we used flux variability analysis [9,25]: for each reaction we computed the maximal and minimal flux values attainable in the space of optimal flux distributions for growth conditions simulating 100 different

---

[1] We use the distance notion instead of the similarity one both in the KF and GF schemes for sake of clarity and for being consistent with commonly used network topology distances.

growth media. Random growth media were generated by setting limiting values to the uptake reactions independently at random. With probability 0.5, the maximal uptake rate was set to 0, *i.e.* only excretion was allowed. Otherwise, uptake rate was limited to a value chosen uniformly at random in the range [0.01, 5], at a resolution of 0.01. A similar sampling method was used in [26]. In addition, in order to ensure enough variability between media, we switched between aerobic and anaerobic growth media with probability 0.5.

For each generated medium we simulated growth conditions similar to [5] and for each reaction checked if it is active across the current growth media. A reaction is considered active in a given flux distribution if its associated flux is non-zero, namely either its maximum or minimum are different than zero. Active genes were denoted by '0' and nonactive ones by '1'. This way we created for each gene a binary vector of its activity across a series of generated media.

We define a measure of metabolic genes co-activity (MGCA) as the Jaccard coefficient [27] between two binary vectors reflecting metabolic genes' activity. The binary Jaccard coefficient measures the degree of overlap between two sets of values, x_a and x_b, and is computed as the ratio between the number of shared attributes of $\mathbf{x}_a$ and $\mathbf{x}_b$ and the number possessed by $\mathbf{x}_a$ or $\mathbf{x}_b$:

$$J(x_a, x_b) = \frac{x_a \cap x_b}{x_a \cup x_b} \tag{4}$$

The pseudo-code of the entire procedure is presented in Figure 2.

Algorithm 1: *FindGenesDist(N)*

Input: *N*: the number of required media.
Output: *results*: matrix *num_genes* × *num_genes* containing the distance
　　　　　between metabolic genes.

for *k=1..N* do
　　for *each external flux f* do
　　　　with probability 0.5, set $f = 0$;
　　　　otherwise f receives a random value chosen uniformly in [0.01, 5];
　　Run FBA to maximize biomass(growth rate)
　　　　and obtain objective value (*wild_growth_rate*);
　　Add constraint: biomass $\geq 0.9 * wild\_growth\_rate$;
　　for *i=1..num_fluxes* do
　　　　Run FBA to maximize flux i, obtain i_{max};
　　　　Run FBA to minimize flux i, obtain i_{min};
　　for *each gene g* do
　　　　if *for one of its related fluxes* $i_{max} = i_{min} = 0$ then
　　　　　　MT[g][k] = 1;
　　　　else
　　　　　　MT[g][k] = 0;
　　for *each gene g1* do
　　　　for *each gene g2 ≠ g1* do
　　　　　　results[g1][g2] = Jaccard_coefficient(MT[g1],MT[g2]);

Fig. 2. The process for computing the GF-based measure

4   Results

Recall that the metabolic state of an organism is controlled by transcriptional regulation which adjusts gene expression levels according to metabolic demands [6]. Thus the experimental pairwise correlations serve as the true benchmark rod to which we compare the computational measures we compute to find out which is the best, *i.e.* closest to reality.

The first computational similarity measure proposed [3] was based on topological properties of the metabolic network . We start by repeating these experiments and then show how our measure can go beyond topological measures.

The obtained metabolic network is used to calculate network distance between genes. We define a pair of directly connected metabolic genes as separated by distance 1, and the network distance between genes X and Y is the length of the shortest path from X to Y in the metabolic network. While any metabolite can be used to establish connections between metabolic genes, the relationships established by the common metabolites and cofactors—such as ATP, water or hydrogen—are not likely to connect genes with similar metabolic functions.

In compiling a metabolic network, we consider a subset of metabolites which excludes the most highly connected metabolic species. An exclusion threshold was determined based on the connectivity of the resulting network. A total of the 10 most highly connected metabolites (ATP, ADP, AMP, CO_2, H, H_2O, NADP, NADPH, phosphate and diphosphate), which compose 1% of all metabolites, and their mitochondrial and external analogs were excluded. We also tried to exclude up to the top 3% of all metabolites, however we found out that the general trends described in this paper are not sensitive to the precise choice of the excluded set of metabolites.

We compared the correlation between the gene functional similarity measure and their expression similarity. To this end, we used Rosetta's "compendium" dataset [28] which measures expression profiles of over 6200 *S. cerevisae* ORFs across 287 deletion strains and 13 chemical conditions. In addition, the dataset contains 63 negative control measurements comparing two independent cultures of the same strain. These were used to establish individual error models for each ORF, providing not only the raw intensity and the ratio measurement values for each experimental data point, but also a *p-value* evaluating the significance of change in expression level. The expression similarity measure between ORFs X and Y was computed according to $1 - Spearman\_rank(p_x, p_y)$ where p_x and p_y are expression profile vectors of X and Y, respectively, and the Spearman rank was calculated as in [29].

As in [3], we observed that the expression distance increases monotonically with network distance ($R^2 = 0.78$, *p*-value $= 1.2 \cdot 10^{-8}$), demonstrating that genes closer to each other in the metabolic network tend to have, on average, higher level of coexpression.

Measuring the correlation between the KF-based distance and those based on the expression data we observed (see Figure 3) a moderate correlation ($R^2=0.36$ in the negatively correlated expressed profiles with a p-value of $8.6 \cdot 10^{-2}$, and $R^2 = 0.45$ in the positively ones with a *p*-value of $\leq 4.6 \cdot 10^{-2}$). Note that the

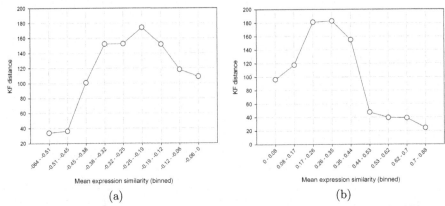

Fig. 3. Correlation between expression levels and genes activities under the KF measure. (a) Negatively expressed pairs. (b) Positively expressed pairs.

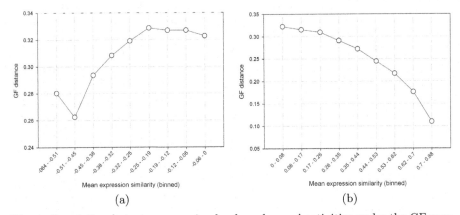

Fig. 4. Correlation between expression levels and genes' activities under the GF measure. (a) Negatively expressed pairs. (b) Positively expressed pairs.

obtained correlation is robust to the initial wild-type metabolic state, as similar correlation levels were observed when starting from different wild-types.

As for the GF-based measure, we observe (see Figure 4) that it exhibits a significant correlation with the expression similarity ($R^2 = 0.78$ in the negatively correlated expressed profiles with a p-value of $5.15 \cdot 10^{-8}$, and $R^2 = 0.94$ in the positively ones with p-value $\leq 1 \cdot 10^{-20}$).

Finally, we observe a significant enhancement of the GF-based measure over the static (topological) metabolic distance indicating that this static distance can explain only partially the demand for common regulation. We use a partial correlation method that describes the relationship between two variables whilst eliminating the effects of another variable on this relationship, namely network distance in our case. Our results show significant partial correlation ($R^2 = 0.65$, with a p-value of $3.8 \cdot 10^{-6}$) between expression levels and our MGCA measure given a metabolic network distance. This higher correlation for our measure supports the fact that the FBA model captures the dynamic metabolic activity of the cell, and that the regulation system indeed works to maximize the growth rate. Moreover, the results stay significant with every thresholds for excluding "currency metabolites" from the metabolic network in the range from 1% to 3%.

In order to evaluate the difference between the MGCA measure and the metabolic network distance measure we analyzed two sets of pairs of genes: one containing pairs of genes that are close under the network distance and distant under the MGCA measure, and vice versa. We observed that the first set is significantly enriched with the GO term *protein biosynthesis* (GO:006412)— 25 annotated genes out of 104 resulting in a p-value ≤ 0.001, as well as with the GO term *nucleobase, nucleoside, nucleotide and nucleic acid metabolism* (GO:006139)—40 annotated genes out of 104 also resulting in a p-value ≤ 0.001. An engrossing result was that the complementary set (genes that are close under the MGCA measure but are distant under the network topology measure) showed no significant enrichment, possibly testifying that such functional similarities occur across a broad and homogeneous span of functional annotations.

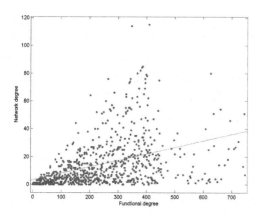

Functional enrichments were computed based on the GO-SLIM process annotations [30] for genes. Yeast GO-SLIM annotations were obtained from SGD [31]. For a given set S and a given term t, the functional enrichment score was computed as follows: suppose S has $n(t)$ genes that are annotated with term t (or with a more specific term). Let $p(t)$ be the hypergeometric probability for observing $n(t)$ or more proteins annotated with the term t in a protein subset of size

Fig. 5. The correlation between functional centrality and network centrality

$|S|$. Having found a term t_0 with minimal probability $p(t_0)$, the score was set to the p-value of the enrichment under term t_0, computed by comparing $p(t_0)$ with the analogous probabilities for 10,000 random sets of proteins of size $|S|$.

In addition we looked at the correlation between the network degree of each gene and its functional degree, *i.e.* the number of functionally highly similar (Jaccard coefficient ≤ 0.3) genes (see Figure 5). As we received a significant correlation of $R^2 = 0.4$ with a p-value ≤ 0.001, it seems that network centrality explains (at least in part) functional centrality. Namely, the more alternative pathways go through a given gene, the more functionally significant it is. We also observed that the correlation is robust to the functionally similarity threshold in the range form 0.01 to 0.3.

5   Conclusions

This paper is the first to show that functional flux-based similarity measures between genes can go beyond previous computational measures based on network topology. We applied two schemes to compute this distance: the knockout-functional (KF) scheme and the growth-functional (GF) scheme. While the former shows a fairly moderate correlation with the experimental results, the latter provides a strong, statistically-significant measure. One possible explanation of this behavior may be that the GF studies probe the natural wild type across a variety of media, whereas the KF method does it in less natural strains and in a sole media. The other reason is the more cumbersome computational method used in the KF case, which is likely to add significant noise to the results obtained.

References

1. Jeong, H., Tombor, B., Albert, R., Oltavi, Z., Barabasi, A.: The large-scale organization of metabolic networks. Nature **407** (2000) 651–654
2. Ravasz, E., Somera, A., Mongru, D., Oltvai, Z., Barabasi, A.: Hierarchical organization of modularity in metabolic networks. Science **297** (2002) 1551–1555
3. Kharchenko, P., Church, G.M., Vitkup, D.: Expression dynamics of a cellular metabolic network. Molecular Systems Biology **1** (2005) E1–E6
4. Chen, L., Vitkup, D.: Predicting genes for orphan metabolic activities using phylogenetic profiles. Genome Biol. **7** (2006) R17
5. Famili, I., Forster, J., Nielsen, J., Palsson, B.Ø.: Saccharomyces cerevisiae phenotypes can be predicted by using constraint-based analysis of a genome-scale reconstructed metabolic network. Proc Natl Acad Sci U S A **100** (2003) 13134–13139
6. Zaslaver, A., Mayo, A., Rosenberg, R., Bashkin, P., Sberro, H., et al: Just-in-time transcription program in metabolic pathways. Nat Genet **36** (2004) 486–491
7. Schuster, S., Dandekar, T., Fell, D.: Detection of elementary flux modes in biochemical networks: a promising tool for pathway analysis and metabolic engineering. Trends Biotechnol **17** (1999) 53–60
8. Schuster, S., Klamt, S., Weckwerth, W., Moldenhauer, F., Pfeiffer, T.: Use of network analysis of metabolic systems in bioengineering. Bioprocess and Biosystems Engineering **24** (2002) 363–372
9. Reed, J., Palsson, B.: Genome-scale in silico models of e. coli have multiple equivalent phenotypic states: assessment of correlated reaction subsets that comprise network states. Genome Res **14** (2004) 1797–1805
10. Bilu, Y., Shlomi, T., Barkai, N., Ruppin, E.: Conservation of expression and sequence of metabolic genes is reflected by activity across metabolic states. PLoS Comp. Bio. (*in press*) (2006)

11. Fell, D., Small, J.: Fat synthesis in adipose tissue. An examination of stoichiometric constraints. Biochem J **238** (1986) 781–786
12. Kauffman, K., Prakash, P., Edwards, J.: Advances in flux balance analysis. Curr Opin Biotechnol **14** (2003) 491–496
13. Price, N.D., Reed, J.L., Palsson, B.Ø.: Genome-scale Models of Microbial Cells: Evaluating the consequences of constraints. Nature Reviews Microbiology **2** (2004) 886–897
14. Varma, A., Palsson, B.: Metabolic capabilities of Escherichia coli: II. Optimal growth patterns. J. Theor. Biol. **165** (1993) 503–522
15. Schilling, C.H., Edwards, J.S., Palsson, B.: Toward metabolic phenomics: analysis of genomic data using flux balances. Biotechnol. Prog **15** (1999) 288–295
16. Fell, D.: Understanding the Control of Metabolism. Portland Press, London (1996)
17. Schilling, C.H., Edwards, J.S., Letscher, D., Palsson, B.Ø.: Combining pathway analysis with flux balance analysis for the comprehensive study of metabolic systems. Biotechnol. Bioeng. **71** (2000) 286–306
18. Vanderbei, R.J.: Linear Programming: Foundations and Extensions. Kluwer Academic Publishers, Boston (1996)
19. Edwards, J., Ibarra, R., Palsson, B.: In silico predictions of Escherichia coli metabolic capabilities are consistent with experimental data. Nat Biotechnol **19** (2001) 125–130
20. Badarinarayana, V., Estep, P.W., Shendure, J., Edwards, J., Tavazoie, S., Lam, F., Church, G.M.: Selection analyses of insertional mutants using subgenic-resolution arrays. Nat. Biotechnol. **19** (2001) 1060–1065
21. Duarte, N., Herrgard, M., Palsson, B.Ø.: Reconstruction and validation of Saccharomyces cerevisiae iND750, a fully compartmentalized genome-scale metabolic model. Genome Res **14** (2004) 1298–1309
22. Segre, D., Vitkup, D., Church, G.: Analysis of optimality in natural and perturbed metabolic networks. Proc. Natl. Acad. Sci. U. S. A. **99** (2002) 15112–15117
23. Shlomi, T., Berkman, O., Ruppin, E.: Regulatory on/off minimization of metabolic flux changes after genetic perturbations. Proc. Natl. Acad. Sci. U. S. A. **102** (2005) 7695–7700
24. Kuepfer, L., Sauer, U., Blank, L.M.: Metabolic functions of duplicate genes in Saccharomyces cerevisiae. Genome Res. **15** (2005) 1421–1430
25. Mahadevan, R., Schilling, C.: The effects of alternate optimal solutions in constraint-based genome-scale metabolic models. Metab Eng **5** (2003) 264–276
26. Almaas, E., Oltvai, Z., Barabasi, A.: The activity reaction core and plasticity of metabolic networks. PLoS Comput Biol **1** (2005) e68
27. Salton, G., McGill, M.J.: Introduction to Modern Information Retrieval. McGraw-Hill, New-York (1983)
28. Hughes, T., et. al.: Flux analysis of underdetermined metabolic networks: the quest for the missing constraints. Cell **102** (2000) 109–126
29. Press, W.H., Teukolsky, S.A., Vetterling, W.T., Flannery, B.P.: Numerical Recipes in C++: the art of scientific computing. Cambridge University Press, Cambridge (2002)
30. Ashburner, M., et. al.: Gene Ontology: tool for the unification of biology. Nat. Genet. **25** (2000) 25–29
31. Issel-Tarver, L., et. al.: Saccharomyces Genome Database. Methods Enzymol **350** (2002) 329–346

Combinatorial Methods for Disease Association Search and Susceptibility Prediction

Dumitru Brinza* and Alexander Zelikovsky**

Department of Computer Science, Georgia State University, Atlanta, GA 30303
dima@cs.gsu.edu, alexz@cs.gsu.edu

Abstract. Accessibility of high-throughput genotyping technology makes possible genome-wide association studies for common complex diseases. When dealing with common diseases, it is necessary to search and analyze multiple independent causes resulted from interactions of multiple genes scattered over the entire genome. This becomes computationally challenging since interaction even of pairs gene variations require checking more than 10^{12} possibilities genome-wide. This paper first explores the problem of searching for *the most disease-associated and the most disease-resistant multi-gene interactions* for a given population sample of diseased and non-diseased individuals. A proposed fast complimentary greedy search finds multi-SNP combinations with nontrivially high association on real data. Exploiting the developed methods for searching associated risk and resistance factors, the paper addresses the *disease susceptibility prediction problem*. We first propose a relevant optimum clustering formulation and the model-fitting algorithm transforming clustering algorithms into susceptibility prediction algorithms. For three available real data sets (Crohn's disease (Daly et al, 2001), autoimmune disorder (Ueda et al, 2003), and tick-borne encephalitis (Barkash et al, 2006)), the accuracies of the prediction based on the combinatorial search (respectively, 84%, 83%, and 89%) are higher by 15% compared to the accuracies of the best previously known methods. The prediction based on the complimentary greedy search almost matches the best accuracy but is much more scalable.

1 Introduction

Disease association studies analyze genetic variation across diseased and non-diseased individuals. The difference between individual DNA sequences occurs at a single-base sites, in which more than one allele is observed across population. Such variations are called single nucleotide polymorphisms (SNPs). Disease association analysis searches for a SNP with frequency among diseased individuals (cases) considerably higher than among non-diseased individuals (controls).

* Partially supported by GSU Molecular Basis of Disease Fellowship.
** Partially supported by NIH Award 1 P20 GM065762-01A1 and US CRDF Award MOM2-3049-CS-03.

P. Bücher and B.M.E. Moret (Eds.): WABI 2006, LNBI 4175, pp. 286–297, 2006.

When dealing with common diseases, it is necessary to search and analyze multiple independent causes each resulted from interaction of multiple SNPs scattered over the entire genome.

Accessibility of high-throughput genotyping technology makes possible genome-wide association studies for common complex diseases. The number of simultaneously typed SNPs for association and linkage studies is reaching 250,000 for SNP Mapping Arrays [1]. High density maps of SNPs as well as massive DNA data with large number of individuals and number of SNPs become publicly available [2].

Several challenges in genome-wide association studies of complex diseases have not yet been adequately addressed [8]: interaction between non-linked genes, multiple independent causes, multiple testing adjustment, etc. The computational challenge (as pointed in [8]) is caused by the dimension catastrophe. Indeed, two-SNP interaction analysis (which can be more powerful than traditional one-by-one SNP association analysis [12]) for a genome-wide scan with 1 million SNPs (3 kb coverage) will afford 10^{12} possible pairwise tests. Multi-SNP interaction analysis reveals even deeper disease-associated interactions but is usually computationally infeasible and its statistical significance drastically decreases after multiple testing adjustment [19,20]. In this paper we explore optimization approach to resolve these issues instead of traditionally used statistics and computational intelligence.

In order to handle data with huge number of SNPs, one can extract informative (indexing) SNPs that can be used for (almost) lossless reconstructing of all other SNPs. Using multiple linear regression based method [11], we have obtained promising results [6]. However, exhaustive searching for all possible SNP combinations is still very slow. A combinatorial search method for finding disease-associated multi-SNP combinations (MSC) applied to index SNPs resulted in finding genetic risk factors which are statistically significant even after multiple testing adjustment, e.g., a few statistically significant MSCs were found (i) for tick-borne encephalitis virus-induced disease [4] and (ii) for Crohn's disease [3] while no single SNP or pair of SNPs show significant association [6].

In this paper we formulate the optimization **problem of finding the most disease-associated multi-SNP combination** for given case-control data. Since it is plausible that common diseases can have also genetic resistance factors, we also search for *the most disease-resistant multi-SNP combination*. Association of risk or resistance factors with the disease can be measured in terms of p-value of the skew in case and control frequencies, risk rates or odds rates. Here we concentrate on so called *positive predictive value* (PPV) which is the frequency of diseased individuals among all individuals with a given multi-SNP combination. This optimization problem is NP-hard and can be viewed as a generalization of the maximum independent set problem. We propose a fast *complimentary greedy search* which we compare with the exhaustive search and combinatorial search method proposed in [6]. Although complimentary greedy search cannot guarantee finding of close to optimum MSCs, in our experiments with real data, it finds MSCs with non-trivially high PPV. For example, for Crohn's disease data [3], complimentary greedy search finds in less than second a case-free MSC containing 24 controls, while exhaustive and combinatorial searches need more than 1 day to find case-free MSCs with at most 17 controls.

We next address the **disease susceptibility prediction problem** (see [16,17,18,21,22]) exploiting the developed methods for searching associated risk and resistance factors. We propose a new optimum clustering problem formulation and suggest a model-fitting method transforming a clustering algorithm into the corresponding model-fitting susceptibility prediction algorithm. Since common diseases can be caused by multiple independent and co-existing factors, we propose association-based clustering of case/control population. The resulted association-based combinatorial prediction algorithm significantly outperforms existing prediction methods. For all three real data sets that were available to us (Crohn's disease [3], autoimmune disorder [10], and tick-borne encephalitis [4]) the accuracy of the prediction based on combinatorial search is higher by 15% compared to the accuracy of all previously known methods implemented in [16,15]. The accuracy of the prediction based on complimentary greedy search almost matches the best accuracy but is much more scalable.

In the next section we will formulate the disease association search problem, overview the searching algorithms and their quality, reformulate the optimization version of disease association search as an independent set problem and propose the complimentary greedy search algorithm. Section 3 is devoted to the disease susceptibility prediction problem. We give the prediction and relevant clustering optimization problem formulations, propose our model-fitting approach of transforming clustering into prediction and describe two new prediction algorithms. Section 4 describes and discusses the results of our experiments with association search and susceptibility prediction on three real data sets.

2   Disease Association Search

In this section we formally describe the search of statistically significant disease-associated multi-SNP combinations. We then formulate the corresponding optimization problem, discuss its complexity, describe combinatorial search introduced in [6] and propose a fast heuristic, so called complementary greedy search.

The typical case/control or cohort study results in a sample population S consisting of n individuals represented by values of m SNPs and the disease status. Since it is expensive to obtain individual chromosomes, each SNP value attains one of three values 0, 1 or 2, where 0's and 1's denote homozygous sites with major allele and minor allele, respectively, and 2's stand for heterozygous sites. SNPs with more than 2 alleles are rare and can be conventionally represented as biallelic. Thus the sample S is an $(0, 1, 2)$-valued $n \times (m + 1)$-matrix, where each row corresponds to an individual, each column corresponds to a SNP except last column corresponding to the disease status (0 stands for disease and 1 stands for non-disease). Let S_0 and S_1 be the subsets of rows with non-disease and disease status, respectively. For simplicity, we assume that there are no two rows identical in all SNP columns.

Risk and resistance factors representing gene variation interaction can be defined in terms of SNPs as follows. A *multi-SNP combination* (MSC) C is a subset of SNP-columns of S (denoted $snp(C)$) and the values of these SNPs, 0, 1, or 2.[1] The subset of individuals-rows of S whose restriction on columns of $snp(C)$ coincide with values of C is denoted $cluster(C)$. A subset of individuals is called a *cluster* if it coincides with $cluster(C)$ for a certain MSC C. For example, if S is represented by an identity matrix I_5, then rows 3, 4, and 5 form a cluster for MSC C with $snp(C) = \{1,2\}$ and both values equal to 0. Obviously, a subset X of rows of S may not form a cluster, but it always can be represented as a union of clusters, e.g., as a union of trivial clusters containing its individual rows. Let $h(C) = cluster(C) \cap S_0$ be the set of non-diseased individuals and $d(C) = cluster(C) \cap S_1$ be the set of diseased individuals in $cluster(C)$.

The association of an MSC C with the disease status can be measured with the following parameters ($h = |h(C)|, d = |d(C)|, H = |S_0|, D = |S_1|$):

- odds ratio $OR = \frac{d \cdot (H-h)}{h \cdot (D-d)}$ (for case-control studies)
- relative risk $RR = \frac{d \cdot (H+D-h-d)}{(D-d)(h+d)}$ (for cohort studies)
- positive predictive value $PPV = \frac{d}{h+d}$ (for susceptibility prediction)
- p-value of the partition of the cluster into diseased and non-diseased:

$$p = \sum_{k=0}^{d} \binom{h+d}{k} \left(\frac{D}{H+D}\right)^k \left(\frac{H}{H+D}\right)^{h+d-k}$$

Since MSCs are searched among all SNPs, the computed p-value requires adjustment for multiple testing which can be done with simple but overly pessimistic Bonferroni correction or computationally extensive but more accurate randomization method.

General disease association searches for all MSCs with one of the parameters above (or below) a certain threshold. The common formulation is to find all MSCs with adjusted p-value below 0.05.

The exhaustive search (ES) checks all 1-SNP, 2-SNP, ..., m-SNP combinations has runtime $O(n^{3m})$ making it infeasible even for small numbers of SNPs m. One either should reduce the depth (number of simultaneously interacting SNPs) or reduce m by extracting informative SNPs from which one can reconstruct all other SNPs. The multiple linear regression based tagging method of [11] has been used in [6]. They choose maximum number of index SNPs that can be handled by ES in a reasonable computational time.

It has been also suggested a *combinatorial search* (CS) which avoids insignificant MSCs or clusters without loosing significant ones. CS searches only for closed MSCs, where closure is defined as follows. The *closure* $\bar{C}$ of MSC C is an MSC with minimum non-diseased elements $h(\bar{C})$ and the same diseased elements $d(\bar{C}) = d(C)$. $\bar{C}$ can be easily found by incorporating into $snp(C)$ all

---

[1] In this paper we restrict ourselves to 0,1, or 2, while in general, the values of MSC can also be negations $\bar{0}, \bar{1}$ or $\bar{2}$, where $\bar{i}$ means that MSC is required to have value unequal to i.

SNP with common values among all diseased individuals in C. Also CS disregards clusters with small number of diseased individuals since they cannot have significant subclusters. CS has been shown much faster than ES and capable of finding more significant MSCs than ES [6].

Here we suggest to consider also optimization formulation corresponding to the general association search problem, e.g., find MSC with the minimum adjusted p-value. In particular, we focus on maximization of PPV. Obviously, the MSC with maximum PPV should not contain non-diseased individuals in its cluster and the problem can be formulated as follows:

Maximum Non-diseased-Free Cluster Problem. (MNFCP) Find a cluster C which does not contain non-diseased individuals and has the maximum number of diseased individuals.

It is not difficult to see that this problem includes the maximum independent set problem. Indeed, given a graph $G = (V, E)$, for each vertex v we put into correspondence a diseased individual v' and for each edge $e = (u, v)$ we put into correspondence a non-diseased individual e' such that any cluster containing u' and v' should also contain e' (e.g., u', v', and e' are identical except one SNP where they have 3 different values 0,1, and 2). Obviously, the maximum independent set of G corresponds to the maximum non-diseased-free cluster and vice versa. Thus one cannot reasonably approximate MNFCP in polynomial time for an arbitrary sample S.

On the other hand, the sample S is not "arbitrary"—it comes from a certain disease association study. Therefore, we may have hope that simple heuristics (particularly greedy algorithms) can perform much better than in the worst arbitrary case. For example in graphs, instead of the maximum independent set we can search for its complement, the minimum vertex cover—repeat picking and removing vertices of maximum degree until no edges left. In our case we minimize the relative cost of covering (or removal) of non-diseased individuals, which is the number of removed diseased individuals. The corresponding heuristic for MNFCP is the following

Complementary Greedy Search
$\quad C \leftarrow S$
$\quad$ Repeat until $h(C) > 0$
$\quad\quad$ For each 1-SNP combination $X = (s, i)$, where s is a SNP and $i \in \{0, 1, 2\}$
$\quad\quad\quad$ find $\bar{d} = d(C) - d(C \cap X)$ and $\bar{h} = h(C) - h(C \cap X)$
$\quad\quad$ Find 1-SNP combination X minimizing $\bar{d}/\bar{h}$
$\quad\quad$ $C \leftarrow C \cap X$

Similarly to the maximum non-diseased-free cluster corresponding to the most expressed risk factor, we can also search for the maximum diseased-free cluster corresponding to the most expressed resistance factor.

Our experiments with three real data sets (see Section 3) show that the complimentary greedy search can find non-trivially large non-diseased-free and diseased-free clusters.

3   Disease Susceptibility Prediction

In this section we show how to apply association search methods to disease susceptibility prediction. We first formulate the problem and discuss cross-validation schemes. We then give a relevant formulation of the optimum clustering problem and propose a general method how any clustering algorithm can be transformed into a prediction algorithm. We conclude with description of two proposed association search-based prediction algorithms.

We start with the formal description of the problem.

Disease Susceptibility Prediction Problem. Given a sample population S (a training set) and one more individual $t \notin S$ with the known SNPs but unknown disease status (testing individual), find (predict) the unknown disease status.

From our point of view, the main drawback of such problem formulation that it cannot be considered as a standard optimization formulation. One cannot directly measure the quality of a prediction algorithm from the given input since it does not contain the predicted status.

A standard way to measure the quality of prediction algorithms is to apply a cross-validation scheme. In the leave-one-out cross-validation, the disease status of each genotype in the population sample is predicted while the rest of the data is regarded as the training set. There are many types of leave-many-out cross-validations where the testing set contains much larger subset of the original sample. Any cross-validation scheme produces a confusion table—see Table 1. The main objective is to maximize prediction accuracy while all other parameters also reflect the quality of the algorithm.

Table 1. Confusion table

	True disease status		
	Diseased	Non-diseased	
predicted diseased	True Positive TP	False Positive FP	Positive Prediction Value PPV= TP/(TP+FP)
predicted non-diseased	False Negative FN	True Negative TN	Negative Prediction Value NPV= TN/(FN+TN)
	Sensitivity TP/(TP+FN)	Specificity TN/(FP+ TN)	Accuracy (TP+TN)/(TP+FP+FN+TN)

In this paper we propose to avoid cross-validation and instead suggest a different objective by restricting the ways how prediction can be made. It is reasonable to require that every prediction algorithm should be able to predict the status inside the sample. Therefore, such algorithms is supposed to be able to partition the sample into subsets based only on the values of SNPs, i.e., partition of S into clusters defined by MSCs. Of course, a trivial clustering where each individual forms its own cluster can always perfectly distinguish between diseased and non-diseased individuals. On the other hand such clustering carries minimum information. Ideally, there should be two clusters perfectly distinguishing diseased from non-diseased individuals. There is a trade-off between number of clusters and the information carried by clustering which results in trade-off between number of errors (i.e., incorrectly clustered individuals) and informativeness which we propose to measure by information entropy instead of number of clusters.

Optimum Disease Clustering Problem. Given a population sample S, find a partition $\mathcal{P}$ of S into clusters $S = S_1 \cup \ldots \cup S_k$, with disease status 0 or 1 assigned to each cluster S_i, minimizing

$$entropy(\mathcal{P}) = -\sum_{i=1}^{k} \frac{|S_i|}{|S|} \ln \frac{|S_i|}{|S|}$$

for a given bound on the number of individuals who are assigned incorrect status in clusters of the partition $\mathcal{P}$, $error(\mathcal{P}) < \alpha \cdot |\mathcal{P}|$.

The above optimization formulation is obviously NP-hard but has a huge advantage over the prediction formulation that it does not rely on cross-validation and can be studied with combinatorial optimization techniques. Still, in order to make the resulted clustering algorithm useful, one needs to find a way ho to apply it to the original prediction problem.

Here we propose the following general approach. Assuming that the clustering algorithm indeed distinguishes *real* causes of the disease, one may expect that the major reason for erroneous status assignment is in biases and lack of sampling. Then a plausible assumption is that a larger sample would lead to a lesser proportion of clustering errors. This implies the following transformation of clustering algorithm into prediction algorithm:

Clustering-based Model-Fitting Prediction Algorithm
Set disease status 0 for the testing individual t and
    Find the optimum (or approximate) clustering $\mathcal{P}_0$ of $S \cup \{t\}$
Set disease status 1 for the testing individual t and
    Find the optimum (or approximate) clustering $\mathcal{P}_1$ of $S \cup \{t\}$
Find which of two clusterings $\mathcal{P}_0$ or $\mathcal{P}_1$ better fits model, and
    accordingly predict status of t,

$$status(t) = arg \min_{i=0,1} error(\mathcal{P}_i)$$

We propose two clustering algorithms based the combinatorial and complementary greedy association searches. Our clustering finds for each individual an MSC or its cluster that contains it and is the most associated according to a certain characteristic (e.g., RR, PPV or lowest p-value)) with disease-susceptibility and disease-resistance. Then each individual is attributed the ratio between these two characteristic values—maximum disease-susceptibility and disease-resistance. Although the resulted partition of the training set S is easy to find, it is still necessary to decide which threshold between diseased and non-diseased clusters should be used. We choose the threshold minimizing the clustering error.

Our *combinatorial search-based prediction algorithm* (CSP) exploits combinatorial search to find the most-associated cluster for each individual. Empirically, the best association characteristic is found to be the relative risk rate RR. Our *complimentary greedy search-based prediction algorithm* (CGSP) exploits complimentary greedy search to find the most-associated cluster for each individual. Empirically, the best association characteristic is found to be the positive predictive value PPV. The leave-one-out cross-validation (see Section 4) show significant advantage of CSP and GCSP over previously known prediction algorithms for all considered real datasets.

4   Results and Discussion

In this section we discuss the results of methods for searching disease-associated multi-SNP combinations and susceptibility prediction on real datasets. We first describe three real datasets, then overview search and prediction methods and conclude with description and discussion of their performance.

Data Sets. The data set Daly *et al* [3] is derived from the 616 kilobase region of human Chromosome 5q31 that may contain a genetic variant responsible for Crohn's disease by genotyping 103 SNPs for 129 trios. All offspring belong to the case population, while almost all parents belong to the control population. In entire data, there are 144 case and 243 control individuals.

The data set of Ueda *et al* [10] are sequenced from 330kb of human DNA containing gene CD28, CTLA4 and ICONS which are proved related to autoimmune disorder. A total of 108 SNPs were genotyped in 384 cases of autoimmune disorder and 652 controls.

The tick-borne encephalitis virus-induced dataset of Barkash *et al* [4] consists of 41 SNPs genotyped from DNA of 21 patients with severe tick-borne encephalitis virus-induced disease and 54 patients with mild disease.

The three datasets have been phased using 2SNP software [5]. The missing data (16% in [3] and 10% in [10]) have been imputed in genotypes from the resulted haplotypes. We have also created corresponding haplotype datasets in which each individual is represented by a haplotype with the disease status inherited from the corresponding individual genotype.

Association Search Methods. We have compared the following 4 methods for search disease-associated multi-SNP combinations.

- Indexed Exhaustive Search (**IES(30)**): exhaustive search on the indexed datasets obtained by extracting 30 indexed SNPs with MLR based tagging method [11];
- Indexed Combinatorial Search (**ICS(30)**): combinatorial search on the indexed datasets obtained by extracting 30 indexed SNPs with MLR based tagging method [11].
- Complementary Greedy Search (**CGS**) (see Section 2)

The quality of searching methods is compared by the sizes of diseased-free and non-diseased-free clusters as well as their statistical significance Table 2. All

Table 2. Comparison of three methods for searching the disease-associated and disease-resistant multi-SNPs combinations with the largest PPV. The starred values refer to results of the runtime-constrained exhaustive search.

Dataset of	Search method	max PPV risk factor				max PPV resistance factor			
		case freq.	control freq.	unadjusted p-value	run-time sec.	case freq.	control freq.	unadjusted p-value	run-time sec.
Crohn's disease [3]	IES(30)	0.09*	0.00	8.7×10^{-7}	21530	0.00	0.07*	3.7×10^{-4}	869
	ICS(30)	0.11	0.00	3.1×10^{-9}	7360	0.00	0.09	5.7×10^{-5}	708
	CGS	0.06	0.00	1.4×10^{-4}	0.1	0.00	0.10	2.2×10^{-5}	0.1
autoimmune disorder [10]	IES(30)	0.04*	0.00	2.5×10^{-8}	7633	0.00	0.04*	4.0×10^{-6}	39
	ICS(30)	0.04	0.00	2.5×10^{-8}	5422	0.00	0.04	4.0×10^{-6}	36
	CGS	0.02	0.00	3.4×10^{-4}	0.1	0.00	0.04	2.5×10^{-5}	0.1
tick-borne encephalitis [4]	ES	0.29*	0.00	4.8×10^{-4}	820	0.00	0.39	1.0×10^{-3}	567
	CS	0.33	0.00	1.3×10^{-4}	780	0.00	0.39	1.0×10^{-3}	1
	CGS	0.19	0.00	6.1×10^{-3}	0.1	0.00	0.32	3.8×10^{-3}	0.1

experiments were ran on Processor Pentium 4 3.2Ghz, RAM 2Gb, OS Linux—the runtime is given in the last column of Table 2.

Prediction Methods. We compare the proposed prediction algorithms based on combinatorial and complimentary greedy searches (see Section 3) with the following three prediction methods. We have chosen these three methods out of 6 compared in [16] and 2 other methods from [15] since they have best prediction results for two real data sets [3] and [10].

Support Vector Machine (SVM). Support Vector Machine (SVM) is a generation learning system based on recent advances in statistical learning theory. SVMs deliver state-of-the-art performance in real-world applications and have been used in case/control studies [18,22]. We use SVM-light [13] with the radial basis function with $\gamma = 0.5$.

Random Forest (RF). A random forest is a collection of CART-like trees following specific rules for tree growing, tree combination, self-testing, and post-processing. We use Leo Breiman and Adele Cutler's original implementation of RF version 5.1 [14]. RF tries to perform regression to generate the suitable model and using bootstrapping produces random trees.

LP-based Prediction Algorithm (LP). This method is based on a graph $X = \{H, G\}$, where the vertices H correspond to distinct haplotypes and the edges G correspond to genotypes connecting its two haplotypes. The density of X is increased by dropping SNPs which do not collapse edges with opposite status. Solving a linear program it assigns weights to haplotypes such that for any non-diseased genotype the sum of weights of its haploptypes is less than 0.5 and greater than 0.5 otherwise. We maximize the sum of absolute values of weights over all genotypes. The status of testing genotype is predicted as sum of its endpoints [15].

Table 3. Leave-one-out cross validation results of four prediction methods for three real data sets. Results of combinatorial search-based prediction (CSP) and complimentary greedy search-based prediction (CGSP) are given when 20, 30, or all SNPs are chosen as informative SNPs.

Dataset	Quality measure	SVM	LP	RF	CGSP			CSP		
					20	30	all	20	30	all
Crohn's disease [3]	sensitivity	20.8	37.5	34.0	28.5	77.1	61.1	68.9	80.0	-
	specificity	88.8	88.5	85.2	90.9	74.1	98.0	79.2	89.7	-
	accuracy	**63.6**	**69.5**	**66.1**	**68.2**	**75.5**	**84.3**	**75.2**	**84.1**	-
	runtime (h)	3.0	4.0	0.08	0.01	0.17	9.0	611	1189	∞
autoimmune disorder [10]	sensitivity	14.3	7.1	18.0	29.4	32.3	51.3	65.9	79.0	-
	specificity	88.2	91.2	92.8	90.7	89.0	94.7	80.0	89.1	-
	accuracy	**60.9**	**61.3**	**65.1**	**68.0**	**68.2**	**82.5**	**74.3**	**83.2**	-
	runtime (h)	7.0	10.0	0.20	0.01	0.32	25.6	9175	17400	∞
tick-borne encephalitis [4]	sensitivity	11.4	16.8	12.7	61.9	52.4	66.7	87.5	80.2	76.2
	specificity	93.2	92.0	95.0	96.2	98.1	94.4	91.2	92.4	94.4
	accuracy	**72.2**	**75.5**	**74.2**	**81.3**	**82.7**	**84.0**	**88.1**	**88.5**	**89.3**
	runtime (h)	0.2	0.08	0.01	0.01	0.01	0.02	1.8	6.3	8.5

Table 3 reports comparison of all considered prediction methods. Their quality is measured by sensitivity, specificity, accuracy and runtime. Since prediction accuracy is the most important quality measure, it is given in bold. Figure 1 shows the receiver operating characteristics (ROC) representing the trade-off between specificity and sensitivity. ROC is computed for all five prediction methods applied to the tick-borne encephalitis data [4].

Discussion. The comparison of three association searches (see Table 2) shows that combinatorial search significantly outperforms the exhaustive search. It always finds the same or larger cluster than exhaustive search and is significantly faster. The search method runtime is a critical in deciding whether it can be used in in clustering and susceptibility prediction. Note that the both exhaustive and combinatorial searches are prohibitively slow on the first two datasets and, therefore, we reduce these datasets to 30 index SNPs while complementary greedy search is fast enough to handle the complete datasets. This resulted in improvement of the complementary greedy over combinatorial search for the first dataset when search for the largest diseased-free cluster - after compression to 30 tags the best cluster simply disappears.

The comparison of the proposed association search-based and previously known susceptibility prediction algorithms (see Table 3) shows a considerable advantage of new methods. Indeed, for the first dataset the best proposed method (CGSP) beats the previously best method (LP) in prediction accuracy 84.3% to 69.5%. For the second dataset, the respective numbers are 83.2% (CSP(30)) to 65.1% (RF), and for the third dataset, they are 89.3% (CSP) to 75.5% (LP). It is important that this lead is the result of much higher sensitivity of new methods—the specificity is almost always very high since all prediction methods tend to be biased toward non-diseased status. The ROC curve also illustrates

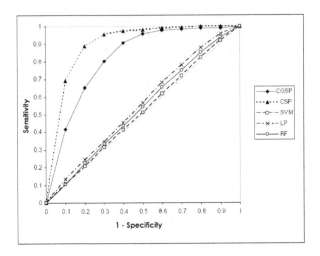

Fig. 1. The receiver operating characteristics (ROC) for the five prediction methods applied to the tick-borne encephalitis data [4]. All SNPs are considered tags for CGSP and CSP.

advantage of CSP and GCSP over previous methods. Indeed the area under ROC curve for CSP is 0.89, for SVM is 0.52 compared with random guessing area of 0.5. Another important issue is how proposed prediction algorithms tolerate data compression. The prediction accuracy (especially sensitivity) is increases for CGSP when more SNPs are made available—e.g., for the second dataset, the sensitivity grows from 29.4% (20 SNPs) to 32.3% (30 SNPs) to 65.9% (all 108 SNPs).

We conclude that the indexing approach, the combinatorial and complementary greedy search methods, and association search-based based susceptibility prediction algorithms are very promising techniques that can possibly help (i) to discover gene interactions causing common diseases and (ii) to create diagnostic tools for genetic epidemiology of common diseases.

References

1. Affymetrix (2005) *http://www.affymetrix.com/products/arrays/*.
2. International HapMap Consortium. (2003) The International HapMap Project. *Nature*, **426**, 789–796, *http://www.hapmap.org*.
3. Daly, M., Rioux, J., Schaffner, S., Hudson, T. and Lander, E. (2001) High resolution haplotype structure in the human genome. *Nature Genetics*, **29**, 229–232.
4. Barkhash, A., Perelygin, A., Brinza, D., Pilipenko, P., Bogdanova, YU., Romaschenko, A., Voevoda, M. and Brinton, M. (2006) Genetic Resistance to Flaviviruses, 5th Conf. on Bioinformatics of Genome Regulation and Structure (BGRS'06), to appear.
5. Brinza, D. and Zelikovsky, A. (2006) 2SNP: Scalable Phasing Based on 2-SNP Haplotypes, *Bioinformatics*, **22(3)**, 371–373.

6. Brinza, D., He, J. and Zelikovsky, A. (2006) Combinatorial Search Methods for Multi-SNP Disease Association, Proc. IEEE Conf. on Engineering in Medicine and Biology (EMBC'06), September 2006, to appear.

7. Clark AG. (2003) Finding Genes Underlying Risk of Complex Disease by Linkage Disequilibrium Mapping, *Curr. Opin. Genet. Dev.*, **13(3)**, 296–302.

8. A.G.Clark et al (2005). Determinants of the success of whole-genome association testing, *Genome Res.*, **15**, 1463–1467.

9. Stephens, M., Smith, N.J., and Donnelly, P. (2001) A New Statistical Method for Haplotype Reconstruction from Population Data, *The American J. of Human Genetics*, **68**, 978-998.

10. Ueda, H., Howson, J.M.M., Esposito, L. et al. (2003) Association of the T Cell Regulatory Gene CTLA4 with Susceptibility to Autoimmune Disease, *Nature*, **423**, 506–511.

11. He, J. and Zelikovsky, A. (2006) Tag SNP Selection Based on Multivariate Linear Regression, *Proc. Int'l Conf. on Computational Science (ICCS'06)*, LNCS **3992**, 750–757.

12. Marchini, J., Donnelley, P. and Cardon, L.R, (2005) Genome-wide strategies for detecting multiple loci that influence complex diseases, *Nature Genetics* **37**, 413–417.

13. Joachims, T. http://svmlight.joachims.org/

14. Breiman, L. and Cutler, A. http://www.stat.berkeley.edu/users/breiman/RF

15. Mao, W., He, J., Brinza, D. and Zelikovsky, A. (2005) A Combinatorial Method for Predicting Genetic Susceptibility to Complex Diseases, *Proc. IEEE Conf. on Engineering In Medicine and Biology (EMBC'05)*, pp. 224–227.

16. Mao, W., Brinza, D., Hundewale, N., Gremalschi, S. and Zelikovsky, A. (2006) Genotype Susceptibility and Integrated Risk Factors for Complex Diseases, *Proc. IEEE Conf. on Granular Computing (GRC 2006)*, pp. 754–757.

17. Kimmel, G. and Shamir R. (2005) A Block-Free Hidden Markov Model for Genotypes and Its Application to Disease Association, *J. of Computational Biology* **12(10)**: 1243–1260.

18. Listgarten, J., Damaraju, S., Poulin B., Cook, L., Dufour, J., Driga, A., Mackey, J., Wishart, D., Greiner,R., and Zanke, B. (2004) Predictive Models for Breast Cancer Susceptibility from Multiple Single Nucleotide Polymorphisms, *Clinical Cancer Research* **10**: 2725–2737.

19. Nelson, M.R., Kardia, S.L., Ferrell, R.E., and Sing, C.F. (2001) A combinatorial partitioning method to identify multilocus genotypic partitions that predict quantitative trait variation, *Genome Res.*, **11**: 458–470.

20. Tahri-Daizadeh, N., Tregouet, D. A., Nicaud, V., Manuel, N., Cambien, F., Tiret L. (2003) Automated detection of informative combined effects in genetic association studies of complex traits, *Genome Res.*, **13**: 1952–1960.

21. Tomita, Y., Yokota, M. and Honda, H. (2005) Classification method for prediction of multifactorial disease development using interaction between genetic and environmental factors, *IEEE Comput. Systems Bioinformatics Conf. CSB'05*, poster.

22. Waddell, M., Page,D., Zhan, F., Barlogie, B., and Shaughnessy, J. (2005) Predicting Cancer Susceptibility from SingleNucleotide Polymorphism Data: A Case Study in Multiple Myeloma, *Proc. BIOKDD'05.*

Integer Linear Programs for Discovering Approximate Gene Clusters

Sven Rahmann[1,2] and Gunnar W. Klau[3,4]

[1] Algorithms and Statistics for Systems Biology group, Genome Informatics,
Technische Fakultät, Bielefeld University, D-33594 Bielefeld, Germany
Sven.Rahmann@cebitec.uni-bielefeld.de
[2] International NRW Graduate School in Bioinformatics and Genome Research
[3] Mathematics in Life Sciences group, Dept. of Mathematics and Computer Science,
Free University Berlin, D-14195 Berlin, Germany
gunnar@math.fu-berlin.de
[4] DFG Research Center MATHEON "Mathematics for key technologies", Berlin

Abstract. We contribute to the discussion about the concept of *approximate conserved gene clusters* by presenting a class of definitions that (1) can be written as integer linear programs (ILPs) and (2) allow several variations that include existing definitions such as common intervals, r-windows, and max-gap clusters or gene teams. While the ILP formulation does not directly lead to optimal algorithms, it provides unprecedented generality and is competitive in practice for those cases where efficient algorithms are known. It allows for the first time a non-heuristic study of large approximate clusters in several genomes. Source code and datasets are available at http://gi.cebitec.uni-bielefeld.de/assb.

1 Introduction

Advances in genome sequencing projects allow to increasingly use methods from comparative genomics to infer gene functions and roles. One approach is based on the following idea: During evolution, genomes rearrange, i.e., whole blocks are cut out, possibly reversed, and moved to different spots in the genome. The rearranged genome will only be fixed in the population if it is viable and its fitness is not significantly lower than that of the presently dominant genome. This is the case only if those genes remain physically close in the genome that need to be expressed together because they act in the same pathway or share an important function. For this reason, conserved gene proximity over long evolutionary times is an indication for a functional relation between the corresponding genes [1]. The above ideas give rise to the problem of finding *approximate conserved gene clusters*, which we loosely define as sets of genes that occur in close proximity in each genome under consideration.

We assume that genes are represented by integers in such a way that paralogous and orthologous genes receive the same number. Homology detection is a delicate procedure, so we must assume that our representation contains errors. As a consequence, we need an error-tolerant formalization of the cluster concept.

P. Bücher and B.M.E. Moret (Eds.): WABI 2006, LNBI 4175, pp. 298–309, 2006.
© Springer-Verlag Berlin Heidelberg 2006

Different ways to formally define gene clusters have been discussed recently; a survey of alternatives has appeared in [2]. Difficulties include:

1. The problem has been attacked from two sides. One philosophy is to specify an algorithm and constructively define the algorithm's results as conserved gene clusters. The drawbacks of this approach are that it is unclear how such an algorithm maps to the biological reality, and that statistical analysis of the results becomes difficult. The other philosophy is to provide a formal specification of what constitutes a gene cluster (modeling step) and then design an algorithm that finds all clusters that satisfy the specification (problem solving step).
2. It is not easy to formally specify what we are looking for. Should we choose a narrow definition at the risk of missing biologically interesting gene sets, or a wide definition and browse through many biologically uninteresting sets?

We think that it is preferable to use separate modeling and solving steps. This allows us to first focus on tuning the model for biological relevance and only then worry about efficient algorithms. Therefore we propose a framework for modeling the *approximate gene cluster discovery problem* (AGCDP), as defined in Problem 1 below, as well as many variants, as an integer linear program (ILP; see [3] for a general introduction).

Contribution and related work. The innovative feature of our approach is its versatility. We are aware that for certain special cases of our ILP formulation, special-purpose algorithms already exist. Using them would solve the corresponding problem more efficiently than using a general ILP solver. However, our approach has the advantage that the objective function and constraints can be easily modified without designing and implementing a new algorithm. The ILP formulation thus allows to test quickly whether a model makes sense from a biological point of view. Incidentally, it also performs well in practice on the known easy formulations. Existing definitions that can be modeled in our framework include (exact) common intervals in permutations [4], (exact) common intervals in arbitrary sequences [5], gene teams or max-gap-clusters [6,7], and r-windows [8], for example.

The paper is structured as follows. Section 2 provides our basic definition of gene clusters and related quantities. Section 3 shows how to formulate the resulting discovery problem as an ILP. Several variations and extensions of the basic model are presented together with the necessary ILP modifications in Section 4, demonstrating the flexibility of our approach. We present computational results in Section 5 and a concluding discussion in Section 6.

2   Basic Problem Specification

Genes and gene sets. Genes are represented by positive integers. If the same integer occurs more than once in the same genome, the genes are paralogs of each other. If the same integer occurs in different genomes, the genes may be orthologs or paralogs. There is also a *special gene* denoted by **0** that represents a different gene at every occurrence and whose purpose it is to model any gene for which no homolog exists in the dataset. The *gene universe* or *gene pool* is denoted by $\mathcal{U} := \{0, 1, \ldots, N\}$ for some integer $N \geq 1$. We are looking for a subset of the gene pool, without the special gene, i.e., $X \subset \mathcal{U}$ with $\mathbf{0} \notin X$, called the *reference gene set*, whose genes occur in close proximity in each genome.

Genomes. A genome is modeled as a sequence of genes; we do not consider intergenic distances. We emphasize that a genome need not be a permutation of the gene pool; each gene (family) can occur zero times, once, or more than once in each genome. Restricting genomes to permutations sometimes allows remarkably efficient algorithms (e.g., [6,4]), but also restricts the modeling power too much for our purposes. To specify the basic problem, we assume that genomes consist of a single linear chromosome. The cases of several chromosomes and of a circular chromosome are discussed in Section 4. We consider m genomes; the length of the i-th genome is n_i: $g^i = (g_1^i, \ldots, g_{n_i}^i)$, $i = 1, \ldots, m$. In the basic model, we look for an approximate occurrence of X in every genome; in Section 4, we describe how to relax this objective.

Genomic intervals and their gene contents. A *linear interval* in a genome $g = (g_1, \ldots, g_n)$ is an index set J, which is either the empty interval $J = \emptyset$, or $J = \{j, j+1, \ldots, k\}$, written as $J = [j : k]$, with $1 \leq j \leq k \leq n$. The *gene content* of $J = [j : k]$ in g is the set $G_J := \{g_j, \ldots, g_k\}$. Note that G_J is a set, and neither a sequence nor a multiset. The *length* of $J = [j : k]$ is $|J| = k - j + 1$. The gene content of $J = \emptyset$ is $G_\emptyset = \emptyset$, and its length is $|J| = |\emptyset| = 0$.

Objective. The goal is to find a gene set $X \subset \mathcal{U}$ without the special gene ($\mathbf{0} \notin X$), and a linear interval J_i for each genome $i \in \{1, \ldots, m\}$, such that, informally $X \approx G_{J_i}^i$ for all i, where $G_{J_i}^i$ denotes the gene content of J_i in the i-th genome.

The agreement of X and the gene content $G_{J_i}^i$ is measured by the number $|G_{J_i}^i \setminus X|$ of genes additionally found in the interval although they are not part of X ("additional genes"), and by the number $|X \setminus G_{J_i}^i|$ of X-genes not found in the interval ("missing genes").

Since gene clusters of different sizes behave differently, it makes sense to parameterize the problem by specifying the size of the reference gene set $|X|$ by enforcing $|X| = D$ or $|X| \geq D$, or a range $D^- \leq |X| \leq D^+$.

Finding an optimal gene cluster. There are several ways to cast the above criteria into an optimization problem: We can let them contribute to the objective function or select thresholds and use them as hard constraints, or both. We start with a formulation with as few hard constraints as possible. A first goal is to find an optimal gene cluster (in terms of the cost function defined below).

Problem 1 (Basic approximate gene cluster discovery problem (AGCDP)). Given
  - the gene pool $\mathcal{U} = \{0, 1, \ldots, N\}$,
  - m genomes $(g^i)_{i=1\ldots m}$, where $g^i = (g_1^i, \ldots, g_{n_i}^i)$,
  - a size range $[D^-, D^+]$ for the reference gene set (possibly $D^- = D^+ =: D$),
  - integer weights $w^- \geq 0$ and $w^+ \geq 0$ that specify the respective cost for each missed and additional gene in an interval,

find $X \subset \mathcal{U}$ with $\mathbf{0} \notin X$ and $D^- \leq |X| \leq D^+$, and a linear interval J_i for each genome in order to minimize

$$c := c(X, (J_i)) = \sum_{i=1}^{m} \left[w^- \cdot |X \setminus G^i_{J_i}| + w^+ \cdot |G^i_{J_i} \setminus X| \right].$$

In Section 3 we show how to write this problem as an ILP; the complexity is discussed in Section 6. In practice, distinct clusters X with the same optimal cost c^* or cost close to c^* may exist, and it is not sufficient to find a single arbitrary optimal one.

Finding all interesting gene clusters. Once we know the optimal cost c^*, we introduce a constraint $c(X, (J_i)) \leq (1+\gamma) \cdot c^*$ with a tolerance parameter $\gamma > 0$ and then enumerate the feasible points (X, J, c) with this additional constraint. The set of feasible points may be redundant in the sense that several solutions lead to similar X, or to different intervals J_i with the same gene content, etc. Therefore we are mainly interested in sufficiently distinct X. After finding one reference gene set X^*, we can force a distinct solution by adding a new constraint $|X \Delta X^*| \geq T$ for a positive threshold T. Here Δ denotes symmetric set difference.

As noted above, the problem is formulated with specific bounds for the reference set size: $|X| \in [D^-, D^+]$ or $|X| = D$. This is useful if we already have an idea of the gene cluster size that we want to discover. Otherwise, we can solve the problem for several values of D. For technical reasons, further discussed below, it is not recommended to choose a large range $[D^-, D^+]$.

3   Integer Linear Programming Formulation

To cast the AGCDP into an ILP framework, we need to represent the reference gene set X, the intervals J_i, and the gene contents $G^i_{J_i}$, as well as several auxiliary variables. Table 1 gives an overview.

We model X as a binary vector $x = (x_0, \ldots, x_N) \in \{0, 1\}^{N+1}$, where we set $x_q = 1$ if and only if $q \in X$. We demand $x_0 = 0$ and $D^- \leq \sum_q x_q \leq D^+$.

To model the selected interval interval J_i in genome i, we use binary indicator vectors $z^i = (z^i_j)_{j=1,\ldots,n_i}$. A linear interval in genome i is characterized by the fact that the ones in z^i occur consecutively. We enforce this property by introducing auxiliary binary vectors $^+z^i = (^+z^i_1, \ldots, ^+z^i_{n_i})$ and $^-z^i = (^-z^i_1, \ldots, ^-z^i_{n_i})$ that model increments and decrements, respectively, in z^i.

We thus set $z^i_1 = {}^+z^i_1 - {}^-z^i_1$, and for $2 \leq j \leq n_i$: $z^i_j = z^i_{j-1} + {}^+z^i_j - {}^-z^i_j$. We forbid a simultaneous increment and decrement at each position: $^+z^i_j + {}^-z^i_j \leq 1$ for all $j = 1, \ldots, n_i$; and we allow at most one increment and decrement: $\sum_{j=1}^{n_i} {}^+z^i_j \leq 1$ and $\sum_{j=1}^{n_i} {}^-z^i_j \leq 1$. Recall that all three vectors z^i, $^+z^i$, and $^-z^i$ are elements of $\{0, 1\}^{n_i}$. It is easy to see that each linear interval can be written in a unique way with this parameterization: For the empty interval, use zero vectors for ^+z and ^-z. For the interval $[j : k]$ with $1 \leq j \leq k < n_i$, set $^+z^i_j = 1$ and $^-z^i_{k+1} = 1$. If $k = n_i$, then ^-z is the zero vector.

Table 1. Overview of variables and expressions representing objects and quantities in the basic ILP formulation. All variables are binary.

Main objects	ILP variables (binary)
reference gene set X	$x = (x_q)_{q=0,\ldots,N}$
interval J_i in i-th genome	$z^i = (z^i_j)_{j=1,\ldots,n_i}, \ i = 1,\ldots,m$
gene content $G^i_{J_i}$ of J_i in g^i	$\chi^i = (\chi^i_q)_{q=0,\ldots,N}, \ i = 1,\ldots,m$

Auxiliary objects	ILP variables (binary)
increments in z^i	$^+z^i = (^+z^i_j)_{j=1,\ldots,n_i}, \ i = 1,\ldots,m$
decrements in z^i	$^-z^i = (^-z^i_j)_{j=1,\ldots,n_i}, \ i = 1,\ldots,m$
intersection $X \cap G^i_{J_i}$	$\iota^i = (\iota^i_q)_{q=0,\ldots,N}, \ i = 1,\ldots,m$

Target quantities	ILP expression
#\{missing genes in g^i\}: $\lvert X \setminus G^i_{J_i} \rvert$	$\sum_{q=0}^{N} x_q - \iota^i_q$
#\{additional genes in g^i\}: $\lvert G^i_{J_i} \setminus X \rvert$	$\sum_{q=0}^{N} \chi^i_q - \iota^i_q$

The gene content $G^i_{J_i}$ in genome i is modeled by another indicator vector $\chi^i = (\chi^i_q)_{q=0,\ldots,N}$: If some position j is covered by the chosen interval J_i, the corresponding gene must be included in the gene content; thus $\chi^i_{g^i_j} \geq z^i_j$ for all $j = 1, \ldots, n_i$ (recall that g^i_j is constant). On the other hand, if some gene $q \in \{1, \ldots, N\}$ is not covered by J_i, it must not be included: $\chi^i_q \leq \sum_{j:g^i_j = q} z^i_j$ for all $q \in \{0, \ldots, N\}$. For each genome i, the above two families of inequalities map the selected intervals exactly to the selected gene contents. Note that if gene q does never appear in genome i, the sum inequality yields $\chi^i_q = 0$, as desired.

To model the target function, we need the intersection between X and the selected gene content $G^i_{J_i}$ in the i-th genome. We define another family of indicator vectors for $i = 1, \ldots, m$: $\iota^i = (\iota^i_q)_{q=0,\ldots,N}$ that we force to model the set intersection $X \cap G^i_{J_i}$ via the inequalities $\iota^i_q \leq x_q$, $\iota^i_q \leq \chi^i_q$, and $\iota^i_q \geq x_q + \chi^i_q - 1$. Then the terms of the target function are

$$|X \setminus G^i_{J_i}| = \sum_{q=0}^{N} (x_q - \iota^i_q); \qquad |G^i_{J_i} \setminus X| = \sum_{q=0}^{N} (\chi^i_q - \iota^i_q).$$

Table 2 presents the whole basic formulation at a glance. After the above discussion, we may state

Theorem 1. *The ILP in Table 2 correctly represents Problem 1 (Basic AGCDP).*

4   Extensions and Variations

Constraining and varying the objective function. The basic ILP in Table 2 always has a feasible solution; an upper bound of the cost is easily obtained by taking any set of size D^- for X, empty intervals in all genomes, and paying the cost of

Table 2. ILP formulation for the basic AGCDP; see Table 1 for variables

Given integers $N \geq 1$, $m \geq 2$, $(n_i)_{i=1,\ldots,m}$ with $n_i \geq 1$, $(g_j^i)_{i=1,\ldots,m;\,j=1,\ldots,n_i}$ from $\{0, 1, \ldots, N\}$, $1 \leq D^- \leq D^+ \leq N$, $w^- \geq 0$ and $w^+ \geq 0$,

$$\text{minimize} \quad \sum_{i=1}^{m} \left[w^- \cdot \sum_{q=0}^{N} (x_q - \iota_q^i) + w^+ \cdot \sum_{q=0}^{M} (\chi_q^i - \iota_q^i) \right] \quad \text{subject to}$$

$$
\begin{aligned}
x_q &\in \{0,1\} & (q = 0, 1, \ldots, N) \\
x_0 &= 0 & \\
\textstyle\sum_{q=0}^{N} x_q &\geq D^- & \\
\textstyle\sum_{q=0}^{N} x_q &\leq D^+ & \\[4pt]
z_j^i, {}^+z_j^i, {}^-z_j^i &\in \{0,1\} & (i = 1, \ldots, m, \,, j = 1, \ldots, n_i) \\
z_1^i &= {}^+z_1^i - {}^-z_1^i & (i = 1, \ldots, m) \\
z_j^i &= z_{j-1}^i + {}^+z_j^i - {}^-z_j^i & (i = 1, \ldots, m, \, j = 2, \ldots, n_i) \\
{}^+z_j^i + {}^-z_j^i &\leq 1 & (i = 1, \ldots, m, \, j = 1, \ldots, n_i) \\
\textstyle\sum_{j=1}^{n_i} {}^+z_j^i &\leq 1 & (i = 1, \ldots, m) \\
\textstyle\sum_{j=1}^{n_i} {}^-z_j^i &\leq 1 & (i = 1, \ldots, m) \\[4pt]
\chi_q^i &\in \{0,1\} & (i = 1, \ldots, m, \, q = 0, 1, \ldots, N) \\
\chi_{g_j^i}^i &\geq z_j^i & (i = 1, \ldots, m, \, j = 1, \ldots, n_i) \\
\chi_q^i &\leq \textstyle\sum_{j : g_j^i = q} z_j^i & (i = 1, \ldots, m, \, q = 0, 1, \ldots, N) \\[4pt]
\iota_q^i &\in \{0,1\} & (i = 1, \ldots, m, \, q = 0, 1, \ldots, N) \\
\iota_q^i &\leq x_q & (i = 1, \ldots, m, \, q = 0, 1, \ldots, N) \\
\iota_q^i &\leq \chi_q^i & (i = 1, \ldots, m, \, q = 0, 1, \ldots, N) \\
\iota_q^i &\geq x_q + \chi_q^i - 1 & (i = 1, \ldots, m, \, q = 0, 1, \ldots, N)
\end{aligned}
$$

$m \cdot D^- \cdot w^-$ for missing all genes in X. In many applications, it makes no sense to consider intervals in which more than a fraction δ^- of the reference genes X are missing or which contain more than a fraction δ^+ of additional genes. Therefore we could restrict the search space by enforcing $\sum_{q=0}^{N} (x_q - \iota_q^i) \leq \lfloor \delta^- \cdot D^+ \rfloor$ and $\sum_{q=0}^{N} (\chi_q^i - \iota_q^i) \leq \lfloor \delta^+ \cdot D^+ \rfloor$. This may, of course, lead to an empty feasible set.

Instead of paying separately for missed and additional genes, we may argue that we should view occurrences of both errors as substitutions to the maximum possible extent. Assuming equal weights $w^- = w^+ = 1$, this leads to a cost contribution of $\max\{|X \setminus G_{J_i}^i|, |G_{J_i}^i \setminus X|\}$ instead of the sum for the i-th genome; see also [9]. More generally, we may replace the objective function by

$$\text{minimize} \quad \sum_{i=1}^{m} \max \left\{ w^- \cdot \sum_{q=0}^{N} (x_q - \iota_q^i), \ w^+ \cdot \sum_{q=0}^{M} (\chi_q^i - \iota_q^i) \right\}$$

by introducing new variables $c_i^- := w^- \cdot \sum_{q=0}^{N} (x_q - \iota_q^i)$ and $c_i^+ := w^+ \cdot \sum_{q=0}^{N} (\chi_q^i - \iota_q^i)$. We let $c_i = \max\{c_i^-, c_i^+\}$ by introducing inequalities $c_i \geq c_i^-$ and $c_i \geq c_i^+$

for $i = 1, \ldots, m$ and writing the objective function as $\min \sum_{i=1}^{m} c_i$, which fixes c_i at the maximum of c_i^- and c_i^+, and not at a larger value.

A single circular chromosome or multiple linear chromosomes. Bacterial genomes usually consist of a single circular chromosome, i.e., any circular permutation of $g = (g_1, \ldots, g_n)$ in fact represents the same genome, and the start and end points are arbitrary. Therefore we need to allow intervals that "wrap around".

Extending the definition of a linear interval from Section 2, we say that an *interval* is either a linear interval or a wrapping interval.

A *wrapping interval* in $g = (g_1, \ldots, g_n)$ is a nonempty index set $J := [j \,|\, k] := \{j, j+1, \ldots, n, 1, \ldots, k\}$, with $1 \leq j, k \leq n$ and $j > k+1$.

The *gene content* of a wrapping interval is $G_J \equiv G_{[j|k]} := \{g_j, \ldots, g_n, g_1, \ldots, g_k\}$, and its *length* is $|J| = n - j + 1 + k$. We specifically disallow $j = k+1$ because this would induce the whole genome, for which we already have the linear interval $[1 : n]$.

As an example, in a genome of length 3, there are seven linear intervals ($\emptyset$, $[1 : 1]$, $[2 : 2]$, $[3 : 3]$, $[1 : 2]$, $[2 : 3]$, $[1 : 3]$), and a single wrapping interval: $[3 \,|\, 1]$.

For a wrapping interval in g^i, the ones in the indicator vector z^i occur in two distinct blocks with the first block starting at position $j = 1$ and the second block ending at position n_i. Therefore there are two points j with $^+z_j^i = 1$, but only if $j = 1$ is one of them. To allow arbitrary intervals (empty, linear, or wrapping), all we need to do is to change the sum constraint for $^+z^i$ from Table 2 into $\sum_{j=2}^{n_i} {}^+z_j^i \leq 1$ $(i = 1, \ldots, m)$.

While the main applications of our work are to genome rearrangements in prokaryotes, we nevertheless show how to allow multiple linear chromosomes: We extend the gene universe by another special number -1 and concatenate the chromosomes of the i-th genome into a single vector g^i as before, representing chromosome borders by -1. We constrain the interval selection variables z_j^i wherever $g_j^i = -1$ to be $z_j^i = 0$; this ensures that the interval J_i does not extend over a chromosome border.

Genome selection. So far we have requested that X occurs in every input genome, or incurred a possibly severe cost of at most $w^- \cdot |X|$ if no gene of X appears in the genome. When we look for approximate gene clusters in a large set of genomes and only require that the cluster occurs in some of them, it is desirable to relax this penalty.

We extend the formulation with an index set $I \subset \{1, \ldots, m\}$ and refer to the genomes indexed by I as the *selected genomes*; these are treated as before, i.e., missing and additional genes in the selected intervals are penalized by w^- and w^+, respectively. For non-selected genomes, we force that J_i is the empty interval, but we only incur a flat penalty $\rho \geq 0$ that should be chosen substantially smaller than $w^- \cdot D^-$. We also specify a minimal number $\mu \leq m$ of genomes to be selected, i.e., we demand that $|I| \geq \mu$. The cost function then becomes

$$c := c(X, I, (J_i)) = \sum_{i \in I} \left[w^- \cdot |X \setminus G_{J_i}^i| + w^+ \cdot |G_{J_i}^i \setminus X| \right] + (m - |I|) \cdot \rho.$$

For the ILP, we model I as another binary vector $y = (y_1, \ldots, y_m) \in \{0, 1\}^m$ with $y_i = 1$ if and only if $i \in I$. We have the constraint $\sum_{i=1}^{m} y_i \geq \mu$. To enforce $J_i = \emptyset$ for $i \notin I$, we use the inequalities $z_j^i \leq y_i$ for all $i = 1, \ldots, m, j = 1, \ldots, n_i$. It remains to properly rewrite the target function. The obvious approach to

$$\text{minimize} \quad \sum_{i=1}^{m} \left[y_i \cdot \left(w^- \cdot \sum_{q=0}^{N} (x_q - \iota_q^i) + w^+ \cdot \sum_{q=0}^{M} (\chi_q^i - \iota_q^i) \right) + (1 - y_i) \cdot \rho \right]$$

does not work, because this function is nonlinear in the variables.

However, a simple solution is available when X is constrained to be of fixed size $D^- = D^+ = D$: If $y_i = 0$, then z^i, χ^i and ι^i are the zero vector and under the old cost function, we would pay $D \cdot w^-$. Now we only pay ρ; therefore we can write the objective function as

$$\text{min.} \quad \sum_{i=1}^{m} \left[w^- \cdot \sum_{q=0}^{N} (x_q - \iota_q^i) + w^+ \cdot \sum_{q=0}^{M} (\chi_q^i - \iota_q^i) + (1 - y_i) \cdot (\rho - Dw^-) \right].$$

If $D^- < D^+$, the above approach does not work, unless we change the flat penalty from ρ into $\rho + |X| - D^-$, which may put larger X at a disadvantage. In that case we can use the same formulation as above with D replaced by D^-.

For the general case of $D^- < D^+$ and a true flat penalty ρ we can use a so-called big-M approach: We write the objective function as

$$c = \rho \cdot \sum_{i=1}^{m} (1 - y_i) + \sum_{i=1}^{m} L_i,$$

where the L_i are new auxiliary variables, which we will force to take values

$$L_i = \begin{cases} \sum_{q=0}^{N} \left(w^- \cdot (x_q - \iota_q^i) + w^+ \cdot (\chi_q^i - \iota_q^i) \right) =: \ell_i & \text{if } y_i = 1, \\ 0 & \text{if } y_i = 0. \end{cases}$$

We achieve this via inequalities $L_i \geq 0$ and $L_i \geq \ell_i - M \cdot (1 - y_i)$ for all $i = 1, \ldots, m$ and a constant M larger than any possible value of ℓ_i. If $y_i = 1$, the inequality becomes $L_i \geq \ell_i$, and since the objective function c is to be minimized, this will lead to $L_i = \ell_i$. If $y_i = 0$, it becomes $L_i \geq -M'$ for some $M' \geq 0$ and is dominated by the non-negativity constraint $L_i \geq 0$. Often, such a big-M approach causes technical problems for the ILP solver, however, as it leads to weak LP relaxations [3].

Using a reference genome. Even in the basic AGCDP, there is a lot of freedom because the reference gene set X need not occur exactly in any of the genomes. In some cases, however, a reference genome may be known and available. This makes the problem much easier, and an ILP formulation would not be required, and the solver could be easily replaced by simpler specialized algorithms. It is reassuring, however, that a reference genome can be easily integrated into the formulation: Without loss of generality, let g^1 be the reference genome. We force $x_q = \chi_q^1 = \iota_q^1$ for $q = 0, \ldots, N$ and possibly $y_1 = 1$ if we are using genome selection.

Modeling common intervals, max-gap clusters and r-windows. By specifying appropriate target functions and constraints, the ILP approach can be used to model existing definitions of gene clusters. For those mentioned here, efficient algorithms exist, and we certainly cannot beat them. It is still convenient that we can treat them in the ILP framework, too.

To model exact common intervals as in [5], we restrict the cost function to take the value zero (i.e., we allow no additional and no missing genes), and set $w^- = w^+ = 1$. Additionally, we can apply genome selection with $\rho = 0$ and a reasonably large value for μ. From the result, we only use the reference set X and disregard the intervals.

The specification of max-gap clusters or gene teams [6] is a generalization of common intervals and demands that between adjacent genes from the reference set X, there are at most δ genes not from X. For $\delta = 0$, we obtain again common intervals. For $\delta > 0$, the max-gap condition states that in each sub-interval of J_i of length $\delta + 1$, we need at least one X-gene: For each $i = 1, \dots, m$ and each $j = 1, \dots, n_i - \delta$ we have that if $z_j^i + \dots + z_{j+\delta}^i = \delta + 1$, then $\iota_{g_j}^i + \iota_{g_{j+1}}^i + \dots + \iota_{g_{j+\delta}}^i \geq 1$ must hold. Each implication can be written as an inequality:

$$\iota_{g_{j+1}}^i + \dots + \iota_{g_{j+\delta}}^i \geq z_j^i + \dots + z_{j+\delta}^i - (\delta + 1) + 1 \quad (i = 1, \dots, m; \ j = 1, \dots, n_i - \delta).$$

We use $w^- = 1$ and $w^+ = 0$ and constrain the target function to zero. To find maximal max-gap clusters, i.e., those not contained in a larger one, we enumerate all max-gap clusters of each size D and subsequently filter out those contained in larger ones.

An r-window cluster for two genomes is defined as a pair of intervals of length r that share at least D genes [8]. To find them, we demand $|X| = D$, set $w^- = 1$, $w^+ = 0$, constrain the target function to zero, and demand that $\sum_{j=1}^{n_i} z_j^i = r$ for each $i = 1, \dots, m$.

5   Computational Results

We have implemented a C++ software tool that reads in a set of genomes, solves one of the integer linear programming formulations presented in Sects. 3 and 4 using the CPLEX optimization library [10], and outputs the list of optimal and close to optimal gene clusters. All experiments were performed on an AMD 2.2 GHz opteron 64 bit processor with 8 GB of main memory using CPLEX 9.03.

Hidden clusters in artificial data. We generate artificial problem instances for benchmarking as follows: We randomly generate 6 genomes of roughly 1,000 genes each ($N = 2000$) with 5% of **0**-genes. For each $D \in \{5, 10, 15, 20, 25\}$, we generate a cluster and hide a perturbed permutation of it in five randomly selected genomes, taking care that the different clusters do not overlap.

Using $w^- = 2$, $w^+ = 1$ and the appropriate value of D, we solve the ILP first with genome selection, setting $\rho = 2D/5$: We retrieve all five clusters in times 29 min, 45 min, 8 min, 113 s, and 14 s, respectively.

Without genome selection, running times are much faster, but we can run into problems because of the high penalty for the genome in which the cluster is missing: We retrieve the clusters of sizes 5, 10, 15, and 25 in 17 min, 7 min, 163 s, and 4 s, respectively. For $D = 20$, we obtain a different cluster than the hidden one that obtains a better target function value without genome selection.

While the running times vary with each instance, the times shown here are representative. This experiment indicates the importance, but also the high complexity of genome selection.

Comparison of two organisms. The genomes of *C. glutamicum* and *M. tuberculosis* consist of 3,057 and 3,991 annotated genes, respectively. The annotated gene set is available at `http://gi.cebitec.uni-bielefeld.de/assb`. We compute the optimal objective function value $c^*(D)$ for each cluster size $D \in [5, 500]$ for the basic formulation with $w^- = w^+ = 1$ (total CPU time: almost 25 days, on average 1:15 hrs per instance). Figure 1 shows the running time per instance as well as the optimal normalized costs $c^*(D)/D$. Local minima correspond to large approximate clusters with comparatively few errors. As Fig. 2 illustrates for $D = 51$, our formulation discovers clusters that cannot be detected by any method that does not consider approximate clusters. The largest exact cluster has size 11.

6   Discussion

We have given several formalizations and corresponding ILP formulations for approximate gene clusters. In contrast to other models, we do not only characterize the set of desired clusters, but also assign a value (the objective function) to them. Our approach allows us to check different gene cluster models for biological relevance before designing optimized algorithms, and to discover *optimal* medium-sized to large approximate clusters that contain no smaller exact ones if they exist and if the ILP solver can handle the problem.

The apparent complexity of our approach comes from the fact that we use a reference set X of genes that need not occur exactly in any genome. While we have not attempted to formally prove the corresponding decision problem NP-hard, the difficulties encountered by the ILP solver and the similarity to the median string problem provide some evidence. The problem becomes even harder (empirically in terms of CPU time) if genome selection is allowed. The situation changes if we require that X occurs in at least one genome without errors. Then a naive polynomial-time algorithm works as follows:

Tentatively set X to the gene set of each interval in each genome. For each genome g except the one where X is taken from, compare X to the character set of each interval J in g, compute the cost according to the number of missing and additional genes, and pick the interval J_g^* in g with minimum cost c_g^*. Now genome selection can be easily applied: Simply remove as many costly genomes as possible. The total cost of X (without genome selection) is $c(X) = \sum_g c_g^*$. Either report the overall best set X, or report each X, where $c(X)$ remains below a given threshold. (Of course, this "algorithm" can be drastically optimized for efficiency.)

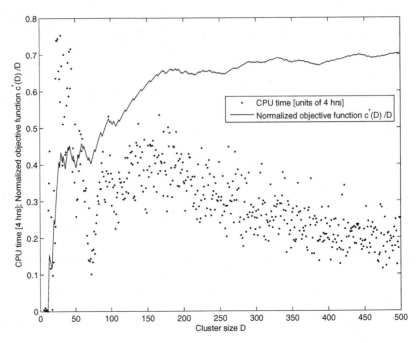

Fig. 1. Comparison of *C. glutamicum* and *M. tuberculosis*: For each cluster size $D \in$ [5, 500], the running time in units of four hours, and the normalized optimal value of the objective function is shown. Note the local minima in the objective function, e.g., at $D = 51$. The apparent correlation between objective function and running time indicates that good approximate clusters are easier to compute than bad clusters.

```
            a
C.glutamicum ( 389 698 33 760 267 267 1156 1 2 55 852 1187 17 321 143 927 372 928 281 0 1739 54
            b                                                                              c
            945 1 979 983 467 524 219 850 914 697 384 1439 648 713 650 268 403 795 124 )

            c
M.tuberculosis ( 124 795 403 268 1 1527 650 0 0 713 648 1439 0 0 384 697 914 850 0 225 9 12 100 4
                                             b             a
            9 1725 84 180 0 9 219 524 467 0 979 9 88 5528 5714 281 928 372 927 143 9 321 9 4
                                             a
            17 3311 852 55 2 1 1156 0 267 760 33 698 389 )
```

Fig. 2. Visualization of an interesting optimal cluster in *C. glutamicum* and *M. tuberculosis* ($D = 51$). Differing genes are marked in grey. Three conserved regions, *a*, *b*, and *c* occur in the cluster.

Important open questions are statistics (significance computations) for gene clusters from our formulations in the spirit of [8], and to formalize a notion of a non-redundant set of gene clusters when enumerating all near-optimal solutions.

Our ILP formulations open a new perspective to the field of approximate gene cluster discovery, and are already usable in practice. We believe that the formulations and the solver can be fine-tuned to solve the same instances even faster, even if the basic AGCDP with or without genome selection is indeed NP-hard.

We are experimenting with alternative methods for genome selection and with an alternative formulation of the consecutive-ones property of the interval indicators z_j^i brought to our attention by Marcus Oswald (Heidelberg).

Another desideratum for the future is to avoid solving the problem for each gene set size D separately. So far this is convenient because it allows simplifications in some formulations, but it seems to slow down the solver drastically (data not shown). Yet a fixed $|X| = D$ is also necessary because optimal objective function values for different $|X|$ do not compare well: Even "good" clusters of size 30 might have higher cost than "bad" clusters of size 5. Normalizing the cost function by $|X|$ seems a promising idea, and we are exploring fractional programming techniques to this end. Overcoming the D-bottleneck would make the ILP approach even more useful in practice and remains an interesting challenge.

Acknowledgments. We thank Thomas Schmidt for providing datasets, Jens Stoye, Yoan Diekmann, Julia Mixtacki, and Markus Oswald for helpful discussions.

References

1. Snel, B., Bork, P., Huynen, M.A.: The identification of functional modules from the genomic association of genes. Proc. Nat'l Acad. Sci. USA **99** (2002) 5890–5895
2. Hoberman, R., Durand, D.: The incompatible desiderata of gene cluster properties. In Proc. 3rd RECOMB Workshop on Comparative Genomics RECOMBCG'05. Volume 3678 of LNCS, Springer Verlaf (2005) 73–87
3. Wolsey, L.A.: Integer programming. Wiley Interscience Series in Discrete Mathematics and Optimization. John Wiley & Sons (1998)
4. Heber, S., Stoye, J.: Algorithms for finding gene clusters. In Proc. 1st Workshop on Algorithms in Bioinformatics WABI'01. Volume 2149 of LNCS, Springer Verlag (2001) 252–263
5. Schmidt, T., Stoye, J.: Quadratic time algorithms for finding common intervals in two and more sequences. In Proc. 15th Symp. on Combinatorial Pattern Matching CPM'04. Volume 3109 of LNCS, Springer Verlag (2004) 347–358
6. Bergeron, A., Corteel, S., Raffinot, M.: The algorithmic of gene teams. In Proc. 2nd Workshop on Algorithms in Bioinformatics WABI'02. Volume 2452 of LNCS, Springer Verlag (2002) 464–476
7. Li, Q., Lee, B.T.K., Zhang, L.: Genome-scale analysis of positional clustering of mouse testis-specific genes. BMC Genomics **6** (2005) 7
8. Durand, D., Sankoff, D.: Tests for gene clustering. J. Comput. Biol. **10** (2003) 453–482
9. Chauve, C., Diekmann, Y., Heber, S., Mixtacki, J., Rahmann, S., Stoye, J.: On common intervals with errors. Technical Report 2006-02, Abteilung Informationstechnik, Technische Fakultät, Universität Bielefeld (2006) ISSN 0946-7831.
10. ILOG, Inc.: CPLEX. http://www.ilog.com/products/cplex (1987–2006)

Approximation Algorithms for Bi-clustering Problems

Lusheng Wang[1], Yu Lin[1,2], and Xiaowen Liu[1]

[1] Department of Computer Science, City University of Hong Kong, Hong Kong
[2] Institute of Computing Technology, Chinese Academy of Sciences, Beijing, China
{lwang, linyu, liuxw}@cs.cityu.edu.hk

Abstract. One of the main goals in the analysis of microarray data is to identify groups of genes and groups of experimental conditions (including environments, individuals and tissues), that exhibit similar expression patterns. This is the so-called bi-clustering problem. In this paper, we consider two variations of the bi-clustering problem: the Consensus Submatrix Problem and the Bottleneck Submatrix Problem. The input of the problems contains a $m \times n$ matrix A and integers l and k. The Consensus Submatrix Problem is to find a $l \times k$ submatrix with $l < m$ and $k < n$ and a consensus vector such that the sum of distance between all rows in the submatrix and the vector is minimized. The Bottleneck Submatrix Problem is to find a $l \times k$ submatrix with $l < m$ and $k < n$, an integer d and a center vector such that the distance between every row in the submatrix and the vector is at most d and d is minimized. We show that both problems are NP-hard and give randomized approximation algorithms for special cases of the two problems. Using standard techniques, we can derandomize the algorithms to get polynomial time approximation schemes for the two problems. To our knowledge, this is the first time that approximation algorithms with guaranteed ratio are presented for microarray analysis.

1 Introduction

In the last several years, microarray technique has been widely used in biological research. Microarray technique has helped to illuminate mechanisms of disease and identify disease subphenotypes, predict disease progression, assign function to previously unannotated genes, group genes into functional pathways, and predict activities of new compounds [1]. Microarray data analysis is an important problem in computational biology [2]. For these large-scale data, classifying genes into different groups under certain conditions is a first step to gain more sophisticated knowledge of different biological pathways or functions. Several clustering or classification techniques, such as k-means [3,4], self-organizing maps [5,6], hierarchical clustering [7,8,9], principal component analysis and singular value decomposition [10,11,12] have been extensively applied to identify groups of similarly expressed genes and conditions from gene expression data.

P. Bücher and B.M.E. Moret (Eds.): WABI 2006, LNBI 4175, pp. 310–320, 2006.

Data errors in microarray are common in the analysis of gene expression data [15,17,18]. The sources of microarray error variability are from various biological and experimental factors, such as biological and individual replication, sample preparation, hybridization and image processing. Moreover, the same gene often shows quite heterogeneous error variability under different biological and experimental conditions [19]. The accurate measurements of absolute expression levels and the reliable detection of low abundance genes are difficult to achieve [16]. For example in mammalian Affymetrix microarrays, an unexpectedly large number of probes (greater than 19% of the probes on each platform) that do not correspond to their appropriate mRNA reference sequences were identified [14]. A lot of work on statistical analysis of gene expression data encourages researchers to consider error and uncertainty in their microarray experiments [13].

It is known that many activation patterns are common to a group of genes only under specific experimental conditions. We should expect subsets of genes to be coregulated and coexpressed only under certain experimental conditions, but to behave almost independently under other conditions, according to our general understanding of cellular processes [21,22,23]. The fact is that we need to discover local patterns in the microarray matrix. The basic model for Bi-clustering is as follows: given an $m \times n$ matrix A, where each element $a_{i,j} \in \{0,1\}$, the problem here is to find a $l \times k$ submatrix with all elements identical to 1 such that $l \times k$ is maximized.

Let $\Sigma = \{\pi_1, \pi_2, \ldots, \pi_{|\Sigma|}\}$ be a fixed size alphabet of symbols. A vector over Σ is a sequence of symbols in Σ. Let A be an $m \times n$ matrix, where each row corresponds to a gene and each column corresponds to a condition. Each element $a_{i,j}$ in A represents the expression level of gene i under condition j. Such a matrix A is defined by its set of m rows, $X = \{x_1, x_2, \ldots, x_m\}$ and its set of n columns, $Y = \{y_1, y_2, \ldots, y_n\}$. Let $P = \{p_1, \ldots, p_l\}$ be a subset of $\{1, 2, \ldots, m\}$ indicating rows in X and $Q = \{q_1, \ldots, q_k\}$ be a subset of $\{1, 2, \ldots, n\}$ indicating columns in Y. The $l \times k$ submatrix $A_{P,Q}$ induced by the pair (P, Q) contains the elements $a_{i,j}$, where $i \in P$ and $j \in Q$. We treat each row in the matrix or submatrix as a vector over Σ. Define $x_i|^Q = a_{i,q_1} a_{i,q_2} \ldots a_{i,q_k}$. Let p and p' be two vectors of the same length over Σ. $d(p, p')$ denotes the number of mismatches between the two vectors. Throughout this paper, we study the following two problems.

The Consensus Submatrix Problem: Given a $m \times n$ matrix A, and integers l and k, find a subset P of l rows, a subset Q of k columns in matrix A and a consensus vector z of length k such that the consensus score $\sum_{i=1}^{l} d(x_{p_i}|^Q, z)$ is minimized.

The Bottleneck Submatrix Problem: Given a $m \times n$ matrix A, and integers l and k, find a subset P of k rows, a subset Q of k columns in matrix A, a center vector z of length k and an integer d such that for every $p_i \in P$ $d(x_{p_i}|^Q, z) \leq d$ and the bottleneck score d is minimized.

In practice, there are errors in microarray data. In the $l \times k$ submatrix, if we assume that the error rate of each row is bounded by a constant, e.g., 10%, then the total consensus score $\sum_{i=1}^{l} d(x_{p_i}|^Q, z)$ is at least $O(lk)$ and the bottleneck score d is at least $O(k)$. Throughout this paper, we assume that for the consensus submatrix problem $\sum_{i=1}^{l} d(x_{p_i}|^Q, z) = O(lk)$ and for the bottleneck submatrix problem $d = O(k)$. Due to technical reasons, in this paper, we consider a special case, where $k = O(n)$.

2   Previous Work

The basic model for biclustering is to find a submatrix $A_{P,Q}$ with all elements identical to a constant value [22]:

$$a_{i,j} = \mu, \quad for \ all \ \ i \in P, j \in Q.$$

If the submatrix is error-free, both the consensus score and the bottleneck score are clearly 0 for the new problems that we proposed in the paper.

In practice, it is interest to find submatrices such that all elements in a row have the same constant value [20,23]. That is,

$$a_{i,j} = c_i, for \ j \in Q.$$

In this case, all the columns in the submatrix are identical. Again, it is clear that both the consensus score and the bottleneck score are 0 if the submatrix is error-free.

A sophisticated approache looks for submatrices in additive model, where

$$a_{i,j} = a_{i',j} + c(i, i'), \quad for \ all \ \ i, i' \in P, \ \ j \in Q \tag{1}$$

[21,22]. That is, for two elements $a_{i,j}$ and $a_{i',j}$ in row i and row i', the difference is a constant $c(i', i)$.

Now we show that our model can also handle the additive model. Let r be a row in the error-free submatrix. We construct a new matrix A' as follows:

$$a'_{i,j} = a_{i,j} - a_{r,j} \quad for \ all \ \ i \in X, \ \ j \in Y.$$

Then, the error-free submatrix is converted into a new submatrix $A'_{P,Q}$ with element

$$\begin{aligned} a'_{i,j} &= a_{i,j} - a_{r,j} \\ &= c(i, r), \quad for \ all \ \ i \in P, \ \ j \in Q. \ \ (\text{From } (1)) \end{aligned}$$

That is, in the resulting submatrix, all elements in a row have the same value. Thus the additive model degenerates to the second case. Therefore, our models can also handle the additive model by trying all rows in A as row $r_{..}$

Cheng and Church proposed the first biclustering model in microarray data analysis [21]. The model introduced a similarity score called the *mean squared residue score H* to measure the coherence of the rows and columns in the sub-matrix.

$$H(P,Q) = \frac{1}{|P||Q|} \sum_{i \in P, j \in Q} (a_{i,j} - a_{i,Q} - a_{P,j} + a_{P,Q})^2$$

where

$$a_{i,Q} = \frac{1}{|Q|} \sum_{j \in Q} a_{i,j}, \quad a_{P,j} = \frac{1}{|P|} \sum_{i \in P} a_{i,j}, \quad and \quad a_{P,Q} = \frac{1}{|P||Q|} \sum_{i \in P, j \in Q} (a_{i,j}).$$

Clearly, the H score is 0 for the first two cases if the submatrix is error-free. We can show that for the additive model, the H score is also 0 if the submatrix is error-free.

In this paper, we design randomized approximation algorithms for both problems. We have an new idea to randomly select $O(\log m)$ columns in the optimal set of columns $Q_{opt} \subseteq Y$ when Q_{opt} is not known. For the bottleneck submatrix problem, we use linear programming and randomized rounding to successfully select a good approximation Q of Q_{opt} and set the letters for the center vector at the columns in Q. Using standard techniques, we derandomize the randomized algorithms to get polynomial time approximation schemes (PTAS) for the two problems. To our knowledge, this is the first time that approximation algorithms with guaranteed ratio are presented for microarray analysis.

The paper is organized as follows. In Section 3, we prove that both problems are NP-hard. In Section 4, we give the algorithm for the consensus submatrix problem, while in Section 5. we give that for the bottleneck submatrix problem.

3   NP-Hardness Result

In this section, we show that both the consensus submatrix problem and the bottleneck submatrix problem are NP-hard. The reduction is from the maximum edge biclique problem. The maximum edge bipartite problem was proved to be NP-hard in [24]. A biclique is a complete bipartite graph where every vertex of the first set is connected to every vertex of the second set.

The Maximum Edge Biclique Problem: Given a biclique graph $G = (V_1 \cup V_2, E)$ and a positive integer X, does G contain a biclique with X edges?

Theorem 1. *The consensus submatrix problem and the bottleneck submatrix problem are NP-hard.*

The proof also suggests that it is NP-hard to decide whether the minimum consensus score is 0 in the consensus submatrix problem and whether the minimum bottleneck score is 0 in the bottleneck submatrix problem. Therefore, there are no approximation algorithms with guaranteed ratio for both problems when the optimal consensus score or the optimal bottleneck score is 0.

4   The Consensus Submatrix Problem

In this section, we present the randomized approximation algorithm for the consensus submatrix problem. Let P_{opt}, Q_{opt} and z_{opt} be the set of rows , the set of columns and the consensus vector of an optimal solution. The optimal consensus score is H_{opt}. By assumption, $H_{opt} = O(kl)$, i.e., there is a constant c' such that $H_{opt} \times c' = kl$. Again, by assumption, $k = O(n)$, i.e., there is a constant c such that $k \times c = n$.

Before we present the algorithm, we first introduce the basic ideas of the algorithm. By enumerating all size k subsets of Y and all length k vectors, we could know Q_{opt} and z_{opt} at some moment. It is easy to see that if we know exactly Q_{opt} and z_{opt}, then we could find the corresponding P_{opt} in polynomial time to minimize the consensus score. However, this straight forward approach costs exponential time. Here we use a random sampling technique to randomly select $O(\log m)$ columns in Q_{opt}, enumerate all possible vectors of length $O(\log m)$ for those columns. At some moment, we know $O(\log m)$ bits of z_{opt} and we can use the partial z_{opt} to select the l rows which are closest to z_{opt} in those $O(\log m)$ bits. After that we can construct a consensus vector z as follows: for each column, choose the (majority) letter that appears the most in each of the l letters in the l selected rows. Then for each of the n columns, we can calculate the number of mismatches between the majority letter and the l letters in the l selected rows. By selecting the best k columns, we can get a good solution.

The remain difficulty is how to randomly select $O(\log m)$ columns in Q_{opt} while Q_{opt} is unknown. Our new idea is to randomly select a set B of $\lceil (c + 1)(\frac{4\log m}{\epsilon^2} + 1) \rceil$ columns from A and enumerate all size $\lceil \frac{4\log m}{\epsilon^2} \rceil$ subsets of B in time $O(m^{\frac{4(c+1)}{\epsilon^2}})$ which is polynomial in terms of the input size $O(mn)$. We can show that with high probability, we can get a set of $\lceil \frac{4\log m}{\epsilon^2} \rceil$ columns randomly selected from Q_{opt}.

Now we describe the complete algorithm in Figure 1.

The following lemma that is originally from [25] is used in our proofs.

Lemma 1. *Let $X_1, X_2, \ldots, X_n$ be n independent random 0-1 variables, where X_i takes 1 with probability p_i, $0 < p_i < 1$. Let $X = \sum_{i=1}^{n} X_i$, and $\mu = E[X]$. Then for any $0 < \epsilon \le 1$,*

$$\mathbf{Pr}(X > \mu + \epsilon n) < (-\frac{1}{3}n\epsilon^2),$$

$$\mathbf{Pr}(X < \mu - \epsilon n) \le (-\frac{1}{2}n\epsilon^2).$$

Lemma 2. *With probability at most $m^{-\frac{2}{\epsilon^2 c^2(c+1)}}$, no subset R of size $\lceil \frac{4\log m}{\epsilon^2} \rceil$ used in Step 1 of Algorithm 1 satisfies $R \subseteq Q_{opt}$.*

Lemma 3. *Assume $|R| = \lceil \frac{4\log m}{\epsilon^2} \rceil$ and $R \subseteq Q_{opt}$. Let $\rho = \frac{k}{|R|}$. With probability at most m^{-1}, there is a row x_i in X satisfying*

$$\frac{d(z_{opt}, x_i|^{Q_{opt}}) - \epsilon k}{\rho} > d(z_{opt}|^R, x_i|^R).$$

Algorithm 1 for The Consensus Submatrix Problem

Input: one $m \times n$ matrix A, integers l and k, and a small number $\epsilon > 0$
Output: a size l subset P of rows, a size k subset Q of columns and a length k
consensus vector z
Step 1: randomly select a set B of $\lceil (c+1)(\frac{4\log m}{\epsilon^2}+1) \rceil$ columns from A.
  (1.1)**for** every size $\lceil \frac{4\log m}{\epsilon^2} \rceil$ subset R of B **do**
    (1.2) **for** every $z|^R \in \Sigma^{|R|}$ **do**
        (a) Select the best l rows $P = \{p_1, ..., p_l\}$ that minimize
           $d(z|^R, x_i|^R)$.
        (b) **for** each column j **do**
            Compute $f(j) = \sum_{i=1}^{l} d(s_j, a_{p_i,j})$, where s_j is the ma-
            jority element of the l rows in P in column j.
            Select the best k columns $Q = \{q_1, ..., q_k\}$ with mini-
            mum value $f(j)$ and let $z(Q) = s_{q_1} s_{q_2} \ldots s_{q_k}$.
        (c) Calculate $H = \sum_{i=1}^{l} d(x_{p_i}|^Q, z)$ of this solution.
Step 2: Output P, Q and z with minimum H.

Fig. 1. Algorithm 1

With probability at most $m^{-\frac{1}{3}}$, there is a row x_i in X satisfying

$$d(z_{opt}|^R, x_i|^R) > \frac{d(z_{opt}, x_i|^{Q_{opt}}) + \epsilon k}{\rho}.$$

Lemma 4. *When $R \subseteq Q_{opt}$ and $z|^R = z_{opt}|^R$, with probability at most $2m^{-\frac{1}{3}}$, the set of rows $P = \{p_1, \ldots, p_l\}$ selected in Step 1 (a) of Algorithm 1 satisfies $\sum_{i=1}^{l} d(z_{opt}, x_{p_i}|^{Q_{opt}}) > H_{opt} + 2\epsilon kl$.*

Theorem 2. *For any $\delta > 0$, with probability at least $1 - m^{-\frac{8c'^2}{\delta^2 c^2(c+1)}} - 2m^{-\frac{1}{3}}$, Algorithm 1 outputs a solution with consensus score at most $(1 + \delta)H_{opt}$ in $O(nm^{O(\frac{1}{\delta^2})})$ time.*

Proof. When $R \subseteq Q_{opt}$ and $z|^R = z_{opt}|^R$, in step 1 (b), we can construct a Q and $z(Q)$ such that

$$\sum_{i=1}^{l} d(z(Q), x_{p_i}|^Q) \leq \sum_{i=1}^{l} d(z_{opt}, x_{p_i}|^{Q_{opt}}). \tag{2}$$

From Lemma 2, we know that with probability at most $m^{-\frac{2}{\epsilon^2 c^2(c+1)}}$, there is no subset R with size $\lceil \frac{4\log m}{\epsilon^2} \rceil$ in Step 1 of Algorithm 1 such that $R \subseteq Q_{opt}$. Combining with Lemma 4, we know that with probability at most $m^{-\frac{2}{\epsilon^2 c^2(c+1)}} + 2m^{-\frac{1}{3}}$, in the execution of Algorithm 1, any set of rows $P = \{p_1, \ldots, p_l\}$ obtained in Step 1(a) satisfies

$$\sum_{i=1}^{l} d(z_{opt}, x_{p_i}|^{Q_{opt}}) > H_{opt} + 2\epsilon kl.$$

In other words, with probability at least $1-m^{-\frac{2}{\epsilon^2 c^2(c+1)}}-2m^{-\frac{1}{3}}$, in the execution of Algorithm 1, we can get a set of rows $P = \{p_1,\ldots,p_l\}$ in Step 1 (a) that satisfies $\sum_{i=1}^{l} d(z_{opt}, x_{p_i}|^{Q_{opt}}) \leq H_{opt} + 2\epsilon kl$.

From (2), we have

$$\sum_{i=1}^{l} d(z(Q), x_{p_i}|^{Q}) \leq \sum_{i=1}^{l} d(z_{opt}, x_{p_i}|^{Q_{opt}}) \leq H_{opt} + 2\epsilon kl.$$

Recall $H_{opt} \times c' = kl$. Set $\epsilon = \frac{\delta}{2c'}$. So with probability at least $1-m^{-\frac{8c'^2}{\delta^2 c^2(c+1)}} - 2m^{-\frac{1}{3}}$, Algorithm 1 outputs a solution with consensus score at most $(1+\delta)H_{opt}$.

For the time complexity, Step1(a), Step1(b) and Step1(c) take $O(mn)$ time. Step 1.1 is repeated $O(2^{\frac{4(c+1)\log m}{\epsilon^2}}) = O(m^{O(\frac{1}{\epsilon^2})}) = O(m^{O(\frac{1}{\delta^2})})$ times . Step 1.2 is repeated $O(m^{O(\frac{\log|\Sigma|}{\epsilon^2})}) = O(m^{O(\frac{1}{\delta^2})})$ times as $\epsilon = \frac{\delta}{2c'}$ and $|\Sigma|$ is fixed constant. Thus, the total running time is $O(nm^{O(\frac{1}{\delta^2})})$.   □

Theorem 3. *There exists a PTAS for the consensus submatrix problem.*

Proof. Algorithm 1 can be derandomized by standard method. For instance, instead of randomly and independently choosing $O(\log m)$ columns from the n columns in Step 1, we can pick the vertices encountered on a random walk of length $O(\log m)$ on a constant degree expander [26]. Obviously, the number of such random walks on a constant degree expander is polynomial in terms of m. Thus, by enumerating all random walks of length $O(\log m)$, we have a polynomial time deterministic algorithm(Also see [27]).   □

5   The Bottleneck Submatrix Problem

In this section, we present the randomized approximation algorithm for the bottleneck submatrix problem. Let P_{opt}, Q_{opt} and z_{opt} be the set of rows , the set of columns and the consensus vector of an optimal solution. The optimal bottleneck score is d_{opt}. By assumption, $d_{opt} = O(k)$ and $k = O(n)$, i.e., there are constants c'' and c such that $d_{opt} \times c'' = k$ and $k \times c = n$.

Similar to Algorithm 1, we can use a random sampling technique to know $O(\log m)$ bits of z_{opt}. Then we can use the partial z_{opt} to select the l rows which are closest to z_{opt} in those $O(\log m)$ bits as in Step 1(a) of Algorithm 1. From Lemma 3, we know that using $O(\log m)$ bits in R, we can get a good estimation of $d(z_{opt}, x_i|^{Q_{opt}})$ for each x_i in X. Thus, if we can correctly select Q_{opt} from the n given columns, then we can get a good approximation solution. However, Step 1 (b) in Algorithm 1 does not work for the bottleneck score in selecting a good approximation of Q_{opt}. Thus, we use a linear programming and randomized rounding technique to select k columns in the matrix.

Linear Programming Formulation

Given a set of rows $P = \{p_1, ..., p_l\}$, we want to find a set of k columns Q and a vector z such that the bottleneck score is minimized. This problem is equivalent to the following optimization problem:

$$\begin{cases} \min d; \\ d(z, x_{p_i}|Q) \leq d, i = 1, 2, \ldots, l, Q \subseteq Y, |Q| = k, z \in \Sigma^k. \end{cases} \quad (3)$$

Let $\Sigma = \{\pi_1, \pi_2, \ldots, \pi_{|\Sigma|}\}$. We introduce $0 - 1$ variable $y_{i,j}$ ($i = 1, 2, \ldots, n, j = 1, 2, \ldots, |\Sigma|$) to indicate whether column i is in Q and the corresponding bit of z. $y_{i,j} = 1$ if and only if column i is in Q and the corresponding bit in z is π_j. For any $a, b \in \Sigma$, $\chi(a, b) = 0$ if $a = b$ and $\chi(a, b) = 1$ if $a \neq b$. (3) can be formulated as the $0 - 1$ Integer Linear Programming:

$$\begin{cases} \min d; \\ \sum_{i=1}^{n} \sum_{j=1}^{|\Sigma|} y_{i,j} = k, \\ \sum_{j=1}^{|\Sigma|} y_{i,j} \leq 1, i = 1, 2, \ldots, n, \\ \sum_{i=1}^{n} \sum_{j=1}^{|\Sigma|} \chi(\pi_j, x_{p_s,i}) y_{i,j} \leq d, s = 1, 2, \ldots, l. \end{cases} \quad (4)$$

Here $y_{i,j}$ is used to achieve two tasks: (1) decide whether column i is selected and (2) if column i is selected, we have to decide the letter in the consensus vector z at the this column.

We can obtain a fractional solution $y_{i,j} = \overline{y_{i,j}}$($i = 1, 2, ..., n, j = 1, 2, \ldots, |\Sigma|$) for (4) in polynomial time. After we get the fractional solution, we do randomized rounding to get an integer solution.

Given a fractional solution $y_{i,j} = \overline{y_{i,j}}$ ($i = 1, 2, \ldots, n, j = 1, 2, \ldots, |\Sigma|$) with cost $\overline{d}$. For each $1 \leq i \leq n, 1 \leq j \leq |\Sigma|$, randomly select column i to Q with probability $\sum_{j=1}^{|\Sigma|} \overline{y_{i,j}}$ and randomly set the bit of z in this column according to the distribution $\overline{y_{i,j}}$ for $j = 1, 2, \ldots, |\Sigma|$). In terms of programming, we can generate a random number ρ in (0,1), for every column i. If $\rho < \sum_{j=1}^{|\Sigma|} \overline{y_{i,j}}$, we select this column into Q and let the bit of z corresponding to this column be π_t if and only if $\sum_{j=1}^{t-1} \overline{y_{i,j}} \leq \rho < \sum_{j=1}^{t} \overline{y_{i,j}}$. If $\rho \geq \sum_{j=1}^{|\Sigma|} \overline{y_{i,j}}$, this column is not selected. Hence we get a 0/1 solution $y' = \{y'_{1,1}, \ldots, y'_{1,|\Sigma|}, \ldots, y'_{n,1}, \ldots, y'_{n,|\Sigma|}\}$.

In this randomized rounding process, we have to do two things. (1) select k' columns, where $k' \geq k - \delta d_{opt}$. (2) get integers values for $y_{i,j}$ such that the distance (restricted on the k' selected columns) between any row in P and the center vector thus obtained is at most γd_{opt}. Here $\delta > 0$ and $\gamma > 0$ are two parameters used to control the errors.

Lemma 5. *When $\frac{n\gamma^2}{3(cc'')^2} \geq 2 \log m$, for any $\gamma, \delta > 0$, with probability at most $exp(-\frac{n\delta^2}{2(cc'')^2}) + m^{-1}$, the rounding result $y' = \{y'_{1,1}, \ldots, y'_{1,|\Sigma|}, \ldots, y'_{n,1}, \ldots, y'_{n,|\Sigma|}\}$ does not satisfy at least one of the following inequalities,*

$$\sum_{i=1}^{n} (\sum_{j=1}^{|\Sigma|} y'_{i,j}) > k - \delta d_{opt},$$

Algorithm 2 for The Bottleneck Submatrix Problem

Input: one matrix $A \in \Sigma^{m \times n}$, integer l, k, a row $z \in \Sigma^n$ and small numbers $\epsilon > 0$, $\gamma > 0$ and $\delta > 0$.

Output: a size l subset P of rows, a size k subset Q of columns and a length k consensus vector z.

if $\frac{n\gamma^2}{3(cc'')^2} \leq 2 \log m$ then

    try all size k subset Q of the n columns and all z of length k

    to solve the problem.

if $\frac{n\gamma^2}{3(cc'')^2} > 2 \log m$ then

    Step 1: randomly select a set B of $\lceil \frac{4(c+1) \log m}{\epsilon^2} \rceil$ columns from A.

        for every $\lceil \frac{4 \log m}{\epsilon^2} \rceil$ size subset R of B **do**

            for every $z|^R \in \Sigma^{|R|}$ **do**

                (a) Select the best l rows $P = \{p_1, ..., p_l\}$ that minimize $d(z|^R, x_i|^R)$.

                (b) Solve the optimization problem (3) by linear programming and randomized rounding to get Q and z.

    Step 2: Output P,Q and z with minimum bottleneck score d.

Fig. 2. Algorithm 2

and for every row $x_{p_s} (s = 1, 2, \ldots, l)$,

$$\sum_{i=1}^{n} (\sum_{j=1}^{|\Sigma|} \chi(\pi_j, x_{p_s,i}) y'_{i,j}) < \overline{d} + \gamma d_{opt}.$$

When $\frac{n\gamma^2}{3(cc'')^2} < 2 \log m$, we try all subsets of X with size k and all length k vectors in polynomial time and get the best solution.

From Lemma 5, we know that in the randomized rounding process, with high probability, we selected k' columns in Q, where $(1 - \epsilon)k \leq k'$. Our aim is to select exactly k columns. If $k' > k$, we can arbitrarily delete $k' - k$ columns from Q and obtain the set of k columns $Q' \subseteq Q$. If $k' < k$, we can arbitrarily select $k - k'$ columns outside Q and add them to Q to get a set of k columns $Q' \supset Q$. By doing so, the extra error introduced is at most ϵk. Since $d = O(n)$, the error ϵk is small and we still can get a PTAS.

Now we describe the complete algorithm in Figure 2.

Similar to Lemma 4, we have

Lemma 6. *When $R \subseteq Q_{opt}$ and $z|^R = z_{opt}|^R$, with probability at most $2m^{-\frac{1}{3}}$, the set of rows $P = \{p_1, \ldots, p_l\}$ obtained in Step 1(a) of Algorithm 2 satisfies $d(z_{opt}, x_{p_i}|^{Q_{opt}}) > d_{opt} + 2\epsilon k$ for some row $x_{p_i} (1 \leq i \leq l)$.*

From Lemmas 2, 5, and 6, we have

Theorem 4. *With probability at least $1 - m^{-\frac{2}{\epsilon^2 c^2 (c+1)}} - 2m^{-\frac{1}{3}} - exp(-\frac{n\delta^2}{2(cc'')^2}) - m^{-1}$, Algorithm 2 runs in time $O(n^{O(1)} m^{O(\frac{1}{\epsilon^2} + \frac{1}{\gamma^2})})$ and obtains a solution with bottleneck score at most $(1 + 2c''\epsilon + \gamma + \delta)d_{opt}$ for any fixed ϵ, γ, $\delta > 0$.*

Theorem 5. *There exists a PTAS for the bottleneck submatrix problem.*

Proof. For Step 1(b), we can use the technique in [25] to derandomize it. The derandomization of the random sampling step is the same as in Algorithm 1. □

6 Conclusion

We have designed PTAS's for both the consensus submatrix and the bottleneck submatrix problems. To our knowledge, this is the first time that an approximation algorithm with guaranteed performance ratio is presented for microarray analysis. It is conscious to point out that the running time is very high and may not work in practice.

Acknowledgements

This work is fully supported by a grant from the Research Grants Council of the Hong Kong Special Administrative Region, China [Project No. CityU 1070/02E].

References

1. R.B. Stoughton. Applications of DNA microarrays in biology. *Annual Rev. Biochem.*, 74:53–82, 2005.
2. D.B. Allison, X. Cui, G.P. Page, and M. Sabripou. Microarray data analysis: from disarray to consolidation and consensus. *Nature Reviews Genetics*, 7:55–65, 2006.
3. S. Tavazoie, J.D. Hughes, M.J. Campbell, R.J. Cho, and G.M. Church. Systematic determination of genetic network architecture. *Nat. Genet.*, 22:281–285, 1999.
4. F.X. Wu, W.J. Zhang, and A.J. Kusalik. A genetic K-means clustering algorithm applied to gene expression data. *LNAI*, 2671, Springer Verlag (2003), 520–526.
5. P. Tamayo, D. Slonim, J. Mesirov, Q. Zhu, S. Kitareewan, E. Dmitrovsky, E.S. Lander, and T.R. Golub. Interpreting patterns of gene expression with self-organizing maps: methods and application to hematopoietic differentiation. *Proc. Nat'l Acad. Sci. USA*, 96:2907–2912, 1999.
6. H. Ressom, D. Wang, and P. Natarajan. Clustering gene expression data using adaptive double selforganizing map. *Physiol. Genomics*, 14:35–46, 2003.
7. M.B. Eisen, P.T. Spellman, P.O. Brown, and D. Botstein. Cluster analysis and display of genome-wide expression patterns. *Proc. Nat'l Acad. Sci. USA*, 95:14863–14868, 1998.
8. V.R. Iyer, M.B. Eisen, D.T. Ross, G. Schuler, T. Moore, J.C. Lee, J.M. Trent, L.M. Staudt, J. Hudson Jr., M.S. Boguski, D. Lashkari, D. Shalon, D. Botstein, and P.O. Brown. The transcriptional program in the response of human fibroblasts to serum. *Science*, 283:83–87, 1999.
9. J. Qin, D.P. Lewis, and W.S. Noble. Kernel hierarchical gene clustering from microarray expression data. *Bioinformatics*, 19:2097–2104, 2003.
10. O. Alter, P.O. Brown, and D. Botstein. Generalized singular value decomposition for comparative analysis of genome-scale expression data sets of two different organisms. *Proc. Nat'l Acad. Sci. USA*, 100:3351–3356, 2003.

11. N.S. Holter, M. Mitra, A. Maritan, M. Cieplak, J.R. Banavar, and N.V. Fedoroff. Fundamental patterns underlying gene expression profiles: simplicity from complexity. *Proc. Nat'l Acad. Sci. USA* 97:8409–8414, 2000.

12. K.C. Li, M. Yan, and S.S. Yuan. A simple statistical model for depicting the cdc15-synchronized yeast cell-cycle regulated gene expression data. *Statistica Sinica*, 12:141–158, 2002.

13. B. Tjaden. An approach for clustering gene expression data with error Information. *BMC Bioinformatics*, 7:17, 2006.

14. B.H. Mecham, D.Z. Wetmore, Z. Szallasi, Y. Sadovsky, I. Kohane, and T.J. Mariani. Increased measurement accuracy for sequence-verified microarray probes. *Physiol. Genomics* 18:308–315, 2004.

15. D.M. Rocke and B. Dubin. A Model for Measurement Error for Gene Expression Arrays. *J. of Computational Biology*, 8(6):557–569, 2001.

16. S. Draghici, P. Khatri, A.C. Eklund, and Z. Szallasi. Reliability and reproducibility issues in DNA microarray measurements. *Trends in Genetics* 22(2):101–109, 2006.

17. J.P. Brody, B.A. Williams, B.J. Wold, and S.R. Quake. Significance and statistical errors in the analysis of DNA microarray data. *Proc. Nat'l Acad. Sci. USA* 99:12975–12978, 2002.

18. E. Purdom and S.P. Holmes. Error distribution for gene expression data. *Statistical Applications in Genetics and Molecular Biology*, 4(1):16, 2005.

19. H. Cho and J.K. Lee. Bayesian hierarchical error model for analysis of gene expression data. *Bioinformatics*, 20:2016–2025, 2004..

20. G. Getz, E. Levine, and E. Domany. Coupled two–way clustering analysis of gene microarray data. *Proc. Nat'l Acad. Sci. USA* 12079–12084, 2000.

21. Y. Cheng and G.M. Church. Biclustering of expression data. *Proc. 8th Conf. on Intelligent Systems for Molecular Biology ISMB'00*, 93–103, 2000.

22. S.C. Madeira and A.L. Oliveira. Biclustering algorithms for biological data analysis: a survey. *IEEE/ACM Transactions on Computational Biology and Bioinformatics*, 1:24–45, 2004.

23. S. Lonardi, W. Szpankowski, and Q. Yang. Finding biclusters by random projections. *Proc. Symp. on Combinatorial Pattern Matching CPM'04*, LNCS, Springer Verlag (2004), 102–116..

24. R. Peeters. The maximum edge biclique problem is NP-complete. *Discrete Applied Mathematics*, 131(3):651–654, 2003.

25. M. Li, B. Ma, and L. Wang. On the closest string and substring problems. *J. ACM*, 49(2):157–171, 2002.

26. D. Gillman. A Chernoff bound for random walks on expander graphs. *Proc. 34th Symp. on Foundations of Computer Science FOCS'93*, IEEE Computer Society Press, 680–691, 1993.

27. S. Arora, D. Karger, and M. Karpinski. Polynomial-time approximation schemes for dense instances of NP-hard problems. *Proc. 27th ACM Symp. on Theory of Computing STOC'95*, ACM Press, 284–293, 1995.

Improving the Layout of Oligonucleotide Microarrays: Pivot Partitioning

Sérgio A. de Carvalho Jr.[1,2,3] and Sven Rahmann[1,3]

[1] International NRW Graduate School in Bioinformatics and Genome Research
[2] Graduiertenkolleg Bioinformatik, Bielefeld University, Germany
Sergio.Carvalho@cebitec.uni-bielefeld.de
[3] Algorithms and Statistics for Systems Biology group, Genome Informatics,
Technische Fakultät, Bielefeld University, D-33594 Bielefeld, Germany
Sven.Rahmann@cebitec.uni-bielefeld.de

Abstract. The production of commercial DNA microarrays is based on a light-directed chemical synthesis driven by a set of masks or micromirror arrays. Because of the natural properties of light and the ever shrinking feature sizes, the arrangement of the probes on the chip and the order in which their nucleotides are synthesized play an important role on the quality of the final product. We propose a new model called *conflict index* for evaluating the layout of microarrays. We also present a new algorithm, called Pivot Partitioning, that improves the quality of layouts, according to existing measures, by over 6% when compared to the best known algorithms.

1 Introduction

An oligonucleotide microarray is a piece of glass or plastic on which single-stranded fragments of DNA, called *probes*, are affixed or synthesized. Affymetrix GeneChip® arrays, for instance, can contain more than one million spots as small as 11 μm, with each spot accommodating several million copies of a probe. Probes are typically 25 nucleotides long and are synthesized on the chip, in parallel, in a series of repetitive steps. Each step appends the same nucleotide to probes of selected regions of the chip. Selection occurs by exposure to light with the help of a photolithographic mask that allows or obstructs the passage of light accordingly [5].

Formally, we have a set of probes $\mathcal{P} = \{p_1, p_2, ...p_n\}$ that are produced by a series of masks $\mathcal{M} = (m_1, m_2, ...m_T)$, where each mask m_t induces the addition of a particular nucleotide $\mathcal{S}_t \in \{A, C, G, T\}$ to a subset of $\mathcal{P}$. The *nucleotide deposition sequence* $\mathcal{S} = \mathcal{S}_1 \mathcal{S}_2 ... \mathcal{S}_T$ corresponding to the sequence of nucleotides added at each masking step is therefore a supersequence of all $p \in \mathcal{P}$ [10].

In general, a probe can be *embedded* within $\mathcal{S}$ in several ways. An embedding of p_k is a T-tuple $\varepsilon_k = (e_{k,1}, e_{k,2}, ...e_{k,T})$ in which $e_{k,t} = 1$ if probe p_k receives nucleotide $\mathcal{S}_t$ (at step t), or 0 otherwise (Figure 1). In particular, a *left-most embedding* is an embedding in which the bases are synthesized as early as possible (see ε_3 in Figure 1).

P. Bücher and B.M.E. Moret (Eds.): WABI 2006, LNBI 4175, pp. 321–332, 2006.

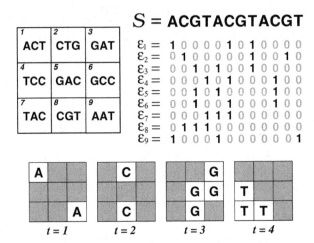

Fig. 1. Synthesis of a hypothetical 3×3 chip. Top left: chip layout and the 3 nt probe sequences. Top right: deposition sequence and probe embeddings. Bottom: first four resulting masks.

The deposition sequence is often taken as a repeated permutation of the alphabet, mainly because of its regular structure and because such sequences maximize the number of distinct subsequences [4].

We distinguish between *synchronous* and *asynchronous* embeddings. In the first case, each probe has exactly one nucleotide synthesized in every cycle of the deposition sequence; hence, 25 cycles or 100 steps are needed to synthesize probes of length 25. In the case of asynchronous embeddings, probes can have any number of nucleotides synthesized in any given cycle, allowing shorter deposition sequences. All Affymetrix chips that we know of can be asynchronously synthesized in 74 steps (18.5 cycles), which is probably due to careful probe selection.

Because of diffraction of light or internal reflection, untargeted spots can be accidentally activated in a certain masking step, producing unpredicted probes that can compromise experimental results. This problem is more likely to occur near the borders between masked and unmasked spots [5]. This observation has given rise to the term *border conflict*.

We are interested in finding an *arrangement* of the probes on the chip together with *embeddings* in such a way that the chances of unintended illumination during mask exposure steps are minimized. The problem appears to be hard because of the exponential number of possible arrangements, although we are not aware of an NP-hardness proof. In a separate work [2], we present a formulation of the above problem as a quadratic assignment problem (QAP), a classical combinatorial optimization problem that is, in general, NP-hard and particularly hard to solve in practice [3]. Optimal solutions are thus unlikely to be found even for small chips and even if we assume that all probes have a single predefined embedding.

If we consider all possible embeddings (up to several million for a typical Affymetrix probe), the problem is even harder. For this reason, the problem has been traditionally tackled in two phases. First, an initial embedding of the probes is fixed and an arrangement of these embeddings on the chip with minimum border conflicts is sought. This is usually referred to as the *placement*. Second, a *post-placement* optimization phase re-embeds the probes considering their location on the chip, in such a way that the conflicts with the neighboring spots are further reduced.

It seems intuitive that better results should be achieved if the placement and embedding phases are considered together, not separately. Because of the generally high number of embeddings of each single probe in the asynchronous setting, it is not easy to design algorithms that make efficient use of this additional freedom and achieve reasonable running times in practice. In fact, so far we know of no single publication that merges the two phases; in this article we propose such a strategy called Pivot Partitioning.

The rest of this paper is structured as follows. Section 2 details two different ways of evaluating computed layouts and embeddings; they form the objective functions that we aim to minimize. As a refinement of the "classical" border length, we introduce the *conflict index* measure. Section 3 reviews existing placement, partitioning, and post-placement strategies. In Section 4 we discuss an extension of the optimal single probe embedding (OSPE) algorithm (that first appeared in [7]) to support our new measure. Our partitioning strategy, that for the first time combines partitioning the chip with embedding the probes, is described in Section 5. Computational results follow in Section 6.

2   Evaluating Layouts and Embeddings

Border length. Hannenhalli and co-workers [6] were the first to give a formal definition of the problem of unintended illumination in the production of microarrays. They formulated the *Border Length Minimization Problem* which aims at finding an arrangement of the probes together with their embeddings in such a way that the number of border conflicts during mask exposure steps is minimal.

The *border length* B_t of a mask m_t is defined as the number of borders shared by masked and unmasked spots at masking step t. The total border length of a given arrangement is the sum of border lengths over all masks. For example, the initial four masks shown in Figure 1 have $B_1 = 4$, $B_2 = 6$, $B_3 = 6$ and $B_4 = 4$. The total border length of that arrangement is 50 (masks 5 to 12 not shown).

Conflict Index. The border length of an individual mask measures the quality of that mask. We are more interested in estimating the risk of synthesizing a faulty probe at a given spot, that is, we need a per-probe measure instead of a per-mask measure. Additionally, the definition of border length does not take into account two important practical considerations [8]:

a) stray light might activate not only adjacent neighbors but also probes that lie as far as three cells away from the targeted spot;

b) imperfections produced in the middle of a probe are more harmful than in its extremities.

This motivates the following definition of the *conflict index* $C(p)$ of a probe of length ℓ_p that is synthesized in T masking steps. First, we define a distance-dependent weighting function, $\delta(p, p', t)$, that accounts for observation a) above:

$$\delta(p, p', t) := \begin{cases} (d(p, p'))^{-2} & \text{if } p' \text{ is unmasked at step } t, \\ 0 & \text{otherwise,} \end{cases} \tag{1}$$

where $d(p, p')$ is the Euclidean distance between the spots of p and p'. This form of weighting function is the same as suggested in [8]. Note that δ is a "closeness" measure between p and p' only if p' is not masked (and thus creates the potential of illumination at p). To limit the number of neighbors that need to be considered, we restrict the support of $\delta(p, p', \cdot)$ to those $p' \neq p$ that are in a 7×7 grid centered around p (see Figure 2 left).

We also define position-dependent weights to account for observation b):

$$\omega(p, t) := \begin{cases} c \cdot \exp(\theta \cdot \lambda(p, t)) & \text{if } p \text{ is masked at step } t, \\ 0 & \text{otherwise,} \end{cases} \tag{2}$$

where $c > 0$ and $\theta > 0$ are constants, and

$$\lambda(p, t) := 1 + \min(b_{p,t}, \ell_p - b_{p,t}) \tag{3}$$

is the distance, from the start or end of the final probe sequence, of the last base synthesized before step t: $b_{p,t}$ denotes the number of nucleotides synthesized within p up to and including step t, and ℓ_p is the probe length (see Figure 2 right).

The motivation behind an exponentially increasing weighting function is that the probability of a successful stable hybridization of a probe with its target should increase exponentially with the absolute value of its Gibbs free energy, which increases linearly with the length of the longest perfect match between probe and target. The parameter θ controls how steeply the exponential weighting function rises towards the middle of the probe. In our experiments, we set $\theta := 5/\ell_p$ and $c = 1/\exp(\theta)$.

We now define the conflict index of a probe p as

$$C(p) := \sum_{t=1}^{T} \left(\omega(p, t) \sum_{p'} \delta(p, p', t) \right), \tag{4}$$

where p' ranges over all probes that are at most three cells away from p. $C(p)$ can be interpreted as the fraction of faulty p-probes (because of unwanted illumination).

We note the following relation between conflict index and border length. Define $\delta(p, p', t) := 1$ if p' is a direct neighbor of p and is unmasked in step t, and $:= 0$ otherwise. Define $\omega(p, t) := 1$ if p is masked in step t, and $:= 0$ otherwise. Then $\sum_s C(p) = 2 \sum_{t=1}^{T} \mathcal{B}_t$, as each border conflict is counted twice, once for p

0.06	0.08	0.10	0.11	0.10	0.08	0.06
0.08	0.13	0.20	0.25	0.20	0.13	0.08
0.10	0.20	0.50	1.00	0.50	0.20	0.10
0.11	0.25	1.00	p	1.00	0.25	0.11
0.10	0.20	0.50	1.00	0.50	0.20	0.10
0.08	0.13	0.20	0.25	0.20	0.13	0.08
0.06	0.08	0.10	0.11	0.10	0.08	0.06

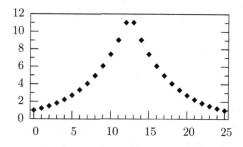

Fig. 2. Ranges of values for both δ and ω on a typical Affymetrix chip where probes of length 25 are synthesized in 74 masking steps. Left: approximate values of the distance-dependent weighting function $\delta(p, p', t)$ for a probe p (shown in the center) and close neighbors p', assuming that p' is unmasked at step t. Right: position-dependent weights $\omega(p, t)$ on the y-axis for each value of $b_{p,t}$ on the x-axis, assuming that p is masked at step t.

and once for p'. Therefore border length and total conflict are equivalent for a particular choice of δ and ω. For our choice (1) and (2), they are not equivalent, but still correlated: a good layout has both low border length and low conflict indices.

3   Previous Work

Up to now, the tasks of probe placement and probe embedding were considered separately. Placement is often handled by (recursively) partitioning the chip into smaller regions before applying a placement algorithm. We now review existing placement algorithms, partitioning algorithms and post-placement strategies.

Placement Algorithms. The border length problem on large oligonucleotide arrays of arbitrary probes was first formally addressed in [6]. The article reports that the first Affymetrix chips were designed using a heuristic for the traveling salesman problem (TSP). The idea consists of building a weighted graph with nodes representing probes, and edges containing the Hamming distance between the probe sequences. A TSP tour is approximated, resulting in consecutive probes in the tour being likely similar. The TSP tour is then *threaded* on the array in a row-by-row fashion. A different threading of the TSP tour, called *1-threading*, is suggested to achieve up to 20% reduction in border length.

A different strategy called *Epitaxial* placement [7] places a random probe in the center of the array and continues to insert probes in spots adjacent to already filled spots. Priority is given to spots with the largest numbers of filled neighbors. At each iteration, it examines all non-filled spots and finds a non-assigned probe with minimum sum of Hamming distances to the neighboring probes, employing a greedy heuristic to select the next spot. A further 10% reduction in border conflict over TSP + 1-threading is claimed.

Both the Epitaxial algorithm and the TSP approach do not scale well to large chips. For this reason, [8] proposes a simpler variant of the Epitaxial algorithm, called *Row-epitaxial*, with two main differences: spots are filled in a pre-defined order, namely from top to bottom, left to right, and only probes of a limited list of candidates are considered when filling each spot. Experiments show that Row-epitaxial is the best large-scale placement algorithm, achieving up to 9% reduction in border length over the TSP + 1-threading.

Partitioning Algorithms. The placement problem can be partitioned by dividing the set of probes into smaller subsets, and assigning these subsets to subregions of the chip. Each subregion can then be treated as an independent chip or recursively partitioned. In this way, algorithms with non-linear time or space complexities can be used to compute the layout of larger chips that otherwise would not be feasible.

The only partitioning that we know of is the Centroid-based Quadrisection [9]. It starts by randomly selecting a probe $c_1 \in \mathcal{P}$. Then, it selects another probe c_2 maximizing $h(c_1, c_2)$, the Hamming distance between their embeddings. Similarly, it selects c_3 and c_4 maximizing the sum of Hamming distance between these four probes that are called centroids. All other probes $p \in \mathcal{P}$ are then compared to the centroids and assigned to a subset $\mathcal{P}_k$ associated with c_k with minimum $h(p, c_k)$. The chip is divided into four quadrants, each being assigned to a subset $\mathcal{P}_k$. The procedure is repeated recursively on each quadrant until a given recursion depth is reached. In the end, the Row-epitaxial algorithm is used to produce the placement of the probes in each final subregion.

Post-placement Optimization. Once the placement is done, further reduction of conflicts can be achieved by re-embedding the probes without changing their locations. The paper [7] presents a dynamic programming algorithm, that we call Optimum Single Probe Embedding (OSPE), for computing an optimum embedding of a probe with respect to the neighboring probes, whose embeddings are considered fixed. Originally, it was developed for border length minimization; in Section 4 we give a slightly more general form that also applies to the conflict index measure.

The OSPE algorithm is the basic operation of several post-placement optimization algorithms: Batched Greedy [7], Chessboard [7] and Sequential [9]. Their main difference lies in the order in which the re-embeddings take place. Since the OSPE never increases the amount of conflicts in a region, all optimization algorithms can be executed several times until a local optimal solution is found, or until the improvements drop below a given threshold.

The Sequential algorithm just proceeds spot by spot, from top to bottom, left to right, re-embedding all probes with the OSPE algorithm. Surprisingly, it achieves the greatest reduction of border conflicts with a running time comparable to Batched Greedy, the fastest among the three.

4   Optimum Single Probe Embedding

The Optimum Single Probe Embedding (OSPE) algorithm finds an optimal embedding of a single probe on a given spot, assuming that all neighboring embeddings are fixed. It can be seen as a special case of a global alignment between the probe sequence p of length ℓ and the deposition sequence $\mathcal{S}$ of length T. We use an $(\ell + 1) \times (T + 1)$ array D, where $D[i, j]$ is defined as the minimum cost of an embedding of $p[1..i]$ into $\mathcal{S}[1..j]$. The cost is the sum of conflicts induced by the embedding of p on its neighbors plus the conflicts suffered by p because of the embeddings of its neighbors.

At every step j of the deposition sequence, the probe p can be either masked or unmasked. Thus, entry $D[i, j]$ is computed as the minimum between the costs resulting from each possible state:

$$D[i, j] = \min(D[i, j-1] + M_{ij}, D[i-1, j-1] + U_j).$$

The costs M_{ij} and U_j depend on probe p and neighboring probes p'. M_{ij} denotes the cost of masking probe p at step j given that base i of p has been synthesized previously. Any unmasked neighbor p' generates a conflict on p with cost $\omega(p, i) \cdot \delta(p, p', j)$; therefore the total cost is

$$M_{ij} = \sum_{p'} \omega(p, i) \cdot \delta(p, p', j).$$

U_j denotes the cost of unmasking probe p at step j, which generates a conflict on each masked neighbor p' with cost $\omega(p', j) \cdot \delta(p', p, j)$; therefore

$$U_j = \sum_{p'} \omega(p', j) \cdot \delta(p', p, j).$$

The first column of D is initialized as follows: $D[0, 0] = 0$ and $D[i, 0] = \infty$ for $0 < i \le \ell$. The first row is $D[0, j] = D[0, j-1] + M_{0j}$ for $0 < j \le T$. The time complexity of the OSPE algorithm is obviously $O(\ell \cdot T)$.

5   Pivot Partitioning

Traditionally, the microarray layout problem has been tackled in two phases: placement, during which an initial embedding of the probes is fixed, and post-placement optimization, when probes are re-embedded using the OSPE algorithm. We believe that better layouts can be produced if the placement phase also considers the various embeddings that a probe can have. In this section we propose a new partitioning algorithm called Pivot Partitioning (PP).

Our algorithm has some similarities with the Centroid-based Quadrisection (CQ) described in Section 3. Its main differences are motivated by the following observation. As mentioned earlier, some probes can have up to several millions different embeddings, while others may have only a few or even only one possible embedding. Probes with more embeddings can better "adapt" to the other

---

Algorithm 1. PivotPartitioning

Input: chip dimension,
             set of probes $\mathcal{P} = \{p_1, p_2, ...p_n\}$,
             maximum partitioning depth t_{max}
Output: placement of the probes $p \in \mathcal{P}$ on the chip

1. Select probes p with minimum number of embeddings, $E(p)$, as pivot candidates:
   (a) Let $\mathcal{Q} = \{p \in \mathcal{P} | E(p) \text{ is minimal}\}$
   (b) Set $\mathcal{P} \leftarrow \mathcal{P} \setminus \mathcal{Q}$
2. Let the region R consist of all rows and columns.
3. Call the Recursive Partitioning with the initial partitioning depth 1:
   return RecursivePartitioning $(1, t_{max}, R, \mathcal{Q}, \mathcal{P})$

---

probes, that is, when placed on a particular spot, they are more likely to have an embedding with fewer conflicts than a probe that has only a limited number of embeddings.

We use the probes with fewer embeddings, which we call "pivots", to drive the partitioning of the probe set and to re-embed the probes just before their placement (as a partitioning algorithm, PP also works in combination with another placement algorithm). Also, we designed our algorithm to work for border length as well as conflict index minimization.

5.1   Pivot Candidates

The first step of the Pivot Partitioning (Algorithm 1), is to select the pivot candidates $\mathcal{Q}$, a set of probes that can later be chosen as pivots. Our pivots are the equivalent of the centroids of the CQ algorithm: they are used to partition the probe set. They are restricted, however, to the probes having fewer embeddings.

The reasons are two-fold. First, less time is spent choosing the pivots since fewer candidates need to be considered. Second, probes with fewer embeddings are better representatives to drive the partitioning. The problem is that some embeddings may have their unmasked steps concentrated in one region of the deposition sequence. This is specially true if the probes are embedded in a left-most or right-most fashion. Some Affymetrix probes, for instance, can be synthesized in the first 37 masking steps, thus using only half of the total 74 steps. Such probes are clearly not good choices for pivots. Probes with fewer embeddings, on the other hand, are guaranteed to cover most (if not all) cycles of the deposition sequence.

In order to guarantee a good partitioning, we limit the size of $\mathcal{Q}$ to a minimum of 1% of the total number of probes[1]. This is achieved by selecting probes with the next minimum number of embeddings. Computing the number of embeddings of a probe takes $O(\ell T)$ time, where ℓ is the length of the probe and T is the

---

[1] Usually, around 1-2% of the probes of an Affymetrix array have only one possible embedding; or two, if we consider that they appear in PM/MM pairs and must be "aligned" in all but the steps that synthesize their middle bases.

Algorithm 2. Recursive Partitioning

Input: current depth t,
          maximum depth t_{max}
          rectangular region R of the chip,
          set of pivot candidates Q,
          set of probes P,
Output: placement of the probes $p \in P$ and $q \in Q$ on the region R of the chip

1. If $t = t_{max}$ then
   (a) Re-embed $p \in P$ optimally with respect to all $q \in Q$
   (b) Return RowEpitaxial $(R, P \cup Q)$
2. Select q' and $q'' \in Q$ such that $h(q', q'')$ is maximal
3. Partition the set of pivot candidates:
   (a) $Q' = \{q \in Q \mid h(q, q') < h(q, q'')\}$
   (b) $Q'' = \{q \in Q \mid h(q, q') > h(q, q'')\}$
       (when $h(q, q') = h(q, q'')$, assignments are made in an attempt to achieve
       balanced partitionings)
4. Partition the set of probes:
   (a) $P' = \{p \in P \mid w(p, q') < w(p, q'')\}$
   (b) $P'' = \{p \in P \mid w(p, q') > w(p, q'')\}$
       (when $h(p, q') = h(p, q'')$, assignments are made in an attempt to achieve
   .   balanced partitionings)
5. Partition R into two subregions R' and R'' proportionally to the number of probes
   in $P' \cup Q'$ and $P'' \cup Q''$
6. return RecursivePartitioning $(t + 1, t_{max}, R', Q', P')$
   $\cup$ RecursivePartitioning $(t + 1, t_{max}, R'', Q'', P'')$

---

length of the deposition sequence. With a few optimizations, however, even a million probes can be examined in a few minutes.

5.2   Recursive Partitioning

The essence of Pivot Partitioning is its recursive procedure (Algorithm 2) that is executed until a given recursion depth t_{max} is reached. If the maximum recursion depth has not been reached yet, we choose a pair of pivots q' and $q'' \in Q$ with maximum Hamming distance between their embeddings, $h(q', q'')$. All other $q \in Q$ are assigned to a subset of Q associated with the pivot whose Hamming distance to q is minimum (step 3).

The next step similarly partitions P into two subsets. A probe $p \in P$ is assigned to the subset associated with the pivot q with minimum weighted distance $w(p, q)$. The weighted distance is computed with the OSPE algorithm, ignoring the location of the probes since they have not been placed yet. In this way, we make the assignments considering all possible embeddings of p.

Step 5 partitions R into two subregions, proportionally to the number of probes in $Q' \cup P'$ and $Q'' \cup P''$, alternating horizontal and vertical divisions. Since we only deal with rectangular regions, sometimes it is necessary to move a few probes from one partition to the other in order to ensure that the probes fit in the subregions.

Each subregion is then processed recursively. Once the maximum partitioning depth t_{max} is reached, the Row-epitaxial [8] algorithm is used to place the probes of $\mathcal{P} \cup \mathcal{Q}$ in the region R. Before that, however, all probes $p \in \mathcal{P}$ are re-embedded optimally with respect to the pivots (again using OSPE ignoring probe locations), which improves the "alignment" of all embeddings in that region.

6   Results and Discussion

We now present the results of running our Pivot Partitioning (PP) algorithm on random chips. Table 1 shows the normalized border length (total border length divided by the number of probes) using our own implementations of Row-epitaxial (for the placement) as well as the Sequential post-placement optimization.

Our results show that, in the first level of partitioning, PP allows for a reduction in border length by as much as 16% when compared to running the Row-epitaxial alone (from 41.27 to 34.69 on 500×500 chips). The total border length for $t_{max} = 2$ on 500×500 is $8\,673\,722$. This represents a reduction of as much as 6.8% over the layout produced by the Centroid-based Quadrisection (CQ) similarly combined with Row-epitaxial and followed by the Sequential optimization, which produced a layout with a border length of $9\,307\,510$ as reported in [9]. In the next levels of partitioning, we observe a small increase in border length but, on the other hand, we also report a significant reduction in running times.

Table 2 shows similar results with the average conflict index. For these experiments, we use a version of Row-epitaxial implemented for conflict index minimization, which fills every spot with a probe p minimizing $\mathcal{C}(p)$. For the post-placement optimization, we use the Sequential algorithm with OSPE for conflict index minimization as described in Section 4. Computing the conflict index of a spot for every probe candidate is not as straight forward as computing

Table 1. Normalized border length of layouts produced by Pivot Partitioning on random chips with dimensions ranging from 100×100 to 500×500, with probes synchronously embedded in a deposition sequence of length 100. Partitioning depths ranges from $t_{max} = 0$ (no partitioning) to $t_{max} = 6$. Row-epitaxial is used for the placement (with $Q = 20\,000$), followed by the Sequential post-placement optimization. Running times are reported in seconds, and do not include the post-placement optimization.

Dim	$t_{max} = 0$		$t_{max} = 2$		$t_{max} = 4$		$t_{max} = 6$	
	Cost	Time	Cost	Time	Cost	Time	Cost	Time
100	42.77	34	39.19	13	40.72	10	42.11	11
200	41.63	429	37.30	155	38.53	62	40.00	85
300	41.38	1 174	36.12	766	37.22	264	38.53	139
500	41.27	3 524	34.69	3 472	35.50	1 996	36.58	713

Table 2. Average conflict index of layouts produced by Pivot Partitioning on random chips of synchronous embeddings. We use versions of the Row-epitaxial (with $Q = 2\,000$) and the Sequential algorithms for conflict index minimization.

Dim	$t_{max} = 0$		$t_{max} = 2$		$t_{max} = 4$		$t_{max} = 6$	
	Cost	Time	Cost	Time	Cost	Time	Cost	Time
100	514.49	45	453.67	37	467.78	19	475.44	15
200	517.07	192	466.22	215	452.41	166	462.55	99
300	518.51	438	475.84	524	452.00	466	448.17	336
500	517.50	1471	481.36	1530	462.33	1472	445.43	1295

the Hamming distance between a probe and its neighbors; thus both versions of Row-epitaxial and Sequential for conflict index minimization are significantly slower. For this reason, we set the limit on the number of probes considered by the Row-epitaxial to $Q = 2\,000$.

We also compare the performance of Pivot Partitioning with the Centroid-based Quadrisection (CQ). Table 3 shows the total border length of layouts produced by CQ as reported in [9]. We run PP on similar input and report the results with equivalent partitioning depths (two levels of PP are equivalent to one level of CQ). The results are shown as a percentage of reduction in border length compared to CQ. For instance, on 500×500 chips, PP produces layouts with 8.95% less conflicts than CQ, on average.

Our results show that PP produces layouts with less conflicts than CQ except for higher partitioning depths on the smaller chips. We suspect that this disadvantage is due to the "borrowing heuristic" used by CQ that permits, during the placement, borrowing probes from neighboring partitions in order to maintain a

Table 3. Comparison between Pivot Partitioning (PP) and Centroid-based Quadrisection (CQ) on random chips with dimensions ranging from 100×100 to 500×500, whose probes are synchronously embedded in a deposition sequence of length 100. The partitioning depths varies from $L = 1$ to $L = 3$ for the CQ algorithm and, equivalently, from $t_{max} = 2$ to $t_{max} = 6$ for PP. Both partitionings use Row-epitaxial for the placement (with $Q = 20\,000$) and are followed by the Sequential post-placement optimization. The data shows the total border length of chips produced by CQ (extracted from [9]), and the results of using PP on similar input, as percentage of the reduction in border length compared to CQ. For instance, PP generates on average 8.95% less border length on 500×500 chips with $t_{max} = 2$.

Dim	CQ $L = 1$	PP $t_{max} = 2$	CQ $L = 2$	PP $t_{max} = 4$	CQ $L = 3$	PP $t_{max} = 6$
100	393 218	0.18%	399 312	-1.89%	410 608	-2.48%
200	1 524 803	2.27%	1 545 825	0.48%	1 573 096	-1.34%
300	3 493 552	7.12%	3 413 316	2.05%	3 434 964	-0.61%
500	9 546 351	8.95%	9 355 231	4.67%	9 307 510	1.03%

high number of probes that can be considered for filling the last spots of a quadrant. We are planning to implement a similar strategy on Pivot Partitioning that could improve the quality of our solutions.

7 Summary

We have presented a new partitioning strategy that for the first time combines the partitioning the chip with embedding of the probes. The main advantages of our approach over previous methods are: faster and better selection of pivots used to drive the assignment of probes to subregions; and improved assignment of probes to regions by considering all valid embeddings of a probe.

Acknowledgments

We thank Ion Mandoiu, Xu Xu and Sherief Reda for providing an implementation of their algorithms.

References

1. Binder, H., Preibisch, S.: Specific and nonspecific hybridization of oligonucleotide probes on microarrays. *Biophysical Journal* (2005) **89** 337–352.
2. de Carvalho Jr., S., Rahmann, S.: Microarray Layout as a Quadratic Assignment Problem. Submitted (2006).
3. Çela,E. (1998) *The Quadratic Assignment Problem: Theory and Algorithms.* Kluwer, Massachessets, USA.
4. Chase, P.: Subsequence numbers and logarithmic concavity. *Discrete Mathematics* (1976) **16** 123–140.
5. Fodor, S., Read, J., Pirrung, M., Stryer, L., Lu, A., Solas, D.: Light-directed, spatially addressable parallel chemical synthesis. *Science* (1991) **251** 767–73.
6. Hannenhalli, S., Hubell, E., Lipshutz, R., Pevzner, P.: Combinatorial algorithms for design of DNA arrays. *Advances in Biochemical Engineering / Biotechnology* (2002) **77** 1–19.
7. Kahng, A., Mandoiu, I., Pevzner, P., Reda, S., Zelikovsky, A.: Border length minimization in DNA array design. In *Proceedings of the Second Workshop on Algorithms in Bioinformatics* (WABI 2002).
8. Kahng, A., Mandoiu, I., Pevzner, P., Reda, S., Zelikovsky, A.: Engineering a scalable placement heuristic for DNA probe arrays. *Proc. 7th Int'l Conf. on Computational Molecular Biology RECOMB'03* (2003) 148–156.
9. Kahng, A., Mandoiu, I., Reda, S., Xu, X., Zelikovsky, A.: Evaluation of placement techniques for DNA probe array layout. *Proc. the IEEE/ACM Conf. on Computer-Aided Design* (2003) 262–269.
10. Rahmann, S.: The shortest common supersequence problem in a microarray production setting. *Proc. 2nd European Conf. on Computational Biology ECCB'03*, *Bioinformatics*, 19(Suppl. 2):ii156–ii161.

Accelerating the Computation of Elementary Modes Using Pattern Trees

Marco Terzer and Jörg Stelling

ETH Zurich, Department of Computer Science, 8092 Zurich, Switzerland
{marco.terzer, joerg.stelling}@inf.ethz.ch

Abstract. Elementary flux modes (EFMs)—formalized metabolic pathways—are central and comprehensive tools for metabolic network analysis under steady state conditions. They act as a generating basis for all possible flux distributions and, thus, are a minimal (constructive) description of the solution space. Algorithms to compute EFMs descend from computational geometry; they are mostly synonymous to the enumeration of extreme rays of polyhedral cones. This problem is combinatorially complex, and algorithms do not scale well. Here, we introduce new concepts for the enumeration of adjacent rays, which is one of the critical and stubborn facets of the algorithms. They rely on variants of k-d-trees to store and analyze bit sets representing (intermediary) extreme rays. *Bit set trees* allow for speed-up of computations primarily for low-dimensional problems. Extensions to *pattern trees* to narrow candidate pairs for adjacency tests scale with problem size, yielding speed-ups on the order of one magnitude relative to current algorithms. Additionally, fast algebraic tests can easily be used in the framework. This constitutes one step towards EFM analysis at the whole-cell level.

1 Introduction

Metabolic networks are characterized by their complexity. Even in simple bacteria, they involve ≈2.000 metabolites and ≈1.000 proteins that catalyze reactions converting external substrates to metabolites and products. For their computational analysis, in particular, stoichiometric or constraint-based approaches have gained popularity because the necessary reaction stoichiometries and reversibilities are usually well–characterized, in contrast to reaction kinetics and associated parameters [1]. For example, genome–scale stoichiometric models have been constructed for several organisms to predict flux distributions in metabolic networks in normal or perturbed conditions as well as optimality and control thereof [2].

Conceptually, the analysis starts from the $m \times q$ stoichiometric matrix $\mathbf{N}$, where m is the number of (internal) metabolites and q the number of reactions. As metabolism usually operates on faster time–scales than other cellular processes, we can assume (quasi) steady–state for the metabolic reactions to derive the fundamental metabolite balancing equation:

$$\boldsymbol{N} \cdot \boldsymbol{r} = \boldsymbol{0} \tag{1}$$

P. Bücher and B.M.E. Moret (Eds.): WABI 2006, LNBI 4175, pp. 333–343, 2006.

where the $(q \times 1)$-vector r represents a *flux distribution*. Additionally, the reaction rates r are subject to thermodynamic feasibility constraints for irreversible reactions (into which any reversible reaction can be decomposed):

$$r \geq 0 \qquad (2)$$

Eqs. (1) and (2) constrain the solution space for valid reaction fluxes to a *convex polyhedral cone P* (see Section 2 for formal definitions). Hence, comprehensively analyzing metabolic network behavior amounts to characterizing P [3]. Metabolic pathways such as elementary flux modes (EFMs) or extreme pathways, which are minimal, linearly independent flux vectors unique for a given network, allow for this because they correspond to extreme rays of P [3].

Thus, computation of EFMs is equivalent to the enumeration of the extreme rays of P, a problem from computational geometry known to be hard for the general case. Current algorithms are variants of the *double description method* (DDM) introduced by Motzkin *et al.* in 1953 [4]. In particular, the *canonical basis approach* [5] and the more efficient *nullspace approach* [6] are used for EFM computation. However, no efficient algorithm is known with time complexity polynomial in the input and output size [7], which currently restricts metabolic pathway analysis to networks of ≈ 100 reactions and metabolites [1].

Here, we propose improved algorithms for EFM computation that address the most critical feature of the DDM, namely the independence tests for (preliminary) extreme rays. We focus on the nullspace approach, but the concepts are readily applicable to the canonical form. After giving fundamental definitions (Section 2) and a detailed description of current algorithms (Section 3), we will present our new approaches relying on k-d trees (Section 4) and experimental results showing their significant impact on performance (Section 5).

2   Fundamentals

Definition 1. *A nonempty set C of points in an Euclidean space is called a (convex) cone if $\lambda x + \mu y \in C$ whenever $x, y \in C$ and $\lambda, \mu \geq 0$.*

Definition 2. *A cone P is polyhedral if $P = \{x \mid A x \geq 0\}$ for some matrix A, i.e. P is the intersection of finitely many linear half-spaces.*

Note that $A = [N^T; -N^T; I]^T$ and $x = r$, with the stoichiometric matrix N, identity matrix I to ensure irreversibility constraints, and the flux distribution r, define the cone in the context of EFM analysis as given by eqs. (1) and (2).

Theorem 1 (Minkowski's Theorem for Polyhedral Cones). *For every cone $P = \{x \mid A x \geq 0\}$ there exists some R such that $P = \{x \mid x = R c$ for some $c \geq 0\}$ is generated by R.*

A is called a *representation matrix* of the polyhedral cone P, R is the *generating matrix* for P. Because both A and R describe the same object P, the pair (A, R) is called *double description pair* or *DD pair* [4,7].

Definition 3. *For any vector $x \in P$, the set $Z(x)$, containing the indices i such that $A_i\, x = 0$, is called the* zero set *of x.*

Definition 4. *A vector r is said to be a* ray *of P if $r \neq 0$ and $\alpha r \in P$ for every $\alpha > 0$. Two rays r and r' are said to be* equivalent, *i.e. $r \simeq r'$, if $r = \alpha r'$ for some $\alpha > 0$.*

Definition 5. *Let r be a ray of P. If one of the following holds, both hold and r is called an* extreme ray:
    (a) $rank(A_{Z(r)}) = rank(A) - 1$
    (b) there is no $r' \in P$ with $Z(r') \supseteq Z(r)$ other than $r' \simeq r$.

If all columns of R are *extreme rays*, R is called a *minimal generating set* for P.

3   Existing Algorithms

3.1   Double Description Method (DDM)

The DDM relies on the definition of *adjacent rays* that is derived from the *extreme ray definition (5)*. Thus, there exist two options to ensure adjacency, (a) sometimes referred to as *algebraic* adjacency test, (b) as *combinatorial* test:

Definition 6. *Let r and r' be two extreme rays of P. If one of the following holds, both hold and r and r' are said to be* adjacent:
    (a) $rank(A_{Z(r) \cap Z(r')}) = rank(A) - 2$
    (b) if $r'' \in P$ with $Z(r'') \supseteq Z(r) \cap Z(r')$ then $r'' \simeq r$ or $r'' \simeq r'$.

The algorithm constructs R from A iteratively as follows:

1. *Initialization Step:* Since P is pointed, i.e. 0 is an extreme point of P, A has full rank d, a nonsingular square sub-matrix A_d exists, and (A_d, A_d^{-1}) is an initial DD pair. As we will see for the *nullspace approach*, other initial pairs are possible.

2. *Iteration Step:* Assume the DD pair (A_j, R_j) with j inequality constraints from $A\,x \geq 0$ already considered. The next DD pair (A_{j+1}, R_{j+1}) is achieved by fulfilling an additional inequality $a_{j+1} := A_{j+1}\, x \geq 0$.

   (a) The hyperplane $H^0_{j+1} = \{x \mid A_{j+1}\, x = 0\}$ separates R_j into 3 parts:
       i. R^0_j, the extreme rays of R_j fulfilling inequality a_{j+1} with equality,
       ii. $R^+_j \subseteq R_j$ fulfilling a_{j+1} with strict inequality and
       iii. $R^-_j \subseteq R_j$ not fulfilling a_{j+1}.

   (b) The matrix R_{j+1} is constructed as the union of
       i. those extreme rays that still fulfill the new condition $(R^0_j \cup R^+_j)$
       ii. together with the rays resulting from the intersection of the separating hyperplane H^0_{j+1} with the hyperplane through the pair of rays (r^-, r^+) where $r^- \in R^-_j$, $r^+ \in R^+_j$ and r^- is *adjacent* to r^+, i.e. the newly created ray is an extreme ray. This step is also known as *Gaussian elimination* with the newly constructed ray r' in H^0_{j+1}:
   $$r' = (A_{j+1}r^+)r^- - (A_{j+1}r^-)r^+.$$

3. Continue with 2 until all inequalities are considered.

3.2   Binary Nullspace Algorithm

Nullspace approach. Wagner [6] proposed to use a well defined form of the kernel matrix K of N as an initial minimal representation matrix, where $K = [I; K^*]^T$. If $N^{m \times q}$ has full rank, i.e. $d = rank(N) = m$, the kernel matrix has dimensions $q \times (q - m)$ and K^* consequently $m \times (q - m)$. Thus, this initialization results in $(q - m) + 2m = q + m$ resolved constraints, leaving m inequalities to be solved in the iteration phase. It can be shown [3] that (A, K) form an initial DD-pair with K being a minimal generating matrix and $A^{(q+m) \times q} = \left[I_{q-m} 0^{(q-m) \times m}; N; -N \right]$. The *nullspace approach* proved to be faster than the original version, removes redundancies (by the nature of the kernel matrix), and simplifies the Gaussian elimination step.

Bit sets. Adjacency tests are the most expensive parts of the algorithm. However, as we only need to know whether or not a ray fulfills a specific inequality with equality, we can use *bit sets* to store this information. Corresponding to the *zero sets* in definition 3, the *bit set zero set* of a given vector x at iteration step j is defined as follows, complementary to [3]:

Definition 7. *For any $x \in P_j$, P_j being the polyhedral cone at iteration step j represented by the double description pair (R_j, A_j), the set*

$$B_j(x) = \{r_1 r_2 \ldots r_j \mid r_i \in [0,1], 1 \le i \le j\} \quad with \quad r_i = \begin{cases} 1 & if\ A_i x = 0 \\ 0 & otherwise \end{cases}$$

is called the bit set representation *of the zero set of x.*

We will use the shorter term *zero set* subsequently for *bit set representation of the zero set*.

   The *bitwise and* operation for zero sets corresponds to the intersection of sets, because for every bit-position in the bit set, the position in the resulting set is 1 iff the position was 1 in both source sets. Accordingly, the subset (or superset) operation can be performed by:

$$B(x) \subseteq B(y) \iff B(x) \wedge B(y) \equiv B(x) \tag{3}$$

Proposition 1. *To derive the zero set of a vector at iteration $j+1$, the following operations are performed:*

$$B_{j+1}(x) = \begin{cases} B_j(x) + 1 & if\ x \in R^0_{j+1} \\ B_j(x) + 0 & if\ x \in R^+_{j+1} \end{cases} \tag{4}$$

for extreme rays which still fulfill the new equation and are kept, and

$$B_{j+1}(x, y) = \{B_j(x) \wedge B_j(y) + 1 \mid x \in R^+_{j+1}, y \in R^-_{j+1}, x\ adj.\ to\ y\} \tag{5}$$

for newly combined rays, where $+$ stands for concatenation, $\wedge$ for the bitwise and operation.

Proof. The proof for (4) and (5) immediately emanates from definition (3).

The bit set representation of zero sets has two main advantages: It demands little space in memory, and set operations (*bitwise and*, subset tests) for adjacency can be performed efficiently. Moreover, storing only one bit for vector elements concerning rows in A which have already been processed is sufficient. The number of zero positions in extreme rays is maximized and the combination of zeros and non-zeros is unique; thus, the original real-valued rays can be reconstructed from the bit set extreme rays after the final iteration step [3].

4   New Approaches

4.1   Bit-Set Trees

The bit sets in definition 7 can be seen as k-tuples of $[0, 1]$ values, and thus search operations on a set of bit sets coincide with queries on a collection of k-dimensional records. For this purpose, k-d-trees have been invented as a structure for storage and retrieval of multidimensional (k-dimensional) data [8].

In the context of EFM-computation, we need to test for the existence of a superset for a given bit set. For 2 adjacent rays r and r' with corresponding zero sets $B(r)$ and $B(r')$, the combinatorial adjacency test as defined in 6(b) bars the existence of a zero set that is superset of $B(r) \cap B(r')$ other than $B(r)$ and $B(r')$. This type of queries can operate on a binary k-d-tree and works similar to the *partial match queries* given in [8].

Tree construction. Given a set of bit sets (our zero sets), the algorithm returns a binary k-d-tree or *bit set tree*. The input of the algorithm is a *set* of bit sets, that is, a collection without duplicates, which conforms to the actual problem. This simplifies step 2 of the algorithm below, where the bit sets are split into two newly created leafs, and we can assure that infinite loops are avoided.

main   Create a leaf node containing all bit sets and invoke **sub** with it. The returned node is the tree's root r.

  sub   1. If the leaf node contains not more elements than some threshold (the maximum leaf size), return it and continue at invoker.
2. Choose some bit j that has not yet been used on prior levels. Separate the leaf's bit sets and create two new leaf nodes *zero* and *one* containing the bit sets with $bit_j = 0$ and $bit_j = 1$, respectively.
3. Recursively invoke **sub** with *zero* and *one*.
4. Create a new intermediary node i with two children *zero* and *one*, the nodes returned by **sub** in 3. Return i and continue at invoker.

Superset existence. Given the root r of a bit set tree t constructed as described above and a bit set s to be tested, where $s = s^+ \cap s^-$ with $s^+ \in t$ and $s^- \in t$, the algorithm returns *true* if a super set of s is contained in t (other than s^+ and s^-), *false* otherwise (i.e. it returns *true* iff s^+ is adjacent to s^-). The functionality of the algorithm is illustrated in Fig. 1.

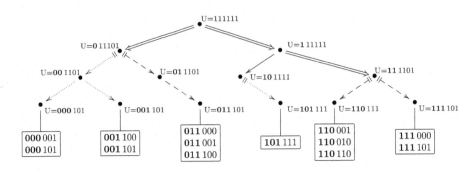

Fig. 1. Superset-Existence algorithm on a *bit-set tree/pattern set tree* with ternary leafs and a test bit set $s = 010011$. Double-lines indicate pointers to child nodes which are traversed in both tree-variants, dotted lines are traversed in neither of them. Dashed lines are only traversed in the bit-set tree, single solid lines only in the pattern-tree. Double-bar arrow-heads highlight truncation by the pattern.

main   Invoke sub with the root node r and return the result from that call.

sub 1. If the current node is a leaf, iterate through the leaf's bit sets and return *true* if any of them is a superset of s (not being s^+ or s^-), *false* otherwise.

2. Let s_i be the bit i of s where i is the bit position corresponding to the current node (this bit has been used to separate the bit sets in child node *zero* from those in *one*).

3. Invoke sub with *one*. If *true* is returned, pass it to the invoker.

4. If $s_i = 0$, call sub with *zero* and return the result, else return *false*.

Correctness and complexity. By the way of constructing the tree, the *zero* child of an intermediary node with selective bit j contains those bit sets that have $bit_j = 0$. Thus, if the set s to be tested contains j, that is, $bit_j = 1$, the bit sets in *zero* cannot be supersets of s and only the bit sets in *one* are superset candidates, conforming with the recursion condition in step 4.

We cannot estimate the number of intermediary modes and, thus, the overall time complexity of the DDM. However, for each step, at least $d - 1$ inequalities are fulfilled with equality, where $d = rank(\boldsymbol{A})$ (definition 5(a)). Since $\boldsymbol{A}$ contains $\boldsymbol{I}$, $d = q$ equals the number of irreversible reactions, and due to the nature of the nullspace, all equality constraints are fulfilled. They correspond to $2m$ rows in $\boldsymbol{A}$ with rank m (assuming independent rows in $\boldsymbol{N}$), thus $q - 1 - m$ positions are left to be fulfilled with equality. That is, the bit sets in t have at least $q - m - 1$ 1-bits, and due to definition 6(a) s at least $q - m - 2$ respectively. With bit set length l ($q - m \leq l \leq q$), the probabilities of a 1 in s and in the tree's bit sets can be estimated:

$$\begin{cases} n \cdot \frac{q-m-1}{l} & \text{remaining bit sets with probability} \quad \frac{q-m-2}{l} \\ n & \text{remaining bit sets with probability} \quad 1 - \frac{q-m-2}{l} \end{cases} \tag{6}$$

We assume a well balanced tree of depth $log_2(n)$ and set $\epsilon_1 = \frac{q-m-1}{l}$ and $\epsilon_2 = \frac{q-m-2}{l}$. The time complexity at step j is proportional to the number of considered bit sets per adjacency test, approximated by

$$n \cdot (1 - \epsilon_2 + \epsilon_1\epsilon_2)^{log_2(n)} = n^{1+log_2(1-\epsilon_2+\epsilon_1\epsilon_2)} \tag{7}$$

Note that the sublinear function in eq. 7 has an optimum at $\epsilon_{1/2} \approx 1/2$. It is relatively insensitive to perturbations in $\approx [0.2, 0.8]$, especially for large n. For real problems, eq. 7 is a good (and conservative) approximation.

In a well-balanced tree, we have $n/2$ nodes holding n unary leafs, requiring $c \cdot 2n$ additional memory space for a total of n intermediary nodes and n leafs, where c is a small constant. Optimizations could be applied, but these memory demands are far from being critical for our purposes. In [8], an algorithm is presented which constructs a balanced tree based on the median of a collection of elements. With binary values, this approach cannot be applied, but we can adjust the selective bit at step 2 of the tree construction. Either a *static* bit order is calculated before constructing the tree, or the most selective bit is chosen *dynamically* when the leaf's bit-sets are subdivided. We get closer to optimally balanced trees with dynamic choice, but loose the property of having the same selective bit for nodes on the same level. Here, we used *static* and *dynamic* heuristics, leaving space for subsequent explorations.

4.2   Pattern Trees

The general idea of *pattern trees* ties up to the *bit set trees*, where bit sets are separated into two child nodes in every intermediary node, taking some designated *selective bit* as criterion for partitioning. In pattern trees, additionally, all intermediary and leaf nodes account for the bit sets of their children by a *union pattern* of all bit sets contained in the subtree. At least the selective bits of the node and its predecessors are common for all bit sets in the subtree. However, since the actual bit sets constitute only a small fraction of all possible values, it is likely that other common 0's will occur in the pattern. This allows a more restrictive pre-rejection of test sets.

Proposition 2. *Let s be a set, E a collection of sets and $U = \{\bigcup e \mid e \in E\}$ the union of all sets in E. Then $s \subseteq U$ is a necessary condition for $\{e \mid s \subseteq e, e \in E\} \neq \emptyset$, i.e. that a superset of s exists in E.*

Proof. If $s \not\subseteq U$, s contains at least some $j \notin U$. Thus, for all $e \in E$, $j \notin e$ and consequently $s \not\subseteq e$ hold.

Tree construction. In addition to the algorithm for constructing *bit set trees*, we calculate the *union pattern* U when a new leaf node (containing the set E of bit sets) is created (in `main` and at step 2) as $U = (\vee e \mid e \in E)$ where $\vee$ stands for the *bitwise or* operation.

Superset existence. The algorithm works very similarly to that given for *bit set trees*, passing the root node r of a pattern tree. Note that no bit tests are performed in step 4 and the recursion is always invoked with both children. Step 1 avoids descending the *zero*-subtree if the test bit of s was 1 since the pattern of the *zero*-child contains 0 at the respective bit position (Fig. 1).

main   Invoke sub with the root node r and return the result from that call.

sub 1. If $s \not\subseteq U$, U being the *union pattern* of the node, return *false*.
   2. If the node is a leaf, iterate through the leaf's bit sets and return *true* if a superset of s exists (not being s^+ or s^-), *false* otherwise.
   3. Invoke sub with *one*. If *true* is returned, pass it to the invoker.
   4. Invoke sub with *zero* and return the result.

Correctness and complexity. The only point where we decide to disregard some superset candidates is at step 1, where sets are excluded if their union pattern does not fulfill the condition introduced in proposition 2 necessary for the existence of supersets in the corresponding subtree.

Both time and memory demands for pattern trees are very similar to those of bit set trees, and it is beyond the scope and objectives of this work to achieve more precise estimates for time complexity.

4.3   Narrowing Adjacent Pair Candidates

In the previous sections, we have addressed *adjacency testing*. We will now focus on narrowing the ray-pairs being candidates for adjacency even before testing.

Proposition 3. *Let E_1, E_2 and E_3 be collections of sets with corresponding union patterns $U_j = \{\bigcup e \mid e \in E_j, 1 \leq j \leq 3\}$, and let $S = \{s_1 \cap s_2 \mid s_1 \in E_1, s_2 \in E_2\}$. Then*

$$\exists s_3 : s_3 \in E_3 \text{ with } U_1 \cap U_2 \subseteq s_3 \implies \exists s_3 : s_3 \in E_3 \text{ with } s \subseteq s_3 \qquad (8)$$

holds for every $s \in S$.

Proof. By definition, $U_1 \cap U_2 = ((s_{11} \cup \cdots \cup s_{1n}) \cap (s_{21} \cup \cdots \cup s_{2m}))$. Applying the distributive law, we get $((s_{11} \cap s_{21}) \cup (s_{11} \cap s_{22}) \cup \cdots \cup (s_{11} \cap s_{2m}) \cup \cdots \cup (s_{1n} \cap s_{2m}))$ being the union of all elements of S. Thus, $s \subseteq U_1 \cap U_2$, and consequently $U_1 \cap U_2 \subseteq U_3 \implies s \subseteq U_3$. From this, eq. (8) follows by replacing U_3 by s_3. If an s_3 exists (left hand of 8), the same s_3 exists on the right hand side.

We can use eq. 8 as necessary preconditions for the all-pair combinations. The success of this shortlisting of candidates highly depends on the relations between the sets or their union patterns. Higher similarities of patterns U_1 and U_2 enhance the probability of the precondition being true, while larger sets E_1 and E_2 are more desirable since more pairs could be eliminated. *Pattern trees* comply with these requirements well since they constitute subtrees with union patterns. Nodes in the upper part of the tree have many, but barely similar entries; descending the tree means lowering the number of entries and increasing the similarity. This characteristic can be used to find the optimal balance between number and similarity of entries in a set.

Here, we implemented an algorithm that tests the *cut-pattern* $U_1 \cap U_2$ for two leaf nodes. We used heuristics to calculate the optimal leaf size, that is the number of bit sets per leaf, in a manner of *statically* balancing similarity and exclusion. Future development should also consider *dynamic* balancing by calculating the *cut pattern* for all nodes, not only leaf nodes, to reject candidates at different tree levels.

In Fig. 1, the pattern combination of the left-most leaf nodes leads to the *cut-pattern* of 000101 (000101 $\wedge$ 001101), for which we find a superset 111101 in the right-most leaf node of the tree. Thus, no pair from the left-most leafs form an adjacent pair, which has been found by one test instead of four for all combinations.

Algorithmic extensions. At iteration j of the double description algorithm, we construct a *pattern tree* t_j as described above with the following extensions:

1. The collections of zero sets in the leafs of the pattern tree are divided into three subsets S^0, S^+, and S^- corresponding to the separation of the rays by the hyperplane H_j^0 at step 2a of the DD-algorithm.

2. We calculate three *union patterns* for every leaf l:

$$
\begin{aligned}
l.U &= \{\bigvee s \mid s \in l.S^0 \cup l.S^+ \cup l.S^-\} \\
l.U^+ &= \{\bigvee s \mid s \in l.S^+\} \\
l.U^- &= \{\bigvee s \mid s \in l.S^-\}
\end{aligned}
$$

To create the extreme rays for the next iteration step, we iterate through the tree's leafs L, and initialize $L_A = L$.

loop$_A$ 1. Choose some leaf l_A from L_A and initialize $L_B = L_A$.

2. Collect the zero sets in S^0 and S^+ of l_A and apply eq. 4.

    loop$_B$ i. Pick some leaf $l_B \in L_B$ (possibly again l_A) and calculate the
$$
\text{cut-patterns} \begin{cases} C^{+-} = l_A.U^+ \wedge l_B.U^- \\ C^{-+} = l_A.U^- \wedge l_B.U^+ \end{cases}
$$

ii. If a superset s for C^{+-} exists in the tree with $s \notin l_A.S^+, s \notin l_B.S^-$, no pair $(s_A^+, s_B^-) \in (l_A.S^+, l_B.S^-)$ is an *adjacent* pair according to eq. 8, thus continue at (iv).

iii. Test every pair $(s_A^+, s_B^-) \in (l_A.S^+, l_B.S^-)$ as usual, i.e. by testing the intersection $s_A^+ \wedge s_B^-$, and apply eq. 5 for adjacent pairs.

iv. Repeat (ii) and (iii) with C^{-+} accordingly.

v. Remove l_B from L_B and continue at **loop$_B$** if L_B is nonempty.

3. Remove l_A from L_A and continue at **loop$_A$** if L_A is nonempty.

5   Experimental Results

As realistic examples, we used variants of a stoichiometric model for the central metabolism of *Escherichia coli* [9]. Network compression techniques mostly

Table 1. Computation of the elementary modes for variants of the central metabolism of *Escherichia coli*. Abbreviations: Glc = glucose, Ac = acetate, Form = formiate, Eth = ethanol, Lac = lactate, CO2 = carbon dioxide, Succ = succinate, Glyc = glycerol. The products considered were Ac, Form, Eth, Lac, CO2. Relative speed figures relate to the bit set tree version. Note that due to the current implementation, the absolute time measurements are somewhat above those given in [10].

	S0		S1		S2	
substrates	Suc		Glc		Glc, Succ, Glyc, Ac	
network size	97 × 114 (28 rev.)		97 × 114 (28 rev.)		97 × 118 (28 rev.)	
compressed size	34 × 45 (16 rev.)		35 × 46 (17 rev.)		37 × 52 (17 rev.)	
iteration steps	28		29		31	
elementary modes	7,055		27,100		507,632	
	time	rel. speed	time	rel. speed	time	rel. speed
iterate all	6.6s	0.5	453s	0.1	-	-
bit set tree	3.6s	1.0	38s	1.0	451min	1.0
pattern tree	3.4s	1.1	28s	1.4	431min	1.1
candidate narrowing	2.6s	1.4	14s	2.7	28min	16.1

identical to those presented in [3] were applied. The algorithm was entirely implemented in Java and the tests were run on a Linux machine with an AMD Opteron(tm) 250 processor with 2.4 GHz, using a Java 5 virtual machine with max. 4 GB memory. The results summarized in table 1 show major improvements from the primitive combinatorial adjacency test to those with bit-set or pattern trees for small problem sizes, where adjacent candidate pair narrowing yields lower advances. With higher dimensional problems, candidate narrowing scales much better than merely improving testing and in fact becomes essential with regard to whole-cell metabolic networks.

6   Conclusions and Prospects

Determining elementary flux modes constitutes an important problem for bioinformatics. In addition, it is relevant for other domains of computer science / applied mathematics due to the nature of the underlying problem: the enumeration of all extreme rays of convex polyhedral cones. In this work, we focused on one aspect of the double description method that is critical for its performance, namely the independence tests for (preliminary) extreme rays. Conceptually, we introduced variants of k-d trees—*bit-set trees* and *pattern trees*—to implement the search for a superset of a given test set in the combinatorial adjacency test, and for effectively restricting the search scopes. Implementations and applications of the algorithms to real-world metabolic networks confirmed performance gains on the order of one magnitude compared to currently employed algorithms. In particular, refined searches using pattern trees scale well with problem size, which is important for ultimately analyzing whole-cell networks.

In perspective, further improvements are expected by exploiting pattern trees for pre-rejection of test candidate pairs in a more sophisticated manner, by adaptive methods for tree-balancing, and by employing the more efficient rank test for adjacency that does not depend on the number of modes to be tested. It is quite simple to combine rank test and candidate narrowing with pattern trees, by demanding that *cut patterns* pass the rank test. All these aspects require further efforts in theory, for instance, to determine optimal balancing schemes. Porting efficiency-sensitive code parts from Java to a high-performance language will certainly help fully realize the algorithms' potential; another relevant and attractive topic regarding applicability to larger networks is parallelization. Overall, we anticipate these approaches to finally enable enumeration of elementary modes for genome-scale metabolic networks. This, of course, still has to be proven. The subsequent interpretation of huge sets of metabolic pathways is yet another challenging and interesting problem.

Acknowledgments

We thank Gaston Gonnet for algorithmic ideas and for comments on the manuscript.

References

1. Klamt, S., Stelling, J.: Stoichiometric and constraint-based modeling. In Szallasi, Z., Stelling, J., Periwal, V., eds.: System Modeling in Cellular Biology. MIT Press (Cambridge / MA) (2006) 73–96
2. Price, N., Reed, J., Palsson, B.: Genome-scale models of microbial cells: Evaluating the consequences of constraints. Nat. Rev. Microbiol. **2** (2004) 886–897
3. Gagneur, J., Klamt, S.: Computation of elementary modes: A unifying framework and the new binary approach. BMC Bioinformatics **5** (2004) 175
4. Motzkin, T.S., Raiffa, H., Thompson, G., Thrall, R.M.: The double description method. In Kuhn, H., Tucker, A., eds.: Contributions to the Theory of Games II. Volume 8 of Annals of Math. Studies., Princeton University Press (Princeton / RI) (1953) 51–73
5. Schuster, S., Hilgetag, C.: On elementary flux modes in biochemical reaction systems at steady state. J. Biol. Syst. **2** (1994) 165–182
6. Wagner, C.: Nullspace approach to determine the elementary modes of chemical reaction systems. J. Phys. Chem. B **108** (2004) 2425–2431
7. Fukuda, K., Prodon, A.: Double description method revisited. In: Combinatorics and Computer Science. (1995) 91–111
8. Bentley, J.L.: Multidimensional binary search trees used for associative searching. Commun. ACM **18** (1975) 509–517
9. Stelling, J., Klamt, S., Bettenbrock, K., Schuster, S., Gilles, E.: Metabolic network structure determines key aspects of functionality and regulation. Nature **420** (2002) 190–193
10. Klamt, S., Gagneur, J., von Kamp, A.: Algorithmic approaches for computing elementary modes in large biochemical reaction networks. IEE Proc. Systems Biol. **152** (2005) 249–55

A Linear-Time Algorithm for Studying Genetic Variation

Nikola Stojanovic[1] and Piotr Berman[2]

[1] Department of Computer Science and Engineering, The University of Texas at Arlington, Arlington, Texas, USA
[2] Department of Computer Science and Engineering, The Pennsylvania State University, University Park, Pennsylvania, USA

Abstract. The study of variation in DNA sequences, within the framework of phylogeny or population genetics, for instance, is one of the most important subjects in modern genomics. We here present a new linear-time algorithm for finding maximal k-regions in alignments of three sequences, which can be used for the detection of segments featuring a certain degree of similarity, as well as the boundaries of distinct genomic environments such as gene clusters or haplotype blocks. k-regions are defined as these which have a center sequence whose Hamming distance from any of the alignment rows is at most k, and their determination in the general case is known to be NP-hard.

1 Introduction

The characterization of inter and intra-species genetic variation is in the core of modern genomics. Studies of homologous regions in different species can provide clues about the evolution, help us identify important loci in human DNA, understand the regulation of complex genetic traits and describe the small number of changes in DNA which have led to the development of consciousness and civilization. The understanding of genetic variation within the human population would help us understand which features in an individual's DNA make that individual more or less susceptible to complex genetic diseases. This would lead to the development of new treatments, and in particular different treatments likely to be effective in different individuals.

Each of these issues has been addressed by major initiatives by the US and other governments, as well as private and nongovernmental organizations. The ENCODE project [10] has started in September of 2003, with the goal of developing high-throughput technologies for cataloguing all functional elements in DNA on a target region comprising approximately 1% of the human genome. The approach so far included extensive sequencing of homologous regions from a large number of species, as well as the development of new computational methods necessary for the analysis of these regions. The HapMap consortium [11] has concentrated on cataloging and analysis of single nucleotide polymorphisms in human DNA, looking at a large number of sequences from individuals belonging to diverse ethnic and racial groups. Both the identification of SNPs [12] and

P. Bücher and B.M.E. Moret (Eds.): WABI 2006, LNBI 4175, pp. 344–354, 2006.

the subsequent determination of haplotype boundaries in different populations required intensive computing and new fast and reliable algorithms [1].

Several years ago, while studying the conservation patterns of short functional sites in DNA, we formulated the 1-*mismatch problem* [8], defined as finding a maximal region in a set of N aligned sequences S_i of length L which have a consensus "center" sequence C whose Hamming distance from each of the S_i, $d(S_i, C)$, is at most 1. We have shown that 1-mismatch problem has a solution running in time $O(NL)$. A natural generalization of this problem is a k-*mismatch problem*, where $d(S_i, C) \leq k$, for any fixed $k > 0$ and all $i \in [1, N]$. However, in the context of the search for regulatory sites in DNA the k-mismatch problem was of lesser interest, as such sites are usually short, ranging between 5 and 25 bases in length. Even as DNA binding proteins are non-specific, interacting with sequences which can have very different composition [2], it is probably not productive to look for more than 1 mismatch per site/sequence in the context of short sites within multiple alignments. However, for other types of problems involving genetic variation, such as the determination of the boundaries of genomic regions, the k-mismatch approach would be appropriate.

The general k-mismatch problem is NP-hard, as shown by independent studies in the coding theory [3], follow-up work on our original paper by other groups [6] [7], and our own analysis (unpublished). In particular, the report in [6] has devised an approximation scheme for this problem. In this paper we show that an important sub-class of the k-mismatch problem, when a Hamming center of exactly 3 sequences is sought, has a linear-time solution[1]. As simultaneous consideration of 3 sequences is often an issue in computational biology (two species in a study, plus an outgroup, for instance) an efficient algorithm for this special case can be of substantial interest. Moreover, since the solution to this problem leads to plausible ancestral sequences it can be used as an aid in finding them as a part of an initiative such as one currently taking place within the framework of the UCSC Genome Browser [4].

2   Algorithm

When three sequences are compared and placed in an alignment, it can have only five possible types of columns: *trivial columns* are these which contain occurrences of one character in all rows, and *nontrivial* columns contain at least two different symbols. Among the latter, it is possible that all three characters in the column are different (we refer to these as *type 0 columns*) or it may be that two characters are same (so we call it the *majority character*) and one occurs only once (*minority character*). If the minority character is in alignment row 1, we call that column *type 1*; if it is in row 2, it is *type 2*; finally, if it is in row 3, we call it *type 3*. By keeping track of the number of columns of each type seen in an interval, it is possible to calculate the distribution of their center characters

---

[1] After an alignment of these sequences has been built, and not counting the time necessary for reporting the overlapping center sequences.

Table 1. Symbols and variables used in the description of the algorithm

Symbol	Meaning
n	Total length of the 3-sequence alignment
n_i	Number of columns of type $i, i \in [0, 3]$
$\mathcal{N}$	Set of all n_i
m_j	Number of mismatches alignment row j has with the center, $j \in [1, 3]$
r_0^j	Number of type 0 columns whose center is from row $j, j \in [1, 3]$
r_i	Number of type i ($i \in [1, 3]$) columns with minority character at center
$\mathcal{R}$	Set of all r_0^j and r_i
start	Beginning of the currently considered interval
stop	End of the currently considered interval (column to the right)
bound	Position which terminated the last reported k-region

so that they create a center sequence for the region, or conclude that a center does not exist, i.e., that the interval is not a k-region.

We shall refer to a consecutive run between alignment columns a and b, inclusive, as to interval $[a, b]$. Within an interval we shall denote the number of columns of type 0, 1, 2 and 3 by n_0, n_1, n_2 and n_3, respectively. We shall refer to all n_i as set $\mathcal{N}$. The number of mismatches alignment rows 1, 2 and 3 have with the center sequence in the current interval will be denoted by m_1, m_2 and m_3, respectively. We shall use symbols $r_0^j, j \in [1, 3]$ to denote the number of type 0 columns whose center is from row j and $r_i, i \in [1, 3]$ to denote the number of type i columns whose center is the minority character. All r_0^j and r_i values will be referred to as set $\mathcal{R}$. Finally, we have referred to the length of the sequences under consideration as L (or rather L_j), but since we are interested in the length of their alignment, i.e., its total number of columns, we shall denote it by n. This notation is summarized in Table 1. The algorithm for locating all maximal k-regions in three aligned sequences is based on the idea that one can proceed through the alignment, from a specified starting point, keeping track of the number of nontrivial columns of each type seen between the start and the current position, and assign center characters to the columns in accordance with the following scheme: for trivial columns, center is always the unique character of the column; for columns of type 0 we choose the character in row 1, and for columns of type 1, 2 and 3 we choose the majority character. This process can continue until some row accumulates $k + 1$ mismatches with the tentative center. At that point, a re-distribution of center characters has to be done.

According to the scheme, $m_1 = n_1$ (the number of type 1 columns). If $m_1 > k$, then some type 1 columns have to have the minority character elected. As each selection of row 1 character for the center of a type 1 column causes an additional mismatch with both rows 2 and 3, this process can be done only as long as both $m_2 \le k$ and $m_3 \le k$. In this case the search for the end of the current maximal k-region can proceed, otherwise the region concluded and it can be reported.

Algorithm 2.1: K3($Al, n, k, threshold$)

$start \leftarrow 1; stop \leftarrow 0; bound \leftarrow 0$
$n_0 \leftarrow 0; n_1 \leftarrow 0; n_2 \leftarrow 0; n_3 \leftarrow 0$
while $bound < n + 1$

do $\begin{cases} \text{if } stop = n + 1 \\ \quad \text{then } halt \leftarrow \text{ true} \\ \quad \text{else } halt \leftarrow \text{CALCULATE\_SCORES}(\mathcal{N}, \mathcal{R}, k) \\ \text{if } halt = \text{ true} \\ \\ \text{then} \begin{cases} \text{if } stop - start + 1 > threshold \text{ and } stop > bound \quad (1) \\ \quad \text{then REPORT\_REGION}(Al, start, stop, \mathcal{R}) \quad\quad (2) \\ bound \leftarrow stop \\ \text{repeat} \\ \quad next \leftarrow \text{ADVANCE\_START}(Al, \mathcal{N}, start) \\ \text{until } next \neq \texttt{TRIVIAL} \end{cases} \\ \\ \text{else} \begin{cases} \text{repeat} \\ \quad next \leftarrow \text{ADVANCE\_STOP}(Al, \mathcal{N}, stop) \\ \text{until } next \neq \texttt{TRIVIAL} \end{cases} \end{cases}$

Fig. 1. Pseudocode for the K3 main loop. The program receives the alignment Al, its length n, the number of permitted mismatches k, and the *threshold* length of the regions to be reported. Other variables are as described in the text.

The situation is more complex if $m_2 > k$ or $m_3 > k$ when, under the original scheme, $m_2 = n_0 + n_2$ and $m_3 = n_0 + n_3$. Without the loss of generality, we shall consider the case of row 2, as the other is symmetrical. As $n_0 + n_2 > k$ the number of mismatches caused by columns of either type 0 or 2 has to decrease, if possible. Decreasing the number of mismatches in columns of type 2 would introduce additional mismatches with both row 1 and row 3, while decreasing the number of mismatches in columns of type 0 would increase m_1 only (as row 3 already has a mismatch there), so we proceed by electing characters from row 2 as centers of these columns, as long as necessary, and possible. It stops being possible when either $m_1 > k$, when there is no way to extend the current maximal k–region, or the character from row 2 has been elected center of all n_0 type 0 columns. In the latter case, and if it is still $m_2 > k$ (for the revised center), the minority character of type 2 columns has to be selected in as many of them as necessary. Since it introduces a corresponding number of new mismatches with both rows 1 and 3, it is possible to be done as long as both $m_1 \leq k$ and $m_3 \leq k$. If this does not hold then the region cannot be extended, and it can be reported, otherwise we continue scanning the alignment for an extension with the revised center sequence. Once the default settings have been modified it would be cumbersome to maintain the distribution of center characters and revise it every time a new nontrivial column is seen. It is much simpler to re-calculate the distribution of the characters for each column type every time when it is determined that a center exists for the current region, and then use the calculated numbers for reporting the sequence. This approach is

Algorithm 2.2: CALCULATE_SCORES($\mathcal{N}, \mathcal{R}, k$)

$halt \leftarrow$ **false**
if $n_1 > k$ **and** $(n_0 + n_2 \geq k$ **or** $n_0 + n_3 \geq k)$
  then $halt \leftarrow$ **true**
  else if $n_1 \leq k$ **and** $n_0 + n_2 \leq k$ **and** $n_0 + n_3 \leq k$
  then RESET($\mathcal{N}, \mathcal{R}$)
  else if $n_1 > k$

$\text{\textbf{then}} \begin{cases} \textbf{if } n_1 - k > k - n_0 - n_2 \textbf{ or } n_1 - k > k - n_0 - n_3 \\ \quad \textbf{then } halt \leftarrow \textbf{ true} \\ \quad \textbf{else } \begin{cases} \text{RESET}(\mathcal{N}, \mathcal{R}) \\ r_1 \leftarrow n_1 - k \end{cases} \end{cases}$

  else if $n_0 + n_2 > k$ **and** $n_0 + n_3 > k$

$\text{\textbf{then}} \begin{cases} \textbf{if } n_0 + n_2 - k + n_0 + n_3 - k > n_0 \\ \quad \textbf{then } halt \leftarrow \textbf{ true} \\ \quad \textbf{else if } n_1 + n_0 + n_2 - k + n_0 + n_3 - k > k \\ \quad \textbf{then } halt \leftarrow \textbf{ true} \\ \quad \textbf{else } \begin{cases} \text{RESET}(\mathcal{N}, \mathcal{R}) \\ r_0^1 \leftarrow 2k - n_0 - n_2 - n_3 \\ r_0^2 \leftarrow n_0 + n_2 - k \\ r_0^3 \leftarrow n_0 + n_3 - k \end{cases} \end{cases}$

  else if $n_0 + n_2 > k$

$\text{\textbf{then}} \begin{cases} \textbf{if } n_0 + n_2 - k > k - n_1 \\ \quad \textbf{then } halt \leftarrow \textbf{ true} \\ \quad \textbf{else if } n_2 \leq k \\ \quad \textbf{then } \begin{cases} \text{RESET}(\mathcal{N}, \mathcal{R}) \\ r_0^1 \leftarrow k - n_2; r_0^2 \leftarrow n_0 + n_2 - k \end{cases} \\ \quad \textbf{else if } n_2 - k > k - n_0 - n_3 \\ \quad \textbf{then } halt \leftarrow \textbf{ true} \\ \quad \textbf{else } \begin{cases} \text{RESET}(\mathcal{N}, \mathcal{R}) \\ r_0^1 \leftarrow 0; r_0^2 \leftarrow n_0; r_2 \leftarrow n_2 - k \end{cases} \end{cases}$

$\text{\textbf{else}} \begin{cases} \textbf{if } n_0 + n_3 - k > k - n_1 \\ \quad \textbf{then } halt \leftarrow \textbf{ true} \\ \quad \textbf{else if } n_3 \leq k \\ \quad \textbf{then } \begin{cases} \text{RESET}(\mathcal{N}, \mathcal{R}) \\ r_0^1 \leftarrow k - n_3; r_0^3 \leftarrow n_0 + n_3 - k \end{cases} \\ \quad \textbf{else if } n_3 - k > k - n_0 - n_2 \\ \quad \textbf{then } halt \leftarrow \textbf{ true} \\ \quad \textbf{else } \begin{cases} \text{RESET}(\mathcal{N}, \mathcal{R}) \\ r_0^1 \leftarrow 0; r_0^3 \leftarrow n_0; r_3 \leftarrow n_3 - k \end{cases} \end{cases}$

  return ($halt$)

Fig. 2. Pseudocode for CALCULATE_SCORES function, returning the value of $halt$ as either **false** or **true**, depending on whether the current k-region can be extended with the current position (alignment column) or not. If the k-region is extended, it also recalculates the distribution of the characters in the center sequence.

implemented by procedure K3, whose pseudocode is given in Figure 1, and its supplementary routines, CALCULATE_SCORES, shown in Figure 2, and RESET, ADVANCE_START, ADVANCE_STOP and REPORT_REGION. The last four procedures are simple, so we shall omit their pseudocode. Briefly, ADVANCE_START and ADVANCE_STOP move the current region starting and ending column, respectively, and update the values of $\mathcal{N}$ depending on the type of the encountered column. RESET sets the values of $\mathcal{R}$, r_0^1 to n_0 and all others to 0.

The calculation of the values from the set $\mathcal{R}$ is not strictly a part of the determination of the boundaries of maximal k-regions, but they facilitate the reporting, if we are interested in the actual center sequences rather than just their boundaries. This is done by the procedure REPORT_REGION, whose pseudocode is omitted as relatively straightforward.

The main loop of the K3 algorithm scans through the alignment, attempting to extend the k-region starting at the current *start* position, until it is not possible any more. At that point the newly discovered maximal k-region is reported, if it satisfies other criteria, such as the minimal reportable length. After that, it moves the *start* to the first possible starting position for the next k-region. In doing this the use of CALCULATE_SCORES, implementing the scheme described above, is instrumental. In addition to estimating whether the new (nontrivial) column can be added to the current k-region, this function also recalculates the allocation of the column-to-center characters, to be used in reporting. In an implementation, the alignment can also be extended by artificial columns at positions 0 and $n + 1$, which makes reasoning easier.

2.1   Algorithm Correctness

In order to prove the algorithm correct, we must show that it reports all maximal k-regions in the alignment only once, and that only such regions are reported. We start by briefly arguing several properties of K3, as their full formal proofs would be too long to be discussed here:

1. The interval between *start* and *stop* is different in every iteration of K3. The first column to the left of the *start* (if any, and if *start* is not set beyond the end of the alignment) is nontrivial, as well as one at the *stop* position.
   Rationale: Every iteration of K3 advances either *start* or *stop*, so the interval between them must be different in every iteration. Loops advancing *start* and *stop* halt only when they pass (*start*) or get positioned at (*stop*) a nontrivial column.

2. Counters n_0, n_1, n_2 and n_3 always contain the correct number of columns of types 0, 1, 2 and 3, respectively, in [*start*, *stop*].
   Rationale: Trivial. These counts are updated by ADVANCE_START and ADVANCE_STOP, which keep track of the type of the current column.

3. If the value of *halt* determined by CALCULATE_SCORES is **false**, then there exists a k-region containing all columns in [*start*, *stop*].
   Rationale: By the scheme applied, if CALCULATE_SCORES returned **false** it means that it could find a satisfactory center assignment.

4. If the value of *halt* determined by CALCULATE_SCORES is **true** then there does not exist a *k*-region containing all columns in [*start, stop*].

   Rationale: Same as above. If *halt* was set to **true**, a satisfying assignment of center letters could not have been found.

Since the first statement procedure K3 reaches after *halt* becomes **true** is that in line (1) it follows that if *start* < *stop* when (1) is reached the interval [*start, stop* − 1] is indeed a *k*-region. Moreover, it is *right-maximal*, in the sense that it cannot be extended by adding more columns to its right end.

Not every column of the alignment can start a maximal *k*-region: it can be either trivial or non-trivial, but, unless it is the very first one in the alignment, it has to be immediately to the right of a nontrivial column. We refer to these as *region-starting* columns, and proceed to show that each such column is indeed considered by K3 as a potential start.

Lemma 1. *Every region-starting column, except these to the right of the* start *at the time* bound *becomes* $n+1$, *is at a position pointed to by* start *exactly once when the algorithm reaches line* (1) *of procedure* K3.

Proof. We prove this by induction on the number of times line (1) is reached.

Induction base Line (1) cannot be reached unless *halt* = **true**, however *start* cannot be increased before *halt* becomes **true**. Therefore the first column of the alignment is pointed to by the *start* variable the first time line (1) is reached. As *start* immediately increases for at least one position, and never decreases, it cannot point to the first column at any later time.

Induction hypothesis. Assume that when line (1) is reached the *m*th time *start* already pointed (once) to the first *m* region-starting columns. Assume further that this iteration of K3 does not set *bound* to $n+1$. It has to be shown that this implies that next time line (1) is reached *start* must point to the region-starting column $m + 1$, and that *start* will never point to that column afterwards.

Induction step. As *start* is increased in the same iteration when line (1) is reached, the *m*th region-starting column can never be at the *start* position again. As soon as *start* passes over a non-trivial column (or reaches the alignment end) it stops increasing, thus the next column it gets positioned to is exactly one that has a nontrivial column to its left, i.e., it is region-starting. If it is immediately to the right of the one previously pointed to by *start*, then it is obviously the next, otherwise the *m*th region-starting column must have been trivial, possibly followed by a run of other trivial columns, so the nontrivial column skipped was not region-starting. In consequence, the column to which *start* is positioned in the same iteration when line (1) was reached for the *m*th time is exactly the next region-starting one. Variable *start* will remain unchanged in all iterations in which *halt* = **false**, so the next time line (1) is reached it will still point to the region-starting column $m + 1$. As *start* then immediately increases, it will never point to this column again.

Lemma 2. *Every maximal k-region in the alignment equals* [*start, stop* − 1] *exactly once when the algorithm reaches line* (1) *of procedure* K3.

Proof. Assume the opposite, i.e., that there is a maximal k-region R which is never equal to $[start, stop - 1]$ when line (1) is reached.

A maximal k-region must start at a region-starting column, each being pointed to by *start* exactly once when line (1) is reached, by Lemma 1. Thus it must also hold for the column which starts R. If R is not equal to $[start, stop - 1]$, it must either be properly contained in this interval or properly contain it.

If R is contained in $[start, stop - 1]$ then it cannot be a maximal k-region. If R properly contains $[start, stop - 1]$ then, as *start* points to the starting column of R, it must be that it contains the column pointed to by *stop*. However, line (1) cannot be reached if *halt* $\neq$ **true**, and if it is **true** then $[start, stop]$ cannot be a k-region (due to the impossible-to-resolve last column). It follows that R, which is at least equal to $[start, stop]$ cannot be a k-region—a contradiction.

Lemma 3. *Every maximal k-region in the alignment which is at least as wide as the threshold is reported by the algorithm exactly once.*

Proof. By Lemma 2, every maximal k-region in the alignment is equal to the interval $[start, stop - 1]$ exactly once when the algorithm reaches line (1) of K3. If its length is at least equal to *threshold* then it would pass the first condition in that line, so it will be reported if $stop > bound$. It thus has to be shown that this condition holds for every maximal k-region.

The value of *bound* is set after line (1) executes. Thus if $bound \geq stop$ at the time the algorithm reaches line (1) with a new interval, then there was some interval $[b, e]$, where $b \leq start$ and $e \geq stop$, which reached line (1) in some previous iteration, as $[start, stop]$ at that time, and it must have been a k-region, too. The current interval $[start, stop - 1]$ is equal or contained in $[b, e - 1]$ and, by Lemma 1, it cannot be equal, so then it cannot be maximal. Therefore, if the current interval $[start, stop - 1]$ is a maximal k-region then $bound < stop$, and the second condition in line (1) must pass.

By Lemma 2 for every maximal k-region in the alignment line (1) is indeed reached, and if its length is at least equal to *threshold* both conditions there must pass, thus every such region is reported. As every region-starting column is pointed to by *start* exactly once when lines (1)–(2) are reached, by Lemma 1, we conclude that every such region is reported only once.

Lemma 4. *Only maximal k-regions whose length is greater than threshold are reported by the algorithm.*

Proof. The first condition in line (1) assures that nothing shorter than *threshold* is reported, and by observations 3. and 4. above if $start < stop$ then $[start, stop - 1]$ must be a k-region. It remains to be shown that no k-region properly contained in another is reported.

Assume the opposite, i.e., that there is a k-region $[b, e]$ which is not maximal, and is yet reported by the algorithm. Interval $[b, e]$ is then a proper subinterval of some maximal k-region $[b', e']$, where $b' \leq b$ and $e \leq e'$. In order to be reported, $[b, e]$ must reach line (1) of K3 as $[start, stop - 1]$. Both $[b, e]$ and $[b', e']$ are k-regions and they must start with a region-starting column. As $b' \leq b$, by Lemma

1 b' must have been pointed to by *start* either now or at some earlier time. If it is now it must also be $e = e'$ as the current region cannot be extended to the right. In the latter case, by Lemma 2 at that time it must have been that $[start, stop - 1]$ was exactly $[b', e']$, setting *bound* to $e' + 1$. The value of *bound* can never decrease, as *stop* can never decrease, thus it must be at least $e' + 1$ at the time $[b, e]$ is to be reported. However, as $bound \geq e' + 1 \geq e + 1 = stop$ it follows that the second condition in line (1) must fail now, preventing the reporting of $[b, e]$, which contradicts the assumption.

We shall omit the proof that the reported center sequence is correct, and instead state some observations to that effect:

1. Line (2) of procedure K3 cannot be reached in two successive iterations. Moreover, every time line (2) is reached it must be *halt* = **true**, and it must have became **false** at least once between two executions of line (2).
2. If CALCULATE_SCORES sets *halt* = **false**, then the $\mathcal{R}$ variables are set to provide a possible center sequence for the k-region contained in $[start, stop)$.
3. Every time the reporting is done, the $\mathcal{R}$ variables contain a correct distribution for the choice of center characters of nontrivial columns, leading to a possible center sequence for the k-region starting at *start* and ending immediately to the left of *stop*.

Theorem 1. *The algorithm implemented by procedure K3 is correct.*

Proof. Lemma 3 guarantees that every maximal k-region in the alignment which is at least as wide as the *threshold* is reported exactly once. By Lemma 4 only these regions are reported and from the observation 3 above it follows that a plausible center sequence is generated, too. We thus conclude that the algorithm implemented by procedure K3 is correct.

2.2   Algorithm Performance

Time complexity. Procedure CALCULATE_SCORES and all supplementary code execute in constant time. Procedure K3 executes in a loop which has either *start* or *stop* advanced in every iteration. Variable *start* can never assume value greater than $stop + 1$, as *stop* always points to a non-trivial (or artificial 0 or $n+1$) column, terminating the advancement of *start* even if it reached it. As the *start* shift always enables further search for the next k-region, it has to eventually cause the advancement of *stop*. In consequence, the K3 loop cannot execute more than $2n + 2$ times (and it must terminate when *stop* becomes $n + 1$). Thus the total time complexity of the algorithm, without center reporting, is $\Theta(n)$.

With reporting, a factor encountering for the output of the centers of the alignment columns from overlapping k-regions, within the areas of overlap, must be added. A single report can take $\Theta(n)$ time, however no single column can be reported more than $3k + 1$ times as a part of any maximal k-region. This is because all such regions must be distinct, and no one can contain more than $3k$ non-trivial columns. When scanning the alignment left-to-right all trivial columns at the left end of the previous reported k-region cannot be contained in the next, and thus they cannot be included in more than $p+1$ reports, where p is the number of maximal k-regions which include the nontrivial column at the right flank of the run of trivial columns. The total cumulative time with the reporting is thus of $O(kn)$. This estimate applies to the setting when an alignment of the sequences in the input has already been built.

Space requirements. Except the alignment itself, all variables used by the algorithm are scalar, thus the program has only constant extra space requirements.

3   Applications

In this work we have concentrated on the core algorithm for the 3-way k-mismatch restriction, rather than on its applications. We are currently working on the study of several genomic regions using the software based on this algorithm, and these findings will be reported in a future manuscript. However, in order to estimate the effectiveness of this approach we have preliminary ran the software implementing the K3 algorithm on a 3-way alignment of *HoxA* region in human, mouse and cow genomes.

Hox genes code for transcription factors involved in the early vertebrate development, and they are well conserved throughout the evolution. Actually, several regions from *Hox* are known to be ultra-conserved in vertebrates [5], and as such they feature multiple long k-regions, for very small k. We have run our software on *HoxA* with k varying between 1 and 5, looking for regions of minimal length between 150 and 400 base pairs. The results were consistent with our previous analysis by other methods, published elsewhere [9], in that, somewhat surprisingly, the conservation appears to be the best in several 5′ UTRs of *Hox* genes. Indeed, although long 3, 4 and 5-regions were present in front of, within and 3′ to *HoxA5* gene, its upstream sequence (plus the start of exon 1) featured the strongest conservation, including 2 overlapping 1-regions of more than 300 bp. 3-regions of more than 300 bp have also been found similarly positioned to *HoxA6* and several long 4 and 5-regions overlapped the first exon of *HoxA11*.

This study served more to demonstrate the workings of the K3 algorithm than to lead to biological discovery. Yet it has shown that K3 can be effective in discovering segments whose conservation need not be obvious, for larger values of k.

Another notable feature of *Hox* regions is the remarkable absence of repeated sequences common almost everywhere else in the human and other genomes, including the areas immediately flanking *Hox* (unpublished study by Ken Dewar, personal communication). This results in a dramatic change of the overall conservation patterns inside and outside *Hox* clusters, which can be effectively captured by this algorithm. In general, it promises to be useful when region boundaries need to be determined, such as in the location of haplotype blocks. For such applications one would often need to combine the centers from multiple 3-sequence sets, and we are currently developing methods for doing that.

Acknowledgments

The ideas described in this manuscript have matured through discussions with Dr. Webb Miller of Penn State University. The sequences we have used have been collected and aligned by Dr. Ken Dewar of McGill University and Genome Quebec Innovation Centre. The authors would like to thank both for their help.

References

1. D.C. Crawford and D.A. Nickerson. Definition and clinical importance of haplotypes. *Ann. Rev. Med.*, 56:303–320, 2005.
2. J.P. Balhoff and G.A. Wray. Evolutionary analysis of the well characterized *endo16* promoter reveals substantial variation within functional sites. *PNAS*, 102:8591–8596, 2005.
3. M. Frances and A. Litman. On covering problems of codes. *Theory of Computing Systems*, 30:113–119, 1997.
4. A.S. Hinrichs, D. Karolchik, R. Baertsch *et al.* The UCSC Genome Browser Database: update 2006. *Nucleic Acids Res.*, 34:D590–D598, 2006.
5. A.P. Lee, E.G. Koh, A. Tay, S. Brenner and B. Venkatesh. Highly conserved syntenic blocks at the vertebrate Hox loci and conserved regulatory elements within and outside Hox gene clusters. *PNAS*, 103:6994–6999, 2006.
6. M. Li, B. Ma and L. Wang. Finding Similar Regions in Many Strings. *Proc. of the 31st ACM Symp. on Theory of Computing STOC'99*, ACM Press, 473–482, 1999.
7. M. Li, B. Ma and L. Wang. On the Closest String and Substring Problems. *J. ACM*, 49:157–171, 2002.
8. N. Stojanovic, P. Berman, D. Gumucio, R. Hardison and W. Miller. A linear–time algorithm for the 1–mismatch problem. *Proc. 5th Workshop on Algorithms and Data Structures WADS'97*, Springer LNCS 1272, 126–135, 1997.
9. N. Stojanovic and K. Dewar. A Probabilistic Approach to the Assessment of Phylogenetic Conservation in Mammalian *Hox* Gene Clusters. *Proc. BIOINFO 2005, Int'l Joint Conf. of InCoB, AASBi and KSBI*, 118–123, 2005.
10. The ENCODE Project Consortium. The ENCODE (ENCyclopedia Of DNA Elements) Project. *Science*, 306:636–640, 2004.
11. The International HapMap Consortium. A haplotype map of the human genome. *Nature*, 437:1299–1320, 2005.
12. The International SNP Map Working Group. A map of human genome sequence variation containing 1.42 million single nucleotide polymorphisms. *Nature*, 409:928–933, 2001.

New Constructive Heuristics for DNA Sequencing by Hybridization*

Christian Blum** and Mateu Yábar Vallès

ALBCOM, Dept. Llenguatges i Sistemes Informàtics
Universitat Politècnica de Catalunya, E-08034 Barcelona, Spain
cblum@lsi.upc.edu

Abstract. Deoxyribonucleic acid (DNA) sequencing is an important task in computational biology. In recent years the specific problem of DNA sequencing by hybridization has attracted quite a lot of interest in the optimization community. However, in contrast to the development of metaheuristics, the work on simple constructive heuristics hardly received any attention. This is despite the fact that well-working constructive heuristics are often an essential component of succesful metaheuristics. It is exactly this lack of constructive heuristics that motivated the work presented in this paper. The results of our best constructive heuristic are comparable to the results of the best available metaheuristics, while using less computational resources.

1 Introduction

Deoxyribonucleic acid (DNA) is a molecule that contains the genetic instructions for the biological development of all cellular forms of life. Each DNA molecule consists of two (complementary) sequences of four different nucleotide bases, namely adenine (A), cytosine (C), guanine (G), and thymine (T). In mathematical terms each of these sequences can be represented as a word from the alphabet $\{A, C, G, T\}$. One of the most important problems in computational biology consists in determining the exact structure of a DNA molecule, called *DNA sequencing*. This is not an easy task, because the nucleotide base sequences of a DNA molecule (henceforth called DNA strands or sequences) are usually so large that they cannot be read in one piece. In 1977, 24 years after the discovery of DNA, two separate methods for DNA sequencing were developed: the chain termination method and the chemical degradation method. Later, in the late 1980's, an alternative and much faster method called *DNA sequencing by hybridization* was developed (see [1,2,3]).

DNA sequencing by hybridization works roughly as follows. The first phase of the method consists of a chemical experiment which requires a so-called DNA

---

* Work supported by the Spanish CICYT project OPLINK (grant TIN-2005-08818-C04-01) and by the "Juan de la Cierva" program of the Spanish Ministry of Science and Technology.
** Corresponding author.

P. Bücher and B.M.E. Moret (Eds.): WABI 2006, LNBI 4175, pp. 355–365, 2006.

array. A DNA array is a two-dimensional grid whose cells typically contain all possible DNA strands—called probes—of equal length l. After the generation of the DNA array, the chemical experiment is started. It consists of bringing together the DNA array with many copies of the DNA sequence to be read, also called the target sequence. Hereby, the target sequence might react with a probe on the DNA array if and only if the probe is a subsequence of the target sequence. Such a reaction is called hybridization. After the experiment the DNA array allows the identification of the probes that reacted with target sequences. This subset of probes is called the *spectrum*. Two types of errors may occur during the hybridization experiment:

1. **Negative errors:** Some probes that should be in the spectrum (because they appear in the target sequence) do not appear in the spectrum. A particular type of negative error is caused by the multiple existence of a probe in the target sequence. This cannot be detected by the hybridization experiment. Such a probe will appear at most once in the spectrum.
2. **Positive errors:** A probe of the spectrum that does not appear in the target sequence is called a positive error.

Given the spectrum, the second phase of DNA sequencing by hybridization consists in the reconstruction of the target sequence from the spectrum. Let us, for a moment, assume that the obtained spectrum is perfect, that is, free of errors. In this case, the original sequence can be reconstructed in polynomial time with an algorithm proposed by Pevzner in [4]. However, as the generated spectra generally contain negative as well as positive errors, the perfect reconstruction of the target sequence is in general impossible.

1.1   DNA Sequencing by Hybridization

The computational part of DNA sequencing by hybridization can be modelled as follows (see [5]). Let the target sequence be denoted by s_t. The number of nucleotide bases of s_t shall be denoted by n (i.e., $s_t \in \{A, C, G, T\}^n$). Furthermore, the spectrum—as obtained by the hybridization experiment—is denoted by S. Remember that each $s \in S$ is an oligonucleotide (i.e., a short DNA strand) of length l (i.e., $s \in \{A, C, G, T\}^l$). In general, the length of any oligonucleotide s is denoted by $l(s)$. Let $G = (V, A)$ be the completely connected directed graph defined by $V = S$. To each link $a_{s,s'} \in A$ is assigned a weight $o_{s,s'}$, which is defined as the length of the longest DNA strand that is a suffix of s and a prefix of s'. A directed Hamiltonian path p in G is a directed path without loops. The length of such a path p, denoted by $l(p)$, is defined as the number of vertices (i.e., oligonucleotides) on the path. In the following we denote by $p[i]$ the i-th vertex in a given path p (starting from position 1). In contrast to the length, the cost of a path p is defined as follows:

$$c(p) \leftarrow l(p) \cdot l - \sum_{i=1}^{l(p)-1} o_{p[i],p[i+1]} \tag{1}$$

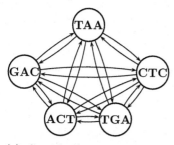

(a) Completely connected directed graph.

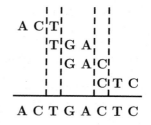

A C T G A C T C

(b) DNA sequence retrieval from a Hamlitonian path.

Fig. 1. (a) The completely connected directed graph with spectrum $S = \{\textbf{ACT,TGA,GAC,CTC,TAA}\}$ as the vertex set. The edge weights (i.e., overlaps) are not indicated for readability reasons. For example, the weight on the edge from **TGA** to **GAC** is 2, because **GA** is the longest DNA strand that is a suffix of **TGA** and a prefix of **GAC**. In (b) is shown how to retrieve the DNA sequence that is encoded by the optimal path $p^* = \langle \textbf{ACT,TGA,GAC,CTC} \rangle$. Note that $c(p^*) = 8$.

The first term sums up the length of the olionucleotides on the path, and the second term (which is substracted from the first one) sums up the overlaps between the neighboring oligonucleotides on p. In fact, $c(p)$ is equivalent to the length of the DNA sequence that is obtained by the sequence of oligonucleotides in p. The problem of DNA sequencing by hybridization consists of finding a directed Hamiltonian path p^* in G with $l(p^*) \geq l(p)$ for all possible paths p that fulfill $c(p) \leq n$. In other words, we aim to find a path p that contains as many oligonucleotides as possible (while respecting the length restriction on the resulting DNA sequence). In the following we refer to this NP-hard optimization problem as *sequencing by hybridization (SBH)*. It can be shown that exist optimal solutions to this problem that have a high probability to resemble the target sequence.

As an example consider the target sequence $s_t = \textbf{ACTGACTC}$. Assuming $l = 3$, the ideal spectrum is $\{\textbf{ACT,CTG,TGA,GAC,ACT,CTC}\}$. Let us assume that the hybridization experiment provides us with the following faulty spectrum $S = \{\textbf{ACT,TGA,GAC,CTC,TAA}\}$. See Figure 1(a) for the corresponding graph G. This spectrum has two negative errors, because **ACT** should appear twice, but can—due to the characteristics of the hybridization experiment—only appear once, and **CTG** does not appear at all in S. Furthermore, S has one positive error, because it includes oligonucleotide **TAA**, which does not appear in the target sequence. An optimal Hamiltonian path in G is $p^* = \langle \textbf{ACT,TGA,GAC,CTC} \rangle$ with $l(p^*) = 4$ and $c(p^*) = 8$. The DNA sequence that is retrieved from this path is **ACTGACTC** (see Figure 1(b)). This sequence is equal to the target sequence.

1.2   Existing Approaches

The first approach to solve the SBH problem was a branch & bound method proposed in [5]. However, this approach becomes unpractical with growing problem size. For example, the algorithm was only able to solve 1 out of 40 different problem instances concerning target sequences with 200 nucleotide bases within one hour. Another argument against this branch & bound algorithm is the fact that an optimal solution to the SBH problem does not necessarily provide a DNA sequence that is equal to the target sequence. This means that the importance of finding *optimal* solutions is not the same as for other optimization problems. Therefore, the research community has focused on (meta-)heuristic techniques for tackling the SBH problem. In addition to two constructive heuristics (see [5,6]), tabu and scatter search approaches [7,8,9] as well as evolutionary algorithms [9,10,11,12,13] were developed. Moreover, a GRASP method proposed in [14] deals with an easier version of the problem in which the first oligonucleotide of each target sequence is given.

The organization of the paper is as follows. In Section 2 we describe our constructive heuristics, and in Section 3 we conduct an experimental evaluation of these heuristics and compare them to the best techniques from the literature. Finally, in Section 4 we offer conclusions and an outlook to the future.

2   New Constructive Heuristics

In this section we first deal with a simple greedy technique from the literature (see [5]). Then, we propose a sensible extension of this heuristic. Finally, we present a conceptionally new heuristic that is based on merging sub-sequences. Before we start the description of the heuristics we introduce some notation. In particular we use

$$\text{pre}(s) \leftarrow \text{argmax}\{o_{s',s} \mid s' \in \hat{S},\ s' \neq s\}\ , \tag{2}$$

$$\text{suc}(s) \leftarrow \text{argmax}\{o_{s,s'} \mid s' \in \hat{S},\ s' \neq s\}\ , \tag{3}$$

where $\hat{S} \subseteq S$ and $s \in \hat{S}$ are given. In words, $\text{pre}(s)$ is the best available predecessor for s in $\hat{S}$, that is, the oligonucleotide that—as a predecessor of s—has the biggest overlap with s. Accordingly, $\text{suc}(s)$ is the best available successor for s in $\hat{S}$. In case of ties, the first one that is found is taken.

LAG (see [5]): Given a graph G and the length n of a target sequence as input, LAG works as shown in Algorithm 1. The construction of a path p in graph G starts by choosing one of the oligonucleotides from S in function Choose_Initial_Oligonulceotide(S). In subsequent construction steps p is extended by appending exactly one oligonucleotide. Finally, the solution construction stops as soon as $c(p) \geq n$, that is, when the DNA sequence derived from the constructed path p is at least as long as the target sequence. In case $c(p) > n$, function Find_Best_Subpath(p) searches for the longest (in terms of the number of oligonucleotdes) subpath p' of p such that $c(p) \leq n$, and replaces p by p'.

In the original version of LAG as presented in [5], the function Choose_Initial_Oligonulceotide(S) chooses a random vertex for starting the path construction. However, in this paper we implemented this function as follows. First, set

---

Algorithm 1. The LAG heuristic

1: **input:** A graph G, and the length of the target sequence n
2: $s^* \leftarrow$ Choose_Initial_Oligonulceotide(S)
3: $p \leftarrow \langle s^* \rangle$
4: $\hat{S} \leftarrow S \setminus \{s^*\}$
5: **while** $c(p) < n$ **do**
6:     $s^* \leftarrow \text{argmax}\{o_{p[l(p)],s} + o_{s,\text{suc}(s)} \mid s \in \hat{S}\}$
7:     Extend path p by adding s^* to its end
8:     $\hat{S} \leftarrow \hat{S} \setminus \{s^*\}$
9: **end while**
10: $p \leftarrow$ Find_Best_Subpath(p)
11: **output:** DNA sequence s that is obtained from p

$S_{bs} \subset S$ is defined as the set of all oligonucleotides in S whose best successor is better or equal to the best successor of all the other oligonucleotides in S.

$$S_{bs} \leftarrow \{s \in S \mid o_{s,\text{suc}(s)} \geq o_{s',\text{suc}(s')}, \ \forall s' \in S\} \tag{4}$$

Then, set $S_{wp} \subseteq S_{bs}$ is defined as the set of all oligonucleotides in S_{bs} whose best predecessor is worse or equal to the best predecessor of all the other oligonucleotides in S_{bs}: $S_{wp} \leftarrow \{s \in S_{bs} \mid o_{\text{pre}(s),s} \leq o_{\text{pre}(s'),s'}, \ \forall s' \in S_{bs}\}$. As starting oligonucleotide we choose the one (from S_{wp}) that is found first. The idea hereby is to start the path construction with an oligonucleotide that has a very good successor and at the same time a very bad predecessor. Such an oligonucleotide has a high probability to coincide with the start of the target sequence.

FB-LAG: A simple extension of the LAG heuristic is obtained by allowing the path construction not only in forward direction but also in backward direction. We call this heuristic henceforth forward-backward lock-ahead greedy (FB-LAG) heuristic. At each construction step the heuristic decides (with the same criterion as LAG) to extend the current path either in forward direction or in backward direction. FB-LAG is obtained from Algorithm 1 by exchanging lines 6, 7, and 8 with the following lines:

$s_r \leftarrow \text{argmax}\{o_{p[l(p)],s} + o_{s,\text{suc}(s)} \mid s \in \hat{S}\}$
$s_l \leftarrow \text{argmax}\{o_{\text{pre}(s),s} + o_{s,p[1]} \mid s \in \hat{S}\}$
if $o_{p[l(p)],s_r} + o_{s_r,\text{suc}(s)} > o_{\text{pre}(s),s_l} + o_{s_l,p[1]}$ **then**
    Extend path p by adding s_r to its end; $\hat{S} \leftarrow \hat{S} \setminus \{s_r\}$
else
    Extend path p by adding s_l to its beginning; $\hat{S} \leftarrow \hat{S} \setminus \{s_l\}$

A second change with respect to LAG concerns the implementation of function Choose_Initial_Oligonulceotide(S). As the path construction allows forward and backward construction it is not necessary to start the path construction with an oligonucleotide that has a high probability of being the beginning of the target sequence. It is more important to start with an oligonucleotide that has a high probability of being part of the target sequence:

$$s^* \leftarrow \text{argmax}\{o_{\text{pre}^2(s),\text{pre}(s)} + o_{\text{pre}(s),s} + o_{s,\text{suc}(s)} + o_{\text{suc}(s),\text{suc}^2(s)} \mid s \in S\} , \quad (5)$$

where $\text{pre}^2(s)$ denotes the best predecessor of the best predecessor of s (i.e., $\text{pre}(\text{pre}(s))$), and similar for $\text{suc}^2(s)$.

SM: The idea of the sub-sequence merger (SM) heuristic (see Algorithm 2) is conceptionally quite different to the LAG and FB-LAG heuristics. Instead of constructing only one path, the heuristic starts with a set of $|S|$ paths, each of which only contains exactly one oligonucleotide $s \in S$, and then merges paths until a path of sufficient size is obtained. The heuristic works in two phases. In the first phase, two paths p and p' can only be merged if p' is the unique best successor of p, and if p is the unique best predecessor of p'. The heuristic enters into the second phase if and only if the first phase has not already produced a path of sufficient length. In the second phase, the uniqueness conditions are relaxed, that is, two paths p and p' can be merged if p' is among the best successors of p, and p is among the best predecessors of p'. The reason of having two phases is the following: The first phase aims to produce possibly error free sub-sequences of the target sequence, whereas the second phase (which is more error prone due to the relaxed uniqueness condition) aims at connecting the sub-sequences produced in the first phase in a reasonable way.

In Algorithm 2, given two paths p and p', $o_{p,p'}$ is defined as $o_{p[l(p)],p'[1]}$, that is, the overlap of the last oligonucleotide in p with the first one in p'. In correspondence to the notations introduced in Equations 2 and 3, the following notations are used:

$$\text{suc}(p) \leftarrow \text{argmax}\{o_{p,p'} \mid p' \in P, \ p' \neq p\} , \quad (6)$$

$$\text{pre}(p) \leftarrow \text{argmax}\{o_{p',p} \mid p' \in P, \ p' \neq p\} . \quad (7)$$

Futhermore, $S_{\text{suc}}(p)$ is defined as the set of best successors of p, that is, $S_{\text{suc}}(p) \leftarrow \{p' \in P \mid o_{p,p'} = o_{p,\text{suc}(p)}\}$; and $S_{\text{pre}}(p)$ is defined as the set of best predecessors of p, that is, $S_{\text{pre}}(p) \leftarrow \{p' \in P \mid o_{p',p} = o_{\text{pre}(p),p}\}$. Finally, function Find_Best_Subpath(p) is implemented as described before.

HSM: The hybrid sub-sequence merger (HSM) heuristic is obtained by combining the FB-LAG heuristic with the SM heuristic. This combination is based on the following observation: At each stage of the SM heuristic, the FB-LAG heuristic can be applied to the problem instance that is obtained as follows. Given the current path set P of the SM heuristic, a spectrum $\hat{S}$ is created that contains the DNA sequences retrieved from the paths in P.[1] The result of the FB-LAG heuristic when applied to this problem instance can (of course) be regarded as a result for the original problem instance. It remains to specify at which stages of the SM heuristic the FB-LAG heuristic is applied. The first application of FB-LAG

---

[1] Note that the oligonucleotides of such a spectrum might have different lengths.

Algorithm 2. The SM heuristic

1: **input:** A graph G, and the length of the target sequence n
2: $P \leftarrow \{\langle s \rangle \mid s \in S\}$
3: **PHASE 1:**
4: $stop =$ FALSE
5: **for** $overlap = l - 1, \ldots, 1$ **do**
6:     **while** $\exists\, p, p' \in P$ **s.t.** $o_{p,p'} = overlap$ **&** $|S_{\mathrm{suc}}(p)| = 1$ **&** $|S_{\mathrm{pre}}(p')| = 1$ **&**
        $\mathrm{suc}(p) = p'$ **&** $\mathrm{pre}(p') = p$ **&** $stop =$ FALSE **do**
7:         Add path p' to the end of path p
8:         $P \leftarrow P \setminus \{p'\}$
9:         **if** $c(p) \geq n$ **then**
10:             $stop =$ TRUE
11:         **end if**
12:     **end while**
13: **end for**
14: **PHASE 2:**
15: **for** $overlap = l - 1, \ldots, 1$ **do**
16:     **while** $\exists\, p, p' \in P$ **s.t.** $o_{p,p'} = overlap$ **&** $p' \in S_{\mathrm{suc}}(p)$ **&** $p \in S_{\mathrm{pre}}(p')$ **&**
        $stop =$ FALSE **do**
17:         Choose p and p' such that $l(p) + l(p')$ is maximal
18:         Add path p' to the end of path p
19:         **if** $c(p) \geq n$ **then**
20:             $stop =$ TRUE
21:         **end if**
22:     **end while**
23: **end for**
24: Let p be the path in P with maximal cost
25: $p \leftarrow$ Find_Best_Subpath(p)
26: **output:** DNA sequence s that is obtained from p

---

is the one to the original problem instance, that is, before the first phase of SM
has started. Then, in the first as well as in the second phase of SM, FB-LAG is
applied at the end of the respective for-loop (i.e., after line 12 and after line 21
in Algorithm 2). However, FB-LAG is only applied if the while-loop before was
executed at least once. Note that in case the while-loop is not even executed a
single time, the problem instance derived from the path set P has not changed
since the previous application of FB-LAG. Finally, the output of HSM is the best
result among the different applications of FB-LAG and the final result of SM.

3   Results

We implemented the 4 heuristics outlined in the previous section in ANSI C++
using GCC 3.2.2 for compiling the software. Our experimental results were obtained on a PC with an AMD64X2 4400 processor and 4 Gb of memory.

Table 1. Results of our constructive heuristics for the instances by Błażewicz et al. [5]

Spectrum size	100	200	300	400	500
Average solution quality	76.98	153.53	230.68	309.03	383.08
Solved instances	23	15	12	7	4
Average similarity score (global)	77.05	133.63	171.78	206.80	218.60
Average similarity score (local)	91.83	152.43	209.33	272.40	293.48
Average computation time (sec)	0.0035	0.016	0.037	0.076	0.13

(a) Results of LAG

Spectrum size	100	200	300	400	500
Average solution quality	78.38	155.70	234.95	310.03	386.20
Solved instances	32	17	18	7	1
Average similarity score (global)	99.78	153.03	225.45	241.00	221.83
Average similarity score (local)	102.38	174.15	253.63	284.58	290.13
Average computation time (sec)	0.0051	0.022	0.054	0.11	0.19

(b) Results of FB-LAG

Spectrum size	100	200	300	400	500
Average solution quality	79.75	157.80	234.90	306.90	367.38
Solved instances	38	31	30	28	18
Average similarity score (global)	106.33	195.85	284.68	357.98	376.25
Average similarity score (local)	107.20	203.03	293.75	377.00	416.68
Average computation time (sec)	0.005	0.02	0.046	0.082	0.13

(c) Results of SM

Spectrum size	100	200	300	400	500
Average solution quality	80.00	159.68	239.90	319.38	398.88
Solved instances	40	36	39	35	31
Average similarity score (global)	108.40	204.78	300.00	396.90	469.55
Average similarity score (local)	108.70	206.85	305.35	399.85	479.88
Average computation time (sec)	0.012	0.048	0.11	0.21	0.35

(d) Results of HSM

A broad set of benchmark instances for DNA sequencing by hybridization was introduced by Błażewicz et al. in [5]. It consists of 40 real DNA target sequences of length 109, 209, 309, 409, and 509 (alltogether 200 instances). Based on real hybridization experiments, the spectra were generated with probe size $l = 10$. All spectra contain 20% negative errors as well as 20% positive errors. For example, the spectra concerning the target sequences of length 109 contain 100 oligonucleotides of which 20 oligonucleotides do not appear in the target sequences.

We applied the 4 heuristics outlined in the previous section to all problem instances. The results are shown in Table 1. The second row of each sub-table

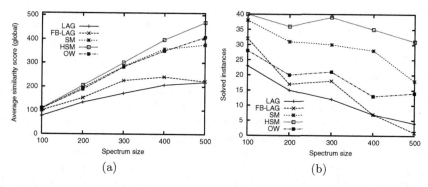

Fig. 2. Comparison of all existing constructive heuristics concerning (a) the global average similarity score obtained, and (b) the number of optimally solved instances. The comparison concerns the instances of Błażewicz et al. [5].

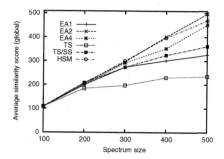

Fig. 3. Comparison of HSM with existing meta-heuristics concerning the instances by Błażewicz et al. [5]

contains the average solution quality (that is, the average number of oligonucleotides in the constructed paths), which must be maximized. The third row provides the number (out of 40) of solved problem instances, that is, the number of instances for which a path of maximal length could be found. The fourth and fifth row provide average similarity scores obtained by comparing the computed DNA sequences with the target sequences. The average scores in the fourth row are obtained from the Needleman-Wunsch algorithm (global alignment), and the average scores that are displayed in the fifth row are obtained from the Smith-Waterman algorithm (local alignment). Both algorithms were applied with the following parameters: +1 for a match of oligonucleotides, -1 for a mismatch or a gap. Finally, the sixth row provides the average computation times for solving one instance (in seconds).

From the results that are displayed in Table 1 we can draw the following conclusions. First, the results of FB-LAG improve in general over the results of LAG. This means that it is beneficial to allow the path construction in two directions (forward as well as backward). Second, the results of the SM heuristic are clearly better than both the results of LAG and the results of FB-LAG. However, the best results are obtained by the HSM heuristic, which is a hybrid between FB-LAG and SM. Even for the largest problem instances, the HSM heuristic produces sequences with very high similarity scores. In order to provide a comparison of all existing constructive heuristics we added the OW heuristic (see [6]) to this comparsion. This comparison is shown graphically in Figure 2. The results clearly show that HSM is currently the best available constructive heuristic.

Finally, in Figure 3 we present a comparison between HSM and available metaheuristic approaches from the literature. The results are surprising: HSM is clearly better than the 4 metaheuristic approaches EA1, EA4, TS, and TS/SS.[2] Furthermore, the results of HSM are—except for the problem instance of target sequence size 509—comparable to the results of the best metaheuristic approach EA2. Taking into account the advantage in computation time (i.e., HSM needs not even half a second to compute its results for the largest problem instances, while EA2—which is the fastest existing metaheuristic—needs several seconds) the HSM heuristic seems to be a good choice even when compared to metaheuristic approaches.

4   Conclusions and Future Work

In this work we have proposed new constructive heuristics for the problem of DNA sequencing by hybridization. First, we extended an existing heuristic. Then, we proposed a conceptionally new heuristic that is based on merging shorter DNA strands into bigger ones until a DNA strand of sufficient size is obtained. Finally we proposed a hybrid between both types of constructive heuristics. The results show that our hybrid method is the best constructive heuristic available to date. Moreover, concerning the results our hybrid method is comparable to state-of-the-art metaheuristics. Only concerning the biggest problem instances our hybrid method has slight disadvantages. On the other side, our constructive heuristics need less computation time.

We believe that our new constructive technique (as implemented by the SM heuristic) can be used to develop metaheuristics that are superior to exisiting metaheuristics for DNA sequencing by hybridization. As a first step, existing metaheuristics might be applied to problem instances resulting from intermediate stages of the SM heuristic. This might improve their results and save much computation time.

References

1. Bains, W., Smith, G.C.: A novel method for nucleid acid sequence determination. J. of Theoretical Biology **135** (1988) 303–307
2. IuP, Y.P.L., Florentiev, V.L., Khorlin, A.A., Khrapko, K.R., Shik, V.V.: Determination of the nucleotide sequence of DNA using hybridization with oligonucleotides. a new method. Doklady Akademii Nauk SSSR **303** (1988) 1508–1511

---

[2] Note that EA3 is not included because from [13] it is not clear which allignment algorithm was used by the authors.

3. Drmanac, R., Labat, I., Brukner, R., Crkvenjakov, R.: Sequencing of megabase plus DNA by hybridization: Theory of the method. Genomics **4** (1989) 114–128
4. Pevzner, P.A.: l-tuple DNA sequencing: Computer analysis. Journal of Biomulecular Structure and Dynamics **7** (1989) 63–73
5. Błażewicz, J., Formanowicz, P., Kasprzak, M., Markiewicz, W.T., Weglarz, J.: DNA sequencing with positive and negative errors. J. of Computational Biology **6** (1999) 113–123
6. Błażewicz, J., Formanowicz, P., Guinand, F., Kasprzak, M.: A heuristic managing errors for DNA sequencing. Bioinformatics **18**(5) (2002) 652–660
7. Błażewicz, J., Formanowicz, P., Kasprzak, M., Markiewicz, W.T., Weglarz, J.: Tabu search for DNA sequencing with false negatives and false positives. European J. of Operational Research **125** (2000) 257–265
8. Błażewicz, J., Glover, F., Kasprzak, M.: DNA sequencing—Tabu and scatter search combined. INFORMS J. on Computing **16**(3) (2004) 232–240
9. Błażewicz, J., Glover, F., Kasprzak, M.: Evolutionary approaches to DNA sequencing with errors. Annals of Operations Research **138** (2005) 67–78
10. Błażewicz, J., Kasprzak, M., Kuroczycki, W.: Hybrid genetic algorithm for DNA sequencing with errors. J. of Heuristics **8** (2002) 495–502
11. Endo, T.A.: Probabilistic nucleotide assembling method for sequencing by hybridization. Bioinformatics **20**(14) (2004) 2181–2188
12. Bui, T.N., Youssef, W.A.: An enhanced genetic algorithm for DNA sequencing by hybridization with positive and negative errors. In Deb et al., eds.: Proc. GECCO'04. Volume 3103 of LNCS, Springer Verlag, (2004) 908–919
13. Brizuela, C.A., González, L.C., Romero, H.J.: An improved genetic algorithm for the sequencing by hybridization problem. Proc. EvoWorkshops. Volume 3005 of LNCS, Springer Verlag, (2004) 11–20
14. Fernandes, E.R., Ribeiro, C.C.: Using an adaptive memory strategy to improve a multistart heuristic for sequencing by hybridization. Proc. WEA'05. Volume 3503 of LNCS, Springer Verlag, (2005) 4–15

Optimal Probing Patterns for Sequencing by Hybridization

Dekel Tsur

Department of Computer Science, Ben-Gurion University of the Negev
dekelts@cs.bgu.ac.il

Abstract. Sequencing by Hybridization (SBH) is a method for reconstructing a DNA sequence based on its k-mer content. This content, called the *spectrum* of the sequence, can be obtained from hybridization with a universal DNA chip. The main shortcoming of SBH is that it reliably reconstructs only sequences of length at most square root of the size of the chip. Frieze et al. [9] showed that by using gapped probes, SBH can reconstruct sequences with length that is linear in the size of the chip. In this work we investigate the optimal placement of the gaps in the probes, and give an algorithm for finding nearly optimal gap placement. Using our algorithm, we obtain a chip design which is more efficient than the chip of Frieze et al.

1   Introduction

Sequencing by Hybridization (SBH) [3, 16] is a method for sequencing DNA molecules. In this method, the target sequence is hybridized to a universal chip containing all 4^k sequences of length k. For each k-long sequence (or *probe*) in the chip, if its reverse complement appears in the target, then the two sequences will bind (or hybridize), and this hybridization can be detected. Thus the hybridization experiment gives the set of all k-long substrings of the target sequence. This set is called the *spectrum* of the target.

Currently, SBH is not considered competitive in comparison with standard gel-based sequencing technologies. The main shortcoming of SBH is that several sequences can have the same spectrum. Thus, if, for example, we wish to reconstruct at least 0.9 fraction of the sequences of length n, then n must be less than roughly 2^k [20, 8, 2, 23]. Several methods for overcoming this limitation of SBH were proposed: interactive protocols [25, 17, 10, 28], using location information [1, 5, 11, 4, 7, 23], using a known homologous string [19, 18, 27], and using restriction enzymes [26, 24].

Another method for enhancing SBH was proposed by Pevzner et al. [20]. They suggested using *gaps* (or *universal bases*) in the probes that can match to any of the four bases. For example, the probe AϕGϕG matches the sequences TAAC, TACC, TAGC, etc. A gapped probe can be implemented using a uniform mixture of the sequences that match the probe with length the same as the probe. Frieze et al. [9] showed that for every k, there is a chip with 4^k probes

P. Bücher and B.M.E. Moret (Eds.): WABI 2006, LNBI 4175, pp. 366–375, 2006.

that can reconstruct sequences of length $\Theta(4^k)$. This result is optimal up to constants. For a fixed k, the chip of Frieze et al. consists of all probes of the form $X^{\lceil k/2 \rceil} \left(\phi^{\lceil k/2 \rceil - 1} X \right)^{\lfloor k/2 \rfloor}$, where the X symbols represent definite bases (namely, each X symbol is replaced by one of the four bases). Preparata and Upfal [22] considered the same probing pattern of Frieze et al., and gave an improved algorithm for reconstructing the sequence from its spectrum. This algorithm allows reconstructing longer sequences than the algorithm of Frieze et al. An even more efficient algorithm was given in [12]. Heath et al. [13] showed that chips containing probes of the form $X^{\lceil k/2 \rceil} \left(\phi^{\lceil k/2 \rceil - 1} X \right)^{\lfloor k/2 \rfloor}$ and of the form $\left(X \phi^{\lceil k/2 \rceil - 1} \right)^{\lfloor k/2 \rfloor} X^{\lceil k/2 \rceil}$ are more efficient than the Frieze et al. chips. A *semi-degenerate base* is a base that matches either A/T or G/C. Probes containing semi-degenerate bases were studied in [20, 21].

In this paper we study the problem of designing optimal probing patterns. More precisely, for a given length and number of definite bases, the goal is to find the probing pattern that allows reconstructing longest sequences (it is desired to use probes with small length and small number of definite bases, since each of these parameters affects the number of molecules on the chip). We give an algorithm for this problem, and show that the probing patterns obtained by this algorithm are about 3 times more efficient than the probing patterns of Frieze et al. For simplicity, we shall restrict our study to a single probing pattern consisting of definite bases and gaps. However, our method can also be used for multiple probing patterns, and for probes containing semi-degenerate bases. We note that our work is somewhat similar to the research on seed design for similarity search (see, for example, [6, 14, 15]).

Due to lack of space, some details are omitted from this extended abstract.

2   Finding Optimal Probing Patterns

We first give some definitions. A *probing pattern* is binary string whose first and last characters are 1. The *weight* of a probing pattern P is the number of ones in P. The set of *probes* that corresponds to a probing pattern P is the set of all strings that are obtained by replacing every 0 in P by the character ϕ, and every 1 in P by one of the characters from $\Sigma = \{A, C, G, T\}$. A probe Q appears in a string S if the string Q matches to a substring of S, where the character ϕ is a don't care symbol (namely, it matches to every letter of Σ). The P-spectrum of a string S is the set of all probes from the set of probes of P that appear in S.

As an example for the definitions above, consider the probing pattern $P = 1101$. The set of probes corresponding to P is $\{AA\phi A, AA\phi C, AA\phi G, AA\phi T, AC\phi A, \ldots, TT\phi T\}$, and the P-spectrum of the string ACGATAC is $\{AC\phi A, AT\phi C, CG\phi T, GA\phi A\}$.

Our goal is to find optimal probing patterns, so we first need to define the notion of optimality. A string S is *unambiguously reconstructable* from its P-spectrum if there is no $S' \neq S$ whose P-spectrum is equal to the P-spectrum of S. For a probing pattern P, the *resolution power* $r(P, n)$ of P is the fraction

of the strings of length n that are unambiguously reconstructable from their
P-spectra. The resolution power is a natural measure for comparison between
different probing patterns. However, this measure ignores the issue of the time
complexity of reconstructing a string from its spectrum. Therefore, instead of the
resolution power, we will use the following measure: Let R be a reconstruction
algorithm, that is, R receives as input a P-spectrum of a string A and outputs
either the string A or 'failure'. The *success probability* of algorithm R on a
probing pattern P, denoted $\mathrm{sp}(P, n, R)$, is the fraction of the strings of length
n that are reconstructed correctly from their P-spectra by algorithm R. The
failure probability of R is $1 - \mathrm{sp}(P, n, R)$.

In this extended abstract, we will concentrate on the reconstruction algorithm
of Preparata and Upfal [22], which will be denoted R_{PU}. Our goal is to find a
probing pattern P of a given length and weight, that maximizes $\mathrm{sp}(P, n, R_{\mathrm{PU}})$
for some given n. Efficiently computing $\mathrm{sp}(P, n, R_{\mathrm{PU}})$ seems a difficult task, so
we will show how to compute a value $\widetilde{\mathrm{sp}}(P, n)$ that approximates $\mathrm{sp}(P, n, R_{\mathrm{PU}})$.
A probing pattern P^{OPT} that maximizes $\widetilde{\mathrm{sp}}(P, n)$ is almost optimal with respect
to $\mathrm{sp}(P, n, R_{\mathrm{PU}})$. An alternative way to approximate $\mathrm{sp}(P, n, R_{\mathrm{PU}})$ is by Monte
Carlo simulations (running algorithm R_{PU} on a large set of random strings and
computing the fraction of the runs in which the algorithm succeeds). However,
this approach is much more computationally intensive than our approach, which
makes it infeasible if the number of probing patterns that needs to be considered
is large. Moreover, our approach gives insight on what makes a probing pattern
efficient.

In the following, we will use $A = a_1 \cdots a_n$ to denote the target string. Let P
be some probing pattern of length L and weight k, and let H be some constant.
We assume that the first and last $L - 1$ letters of A are known. Algorithm R_{PU}
reconstructs the first $n - H + 1$ letters of A as follows (reconstructing the last
$H - 1$ letters is performed in a similar manner by reconstructing the sequence
backwards):

1. Let $s_1, \ldots, s_{L-1}$ be the first $L - 1$ letters of A.
2. For $t = L, L + 1, \ldots, n - H + 1$ do:
   (a) Let $\mathcal{B}_t$ be the set of all strings B of length H such that the string
       $s_1 \cdots s_{t-1} B$ is consistent with the P-spectrum of A (i.e., the P-spectrum
       of $s_1 \cdots s_{t-1} B$ is a subset of the P-spectrum of A).
   (b) If all the strings in $\mathcal{B}_t$ have a common first letter a, then set $s_t \leftarrow a$.
       Otherwise, return 'failure'.
3. Return $s_1 \cdots s_{n-H+1}$.

For the rest of this section, we show how to compute an approximation $\widetilde{\mathrm{fp}}(P, n) = 1 - \widetilde{\mathrm{sp}}(P, n)$ of the failure probability of algorithm R_{PU}. Our analysis is similar
to the analysis of Heath and Preparata [12]. However, the analysis of Heath and
Preparata is specific to the probing pattern of Frieze et al. In particular, they
omitted several cases from the analysis which are negligible for that probing
pattern. In our analysis, we consider more cases. Moreover, we handle the time
complexity for computing $\widetilde{\mathrm{fp}}(P, n)$, which is not done in [12].

It is easy to verify that if algorithm R_{PU} does not return 'failure', then $s_1 \cdots s_{n-H+1} = a_1 \cdots a_{n-H+1}$. Moreover, the algorithm stops at some t if and only if there is a string $B \in \mathcal{B}_t$ whose first letter is not equal to a_t. Such a string B will be called a *bad extension*. The string $a_t \cdots a_{t+H-1} \in \mathcal{B}_t$ is called the *correct extension*.

Suppose that the algorithm failed at some t, and let $B = b_1 \cdots b_H$ be the corresponding bad extension. Denote $B' = a_{t-l+1} \cdots a_{t-1} b_1 \cdots b_H$. By definition, the H probes that appear in B' also appear in A. That is, for every $i = 1, \ldots, H$, there is an index r_i such that $B'[i+j-1] = A[r_i+j-1]$ for all $1 \le j \le L$ for which $P[j] = 1$. The probe that corresponds to the index r_i is called *supporting probe i*. Supporting probe i is called a *fooling probe* if $r_i \ne t - L + i$. Fooling probe i is called *close* if $r_i \in \{t - L + 2, \ldots, t\}$. Two supporting probes i and j will be called *adjacent* if $r_j - r_i = j - i$, and they will be called *overlapping* if $|r_j - r_i| < L$ and they are not adjacent. Fooling probe i is *simple* if it is not close, and it is not adjacent or overlapping with another fooling probe.

Let J be the minimum index such that the probes $J, J+1, \ldots, H$ are pairwise adjacent. Note that some of the supporting probes can appear more than once in A, and therefore, there may be several ways to choose the values of $r_1, \ldots, r_H$. We assume that these values are chosen in a way that minimizes the number of fooling probes and the value of J.

2.1   Simple Probes

We first assume that fooling probes $1, \ldots, J - 1$ are simple, and that probe J is a fooling probe. We will analyze the case of non-simple fooling probes in Section 2.2. Let α denote the probability that a random probe appears in the string A. Using the Chen-Stein method, it is easy to show that the number of occurrences of a random probes in A is approximated by a Poisson distribution with mean $(n - L + 1)/4^k$. In particular, we have that $\alpha \approx 1 - e^{-(n-L+1)/4^k}$. We consider several cases:

Case 1. $J = 1$. In this case we have that $a_{r_1} \cdots a_{r_1+L-1} = a_{t-L+1} \cdots a_{t-1} b_1$. This event is composed of $L - 1$ character equalities in A ($a_{r_1+j-1} = a_{t-L+j}$ for $j = 1, \ldots, L - 1$), and one character inequality ($a_{r_1+L-1} = b_1 \ne a_t$). Thus, this event happens with probability $3/4^L$ for fixed t and r_1. Since there are approximately $\binom{n}{2}$ ways to choose t and r_1, it follows that the contribution of case 1 to $\widetilde{\mathrm{fp}}(P, n)$ is by

$$b_1 = \frac{n^2}{2} \cdot \frac{3}{4^L}.$$

Case 2. $2 \le J \le L$. In this case we have $a_{r_J} \cdots a_{r_J+L-J-1} = a_{t-L+J} \cdots a_{t-1}$, and $a_{r_J+L-J} \ne a_t$. Moreover, from the minimality of J we have that $a_{r_J-1} \ne a_{t-L+J-1}$. Which of the probes $1, \ldots, J - 1$ are fooling probes? If for a probe i, $b_{i-L+j} = a_{t+i-L+j-1}$ for all j for which $L - i + 1 \le j \le L$ and $P[j] = 1$ (in words, the characters sampled by probe i are equal in the bad extension and the correct extension) then the probe is not a fooling probe. Therefore, the number of fooling probes depends on the mismatches between the strings $a_t \cdots a_{t+J-2}$ and $b_1 \cdots b_{J-1}$. Let C be a binary string of length $J - 2$, where $C[i] = 1$ if $a_{t+i} \neq b_{i+1}$, and $C[i] = 0$ otherwise. Let $\hat{C} = 0^{L-1}1C$, namely, a string with $L - 1$ zeros followed by one is concatenated to C (the leftmost 1 is due to the fact that we always have $a_t \neq b_1$). We say that the pattern P *hits* a string S of length L if there is an index i such that $P[i] = S[i] = 1$. Thus, the number of fooling probes among probes $1, \ldots, J - 1$ is equal to the number of substrings of $\hat{C}$ of length L that are hit by P, which will be denoted $\text{hits}(\hat{C})$.

Since probes $J, J+1, \ldots$ are pairwise adjacent, we have that $b_i = a_{r_J+L-j+i-1}$ for $i = 1, \ldots, L - 1$. Therefore, $C[i] = 0$ forces the equality $a_{t+i} = a_{r_J+L-J+i}$ (which happens with probability $1/4$), and $C[1] = 1$ forces the inequality $a_{t+i} \neq a_{r_J+L-J+i}$ (which happens with probability $3/4$). Thus, for fixed t, r_i, and C, the probability that the equalities between the symbols a_{t+i} and b_{i+1} are according to C is $3^{\text{ones}(C)}/4^{J-2}$, where $\text{ones}(C)$ is the number of ones in C. It follows that the contribution of case 2 to $\widetilde{\text{fp}}(P, n)$ is $\sum_{J=2}^{L} b_J$, where

$$b_J = n^2 \left(\frac{3}{4}\right)^2 \frac{1}{4^{L-J}} \sum_{C \in \{0,1\}^{J-2}} \frac{3^{\text{ones}(C)} \alpha^{\text{hits}(0^{L-1}1C)}}{4^{J-2}}$$

$$= n^2 \frac{9}{4^L} \sum_{C \in \{0,1\}^{J-2}} 3^{\text{ones}(C)} \alpha^{\text{hits}(0^{L-1}1C)}.$$

Case 3. $L + 1 \leq J \leq H - L + 2$. This case is similar to case 2, so we omit the details. The contribution of this case to $\widetilde{\text{fp}}(P, n)$ is $\sum_{J=L+1}^{H-L+2} b_J$, where

$$b_J = n^2 \frac{9}{4^L} \sum_{C \in \{0,1\}^{J-2}} 3^{\text{ones}(C)} \alpha^{\text{hits}(0^{L-1}1C)}.$$

Case 4. $J > H - L + 2$. The contribution of this case to $\widetilde{\text{fp}}(P, n)$ is negligible (we omit the details).

We now handle the time complexity of computing $b_2, \ldots, b_{H-L+2}$. A straightforward computation of $b_2, \ldots, b_{H-L+2}$ takes $O(\sum_{J=2}^{H-L+2} LJ \cdot 2^J) = O(LH \cdot 2^H)$ time. We now show a dynamic programming algorithm for computing $b_2, \ldots, b_{H-L+2}$ in $O(H \cdot 2^L)$ time. For $J = 2, \ldots, H - L + 2$ and a binary string C of length $\min(J - 2, L - 1)$, define

$$b(J, C) = \sum_{C' \in \{0,1\}^{J-2}:C' \text{ is a suffix of } C} 3^{\text{ones}(C')} \alpha^{\text{hits}(0^{L-1}1C')}.$$

Clearly,

$$b_J = n^2 \frac{9}{4^L} \sum_{C \in \{0,1\}^{\min(J-2,L-1)}} b(J, C),$$

and the following recurrence is used to compute $b(J, C)$: For $J < L + 2$,

$$b(J, C) = b(J - 1, C[1]C[2] \cdots C[|C| - 1]) \cdot 3^{C[|C|]} \cdot \alpha^{\text{hits}(0^{L-|C|-1}1C)},$$

and for $J \geq L + 2$,

$$b(J, C) = \sum_{x \in \{0,1\}} b(J - 1, xC[1]C[2] \cdots C[|C| - 1]) \cdot 3^{C[|C|]} \cdot \alpha^{\text{hits}(xC)}.$$

The computation of $\text{hits}(0^{L-|C|-1}1C)$ or $\text{hits}(xC)$ is done in $O(1)$ time by computing a table that stores the value of $\text{hits}(C')$ for every string C' of length L.

2.2   Nonsimple Probes

Consider cases 2 and 3 above.

Case 2. Fix some $2 \leq J \leq L$. In this extended abstract, we only handle the case when some of the probes $1, \ldots, J - 1$ are adjacent to probe J, and the rest of the probes from $1, \ldots, J - 1$ are pairwise non-adjacent. Consider some fixed $C \in \{0, 1\}^{J-2}$, and let I_C be the set of fooling probes that correspond to substrings of $0^{L-1}1C$ that are hit by P. We say that a probe $i \in I_C$ *samples* position $r_J - j$ if $1 \leq j \leq J - i$ and $P[(J - i) + 1 - j] = 1$, or in other words, probe i contains the character $a_{r_J - j}$ if it is adjacent to probe J.

By the definition of J, $a_{r_J - 1} \neq a_{t - L + J - 1}$. Therefore, the probes that sample position $r_J - 1$ cannot be adjacent to probe J. Let I_C' be the set of the probes in I_C that do not sample $r_J - 1$. Let $S_C = \{r_J - j_1, \ldots, r_J - j_{|S_C|}\}$ be the set of all the positions $r_J - j$ that are sampled by at least one probe from I_C'. If a probe $i \in I_C'$ is adjacent to probe J then $a_{r_J - j} = a_{t - L + J - j}$ for every position $r_J - j$ that is sampled by probe i. For a target string A, the equalities of the form $a_{r_J - j} = a_{t - L + J - j}$ that are satisfied for positions $r_J - j \in S_C$ can be represented by a binary string C' of length $|S_C|$: $C'[l] = 1$ if $a_{r_J - j_l} \neq a_{t - L + J - j_l}$ and $C'[l] = 0$ otherwise. The probes in I_C' that are adjacent to probe J can be determined from the string C': For each such probe, $C'[l] = 0$ for every l such that position $r_J - j_l$ is sampled by the probe. We define $\text{fooling}(P, I_C, C')$ to be the number of probes in I_C' that sample some position $r_J - j_l$ with $C'[l] = 1$. To account for non-simple probes, we change the definition of b_J from Section 2.1 to

$$b_J = n^2 \frac{9}{4^L} \sum_{C \in \{0,1\}^{J-2}} \beta(P, I_C),$$

where

$$\beta(P, I_C) = 3^{\text{ones}(C)} \alpha^{|I_C - I_C'|} \frac{1}{4^{|S_C|}} \sum_{C' \in \{0,1\}^{|S_C|}} 3^{\text{ones}(C')} \alpha^{\text{fooling}(P, I_C, C')}.$$

A naive computation of $\beta(P, I_C)$ is time consuming. To compute $\beta(P, I_C)$ more efficiently, we use the following idea: Let $S \subseteq S_C$ be the set of all positions $r_J - j_l \in S_C$ such that the set of probes that sample $r_J - j_l$ is equal to the set of probes that sample $r_J - j_1$. The positions in S can be collapsed into a single positions, namely instead of representing a configuration by a binary string C' of length S_C, we can represent a configuration using a string C'' of length $1 + |S_C - S|$. This can be repeated with the other positions in S_C.

Another speedup follows from the following observation:

Claim. For two probing patterns P and P', if $P[i] \geq P'[i]$ for all i, then $\beta(P, I_C) \leq \beta(P', I_C)$ for every set I_C.

We use the claim as follows. When searching for the optimal probing pattern of length L and weight k, we first compute $\beta(P', I_C)$ for all sets I_C and for all probing patterns with length L and weight at most k, in which all the ones are in first 8 positions of the pattern or the last position. Then, when computing the failure probability for some pattern P, we choose the pattern P' whose prefix of length 8 is equal to the prefix of length 8 of P, and we use $\beta(P', I_C)$ instead of $\beta(P, I_C)$.

Case 3. In this we need to consider two sub-cases. The first case is when probe J is a fooling probe. The analysis of this case is similar to the analysis of the previous case. The second case is when probe J is not a fooling probe, namely $r_J = t - L + J$.

We have that $b_1 \neq a_t$ and from the minimality of J, $b_{J-L} \neq a_{t-1+J-L}$. Recall that C is a binary string of length $J - 2$, where $C[i] = 1$ if $a_{t+i} \neq b_{i+1}$, and $C[i] = 0$ otherwise. From the fact that $r_{J+i} = t - L + J + i$ for $i \geq 0$ it follows that $C[J - L] = C[J - L + 1] = \cdots = C[J - 2] = 0$. From the minimality of J, $C[J - L - 1] = 1$ when $J > L + 1$. Therefore, for $J > L + 1$, we add the term

$$b'_J = 9n \sum_C 3^{\text{ones}(C)} \alpha^{\text{hits}(0^{L-1}1C)}$$

to b_J, where the sum is over all strings $C \in \{0, 1\}^{J-2}$ that satisfy $C[J - L] = C[J - L + 1] = \cdots = C[J - 2] = 0$ and $C[J - L - 1] = 1$. For $J = L + 1$ we have that only one string C satisfies the requirements (the string $C = 0^{J-2}$) and we have the term

$$b'_{L+1} = 3n \cdot 3^{\text{ones}(0^{J-2})} \alpha^{\text{hits}(0^{L-1}10^{J-2})} = 3n \cdot \alpha^k.$$

The case of probe J not being a fooling probe for $J = L + 1$ was called "Mode 1" in [12].

3   Results

The s, r-*probing pattern* of Frieze et al. [9] is the pattern $1^s(0^{s-1}1)^r$. For a fixed weight k, the optimal s, r-probing pattern is the pattern with $s = \lceil k/2 \rceil$ (and $r = \lfloor k/2 \rfloor$). Denote this pattern by P_k^{FPU}. We run the algorithm of Section 2 with $k = 7$, $L = 15, 16, 17$, and $n = 4000$. The best patterns found by the the algorithm for $L = 15, 16, 17$ are $P_{15,7}^{\text{OPT}} = 111001000001011$, $P_{16,7}^{\text{OPT}} = 1101000100001011$, and $P_{17,7}^{\text{OPT}} = 11010001000001011$, respectively. For each probing pattern, we ran algorithm R_{PU} on 1000 random target strings (with the probing patterns P_7^{FPU}, $P_{15,7}^{\text{OPT}}$, $P_{16,7}^{\text{OPT}}$, and $P_{17,7}^{\text{OPT}}$), and computed the success rate of the algorithm. The results are given in Figures 1 and 2. The failure rate of R_{PU} for the pattern $P_{16,7}^{\text{OPT}}$ (whose length is the same as the length of P_7^{FPU}) is about 3 times smaller than the failure rate for P_7^{FPU}.

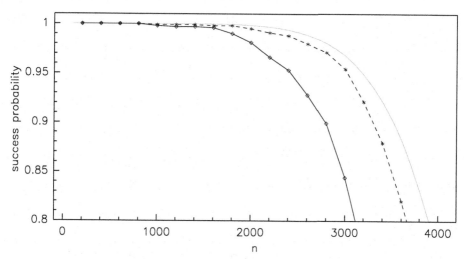

Fig. 1. Success probability of algorithm R_{PU} on the patterns P_7^{FPU} (solid line) and $P_{16,7}^{\text{OPT}}$ (dashed line) for various values of n. The gray solid line gives the probability that Mode 1 does not occur, which is an upper bound on the success probability for any pattern.

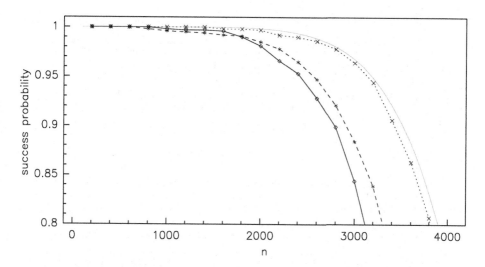

Fig. 2. Success probability of algorithm R_{PU} on the patterns P_7^{FPU} (solid line), $P_{15,7}^{\text{OPT}}$ (dashed line), and $P_{17,7}^{\text{OPT}}$ (dotted line) for various values of n.

Recall that Mode 1 refers to the case when the bad extension differs from the correct extension only in the first letter. The bad extension is supported by k fooling probes. Using Poisson approximations, the probability that a failure due to Mode 1 occurs is approximately $1 - e^{-3(n-L+1)\alpha^k}$. Note that this probability depends only on the length L of the probing pattern, but not on the pattern itself. Therefore, $e^{-3(n-19)\alpha^k}$ is an upper bound on the success probability of all probing patterns of length at most 20. This upper bound is shown as a gray solid line in Figures 1 and 2. An analysis of the failures show that most (about two thirds) of the failures in the runs of $P_{16,7}^{\mathrm{OPT}}$ are due to Mode 1, while only small part of the failures in the runs of P_7^{FPU} are due to Mode 1. Since Mode 1 failure is unavoidable, we have that the probing pattern $P_{16,7}^{\mathrm{OPT}}$ is very close to optimal.

It is clear from the analysis of Section 2 that longer probing patterns can achieve smaller failure probability. Indeed, the pattern $P_{17,7}^{\mathrm{OPT}}$ performs better than $P_{16,7}^{\mathrm{OPT}}$, and its failure probability is very close to the lower bound of Mode 1 failure probability. Moreover, while the pattern $P_{15,7}^{\mathrm{OPT}}$ is shorter than P_7^{FPU}, it has a smaller failure probability than P_7^{FPU} (for $n \geq 2000$).

References

1. L.M. Adleman. Location sensitive sequencing of DNA. Technical report, University of Southern California, 1998.
2. R. Arratia, D. Martin, G. Reinert, and M.S. Waterman. Poisson process approximation for sequence repeats, and sequencing by hybridization. *J. of Computational Biology*, 3(3):425–463, 1996.
3. W. Bains and G.C. Smith. A novel method for nucleic acid sequence determination. *J. Theor. Biology*, 135:303–307, 1988.
4. A. Ben-Dor, I. Pe'er, R. Shamir, and R. Sharan. On the complexity of positional sequencing by hybridization. *J. Theor. Biology*, 8(4):88–100, 2001.
5. S.D. Broude, T. Sano, C.S. Smith, and C.R. Cantor. Enhanced DNA sequencing by hybridization. *Proc. Nat'l Acad. Sci. USA*, 91:3072–3076, 1994.
6. J. Buhler, U. Keich, and Y. Sun. Designing seeds for similarity search in genomic DNA. *J. of Computer and System Sciences*, 70(3):342–363, 2005.
7. R. Drmanac, I. Labat, I. Brukner, and R. Crkvenjakov. Sequencing of megabase plus DNA by hybridization: theory of the method. *Genomics*, 4:114–128, 1989.
8. M.E. Dyer, A.M. Frieze, and S. Suen. The probability of unique solutions of sequencing by hybridization. *J. of Computational Biology*, 1:105–110, 1994.
9. A. Frieze, F.P. Preparata, and E. Upfal. Optimal reconstruction of a sequence from its probes. *J. of Computational Biology*, 6:361–368, 1999.
10. A.M. Frieze and B.V. Halldórsson. Optimal sequencing by hybridization in rounds. *J. of Computational Biology*, 9(2):355–369, 2002.
11. S. Hannenhalli, P.A. Pevzner, H. Lewis, and S. Skiena. Positional sequencing by hybridization. *Computer Applications in the Biosciences*, 12:19–24, 1996.
12. S.A. Heath and F.P. Preparata. Enhanced sequence reconstruction with DNA microarray application. *Proc. 7th Conf. on Combinatorics and Computing CO-COON'01*, pages 64–74, 2001.
13. S.A. Heath, F.P. Preparata, and J. Young. Sequencing by hybridization using direct and reverse cooperating spectra. *J. of Computational Biology*, 10(3/4):499–508, 2003.
14. U. Keich, M. Li, B. Ma, and J. Tromp. On spaced seeds for similarity search. *Discrete Applied Mathematics*, 138(3):253–263, 2004.
15. G. Kucherov, L. Noé, and M. Roytberg. A unifying framework for seed sensitivity and its application to subset seeds. *Proc. 5th Workshop on Algorithms in Bioinformatics WABI'05*, LNCS 3692, pages 251–263, 2005.

16. Y. Lysov, V. Floretiev, A. Khorlyn, K. Khrapko, V. Shick, and A. Mirzabekov. DNA sequencing by hybridization with oligonucleotides. *Dokl. Acad. Nauk USSR*, 303:1508–1511, 1988.

17. D. Margaritis and S. Skiena. Reconstructing strings from substrings in rounds. *Proc. 36th Symp. on Foundations of Computer Science FOCS'95*, pages 613–620, 1995.

18. I. Pe'er, N. Arbili, and R. Shamir. A computational method for resequencing long DNA targets by universal oligonucleotide arrays. *Proc. Nat'l Acad. Sci. USA*, 99:15497–15500, 2002.

19. I. Pe'er and R. Shamir. Spectrum alignment: Efficient resequencing by hybridization. *Proc. 8th Conf. on Intelligent Systems in Molecular Biology ISMB'00*, pages 260–268, 2000.

20. P.A. Pevzner, Y.P. Lysov, K.R. Khrapko, A.V. Belyavsky, V.L. Florentiev, and A.D. Mirzabekov. Improved chips for sequencing by hybridization. *J. Biomolecular Structure and Dynamics*, 9:399–410, 1991.

21. F.P. Preparata and J.S. Oliver. DNA sequencing by hybridization using semi-degenerate bases. *J. of Computational Biology*, 11(4):753–765, 2004.

22. F.P. Preparata and E. Upfal. Sequencing by hybridization at the information theory bound: an optimal algorithm. *J. of Computational Biology*, 7:621–630, 2000.

23. R. Shamir and D. Tsur. Large scale sequencing by hybridization. *J. of Computational Biology*, 9(2):413–428, 2002.

24. S. Skiena and S. Snir. Restricting SBH ambiguity via restriction enzymes. *Proc. 2nd Workshop on Algorithms in Bioinformatics WABI'02*, LNCS 2452, pages 404–417, 2002.

25. S. Skiena and G. Sundaram. Reconstructing strings from substrings. *J. of Computational Biology*, 2:333–353, 1995.

26. S. Snir, E. Yeger-Lotem, B. Chor, and Z. Yakhini. Using restriction enzymes to improve sequencing by hybridization. Technical Report CS-2002-14, Technion, Haifa, Israel, 2002.

27. D. Tsur. Bounds for resequencing by hybridization. *Proc. 3rd Workshop on Algorithms in Bioinformatics WABI'03*, LNCS 2812, pages 498–511, 2003.

28. D. Tsur. Sequencing by hybridization in few rounds. *Proc. 11th European Sym. on Algorithms ESA'03*, LNCS 2832, pages 506–516, 2003.

Gapped Permutation Patterns
for Comparative Genomics

Laxmi Parida

Computational Biology Center, IBM T. J. Watson Research Center,
Yorktown Heights, New York 10598, USA
parida@us.ibm.com

Abstract. The list of species whose complete DNA sequence have been read, is growing steadily and it is believed that comparative genomics is in its early days [12]. Permutations patterns (groups of genes in some "close" proximity) on gene sequences of genomes across species is being studied under different models, to cope with this explosion of data. The challenge is to (intelligently and efficiently) analyze the genomes in the context of other genomes. In this paper we present a generalized model that uses three notions, *gapped* permutation patterns (with gap g), genome clusters, via *quorum*, $K > 1$, parameter, and, possible multiplicity in the patterns. The task is to automatically discover all permutation patterns (with possible multiplicity), that occur with gap g in at least K of the given m genomes. We present $\mathcal{O}(\log m N_I + |\Sigma| \log |\Sigma| N_O)$ time algorithm where m is the number of sequences, each defined on Σ, N_I is the size of the input and N_O is the size of the maximal gene clusters that appear in at least K of the m genomes.

Keywords: Pattern discovery, data mining, labeled trees, clusters, patterns, motifs, permutation patterns, comparative genomics, whole genome analysis, evolutionary analysis.

1 Introduction

As research in genomic science evolves, there is a rapid growth in the number of available complete genome sequences. To date about a dozen species of animals have had their complete DNA sequence determined and the list is growing steadily [12]. The knowledge of gene positions on a chromosome, combined with the strong evidence of a correlation between the position and the function of a gene makes the discovery of common gene *clusters* invaluable firstly for understanding and predicting gene/protein functions and secondly for providing insight into ancient evolutionary events.

In recent years, this problem has be modeled and studied intensively by the research community. The first model of a genome sequence allowing only orthologous[1] genes was introduced by Uno and Yagiura [14]. By modeling genomes as

---

[1] These genes appear in different organisms, and are believed to have the same evolutionary origin (generated during speciation).

P. Bücher and B.M.E. Moret (Eds.): WABI 2006, LNBI 4175, pp. 376–387, 2006.
© Springer-Verlag Berlin Heidelberg 2006

permutations, they defined a *common interval* to be a pair of intervals of two given permutations consisting of the same set of genes. The number of such intervals, N_O, can be quadratic in the size of the input, N_I. They gave an optimal $\mathcal{O}(N_I + N_O)$ time algorithm based on a clever observation of monotonicity of certain integer functions.

Heber and Stoye [9] extended this result to $m \geq 2$ permutations. They introduced the idea of *irreducible* intervals, whose number can be only linear in the size of the input. They offered an optimal $\mathcal{O}(N_I)$ time algorithm to detect all the irreducible intervals and based on this they gave an optimal $\mathcal{O}(N_I + N_O)$ algorithm for the general problem with $m \geq 2$ permutations. The problem of identification of common intervals has been revisited in [3].

Bergeron, Corteel and Raffinot [5] formalized the notion of *distance-based clusters*: this allows the genes in a cluster to be separated by gaps that do not exceed a pre-defined threshold. Such clusters are termed *gene teams* and they are constrained to appear in *all* given m sequences. The authors present an $\mathcal{O}(mn \log^2 n)$ algorithm to detect the gene teams that occur in m sequences defined on n genes (alphabet).

A slightly modified model of a genome sequence, that allows paralogs[2] was introduced in [7] and a pattern discovery framework was formalized for this setting: a pattern is a gene cluster, that allows for multiplicity, i.e., paralogous genes within a cluster appearing quorum $K > 1$ times. K is the quorum parameter, usually used in the context of patterns. Formally, the following problem was addressed: *Given m strings (with possible multiplicity) on Σ of length n_i each, and a quorum K, the task is obtain all permutation patterns p that occur in at least K sequences.* The algorithm presented in [7] is based on *Parikh-maps* and has a time complexity of $\mathcal{O}(Ln \log |\Sigma| \log n)$, where L ($< \max_i(n_i)$) is the size of the largest cluster and $n = \Sigma_i(n_i)$. Here, we introduced the notion of *maximal* permutation patterns or clusters. In this context, maximality is a non-trivial notion requiring a special notation. In [11], this notation was shown to be a PQ tree structure [10] and a linear time algorithm was presented to compute it for each pattern (cluster).

Using a similar model of genomes as sequences rather than permutations, Schmidt and Stoye [13] devised a $\Theta(n^2)$ algorithm for extracting all common intervals of two sequences. In [8], He and Goldwasser extend the notion of gene teams to COG (clusters of orthologous genes) teams by allowing any number of paralogs and orthologs, and devised an $\mathcal{O}(mn)$ time algorithm to find such COG teams for *pairwise* chromosome comparison where m and n are the number of orthologous genes in the two chromosomes.

As the number of genomes under study grows in number, it becomes important to handle not only distance based clusters (or gapped clusters) but also its co-occurrence in some subset and not necessarily *all* the m genomes. Again, maximality is an important idea that potentially cuts down the number of clusters without significant loss of information.

---

[2] Paralogous genes appear in the same organism and caused by the duplication of ancestor genes.

Contributions of this paper. We formalize a generalized model that uses three notions explored independently before, as we discussed above: 1) *gapped* gene clusters, with gap g, and, (2) genome clusters, via a *quorum*, $K > 1$, parameter and (3) multiplicity in the gene clusters. The task is to automatically discover all permutation patterns (with possible multiplicity) occurring with gap at most g in at least K of the given m sequences. let P_g denote the patterns that with gap g (see precise definition in the next section) on K of the m input sequences. Note that the use of (1) the quorum parameter ($m \geq K > 1$) with multiplicity in the input and (2) multiple sequences ($m \geq 2$) distinguishes it from the previous gapped cluster models [5,8]: in the first, K is fixed at m, without multiplicity, and in the second, m is fixed at 2. Although, in this paper gap g is defined in terms of number of intervening genes, it can be simply generalized to other definition of gap (such as actual location on the chromosome or "distance" from a target gene and so on) and one of these variations will be discussed elsewhere.

Overview of the approach. When $g = 0$, the problem has a $\mathcal{O}(Ln \log |\Sigma| \log n)$ time solution, where L is the size of the pattern. When $g = 0$ the size of the output is no more than $\Sigma_i(n_i^2)$. We also note that when $g > 0$, multiple patterns may occur with the same imprint and thus the output size could be potentially exponential in the input parameter (say m). When g is very large (or gene clusters), the problem has an output-sensitive algorithm (discussed in Section 3.1). Note that if $p \in P_g$, it is possible that $p \notin P_0$ and there is no $p' \in P_0$ such that p can be deduced from p'. But if $p \in P_g$, then there must exist $p'' \in P_\infty$ such that $p \subseteq p''$ and the occurrences of p can be deduced from the occurrences of p''. We use this as a handle to solving the $0 < g < \infty$ case. We solve the problem in two stages: in the first stage we solve the problem for large gaps ($g = \infty$). In fact since the number of patterns is very large, we compute only the maximal permutation patterns. We use the solution of the first stage to construct the solutions for the given gap g. The overall time complexity of this two-stage algorithm is $\mathcal{O}(\log m N_I + |\Sigma| \log |\Sigma| N_O)$ where N_I is the size of the input and N_O is the the number of maximal gene clusters that appear in at least K of the m genomes. For the sake of completeness, we give a method to extract all the non-maximal patterns out of the maximal ones of the last stage.

2 Notation

A genome or chromosome is denoted by a sequence s which is defined on the genes or a finite alphabet Σ. $s[i \ldots j]$ denotes the subsequence of s from ith to the jth element. Also, a gene cluster is referred to as a pattern (or permutation pattern or πpattern) in this paper.

Definition 1. *($\Pi(s)$), $\Pi'(s)$) Given a sequence s on Σ, $\Pi(s) = \{\sigma \in \Sigma \mid \sigma = s[i] \text{ for some } 1 \leq i \leq |s|\}$. $\Pi'(s) = \{\sigma(l) \mid \sigma \in \Pi(s), \sigma \text{ appears exactly } l \text{ times in } s\}$.*

For example if $s = abcda$, $\Pi(s) = \{a, b, c, d\}$. If $s = abbccdac$, $\Pi'(s) = \{a(2), b(2), c(3), d(1)\}$. The latter set is said to have *multiplicity*, i.e., it has multiple "copies" of one or more elements of the set.

Definition 2. *($p_1 \subseteq p_2$) Let p_1 and p_2 be sets with multiplicity. Then, $p_1 \subseteq p_2$ if and only if for each $\sigma(l) \in p_1$, $\sigma(l') \in p_2$ holds where $l' \geq l$.*

For convenience, sometimes the $l = 1$ annotation is omitted. For example $\{a(2), b(2), c(3), d(1)\}$ is written as $\{a(2), b(2), c(3), d\}$.

Definition 3. *(occurs with gap g, imprint) Given a sequence s on Σ, a set $p \in 2^\Sigma$ occurs at position i on s with gap g if all the following hold:*

1. *$s[i] \in p$ (this ensures that the very first character is not a gap),*
2. *l is the smallest number such that $p \subseteq \Pi(s[i \ldots (i+l-1)])$ (this ensures that the very last character is not a gap). and,*
3. *if for every pair $i \leq i_1 < i_2 \leq (i+l-1)$ with $s[i_1], s[i_2] \in p$, and for all j, $i_1 < j < i_2$, $s[j] \notin p$, the distance between i_1 and i_2 is no more than g i.e., $(i_2 - i_1 - 1) \leq g$ holds. In other words, there are at most g gaps between any two non-gap (solid) characters.*

Further, the subsequence $s[i \ldots (i+l-1)]$ is an imprint of p at location i.

For example, if $p = \{a, b, c\}$ with $s = c\,e\,b\,a\,h\,r\,s\,c\,b\,e\,a\,g$, then p occurs at locations 1 and 8 with gap $g = 1$ with the imprint of the occurrence shown boxed in $s = \boxed{c\,e\,b\,a}\,h\,r\,s\,\boxed{c\,b\,e\,a}\,g$. However, if gap $g \geq 3$, then there is a third occurrence of p as well at location 4 whose imprint is shown as $s = c\,e\,\boxed{b\,a\,h\,r\,s\,c}\,b\,e\,a\,g$

Definition 4. *(permutation pattern, πpattern p, location list $\mathcal{L}_p^g$) Given m sequences s_i ($1 \leq i \leq m$) each on alphabet Σ, a quorum $1 < K \leq m$ and a gap $0 \leq g$, $p \in 2^\Sigma$ is a permutation pattern or a πpattern with quorum K, if location list $\mathcal{L}_p^g = \{(i, l, u) \mid p \text{ occurs with gap } g \text{ and imprint } s_i[l, u]\}$ is such that $|\mathcal{L}_p^g| \geq K$.*

In the following, assume that $\mathcal{P}$ is the set of all πpatterns on the given m sequences. Note that if $p \in \mathcal{P}$, it does not imply that any $(p' \subset p) \in \mathcal{P}$. So, maximality in permutation patterns is not straightforward and is defined formally as follows [7].

Definition 5. *[7](maximal p) Given $\mathcal{P}$, $p_a \in \mathcal{P}$ is non-maximal if there exists $p_b \in \mathcal{P}$ such that: (1) the imprint of each occurrence of p_a is contained in the imprint of an occurrence of p_b, and (2) the imprint of each occurrence of p_b contains $l \geq 1$ imprints of occurrence(s) of p_a. A pattern p_b that is not non-maximal is maximal.*

In fact, it was shown in [7] that when the gap $g = 0$, there is a concise notation to represent all the non-maximal patterns of p and in [11] it was shown that these non-maximal patterns can be arranged as a PQ tree structure. It was also demonstrated in [11] that when gap $g = 0$, with multiplicity in the patterns, a single PQ tree may fail to capture all the non-maximal patterns.

   We demonstrate here that when the gap $g > 0$, there is no straightforward representation using a single PQ structure. Consider the following example.

Example 1. Let $s_1 = a\,b\,c\,d\,e\,f\,g\,h\,j$ and $s_2 = b\,j\,d\,g\,f\,e\,h\,c\,l$ with quorum $K = 2$ and gap $g = 1$. Consider the following: $p_0 = \{c, e, g, j, b, d, f, h\}$, $p_1 = \{b, d, f, h\}$, $p_2 = \{c, e, g, j\}$, $p_3 = \{e, g, d, f\}$, $p_4 = \{c, e, d, f\}$, $p_5 = \{g, j, b, d\}$, $p_6 = \{c, e, b, d\}$, $p_7 = \{c, e, j, b\}$, $p_8 = \{c, b, d, h\}$. For $g = 1$, p_0, p_1, p_2, p_3 and p_4 are πpatterns but p_5, p_6, p_7 and p_8 are not. Note that p_1, p_2, p_3, p_4 are non-maximal with respect to p_0. It is easy to see that a single PQ structure cannot succinctly represent the πpatterns.

The reason for using maximal patterns although there is no convenient structure capturing the non-maximal patterns is that, the number of patterns are smaller and the overcounting caused by non-maximal patterns can be avoided. We use the following definition of maximality:

Definition 6. (maximal p) *Let $\mathcal{P}$ be the set of all πpatterns on a set of given m sequences. Then $p_a \in \mathcal{P}$ is non-maximal if there exists $(p_b \supset p_a) \in \mathcal{P}$ such that the imprint of each occurrence of p_a is contained in the imprint of an occurrence of p_b. A pattern p_b that is not non-maximal is maximal.*

For the remainder of the discussion input size, N_I, and output size, N_O, are defined as follows:

$$N_I(s, m) = \sum_{i=1}^{m} |s_i| \quad \text{and} \quad N_O(P) = \sum_{p \in P} (|p| + |\mathcal{L}_p|) \tag{1}$$

Note that $Lst(\mathcal{L}_p)$ is simply a collection of the sequence indices.

Let P_g denote the maximal patterns with gap g. The following example demonstrates the case when the output size can be exponential.

Example 2. Let $s_1 = 2\,3\,4\,5$, $s_2 = 1\,3\,4\,5$, $s_3 = 2\,6\,4\,5$, $s_4 = 2\,3\,7\,5$, $s_5 = 2\,3\,4\,8$ and $K = 2$ and gap $g = 1$. Then $P_1 = \{\ \{2\}, \{3\}, \{4\}, \{5\}, \{2, 3\}, \{2, 4\}, \{2, 5\}, \{3, 4\}, \{3, 5\}, \{4, 5\}, \{2, 3, 4\}, \{2, 3, 5\}, \{3, 4, 5\}, \{2, 4, 5\}\ \}$. Thus $P_1 = \mathcal{O}(2^m)$.

Next we define a collection of sequence indices of a pattern p as follows used in the next section.

Definition 7. ($Lst(\mathcal{L}_p^g)$) *Given $\mathcal{L}_p^g$, $Lst(\mathcal{L}_p^g) = \{i \mid (i, -, -) \in \mathcal{L}_p^g\}$.*

3   The Gapped Permutation Pattern Problem

Problem 1. (*g-**Gap πPattern Problem** g-$GPP(s, m, K, P)$*) Given m sequences s_i defined on a finite alphabet Σ and some integer $g \geq 0$ and $K > 1$, the task is to find P, the collection of maximal πpatterns that occur on s with gap g and quorum K.

One of the main problems with permutation patterns is that it cannot be built from smaller units (as in sequence patterns, say). In other words p_1 and p_2 may not, but $p_1 \cup p_2$ could be a pattern. For example consider $s_1 = 7\,1\,2\,3\,4\,5\,7\,8$, $s_2 = 6\,2\,4\,1\,3\,9$ with $K = 2$. $p_1 = \{1, 2\}$ and $p_2 = \{3, 4\}$ are not but $p = p_1 \cup p_2 = \{1, 2, 3, 4\}$ is a permutation pattern.

∞-GapMaxπPattern(s, m, K, P_∞)
$Dic[j] = \{i(\ell) \mid \sigma_j$ occurs exactly ℓ times in $s_i\}$ //dictionary
$\mathcal{S}[1] \leftarrow \{0, 1, 2, \ldots, m\}; \mathcal{S}[2] \leftarrow \phi$ //dummy 0 in $\mathcal{S}[1]$
MineMaxπpat$(K, 1, 1, \mathcal{S})$ //main call

            //σ_j appears $l_{j_1} < l_{j_2} < \ldots < l_{j_n}$ times in input s
MineMaxπpat$(K, j, r, \mathcal{S})$
BEGIN
IF $(j > |\Sigma|)$ EXIT
$\ell \leftarrow l_{j_r}$
$\mathcal{S}_{new}[1] \leftarrow \{i \mid (i \in \mathcal{S}[1])$ AND $(i(\ell') \in Dic[j], \ell' \geq \ell)\}$
$\mathcal{S}_{new}[2] \leftarrow \mathcal{S}[2] \cup \{\sigma_j(\ell)\}$
IF $(|\mathcal{S}_{new}[1]| \geq K)$
    $MAXML \leftarrow FALSE$
    IF $\mathcal{S}_{old} = \mathbf{Exists}(\mathcal{T}, \mathcal{S}_{new}[1])$
        IF $\mathcal{S}_{old}[2] \subset \mathcal{S}_{new}[2]$ $\mathbf{Update}(\mathcal{T}, \mathcal{S}_{old}[2], \mathcal{S}_{new}[2])$
        ELSE $MAXML \leftarrow TRUE$
    ELSE $\mathbf{Add}(\mathcal{T}, \mathcal{S}_{new})$
    IF $MAXML = FALSE$
        $j' \leftarrow (j_r = j_n)? (j+1) : j; r' \leftarrow (j_r = j_n)? 1 : (r+1)$
        MineMaxπpat$(K, j', r', \mathcal{S}_{new})$
MineMaxπpat$(K, j+1, 1, \mathcal{S})$
END

Fig. 1. The pseudocode for finding maximal patterns (with possible multiplicity) that occur with ∞ gap in at least K of the m sequences

This problem is compounded with gapped occurrences of the patterns. In [7] we presented a Parikh-mapping based solution that used a fixed window size (pattern size). However this approach cannot be used when a non-zero gap is defined (since it is unclear which subset of the window are the solid characters of the pattern).

3.1   (Stage 1) ∞-Gap Maximal πPattern Problem

At this stage, we seek all collection of characters that co-occur in at least K sequences. The problem is formally defined as follows.

Problem 2. (∞-**Gap πPattern Problem**, ∞-**GPP**(s, m, K, P)) The input is m sequences s_i, each of length n_i, $1 \leq i \leq m$, defined on a finite alphabet Σ, and a quorum K. The output is all maximal permutation patterns (with possible multiplicity), p, with its location list $\mathcal{L}_p^\infty$.

In the worst case, the size of the output could be very large, i.e., $|P_\infty| = \mathcal{O}(2^m)$, assuming $m \ll n_i$. So there is no hope for a worst-case polynomial time algorithm and we explore an output sensitive method to compute the maximal patterns P_∞ in the following discussion.

Overview of the approach. Without loss of generality, assume an ordering on the alphabet set as $\sigma_1 < \sigma_2 < \ldots < \sigma_l$, where $|\Sigma| = l$. We next define the following.

Definition 8. *(s_p^O) Given p on a finite alphabet with ordering O given as $\Sigma = \{\sigma_1 < \sigma_2 < \ldots < \sigma_l\}$, s_p^O is a sequence of length $|p|$ such that $\Pi(s_p^O) = p$ and for each pair $i_1 < i_2$, $s_p^O[i_1] < s_p^O[i_2]$ holds.*

Let $\mathcal{X}$ be $(2^\Sigma \setminus \phi)$. In other words, $\mathcal{X}$ is the collection of all non-empty subsets of Σ. Now consider the trie [2] of the following collection of sequences $\{s_X^O \mid X \in \mathcal{X}\}$. Given a node A in the trie, the unique path from the root to A defines a sequence s_p denoted as $\pi path(A)$. Clearly for an instance of problem any $p \in P_\infty$, $p = \pi path(A)$ holds for some node A in the trie. However, the converse is not true. So our interest is in the following modification of this trie.

Definition 9. *($S_\infty^O, \mathbf{T}^O(P_\infty)$) Given an instance of ∞-GPP(s, m, K, P_∞), let $S_\infty^O = \{s_p^O \mid p \in P_\infty\}$ for some ordering O on the elements of Σ. Then the trie on S_∞^O is defined as $\mathbf{T}^O(P_\infty)$.*

Thus different orderings give different trie's, but they denote the same sets. It is easy to see that if O_1 and O_2 are two distinct orderings, then, it is possible that $\mathbf{T}^{O_1}(P_\infty) \neq \mathbf{T}^{O_1}(P_\infty)$ but $\{\Pi(s_p^{O_1}) \mid p \in P_\infty\} = \{\Pi(s_p^{O_2}) \mid p \in P_\infty\} = P_\infty$.

The algorithm & its correctness. The algorithm is best described through the pseudocode that is presented in Figure 1 and a complete example is discussed in Figure 2. We first reorganize the input data s, by constructing the dictionary Dic. The dictionary is an array, (indexed on σ_j) of ordered lists of two tuples, written as $i(k)$, sorted by i. k is the number of times σ_j appears in s_i. It is easy to see that this construction takes linear time and Dic takes linear space. The other initialization details are shown in Figure 1. To avoid clutter, in Figure 2(2), k is omitted from the dictionary.

The main routine is *MineMaxπPat()* whose pseudocode is shown in Figure 1. It is easy to verify that this recursive procedure implicitly constructs and traverses $\mathbf{T}(P_\infty)$ depth first. The left-to-right ordering of the children is defined by the order of the elements of Σ (that label the branches).

In the following we give the correspondence between the recursive call and $\mathbf{T}(P_\infty)$. Each recursive call is a node, labeled with $\mathcal{S}[1]$ and a (descending) branch, labeled with $\sigma_j(l)$ in Figure 2(c). In Figure 2(c), the solid edges denote the edges of the tree and the dashed edges denote the termination of the calls or the backtracked edges. This shows that the algorithm does indeed construct and traverse the tree $\mathbf{T}(P_\infty)$ and the correctness of the algorithm follows from the lemma.

Lemma 1. *The number of edges in $\mathbf{T}(P_\infty)$, constructed by MineMaxπPat(), including the backtracked edges is no more than $|\Sigma| \sum_{p \in P_\infty} |p|$.*

Proof. The total number of prefixes of p, $p \in P_\infty$ is $\sum_{p \in P_\infty} |p|$ which is the number of edges in the trie $\mathbf{T}(P_\infty)$. Also, the number of backtracked edges (shown by dashed edges in the figures) for each node can be no more than $|\Sigma|$. Hence the result. $\square$

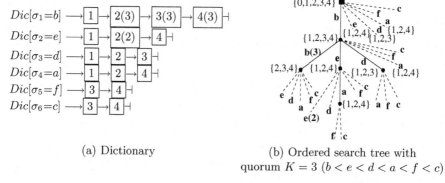

(a) Dictionary

(b) Ordered search tree with
quorum $K = 3$ ($b < e < d < a < f < c$)

Fig. 2. The algorithm described in Figure 1 for a sample data: $s_1 = beda$, $s_2 = edbabeb$, $s_3 = bbfcdb$ and $s_4 = cafbebb$, with $K = 3$. The dictionary Dic which is an array of ordered lists is shown in (2). (3) displays the implicit tree generated by the recursive routine $MineMax\pi Pat()$. Each call is shown as a node, labeled with $\mathcal{S}[1]$ and a (descending) branch, labeled with $\sigma_j(\ell)$ (when $\ell = 1$, the annotation is omitted). Also, if $|\mathcal{S}[1]| < (K = 3)$, it is not shown, to avoid clutter.

Time complexity. The generated location lists are stored on a data structure (say balanced trees) that allows for efficient retrieval and insertion. A pattern p is stored in this data structure, indexed by $Lst(\mathcal{L}_p)$ (see Definition 7). The routines shown in bold in the figure, $Update(\mathcal{T}, \mathcal{S}_{old}[2], \mathcal{S}_{new}[2])$, $Add(\mathcal{T}, \mathcal{S}_{new})$ and $Exist(\mathcal{T}, \mathcal{S}[1])$ are the update, insertion and retrieval routines respectively on this data structure. $Exist(\mathcal{T}, \mathcal{S}[1])$ returns the pattern denoted by $\mathcal{S}_{old}$ that already exists in the data structure. $Add(\mathcal{T}, \mathcal{S}_{new})$ stores $\mathcal{S}_{new}$ in the data structure. However, if $\mathcal{S}_{old}$ already exists, the associated pattern is updated. Since $K' \leq m$, each of the routine takes $\mathcal{O}(m \log m)$ time.

At each recursive call (or edge of $\mathbf{T}(P_\infty)$), the routine (except for $Update()$, $Add()$ and $Exist()$ takes time $\mathcal{O}(m)$. This is possible since $\mathcal{S}_{new}[1]$ can be efficiently computed as the merge of two ordered lists $\mathcal{S}[1]$ and $Dic[j]$. Thus, each call takes $\mathcal{O}(m \log m)$ time. The number of calls is $|\Sigma| \sum_{p \in P_\infty} |p|$ by Lemma 1. Hence the algorithm takes $\mathcal{O}(N_I(s, m) + |\Sigma| N_O(P_\infty) m \log m)$ time where $N_I(s, m)$ and $N_O(P_\infty)$ are the sizes of the input and output respectively (see Equation (1) for N_I and N_O).

3.2 (Stage 2) g-Gap Maximal πPattern Problem

At this stage, using the results of the last section, we extract the maximal permutation patterns that occur with gap $g(< \infty)$.

At the end of Stage 1, the maximal permutation patterns, p that occur with ∞ gaps is stored in a balanced binary tree data structure $\mathcal{T}$. A maximal permutation pattern p' (see Definition 6) that occurs with gap g is obtained from a maximal pattern p with $p' \subseteq p$ where p occurs with ∞ gap as follows:

Let the imprint of p in s_i be $s_i[j_{i1_\infty}, j_{i2_\infty}]$, $i \in Lst(\mathcal{L}_p^\infty)$. Let $p' \subseteq p$ be a maximal set (permutation pattern) such that imprint of p' with gap g on s_i is given as $s_i[j_{i1_g}, j_{i2_g}]$ where $j_{i1_\infty} \leq j_{i1_g} \leq j_{i2_g} \leq j_{i2_\infty}$, for some $i \in Lst(\mathcal{L}_p^\infty)$. Following this notation, the collection of i's for a p' is defined as follows (i.e., the i's where p' occurs with gap g):

$$\mathcal{L}' = \{(i \in Lst(\mathcal{L}_p^\infty)) \mid j_{i1_\infty} \leq j_{i1_g} \leq j_{i2_g} \leq j_{i2_\infty}\}$$

Then, the following holds.

Observation 1. *If $p' \in P_\infty$, then $p' \in P_g$ with $\mathcal{L}_p^\infty = \mathcal{L}_{p'}^g$. If $p' \notin P_\infty$, then there exists a unique $p'' \in P_\infty$, which is the smallest (cardinality) set such that $p' \subseteq p'' \subseteq p$. (See above for notation.)*

Proof. Assume the contrary that there exists at least two maximal patterns $p_1, p_2 \in P_\infty$ such that $p' \subseteq p_1$ and $p' \subseteq p_2$. If p_1 and p_2 are distinct and neither is included in the other, then $p_1 \cap p_2 = p_3$ must be maximal. Also $p' \subseteq p_3$ contradicting the fact that p_1 and p_2 are the smallest cardinality sets. Thus $p_1 = p_2 = p''$.

Next we show that $p'' \subset p$. Again, assume the contrary, i.e., $p'' \not\subset p$ (but $p' \subset p$ and $p' \subset p''$), then $p_3 = (p \cap p'')$ must be maximal and $p_3 \in P_\infty$. Then $p_3 \subseteq p''$ and p'' cannot be the smallest cardinality maximal set, leading to a contradiction. Hence the result. $\qquad\square$

Corollary 1. *For a $p \in P_\infty$ and $p' \subseteq p$ such that $p' \in P_g$, as above, if $p'' = p$, then $Lst(\mathcal{L}_{p'}^g) = \mathcal{L}'$.*

This observation gives a very efficient (output-sensitive) algorithm to construct P_g from P_∞, along with the location lists. Next, $\mathcal{L}_p^g$ is computed from L'' (recall $L'' = Lst(\mathcal{L}_p^g)$).

If our interest was only in maximal permutation patterns, the process stops here. However, to obtain *all* permutation patterns, we process each maximal permutation pattern to obtain the non-maximal patterns in Stage 3.

3.3   (Stage 3) Restricted g-Gap πPattern Problem

Problem 3. **(Restricted g-Gap πPattern Problem g-RGPP(r, l, L, K, P))** The input is m sequences s_i each of length l, where $\Pi'(s_i) = \Pi'(s_j)$ for each $1 \leq i, j \leq L$. The output is all permutation patterns that occur with gap g in at least K sequences.

Since $\Pi'(s_i) = \Pi'(s_j)$, for all $1 \leq i, j \leq L$, each sequence is a permutation (albeit with multiplicity) of s_1. For convenience, we can assign integers to the characters. First we seek all those patterns that occur in s_1. We convert the input to integers as follows. So, let s_1 be the reference sequence. Since there is possible multiplicity in s_1, the mapping is not necessarily one-to-one and if $\sigma(l) \in \Pi'(s_1)$, then σ is mapped to a set of l integers. In other words, $F(\sigma) = \{j \mid s_1[j] = \sigma\}$.

The size of a pattern p is defined to be $sz(p) = \Sigma_{\sigma(l) \in \Pi'(p)}(l)$. Note that when p has no multiplicities, $sz(p) = |\Pi(p)|$.

Observation 2. *Any pattern p that occurs with gap g on the reference sequence (say s_1) is such that there exists an ordering $f(\sigma_1) < f(\sigma_2) < \ldots < f(\sigma_{sz(p)})$ satisfying the condition $(f(\sigma_{i+1}) - f(\sigma_i)) \leq (g-1)$, for each $1 \leq i < sz(p)$, where $\sigma_i \in \Pi(p)$ (σ_i not necessarily distinct from σ_j when $i \neq j$), and $f(\sigma_i) \in F(\sigma_i)$ for $1 \leq i \leq sz(p)$.*

Example 3. Let $s_1 = a\,b\,c\,d\,b$ and $s_2 = d\,b\,b\,a\,c$. Using s_1 as the reference sequence, the integer mappings are as follows: $F(a) = \{1\}$, $F(b) = \{2, 5\}$, $F(c) = \{3\}$, $F(d) = \{4\}$. Note that $p = \{b(2), c(1)\}$ is a pattern that occurs in both s_1 and s_2, and, the ordering

$$(f(b) = 2) < (f(c) = 3) < (f(b) = 5)$$

satisfies the condition of Observation 2 for gap $g = 1$.

The algorithm performs an ordered search by scanning each string s_i, $i > 1$, (s_1 is the reference sequence) from left to right. We first fix a left pointer at j_l and move a right pointer j_r from $j_l + 1$ to the end of the string. At each scan, we are checking for a pattern that occurs on s_i with imprint $s_i[j_l \ldots j_r]$.

This ordered search is best explained through an example. Continuing Example 3, s_2 is written in terms of integers as: $s_2' = \{4\}\{2, 5\}\{2, 5\}\{1\}\{3\}$. Let $j_l = 1$ and $j_r = 2$. The elements of $s_2'[j_l \ldots j_r]$ (4, and, 2 or 5) are shown boxed in the figure below. The two orderings (1) $q_1 = 2 < 4$ and (2) $q_2 = 4 < 5$, satisfy the conditions of Observation 2. The p corresponding to an ordering q is defined as $p = \{\sigma \mid k \in q$ and $k \in F(\sigma)\}$. In the running example, $p = \{b, d\}$ occurs in s_1 with imprints $s_1[2 \ldots 4]$ and $s_1[4 \ldots 5]$. p occurs in s_2 with imprint $s_2[j_l \ldots j_r]$. Thus $\mathcal{L}_p^g = \{(1, 2, 4), (1, 4, 5), (2, 1, 2)\}$. At each scan, there are two mandatory integers, corresponding to $s_2'[j_l]$ and $s_2'[j_r]$ (shown in bold in the figure below).

j_r	ordering	ordering	p	$\mathcal{L}_p^1$
		$s_2' = 4\,\{2,5\}\,\{2,5\}\,1\,3$		
2	**2** **4** **5**	$2 < 4$	$\{b, d\}$	$\{(1, 2, 4),$
		$4 < 5$		$(1, 4, 5), (2, 1, 2)\}$
3	**2** **4** **5**	$2 < 4 < 5$	$\{b(2), d\}$	$\{(1, 2, 5), (2, 1, 3)\}$
4	**1** **2** **4** **5**	$1 < 2 < 4$	$\{a, b, d\}$	$\{(1, 1, 4), (2, 1, 4)\}$
		$1 < 2 < 4 < 5$	$\{a, b(2), d\}$	$\{(1, 1, 5), (2, 1, 4)\}$
5	**1** **2** **3** **4** **5**	$1 < 2 < 3 < 4$	$\{a, b, c, d\}$	$\{(1, 1, 4),$
		$1 < 3 < 4 < 5$		$(1, 1, 5), (2, 1, 5)\}$
		$2 < 3 < 4 < 5$	$\{b(2), c, d\}$	$\{(1, 2, 5), (2, 1, 5)\}$
		$1 < 2 < 3 < 4 < 5$	$\{a, b(2), c, d\}$	$\{(1, 1, 5), (2, 1, 5)\}$

Fig. 3. Consider Example 3. The patterns (and their location lists) generated as j_l is fixed at 1 and j_r moves from 2 to 5 on s_2'.

Any valid subsequence that satisfies conditions of Observation 2 must contain these mandatory elements and must occur on s_2 with gap g at $s_2[j_l \dots j_r]$.

Each pattern that is extracted in the last step is stored in a balanced tree data structure say $\mathcal{T}^g$. At the end of the process, the patterns on the node of this tree are checked to see if they appear in at least K of the given sequences.

Time Complexity. Let $S = j_r - j_l + 1$. All the orderings can be searched using a traversal scheme (such as DFS). Note that there are at most g ways of picking the next character (node) in the traversal. Thus the time taken at each scan to compute all the orderings is $\mathcal{O}(g^2 S)$ including the back edges (excluding the enumeration of the pattern). For a reference sequence, s_i, the scan is repeated for each j_l, $1 \leq j_l < l$ and for each sequence s_j, $j \neq i$. At the end the data structure $\mathcal{T}^g$ needs to be scanned only once. Thus the total time taken is $\mathcal{O}(g^2 l^2 L^2)$.

3.4   Putting It All Together

If the task is to obtain *all* permutation patterns that occur with gap g in at least K of the given m sequences, then there are two options. The first option is to extract the patterns in the three stages: first obtain all the maximal (Definition 6) patterns and then extract the non-maximal patterns from each maximal pattern. Note that in this case the integer encoding of the reference sequence s_i', which is a fragment of the original sequence s_i, will reflect the indices of s_i. For example, let $s_1 = a\,g\,h\,\mathbf{b}\,\mathbf{a}\,\mathbf{c}\,\mathbf{d}\,a$ with the maximal pattern shown in bold. This produces the fragment $b\,a\,d\,a$ and the integer mappings are $F(a) = \{5, 8\}$, $F(b) = \{4\}$, $F(d) = \{7\}$.

The second option is to augment each sequence s_i to $\bar{s}_i$ such that $\Pi'(\bar{s}_i) = \Pi'(\bar{s}_j)$ for each i, j and directly apply the algorithm of Stage 3 to obtain the maximal patterns. However, the first option, in practice, is preferred since it helps weed out a lot of candidate patterns. We use this approach for the experiments in the next section.

4   Conclusion and Ongoing Work

We have formalized a generalized model that uses maximal gapped patterns on a subset of the genomes. The subset of the genomes is dictated by the quorum parameter K. The gaps also help handle noisy and incomplete data. We present an output sensitive algorithm to compute all the permutation patterns. We are currently testing this model on a series of synthetic data and on available public data. The two sets of real data are chloroplast gene order of Campanulaceae data (used in [6]), and, the human and rat data [1,4]. We are applying the results of this model for functional classification of genes/proteins, as well as construction of phylogeny of the genomes. Our preliminary results on the latter application are very encouraging, both on synthetic and chloroplast data.

References

1. M. Alexandersson, S. Cawley, and L. Pachter. SLAM—Cross-species gene finding and alignment with a generalized pair hidden markov model. *Genome Research*, 13:496–502, 2003.
2. A.V. Aho, J.E. Hopcroft, and J.D. Ullman. *Data Structure and Algorithms*. Addison-Wesley Publishing Company, 1983.
3. A. Bergeron, C. Chauve, F. de Montgolfier, and M. Raffinot. Computing common intervals of k permutations, with applications to modular decomposition of graphs. *Proc. 9th European Symp. on Algorithms ESA'05*, volume 3669 of *Lecture Notes in Computer Science*, pages 779–790, Springer Verlag, 2005.
4. N. Bray, O. Couronne, I. Dubchak, T. Ishkhanov, L. Pachter, A. Poliakov, E. Rubin, and D. Ryaboy. Strategies and tools for whole-genome alignments. *Genome Research*, 1:73–80, 2003.
5. A. Bergeron, S. Corteel, and M. Raffinot. The algorithmic of gene teams. *Proc. 2nd Workshop on Algorithms in Bioinformatics WABI'02*, volume 2452 of *Lecture Notes in Computer Science*, pages 464–476, Springer Verlag, 2002.
6. M.E. Cosner, R.K. Jansen, B.M.E. Moret, L.A. Raubeson, L.-S. Wang, T. Warnow, and S. Wyman. An empirical comparison of phylogenetic methods on chloroplast gene order data in Campanulaceae. *Comparative Genomics: Empirical and Analytical Approaches to Gene Order Dynamics, Map Alignment, and the Evolution of Gene Families*, Kluwer, 2000.
7. R. Eres, G. Landau, and L. Parida. A combinatorial approach to automatic discovery of cluster-patterns. *Proc. 3rd Workshop on Algorithms in Bioinformatics WABI'03*, volume 2812 of *Lecture Notes in Bioinformatics*, pages 139–150. Springer Verlag, 2003.
8. X. He and M.H. Goldwasser. Identifying conserved gene clusters in the presence of orthologous groups. *Proc. 8th Conf. on Research in Computational Molecular Biology RECOMB'04*, pages 272–280. ACM Press, 2004.
9. S. Heber and J. Stoye. Finding all common intervals of k permutations. *Proc. 12th Symp. on Combinatorial Pattern Matching CPM'01*, volume 2089 of *Lecture Notes in Computer Science*, pages 207–218. Springer Verlag, 2001.
10. K. Booth and G. Leukar. Testing for the consecutive ones property, interval graphs, and graph planarity using pq-tree algorithms. *J. of Computer and System Sciences*, 13:335–379, 1976.
11. Gad Landau, Laxmi Parida, and Oren Weimann. Using pq trees for comparative genomics. In *Proc. Symp. on Combinatorial Pattern Matching CPM'05*, volume 3537 of *Lecture Notes in Computer Science*, pages 128–143. Springer Verlag, 2005.
12. J. Mulley and P. Holland. Small genome, big insights. *Nature*, 431:916–917, 2004.
13. T. Schmidt and J. Stoye. Quadratic time algorithms for finding common intervals in two and more sequences. In *Proc. Symp. on Combinatorial Pattern Matching CPM'04*, volume 3109 of *Lecture Notes in Computer Science*, pages 347–358. Springer Verlag, 2004.
14. T. Uno and M. Yagiura. Fast algorithms to enumerate all common intervals of two permutations. *Algorithmica*, 26(2):290–309, 2000.

Segmentation with an Isochore Distribution[*]

Miklós Csűrös[1], Ming-Te Cheng[1], Andreas Grimm[2],
Amine Halawani[1], and Perrine Landreau[3]

[1] Department of Computer Science and Operations Research, Université de Montréal
C.P. 6128, succ. Centre-Ville, Montréal, Québec, Canada, H3C 3J7
`csuros@iro.umontreal.ca`
[2] Lehr- und Forschungseinheit für Bioinformatik
Ludwig-Maximilians-Universität München, 80333 München, Germany
[3] Institut Scientifique Polytechnique Galilée—Université Paris XIII
93430 Villetaneuse, France

Abstract. We introduce a novel generative probabilistic model for segmentation problems in molecular sequence analysis. All segmentations that satisfy given minimum segment length requirements are equally likely in the model. We show how segmentation-related problems can be solved with similar efficacy as in hidden Markov models. In particular, we show how the best segmentation, as well as posterior segment class probabilities in individual sequence positions can be computed in $O(nC)$ time in case of C segment classes and a sequence of length n.

1 Introduction

Let $\mathbf{x} = x_1 x_2 \cdots x_n$ be a sequence of characters over a finite alphabet $\mathcal{A}$. A *segmentation* of $\mathbf{x}$ is described as a sequence $\mathbf{z} = z_1 z_2 \ldots z_n$ that assigns a *segment class* to each sequence position. The segmentation is thus a sequence over an alphabet $\mathcal{C}$, where $\mathcal{C}$ is the set of segment classes. A *segment* is a maximal contiguous region of positions that belong to the same class. Many molecular sequence analysis problems can be formulated as segmentation problems [1]. Obvious examples include the identification of isochores [2] in genomic DNA, and identification of charge clusters and hydrophobic profiles for proteins. In principle, all sequence annotation tasks (with non-overlapping segments) fit this general segmentation framework. For example, even such a complex task as eukaryotic gene prediction [3], entails the segmentation of a genomic sequence into classes such as "intergenic" and "exonic." In this work we are interested in generative probabilistic models, when the sequence $\mathbf{x}$ is the observed value of a random variable that depends solely on $\mathbf{z}$, which is also a random instance. Furthermore, we assume independence in the sense that each x_i depends on z_i only. Such probabilistic models include hidden Markov models [4,5], and other notable examples [6,7]. Hidden Markov models (HMMs) have the computational advantage that various segmentation-related problems, including that of finding

---

[*] Work supported by a grant from the Natural Sciences and Engineering Research Council of Canada.

P. Bücher and B.M.E. Moret (Eds.): WABI 2006, LNBI 4175, pp. 388–399, 2006.

the most likely segmentation, can be solved with linear-time algorithms in the sequence length n.

This paper's main goal is to introduce a new class of prior segmentation distributions; namely, a uniform distribution over segmentations in which all segments are longer than some specified threshold. Such a distribution captures usual expectations from segmentation results. We show that it is possible to compute the most likely segmentation in linear time in n, while the minimum segment length does not affect the running time. We show the same asymptotic running times for computing the posterior probabilities for segment class memberships and segment boundaries. In other words, we describe the analogues of the Viterbi and forward-backward algorithms.

An important motivation for our segmentation model comes from the *isochore theory* [8]. It postulates that the genome of warm-blooded vertebrates is composed of *isochores* in a mosaic structure. An isochore is a long contiguous segment of genomic DNA with a "fairly homogeneous" guanine+cytosine (GC) content [9]. The old debate about the theory's utility reemerged at the completion of the human genome draft sequence and persists to this day [9,10,11,12,13]. Eyre-Walker and Hurst [14] review biologically relevant issues in conjunction with isochores. We do not want to settle the question of biological relevance, but rather treat isochores as a technically useful concept describing the "fairly homogeneous" GC content of a region within an environment of at least 50–300 thousand base pairs. Usual isochore computations involve sliding windows of fixed length [10,11,13,14]. Window-less methods usually correspond to the minimization of some segment homogeneity measure [2,15]. To our knowledge, no generative model exists until now that explicitly captures the notions of minimum length and homogeneity at the same time. Here we put forward such a minimalist model, along with relevant computations.

1.1   Model and Model Selection

First we describe a generative framework for defining segmentation problems. A sequence of random variables $\mathbf{X} = (X_i \colon i = 1, \ldots, n)$ is dependent on a sequence of (unknown) segment class memberships $\mathbf{Z} = (Z_i \colon i = 1, \ldots, n)$. Here $X_i \in \mathcal{A}$ are letters from a finite alphabet and $Z_i \in \mathcal{C}$ are segment classes. The possible segment classes $\mathcal{C}$ are known. From an observed sequence $\mathbf{x} = x_1 \cdots x_n$, we want to deduce a segmentation $\mathbf{z} = z_1 \cdots z_n$. The human genome is often analyzed in terms of isochores named L1, L2, H1, H2, H3 with typical GC level cutoffs of 0.37, 0.41, 0.46, 0.53. In our probabilistic framework, a human chromosome sequence forms $\mathbf{x}$, and $\mathcal{C}$ comprises isochore classes.

More or less general versions of this framework were considered in the statistical literature [6,16]. They usually involve $\Omega(n^2)$-time computations for determining optimal segmentations [16,17]. Optimality is measured by some fitness or homogeneity measure. We focus on cases when the optimal segmentation can be found efficiently by some reasonable principle. First of all, we assume *independence*: the distribution of each X_i is completely determined by the probabilities

$$p_z(x) = \mathbb{P}\Big\{X_i = x \mid Z_i = z\Big\}.$$

A direct likelihood maximization approach cannot be used to choose a hypothesis $\mathbf{z}$, since the likelihood is maximized when each $z_i = \max_z p_z(x_i)$, which is rarely a consistent estimation. (For example, in GC content analysis, the best segmentation is a binary sequence of two classes for 100% and 0% GC.) We discuss two main principles that lead to better estimates without overfitting. The first principle is a Bayesian one: by imposing a prior distribution on $\mathbf{Z}$, one can select $\mathbf{z}$ that maximizes the posterior probability $\mathbb{P}\big\{\mathbf{Z} = \mathbf{z} \,\big|\, \mathbf{X} = \mathbf{x}\big\}$. This principle is employed in hidden Markov models. If $\mathbf{Z}$ is a Markov chain with a finite state set $\mathcal{C}$, then the best segmentation can be found in $O(n|\mathcal{C}|)$ time using the Viterbi algorithm [4, 5]. An alternative principle is to incorporate a notion of complexity in the optimization. For instance, the likelihood can be combined with description length [18], which penalizes complicated segmentations. When $\mathcal{C}$ is finite, and the segmentation's complexity is measured by the number of its segments, the best segmentation can be found efficiently in $O(n|\mathcal{C}|)$ time [7]. When $\mathcal{C}$ is the set of all possible distributions over $\mathcal{A}$, then the best segmentation minimizes the entropy with an adequate complexity penalization [2, 15].

The Bayesian approach of imposing a prior distribution on $\mathbf{Z}$ has the methodological advantage that it enables one to define probabilities of the type $\mathbb{P}\big\{\chi(\mathbf{Z}) \,\big|\, \mathbf{X} = \mathbf{x}\big\}$, where $\chi(\cdot)$ is some "interesting" property. Interesting properties include segment boundaries ($\chi(\mathbf{z}) = \{z_{i-1} = z'; z_i = z\}$) and the class of a position ($\chi(\mathbf{z}) = \{z_i = z\}$). Concerning the notation $\chi(\cdot)$, we use events and their indicators interchangeably, and, thus, $\{z_i = z\}$ denotes both the event that position i belongs to class z and the indicator variable which takes the value of 1 or 0, when the event occurs or not, respectively.

2   Isochore Distribution

In what follows, we focus on the case when $\mathbf{Z}$ is uniformly distributed over all segmentations satisfying certain minimum segment length requirements. We call such a distribution an *isochore distribution*. When the segmentation prior is uniform over a set $\mathcal{Z}$, the posterior probabilities can be computed as

$$\mathbb{P}\big\{\chi(\mathbf{Z}) \,\big|\, \mathbf{X} = \mathbf{x}\big\} \propto \sum_{\mathbf{z} \in \mathcal{Z} \cap \chi(\mathbf{z})} \mathbb{P}\big\{\mathbf{X} = \mathbf{x} \,\big|\, \mathbf{Z} = \mathbf{z}\big\}, \tag{1}$$

since $\mathbb{P}\{\mathbf{X} = \mathbf{x}\}$ does not depend on $\mathbf{z}$ and $\mathbb{P}\{\mathbf{Z} = \mathbf{z}\}$ is the same for every choice of $\mathbf{z} \in \mathcal{Z}$. In our case, the main difficulty is the efficient enumeration of segmentations that satisfy the minimum length requirements when the segmentation value is fixed in a position.

We are interested in segmentations where segments of class $z \in \{1, \dots, C\}$ are of minimum length $m_z > 0$. The notion of minimum segment length is captured through the following notation. We define $\mathsf{left}(\mathbf{z}, i)$ as the number of positions to the left of i that belong to the same segment class, and $\text{right}(\mathbf{z}, i)$ as the number of positions to the right that belong to the same segment class. Formally,

$$\text{left}(\mathbf{z}, i) = \left(\min_{d>0}\{d\colon z_{i-d} \neq z_i\}\right) - 1; \qquad \text{right}(\mathbf{z}, i) = \left(\min_{d>0}\{d\colon z_{i+d} \neq z_i\}\right) - 1.$$

We extend the notation so that $z_i = 0$ whenever $i \leq 0$ or $i > n$: if $z_j = z$ for all $j \leq i$ then $\text{left}(\mathbf{z}, j) = j - 1$ for all $j \leq i$, and an analogous statement holds for $\text{right}()$ in the rightmost segment. Clearly, the length of the segment that includes position i is the value $\text{length}(\mathbf{z}, i) = \text{left}(\mathbf{z}, i) + \text{right}(\mathbf{z}, i) + 1$.

Definition 1. *Let* $m_1, \ldots, m_C > 0$ *be the minimum segment lengths for the segment classes. A segmentation* $\mathbf{z}$ *is* valid *if and only if* $\text{length}(\mathbf{z}, i) \geq m_{z_i}$ *for all* $i = 1, \ldots, n$. *A random variable* $\mathbf{Z}$ *has an* isochore distribution *if it is drawn uniformly from the set of valid segmentations.*

2.1   Number of Valid Segmentations

It is useful to compute the number of valid segmentations, since it defines our prior. Let $N_z(n)$ be the number of valid segmentations for a sequence of length n which end with a segment of class z, and let $N(n) = \sum_z N_z(n)$ be the total number of valid segmentations. These values can be computed exactly:

$$N_z(n) = \begin{cases} 0 & \text{if } n < m_z; \\ 1 & \text{if } n = m_z; \\ N_z(n-1) + \sum_{z' \neq z} N_{z'}(n - m_z) & \text{if } n > m_z. \end{cases}$$

For the particular case of $\forall z\colon m_z = m$, i.e., identical segment length thresholds, we have the recursion $N(n) = N(n-1) + (C-1)N(n-m)$ for $n > m$, with the initial values $N(n) = 0$ for $n < m$ and $N(m) = C$. Clearly, $N(n)$ grows exponentially with n. In general, $N(n) = \Theta(\beta^{n/m})$ where β is the root of the characteristic equation $\beta - \beta^{1-1/m} - (C-1) = 0$. The value $N(n)$ provides the normalizing value in Eq. (1) and can be used for normalization in upcoming formulas.

2.2   Computing the Best Segmentation

Finding the best segmentation under the isochore distribution prior is not difficult. The dynamic programming method outlined in [7] for $C = 2$ can be generalized to an arbitrary number C of classes. Define

$$\xi_z(i) = p_z(x_i) \qquad \text{and} \qquad \Xi_z(i, i') = \prod_{j=i}^{i'} \xi_z(j).$$

In other words, $\Xi_z(i', i)$ is the likelihood for a segment $i..i'$ in class z. We derive a dynamic programming algorithm for the variables $V_z(i)$ for all $z \in \{1, \ldots, C\}$

and $i = 1, \ldots, n$. The variable $V_z(i)$ gives the likelihood for the best segmentation that is valid within the prefix $x_{1..i}$ and ends with class $z_i = z$.

$$V_z(i)$$
$$= \begin{cases} 0 & i < m_z; \\ \Xi_z(1, m_z) & i = m_z; \\ \max\left\{\xi_z(i)V_z(i-1), \Xi_z(i - m_z + 1, i)\max_{z'} V_{z'}(i - m_z)\right\} & i > m_z. \end{cases} \quad (2)$$

After carrying out the computations for all z and i, the best segmentation ends with $\arg\max_z V_z(n)$ and previous classes can be found by tracing back the maxima in (2). An advantageous technique is to keep track of letter counts

$$c_a(i) = \sum_{j=1}^{i} \{x_j = a\}$$

for all $a \in \mathcal{A}$ and i and then compute $\Xi_z(i, j) = \prod_{a \in \Sigma} \left(p_z(a)\right)^{c_a(j) - c_a(i-1)}$ (with $c_a(0) = 0$). In order to reduce costly floating-point calculations, $\left(p_z(a)\right)^c$ should be computed beforehand for all $z \in \{1, \ldots, C\}$, $a \in \mathcal{A}$ and $c \in \{0, 1, \ldots, m\}$. One can also work with $\log V_z(i)$ instead to avoid underflow, and to expedite the computations by performing additions instead of multiplications.

Theorem 1. *A segmentation* $\mathbf{z}$ *that maximizes* $\mathbb{P}\left\{\mathbf{Z} = \mathbf{z} \mid \mathbf{X} = \mathbf{x}\right\}$ *can be found in* $O(nC)$ *time when* $\mathbf{Z}$ *has an isochore distribution with* C *segment classes.*

Proof. The recurrences of (2) can be computed in $O(1)$ time for every z and i, by keeping track of the letter counts $c_a(j)$ and the maxima $\max_{z'} V_{z'}(j)$ in every position j. □

2.3   Computing Posteriors

For computing posterior probabilities, we need to be able to sample valid segmentations that are constrained at a position. In order to simplify the formulas, we assume from now on that the minimum segment lengths are identical, i.e., for all z, $m_z = m$, and that the minimum length m is an even number.

In order to derive recurrence relations, consider the following sets of (not necessarily valid) segmentations for $z \in \{1, \ldots, C\}$, $i \in \{1, \ldots, n\}$ and $d \in \{0, \ldots, n\}$:

$$\mathcal{L}_z^{(d)}(i) = \left\{\mathbf{z}\colon z_i = z; \mathsf{left}(\mathbf{z}, i) \geq d; \forall j < i - \mathsf{length}(\mathbf{z}, i)\colon \mathsf{length}(\mathbf{z}, j) \geq m\right\}$$

$$\mathcal{R}_z^{(d)}(i) = \left\{\mathbf{z}\colon z_i = z; \mathsf{right}(\mathbf{z}, i) \geq d; \forall j > i + \mathsf{length}(\mathbf{z}, i)\colon \mathsf{length}(\mathbf{z}, j) \geq m\right\}.$$

In other words, $\mathcal{L}_z^{(d)}(i)$ is the set of segmentations that are restricted only for the prefix $z_1, \ldots, z_i$ so that (a) positions $i - d, \ldots, i$ are in class z, and (b) segments before the segment of i satisfy the minimum length requirements. The sets $\mathcal{R}_z^{(d)}(i)$ are defined analogously for suffixes of $\mathbf{z}$. Now, $\mathcal{L}_{z'}^{(m-1)}(i-1) \cap \mathcal{R}_z^{(m-1)}(i)$ is the set of valid segmentations that have a $z' \to z$ segment boundary at i. Hence, the posterior probability of a boundary at position $i > 1$ can be written as

$$q_{z' \to z}(i) = \mathbb{P}\Big\{ Z_{i-1} = z'; Z_i = z \,\Big|\, \mathbf{X} = \mathbf{x} \Big\}$$
$$\propto \mathbb{P}\Big\{ \mathbf{X} = \mathbf{x} \,\Big|\, \mathbf{Z} \in \mathcal{L}_{z'}^{(m-1)}(i-1) \cap \mathcal{R}_z^{(m-1)}(i) \Big\}$$

when $z' \neq z$. It will be useful to define the posterior probabilities for position $1 < i < n$ being the left or right end of a segment in class z:

$$q_{\to z}(i) = \sum_{z' \neq z} q_{z' \to z}(i); \qquad\qquad 1 < i \leq n; \qquad (3a)$$

$$q_{z \to}(i) = \sum_{z' \neq z} q_{z \to z'}(i+1); \qquad\qquad 1 \leq i < n. \qquad (3b)$$

The posterior probability that position i belongs to class z is denoted by

$$q_z(i) = \mathbb{P}\Big\{ Z_i = z \,\Big|\, \mathbf{X} = \mathbf{x} \Big\}.$$

For the sake of completeness, we extend the notation of Eqs. (3) to the sequence extremities: $q_{\to z}(1) = q_z(1)$ and $q_{z \to}(n) = q_z(n)$.

Theorem 2. *Let* $\mu_z(i) = \mathbb{P}\Big\{ \mathbf{Z} \in \mathcal{L}_z^{(m/2)}(i) \cap \mathcal{R}_z^{(m/2)}(i) \,\Big|\, \mathbf{X} = \mathbf{x} \Big\}$. *For all* $i \in \{1, \dots, n\}$ *and* $z \in \{1, \dots, C\}$, *the probability that position* $1 < i < n$ *belongs to segment class* z *can be written as*

$$q_z(i) = \mu_z(i) + \sum_{\delta=0}^{\max\{i-1, \frac{m}{2}-1\}} q_{\to z}(i-\delta) + \sum_{\delta=0}^{\max\{n-i, \frac{m}{2}-1\}} q_{z \to}(i+\delta).$$

Proof. If $z_i = z$ and $\mathbf{z}$ is a valid segmentation, then exactly one of the following is true

1. $\mathsf{left}(\mathbf{z}, i) \geq m/2$ and $\mathsf{right}(\mathbf{z}, i) \geq m/2$ simultaneously;
2. position i's segment starts at position $i - \delta$ for some $0 \leq \delta < m/2$.
3. position i's segment ends at position $i + \delta$ for some $0 \leq \delta < m/2$.

The probability for Case 1 is $\mu_z(i)$. The probability of Case 2 is $\sum_\delta q_{\to z}(i-\delta)$; the probability of Case 3 is $\sum_\delta q_{z \to}(i+\delta)$. $\qquad\square$

3   Algorithm for Posterior Probabilities

Define the following likelihoods

$$L_z(i) = \sum_{\mathbf{z} \in \mathcal{L}_z^{(m-1)}(i)} \mathbb{P}\left\{X_{1..i-1} = x_{1..i-1} \,\middle|\, \mathbf{Z} = \mathbf{z}\right\}; \tag{4a}$$

$$\lambda_z(i) = \sum_{\mathbf{z} \in \mathcal{L}_z^{(m/2)}(i)} \mathbb{P}\left\{X_{1..i-1} = x_{1..i-1} \,\middle|\, \mathbf{Z} = \mathbf{z}\right\}; \tag{4b}$$

$$R_z(i) = \sum_{\mathbf{z} \in \mathcal{R}_z^{(m-1)}(i)} \mathbb{P}\left\{X_{i+1..n} = x_{i+1..n} \,\middle|\, \mathbf{Z} = \mathbf{z}\right\}; \tag{4c}$$

$$\varrho_z(i) = \sum_{\mathbf{z} \in \mathcal{R}_z^{(m/2)}(i)} \mathbb{P}\left\{X_{i+1..n} = x_{i+1..n} \,\middle|\, \mathbf{Z} = \mathbf{z}\right\}; \tag{4d}$$

$$b_{z' \to z}(i) = \sum_{\mathbf{z} \in \mathcal{L}_{z'}^{(m-1)}(i-1) \cap \mathcal{R}_z^{(m-1)}(i)} \mathbb{P}\left\{\mathbf{X} = \mathbf{x} \,\middle|\, \mathbf{Z} = \mathbf{z}\right\}, \qquad i > 1. \tag{4e}$$

Clearly, $b_{z' \to z}(i) = L_{z'}(i-1)\xi_{z'}(i-1)\xi_z(i)R_z(i)$. whenever $1 < i \le n$. Let

$$b_{\to z}(i) = \sum_{z' \ne z} b_{z' \to z}(i) = \xi_z(i)R_z(i) \sum_{z' \ne z} \xi_{z'}(i-1)L_{z'}(i-1), \qquad i > 1;$$

$$b_{z \to}(i) = \sum_{z' \ne z} b_{z \to z'}(i+1) = \xi_z(i)L_z(i) \sum_{z' \ne z} \xi_{z'}(i+1)R_{z'}(i+1), \qquad i < n.$$

For the sequence extremities,

$$q_z(1) \propto b_{\to z}(1) = \xi_z(1)R_z(1); \tag{5a}$$
$$q_z(n) \propto b_{z \to}(n) = \xi_z(n)L_z(n). \tag{5b}$$

By Theorem 2, the posterior probabilities for segment class memberships can be computed for all $1 < i < n$ as

$$q_z(i) \propto \lambda_z(i)\xi_z(i)\varrho_z(i) + h_z(i), \tag{6}$$

where

$$h_z(i) = \sum_{\delta=0}^{\min\{i-1, \frac{m}{2}-1\}} b_{\to z}(i-\delta) + \sum_{\delta=0}^{\min\{n-i, \frac{m}{2}-1\}} b_{z \to}(i+\delta).$$

The right-hand sides of Eqs. (5) and (6) are normalized by dividing them with $Q = \sum_z \xi_z(1)R_z(1) = \sum_z \xi_z(n)L_z(n)$. In fact, posterior probabilities for segment boundaries are computed by the same normalization:

$$q_{\to z}(i) = Q^{-1}b_{\to z}(i) \qquad \text{and} \qquad q_{z \to}(i) = Q^{-1}b_{z \to}(i).$$

Additionally, since $\mathbb{P}\{\mathbf{Z} = \mathbf{z}\} = 1/N(n)$ for all $\mathbf{z}$, Bayes' theorem gives $\mathbb{P}\{\mathbf{X} = \mathbf{x}\} = \frac{Q}{N(n)}$.

The variables of Eqs. (4) are computed by the following recurrences.

$$\lambda_z(i) = \xi_z(i-1)\lambda_z(i-1) + \Xi_z(i - \frac{m}{2}, i-1) \qquad\qquad i > \frac{m}{2} + 1 \quad (7a)$$
$$\times \sum_{z' \neq z} \xi_{z'}(i - \frac{m}{2} - 1)L_{z'}(i - \frac{m}{2} - 1);$$

$$L_z(i) = \xi_z(i-1)L_z(i-1) \qquad\qquad\qquad\qquad i > m \quad (7b)$$
$$+ \Xi_z(i - m + 1, i - 1)\sum_{z' \neq z} \xi_{z'}(i - m)L_{z'}(i - m);$$

Analogous formulas are used to compute $\varrho_z(i)$ and $R_z(i)$. If $\frac{m}{2} < i \le n - \frac{m}{2} + 1$, then

$$h_z(i) = h_z(i-1) + b_{\to z}(i) - b_{\to z}(i - \frac{m}{2}) + b_{z\to}(i + \frac{m}{2} - 1) - b_{z\to}(i - 1). \quad (8)$$

Obviously, $h_z(1) = b_{\to z}(1)$. For $1 < i \le \frac{m}{2}$ the recurrence of Eq. (8) does not include the subtraction of $b_{\to z}(i - \frac{m}{2})$ and for $i > n - \frac{m}{2} + 1$ the recurrence does not include the term $b_{z\to}(i + \frac{m}{2} - 1)$. The variables of Eqs. (7) are initialized in an obvious manner.

A useful algorithmic technique for computing expressions of the type $A(z) = \sum_{z' \neq z} B(z')$ for all z in $O(C)$ total time is the following. First compute $B_{lo}(z) = \sum_{z' < z} B(z')$ for all z. Then compute $B_{hi}(z) = \sum_{z' > z} B(z')$ for all z. Clearly, this can be done in $O(C)$ time. Now, $A(z) = B_{lo}(z) + B_{hi}(z)$ can be set in $O(1)$ time for each z. Using this technique, all variables can be computed for every i in $O(nC)$ time. Notice that the Ξ_z can be computed in $O(1)$ time for all z, by keeping track of character counts in prefixes and suffixes as described in §2.2.

REMARK.   It may seem that when the minimum lengths differ, $\sum_{z' \neq z} \xi_{z'}(i - m_z)L_{z'}(i - m_z)$ in (7b), for example, needs to be computed for each z separately, resulting in a $\Theta(C^2)$ factor in the running time. The technique, however, can be readily adapted to this case. The appropriate B_{lo} and B_{hi} values need to be kept for recent values of $j = i - m_z$, which again leads to a linear running time in C.

Theorem 3. *All posterior probabilities for segment class memberships and segment boundaries can be computed in $O(nC)$ time when* $\mathbf{Z}$ *has an isochore distribution with C segment classes.*

The posterior probabilities can be used in an Expectation Maximization framework, as in Baum-Welch training for HMMs [4, 5]. Simply, the $p_z(x)$ are estimated as

$$\hat{p}_z(x) = \frac{\sum_{i=1}^{n} q_z(i)\{x_i = x\}}{\sum_{i=1}^{n} q_z(i)}.$$

3.1   Memory Management

Since the recurrences for ϱ and R can be computed from right to left while those for λ, L and h are computed in a left to right direction, a direct implementation would need to first compute and store the ϱ and R values and then proceed from left to right to carry out the posterior computations. The left-to-right computation proceeds in a "lookahead" fashion: for every i, $\lambda_z(i)$, $L_z(i+\frac{m}{2}-1)$, $h_z(i)$ and $q_z(i)$ are computed, in this order. Consequently, an array of size m can store the necessary values $L_z(j)$ for $i - \frac{m}{2} \leq j < i + \frac{m}{2}$ to carry out one step of the left-to-right computations. For λ_z and h_z, only the previous values are needed. It is, however, a good idea to keep track of recent values of $b_{z\rightarrow}$ and $b_{\rightarrow z}$ so that they are not computed twice.

A direct implementation, in which all $\varrho_z(i)$ and $R_z(i)$ are computed before proceeding to the left-to-right computations, may be impractical for large sequences because of large memory requirements. For longer sequences, it is possible to do the computations using a "slicing" or "checkpointing" technique, similar to those employed in pairwise sequence alignment and HMM training [19]. We do not discuss the details here due to space limitations. The technique allows for computing the probabilities on all-purpose desktop computers: our implementation was used to carry out the segmentations with five classes and $m = 50000$ for human chromosome 1 (246 Mbp), with a memory footprint below 2 Gigabytes. A recursive checkpointing technique leads to the following theorem.

Theorem 4. *For C segment classes with minimum length m and a sequence of length n, the posterior probabilities can be computed in $O(LnC)$ time using $O(Cm^{1-1/L}n^{1/L}L)$ workspace, where L is an arbitrary positive integer. In particular, by choosing $L = \Theta\left(\log\frac{n}{m}\right)$, the probabilities are computed in $O\left(Cn\log\frac{n}{m}\right)$ time using $O\left(Cm\log\frac{n}{m}\right)$ workspace.*

4   Experiments

We implemented the described procedure for posterior calculations in a Java package. Figure 1 compares in a simulated experiment the quality of HMM-based

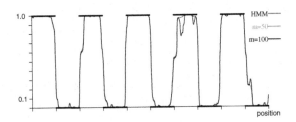

Fig. 1. Posterior segment class membership by HMM and isochore distributions. A random DNA sequence of 1000 characters was generated with alternating 30% and 70% GC level in 100bp segments. The plot compares the posterior segment class membership for the 30% GC class as computed by an HMM (two states, state switching transition probabilities are 0.01), and those computed using isochore distributions with minimum length 50 and 100. The former already gives smoother results (see especially the seventh segment), while the latter finds the true segmentation perfectly.

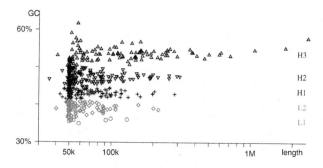

Fig. 2. Segment composition and length in the segmentation of chr19. Segment class levels are as follows: 35%, 39%, 43%, 47%, and 53% GC in L1–H3, respectively.

predictions and our method. The figure illustrates that HMM predictions are more easily affected by random fluctuations in the sequence composition.

For illustrative purposes, we carried out a segmentation of human chromosome 19 [20]. The results of the segmentation can be viewed as a custom annotation track in the UCSC genome browser [21]; the track can be downloaded from http://www.iro.umontreal.ca/~{}csuros/segmentation/hg17/chr19-segments.bed.

There are two principal questions that need to be addressed in this context: whether most of the genome can be classified into isochores, and whether there is a non-arbitrary threshold on homogeneous region lengths. Using five isochore classes, we segmented the sequence into segments within which the class membership can be established with at least 90% probability, using a minimum length of 50000 base pairs. About 85% of the sequence can be classified into one of the isochore classes with more than 90% fidelity. Almost all of the missing 15% fall into the unsequenced centromeric region, and the few percents that remain are mostly in short segment boundaries. This fact does not necessarily reflect the validity of classification, as long segments have a very small chance to fall right between two classes in GC composition. Figure 2 plots the statistics on the segments. This chromosome is unusually GC-rich [20], 1.4%, 9.4%, 15.9%, 22.8% and 35.5% of the positions are classified into the classes L1, L2, H1, H2 and H3, respectively. It is interesting to notice that a large number of the segments have a length very close to the lower bound, which hints at heterogeneity below the minimum length cutoff. Classically, isochores are said to be hundreds of thousand base pairs in length: our segmentation does not reveal such a phenomenon.

5   Conclusion

We presented a novel probabilistic model for segmentations and showed how usual techniques associated with hidden Markov models have their equivalents, including a Viterbi-style algorithm for finding the best segmentation, a forward-backward algorithm for computing posteriors, and expectation maximization for setting class parameters. The model features an explicit minimum segment length parameter, which is not easily captured by an HMM. Our "minimalist" model assumes a uniform distribution among segmentations that obey the segment length constraints. Some additional parameters can be easily incorporated into the model. For instance, one can add conditional probabilities for changing segment classes, or have a segment length distribution that is a shifted geometric one. Using the example of Eq. (7b), write

$$L_z(i) = \tau_0 \xi_z(i-1) L_z(i-1) + \Xi_z(i-m+1, i-1) \sum_{z' \neq z} \tau_{z'} \xi_{z'}(i-m) L_{z'}(i-m).$$

The parameter τ_0 implies that segment length has a thresholded geometric distribution and the parameters $\tau_{z'}$ model different probabilities for the preceding segment class. In fact, such a parametrization is the equivalent of posterior computations for HMMs when the state sequence has to obey some duration thresholds. Hidden Markov models are sometimes used along with some ad hoc thresholding on segment lengths (e.g., [22]). Our results show that such an approach can be implemented in a theoretically sound manner. There are some standard techniques [5], involving extra states or transitions, which can model minimum segment lengths at the price of increased time complexity. In contrast, our algorithms' running time is linear in the number of segment classes (using the equivalent of two states per class), and the time complexity is not affected by the minimum segment length.

Without doubt, many genome features (such as gene density, retrotransposition and replication timing) are linked to regional GC composition, but there is still need for an adequate "isochore theory" that explains genome organization in terms of isochores. A main difficulty in assessing the role of isochores in mammalian genome analysis has been the lack of a widely accepted generative (as opposed to descriptive) model. In our opinion, such a falsifiable model is necessary for a useful scientific discussion, and would open up the path to meaningful hypothesis testing procedures. Refutation attemps [10, 11, 14] have been rebuked on the basis that the employed statistical models do not adequately capture the true nature of isochores [9, 12]. On the other hand, proponents of the theory largely relied on ad hoc segmentation procedures [2, 13], which result in useful genome annotations, but make it difficult to assess statistical validity. We intend to continue our work toward an adequate isochore model, by incorporating positional dependence and other essential features.

We hope that our model and the associated computational results will be useful on their own for "simple" sequence analysis tasks, such as the identification of isochores or CpG islands, or as part of more sophisticated probabilistic models for complicated analysis problems, such as *ab initio* gene prediction.

References

1. Karlin, S.: Statistical signals in bioinformatics. Proc. Nat'l Acad. Sci. USA **102** (2005) 13355–13362
2. Li, W., Bernaola-Galván, P., Haghighi, F., Grosse, I.: Applications of recursive segmentation to the analysis of DNA sequences. Comput. Chem. **26** (2002) 491–510
3. Mathé, C., Sagot, M.F., Schiex, T., Rouzé, P.: Current methods of gene prediction, their strengths and weaknesses. Nucleic Acids Res. **30** (2002) 4103–4117
4. Rabiner, L.R.: A tutorial on Hidden Markov Models and selected applications in speech recognition. Proc. IEEE **77** (1989) 257–286
5. Durbin, R., Eddy, S.R., Krogh, A., Mitchison, G.: Biological Sequence Analysis. Cambridge University Press, UK (1998)
6. Fu, Y.X., Curnow, R.N.: Maximum likelihood estimation of multiple change points. Biometrika **77** (1990) 563–573
7. Csűrös, M.: Maximum-scoring segment sets. IEEE/ACM Trans. Comput. Biol. Bioinf. **1** (2004) 139–150
8. Bernardi, G., Olofsson, B., Filipski, J., Zerial, M., Salinas, J., Cuny, G., Meunier-Rotival, M., Rodier, F.: The mosaic genome of warmblooded vertebrates. Science **228** (1985) 953–958
9. Bernardi, G.: Misunderstandings about isochores: Part I. Gene **276** (2001) 3–13
10. IHGSC: Initial sequencing and analysis of the human genome. Nature **409** (2001) 860–921
11. Cohen, N., Dagan, T., Stone, L., Graur, D.: GC composition of the human genome: in search of isochores. Mol. Biol. Evol. **22** (2005) 1260–1272
12. Clay, O., Bernardi, G.: How not to look for isochores: A reply to Cohen et al. Mol. Biol. Evol. **22** (2005) 2315–2317
13. Constantini, M., Clay, O., Auletta, F., Bernardi, G.: An isochore map of the human genome. Genome Res. **16** (2006) 536–541
14. Eyre-Walker, A., Hurst, L.D.: The evolution of isochores. Nat. Rev. Genet. **2** (2001) 549–555
15. Szpankowski, W., Ren, W., Szpankowski, L.: An optimal DNA segmentation based on the MDL principle. Int. J. Bioinformatics Research and Applications **1** (2005) 3–17
16. Barry, D., Hartigan, J.A.: Product partition models for change point problems. Ann. Statist. **20** (1992) 260–279
17. Auger, I.E., Lawrence, C.E.: Algorithms for the optimal identification of segment neighborhoods. Bull. Math. Biol. **51** (1989) 39–54
18. Rissanen, J.: A universal prior for integers and estimation by minimum description length. Ann. Statist. **11** (1983) 416–431
19. Tarnas, C., Hughey, R.: Reduced space hidden markov model training. Bioinformatics **14** (1998) 401–406
20. Grimwood, J., et al.: The DNA sequence and biology of human chromosome 19. Nature **428** (2004) 529–535
21. Karolchik, D., Baertsch, R., Diekhans, M., Furey, T.S., Hinrichs, A., Lu, Y.T., Roskin, K.M., Schwartz, M., Sugnet, C.W., Thomas, D.J., Weber, R.J., Haussler, D., Kent, W.J.: The UCSC genome browser database. Nucleic Acids Res. **31** (2003) 51–54
22. Klein, R.J., Misulovin, Z., Eddy, S.R.: Noncoding RNA genes identified in AT-rich hyperthermophiles. Proc. Nat'l Acad. Sci. USA **99** (2002) 7542–7547

Author Index

Abul, Osman   197
Agarwal, Pankaj K.   207
Albrecht, Andreas A.   252
Alves, Carlos E.R.   186

Barral, Yves   1
Bergeron, Anne   163
Berman, Piotr   138, 344
Blum, Christian   355
Böcker, Sebastian   12
Brinza, Dumitru   286
Burger, Lukas   44

Cai, Liming   68, 262
Chan, M.Y.   56
Chan, Wun-Tat   56
Cheng, Ming-Te   388
Chin, Francis Y.L.   56
Conner, Glenn   162
Csűrös, Miklós   388

Darling, Aaron E.   126
de Carvalho Jr., Sérgio A.   321
Deng, Zhidong   242
do Lago, Alair Pereira   186
Drabløs, Finn   197

Eidsheim, Lars Andreas   197

Friberg, Markus T.   1
Fung, Stanley P.Y.   56

Gambin, Anna   32
Gonnet, Gaston H.   1
Gonnet, Pedro   1
Gramm, Jens   92
Grimm, Andreas   388

Halawani, Amine   388
Hartman, Tzvika   92
Hein, Jotun   103
Hollan, Barbara   162
Hong, Eun-Jong   219
Hou, Minmei   138

Huber, Katharina T.   162
Huson, Daniel H.   150
Ilinkin, Ivaylo   115
Isom, Adam   115

Janardan, Ravi   115
Jenkins, Paul   103

Kao, Ming-Yang   56
Karczmarski, Jakub   32
Keijsper, Judith   80
Kelk, Steven   80
Klau, Gunnar W.   298
Kluge, Bogusław   32
Kuiken, Carla   126

Landreau, Perrine   388
Letzel, Matthias C.   12
Li, Ming   231
Li, Shuai Cheng   231
Ligeti, Péter   174
Lin, Yu   310
Lipták, Zsuzsanna   12
Liu, Chunmei   68
Liu, Xiaowen   310
Lozano-Pérez, Tomás   219
Łuksza, Marta   32
Lyngsø, Rune   103

Malmberg, Russell L.   68, 262
Messeguer, Xavier   126
Miklós, István   174
Miller, Webb   138
Mixtacki, Julia   163
Moulton, Vincent   162

Nedland, Magnar   197
Nierhoff, Till   92

Ostrowski, Jerzy   32

Paige, Timothy Brooks   174
Parida, Laxmi   376
Perna, Nicole T.   126
Pervukhin, Anton   12
Phillips, Jeff M.   207
Pinter, Ron Y.   274

Rahmann, Sven   298, 321
Rokhlenko, Oleg   274
Rudolph, Johannes   207
Ruppin, Eytan   274

Sandve, Geir Kjetil   197
Schraudolph, Nicol N.   1
Sharan, Roded   92, 274
Shlomi, Tomer   274
Skaliotis, Alexandros   252
Song, Dandan   242
Song, Yinglei   68
Song, Yixu   24
Steel, Mike A.   150
Steinhöfel, Kathleen   252
Stelling, Jörg   333
Stojanovic, Nikola   344
Stougie, Leen   80
Stoye, Jens   163
Syrstad, Øyvind Bø   197

Tantau, Till   92
Terzer, Marco   333
Treangen, Todd J.   126
Tsur, Dekel   366

Vallès, Mateu Yábar   355
van Iersel, Leo   80
van Nimwegen, Erik   44
Vellozo, Augusto F.   186

Wang, Jiaxin   24
Wang, Lusheng   310
Whitfield, Jim   150

Yang, Zehong   24
Ye, Jieping   115

Zelikovsky, Alexander   286
Zhang, Louxin   126, 138
Zhao, Jizhen   262
Zhu, Hongmei   24

Lecture Notes in Bioinformatics

Vol. 4175: P. Bücher, B.M.E. Moret (Eds.), Algorithms in Bioinformatics. XII, 402 pages. 2006.

Vol. 4146: J.C. Rajapakse, L. Wong, R. Acharya (Eds.), Pattern Recognition in Bioinformatics. XIV, 186 pages. 2006.

Vol. 4115: D.-S. Huang, K. Li, G.W. Irwin (Eds.), Computational Intelligence and Bioinformatics, Part III. XXI, 803 pages. 2006.

Vol. 4075: U. Leser, F. Naumann, B. Eckman (Eds.), Data Integration in the Life Sciences. XI, 298 pages. 2006.

Vol. 4070: C. Priami, X. Hu, Y. Pan, T.Y. Lin (Eds.), Transactions on Computational Systems Biology V. IX, 129 pages. 2006.

Vol. 3939: C. Priami, L. Cardelli, S. Emmott (Eds.), Transactions on Computational Systems Biology IV. VII, 141 pages. 2006.

Vol. 3916: J. Li, Q. Yang, A.-H. Tan (Eds.), Data Mining for Biomedical Applications. VIII, 155 pages. 2006.

Vol. 3909: A. Apostolico, C. Guerra, S. Istrail, P. Pevzner, M. Waterman (Eds.), Research in Computational Molecular Biology. XVII, 612 pages. 2006.

Vol. 3886: E.G. Bremer, J. Hakenberg, E.-H.(S.) Han, D. Berrar, W. Dubitzky (Eds.), Knowledge Discovery in Life Science Literature. XIV, 147 pages. 2006.

Vol. 3745: J.L. Oliveira, V. Maojo, F. Martín-Sánchez, A.S. Pereira (Eds.), Biological and Medical Data Analysis. XII, 422 pages. 2005.

Vol. 3737: C. Priami, E. Merelli, P. Gonzalez, A. Omicini (Eds.), Transactions on Computational Systems Biology III. VII, 169 pages. 2005.

Vol. 3695: M.R. Berthold, R.C. Glen, K. Diederichs, O. Kohlbacher, I. Fischer (Eds.), Computational Life Sciences. XI, 277 pages. 2005.

Vol. 3692: R. Casadio, G. Myers (Eds.), Algorithms in Bioinformatics. X, 436 pages. 2005.

Vol. 3680: C. Priami, A. Zelikovsky (Eds.), Transactions on Computational Systems Biology II. IX, 153 pages. 2005.

Vol. 3678: A. McLysaght, D.H. Huson (Eds.), Comparative Genomics. VIII, 167 pages. 2005.

Vol. 3615: B. Ludäscher, L. Raschid (Eds.), Data Integration in the Life Sciences. XII, 344 pages. 2005.

Vol. 3594: J.C. Setubal, S. Verjovski-Almeida (Eds.), Advances in Bioinformatics and Computational Biology. XIV, 258 pages. 2005.

Vol. 3500: S. Miyano, J. Mesirov, S. Kasif, S. Istrail, P. Pevzner, M. Waterman (Eds.), Research in Computational Molecular Biology. XVII, 632 pages. 2005.

Vol. 3388: J. Lagergren (Ed.), Comparative Genomics. VII, 133 pages. 2005.

Vol. 3380: C. Priami (Ed.), Transactions on Computational Systems Biology I. IX, 111 pages. 2005.

Vol. 3370: A. Konagaya, K. Satou (Eds.), Grid Computing in Life Science. X, 188 pages. 2005.

Vol. 3318: E. Eskin, C. Workman (Eds.), Regulatory Genomics. VII, 115 pages. 2005.

Vol. 3240: I. Jonassen, J. Kim (Eds.), Algorithms in Bioinformatics. IX, 476 pages. 2004.

Vol. 3082: V. Danos, V. Schachter (Eds.), Computational Methods in Systems Biology. IX, 280 pages. 2005.

Vol. 2994: E. Rahm (Ed.), Data Integration in the Life Sciences. X, 221 pages. 2004.

Vol. 2983: S. Istrail, M.S. Waterman, A. Clark (Eds.), Computational Methods for SNPs and Haplotype Inference. IX, 153 pages. 2004.

Vol. 2812: G. Benson, R.D. M. Page (Eds.), Algorithms in Bioinformatics. X, 528 pages. 2003.

Vol. 2666: C. Guerra, S. Istrail (Eds.), Mathematical Methods for Protein Structure Analysis and Design. XI, 157 pages. 2003.